THE REALM OF INTERACTING BINARY STARS

ASTROPHYSICS AND SPACE SCIENCE LIBRARY

A SERIES OF BOOKS ON THE RECENT DEVELOPMENTS
OF SPACE SCIENCE AND OF GENERAL GEOPHYSICS AND ASTROPHYSICS
PUBLISHED IN CONNECTION WITH THE JOURNAL
SPACE SCIENCE REVIEWS

Editorial Board

R. L. F. BOYD, *University College, London, England*

W. B. BURTON, *Sterrewacht, Leiden, The Netherlands*

C. DE JAGER, *University of Utrecht, The Netherlands*

J. KLECZEK, *Czechoslovak Academy of Sciences, Ondřejov, Czechoslovakia*

Z. KOPAL, *University of Manchester, England*

R. LÜST, *Max-Planck-Institut für Meteorologie, Hamburg, Germany*

L. I. SEDOV, *Academy of Sciences, Moscow, Russia*

Z. ŠVESTKA, *Laboratory for Space Research, Utrecht, The Netherlands*

VOLUME 177

THE REALM OF INTERACTING BINARY STARS

Edited by

J. SAHADE

*Observatorio Astronómico, La Plata and
Instituto Argentino de Radioastronomica,
Villa Elise (Bs.As.), Argentina*

G. E. McCLUSKEY, JR.

*Dept. of Mathematics, Lehigh University,
Bethlehem, Pennsylvania, U.S.A.*

and

Y. KONDO

*Goddard Space Flight Center, Greenbelt,
Maryland, U.S.A.*

SPRINGER-SCIENCE+BUSINESS MEDIA, B.V.

ISBN 978-94-010-5066-1   ISBN 978-94-011-2416-4 (eBook)
DOI 10.1007/978-94-011-2416-4

Printed on acid-free paper

All Rights Reserved
© 1993 Springer Science+Business Media Dordrecht
Originally published by Kluwer Academic Publishers in 1993
Softcover reprint of the hardcover 1st edition 1993
No part of the material protected by this copyright notice may be reproduced or
utilized in any form or by any means, electronic or mechanical,
including photocopying, recording or by any information storage and
retrieval system, without written permission from the copyright owner.

TABLE OF CONTENTS

J. Sahade, G.E. McCluskey Jr., and Y. Kondo
INTRODUCTION ... 1

A.H. Batten and F.B. Wood
THE DEVELOPMENT OF THE IDEA OF INTERACTING
   DOUBLE STARS ... 3

J. Sahade, G.E. McCluskey Jr., and Y. Kondo
THE INTERACTING BINARY ZOO ... 13

J. Sahade
MASS LOSS IN INTERACTING BINARY SYSTEMS ... 17

S.H. Lubow
SOME ISSUES IN THE THEORY OF MASS TRANSFER ... 25

G.E. McCluskey, Jr.
THE ALGOL-TYPE INTERACTING BINARIES ... 39

E.F. Guinan and A. Giménez
MAGNETIC ACTIVITY IN CLOSE BINARIES ... 51

S.M. Rucinski
CONTACT BINARIES OF THE W UMa TYPE ... 111

E. Meyer-Hofmeister and H. Ritter
ACCRETION DISKS IN CLOSE BINARIES ... 143

A.R. King and J.P. Lasota
MAGNETIC CATACLYSMIC VARIABLES ... 169

A.A. Boyarchuk
SYMBIOTIC STARS ... 189

S. Starrfield
THEORY AND OBSERVATIONS OF CLASSICAL NOVAE
   IN OUTBURST ... 209

K. Nomoto
INTERACTING BINARIES AND TYPE I SUPERNOVAE 247

V. Trimble
BINARY SYSTEMS WITH COMPACT COMPONENTS 271

T.C. Weekes
TeV RADIATION FROM BINARY STARS 297

I.A. Ahmad
ζ AURIGAE ATMOSPHERIC ECLIPSING BINARIES 305

R.H. Koch
POLARIZED RADIATION FROM CLOSE BINARIES 311

N. M. Elias II and R.L. Mutel
RADIO EMISSION FROM ACTIVE LATE-TYPE BINARIES 335

A.P. Boss
FORMATION OF BINARY STARS 355

G.E. McCluskey, Jr. and Y. Kondo
THE EVOLUTION OF INTERACTING BINARY SYSTEMS:
   OBSERVATIONAL ASPECTS 381

J. Sahade, G.E. McCluskey, Jr., and Y. Kondo
PENDING PROBLEMS AND FINAL COMMENTS 399

REFERENCES 403

INDEX 443

INTRODUCTION

More than two centuries have elapsed since the story of interacting binaries began with the rediscovery of the variability of Algol by John Goodricke and the interpretation he proposed for explaining the regular periodic brightness variations which he found.

Over this long span of time our knowledge about these systems has been growing, and we have now reached a fairly good understanding of the structure and behavior of this interesting group of objects.

Early on, interacting binaries – a term coined independently by Paczynski and by Plavec in 1966 – posed problems of a different type and some of them were really an extremely difficult riddle to resolve.

Otto Struve was first in trying to disentangle these puzzles with the cooperation of other colleagues, and his efforts, which covered about a quarter of a century, led to what Popper has called the "Struve revolution".

The concepts of *gaseous stream*; *gaseous ring* (this suggested by A.H. Joy from his observations of RW Tauri), *expanding outer envelope*, *Roche equipotential lobes*, belong to this epoch, and one of Struve's co-workers, Su-shu Huang, introduced the idea of a *thick disk* around the fainter companion in ϵ Aurigae and in β Lyrae, which then was extended by others to describe the behavior of cataclysmic variables.

A number of other people contributed to the growth of our knowledge in this field. F.B. Wood was the first to try to explain period changes as due to mass loss, and J.A. Crawford and Z. Kopal, independently, suggested an explanation for the Algol paradox, which opened up the field of the evolution of close binaries. This took place initially through the work of three separate groups, namely those of Kippenhahn and Weigert, Paczynski, and Plavec.

An important breakthrough resulted from the discoveries of the binary nature of the recurrent nova T Coronae Borealis by Sanford in 1949, of the binary nature of the dwarf novae SS Cygni and AE Aquarii by Joy in 1954 and 1956 respectively, and of the binary nature of the old novae by Kraft in 1964: all this brought to light the fact that all cataclysmic variables are close binary systems. We could add the suggestion by Walker and Herbig in 1954, of the existence of a *hot spot* in the nebular ring around the hot component in UX Ursae Majoris, which later came to be considered as a common feature of the region where the gaseous stream from the companion impacts upon the disk that surrounds the white dwarf component in cataclysmic variables.

In recent years substantial progress has been made in different ways, e.g., through the discovery of radio emissions in a number of interacting binaries or through high time resolution of circumstellar envelopes in Algols, by Kaitchuck and his group.

The application of space technology for observations in the ultraviolet, particularly with the IUE satellite, and at X-ray energies, have provided us with information that has proven to be invaluable in rapidly advancing our knowledge of the structure and dynamics of the gaseous matter in interacting binaries.

In addition, new observational technology on the ground and the use of computers have made available to us means and ways to obtain and to process observational material in a manner that was undreamed of only a few years ago.

It seems timely to try to summarize in a book our present knowledge of interacting binaries, each aspect written by a colleague who has done relevant research in the field. With the contributions of the authors of the different chapters – to whom we express our warm appreciation and thanks for their effort – we have attempted to achieve this aim. We hope to have succeeded.

THE EDITORS,
JORGE SAHADE
GEORGE E. MCCLUSKEY, JR.
YOJI KONDO

THE DEVELOPMENT OF THE IDEA OF INTERACTING DOUBLE STARS

A. H. BATTEN and F. B. WOOD

1. Introduction

The term "interacting binaries" has become increasingly popular over the last decade, but is rarely precisely defined. Perhaps the growing popularity is a consequence of the realization that the older term "close binary systems" means rather different things to the observers of eclipsing and spectroscopic binaries on the one hand, and of visual binaries on the other, but the new term is not completely satisfactory. By definition, any binary system is composed of two stars that interact gravitationally to the extent that they are constrained to move in closed orbits about a common centre of mass. In this sense, therefore, *all* binaries are "interacting". Furthermore, many if not most binaries that are observed either as eclipsing or spectroscopic binaries, show evidence of tidal distortion (another manifestation of gravitational interaction) or the mutual irradiation that we erroneously but conveniently call "reflection". When we refer to "interacting binaries", we usually mean to imply stronger interactions that may significantly affect the course of evolution of the two components and of the system as a whole. By "interacting binary" we mean a system in which at least one of the component stars is unstable and expelling matter, either into the gravitational field of the other, or out of the system altogether. In any given system, of course, both these things may be (and probably are) happening together, and matter expelled from the system does not necessarily leave it immediately. The unstable star may be so *because* of the presence of the companion – as we believe is the case for the secondary components of Algol systems – or it may simply have reached an unstable phase in its own evolution. A massive early-type star, for example, may lose considerable mass through a stellar wind, whether or not it is attended by a companion. If it does have a companion, we will regard the system as an interacting binary. There is not, of course, a hard and fast distinction between these two causes of instability: an intrinsically unstable star may be further disturbed by a companion. Thus the filling of the appropriate Roche lobe by either member of a system is not a necessary condition for interaction, in the sense we are defining it, or for the presence of circumstellar matter in the system. Our definition, therefore, does not permit us to draw a firm distinction, in terms of a single parameter, such as orbital period or major semi-axis, between "interacting" and "non-interacting" binaries. To a large extent, it is the distinction between spectroscopic and eclipsing binaries on the one hand, and visual binaries on the other, but there are exceptions both ways. Popper (1980) and Andersen and colleagues (e.g. Andersen 1991, Andersen *et al.* 1988) have studied many well-behaved main-sequence, and even evolved, spectroscopic binaries that are certainly not yet interacting, while Kopal (1984) has frequently reminded us that the mere existence of Sirius B demonstrates that wide visual binaries can also interact.

If, despite these reservations, we agree to regard the distinction between interacting and non-interacting binaries as roughly equivalent to that between eclipsing and visual pairs, then we can date the beginning of the study of interacting binaries from the recognition of the first eclipsing one. This is the more appropriate since the binary in question, Algol, is interacting. We cannot know for certain which was the first eclipsing variable to be noticed, or who first observed one. The oldest known European observation was of a minimum of Algol, noted by Montanari in 1670. Earlier observations, either by Europeans or by members of other civilizations, may one day come to light, but we can be fairly sure that it was John Goodricke (1783) who was the first to recognize (in 1782) that Algol's variation in light is periodic, and to suggest eclipses in a possible binary system as the cause. Eclipsing and visual binaries, therefore, were recognized at about the same time - towards the end of the eighteenth century, when the "binary hypothesis" suddenly became fashionable. Perhaps the hypothesis became popular because instruments had just developed to the point at which stellar duplicity could be recognized easily, but the popularity might, as M.J. Crowe (private communication) has suggested, have been a manifestation of an early form of SETI. Faint companions of bright stars were thought of as possible planets.

In their book *Interacting Binary Stars* Sahade and Wood (1978) suggested that there had been four different epochs in the study of close binaries, and gave reasons for their division of them. The first epoch ran from the work of Goodricke to the papers by Russell, which provided a theoretical treatment adequate for that time. The second era extended until the advent of photoelectric photometry, while the third, beginning about 1941, featured among other things what has been called "the Struve revolution" (Popper 1970a), which produced marked changes in our interpretation – especially of the spectrographic work. The fourth era was, rather hesitantly, thought of as beginning around 1966, and it may be considered as still continuing. Much of our discussion concentrates on the second and third epochs, but material from the other two is included wherever pertinent.

2. Treading the Royal Road

Although some attempts to develop Goodricke's hypothesis were made in the nineteenth century, and more eclipsing variables were discovered, little progress was made until Vogel (1890) measured spectroscopically the orbital motion of the brighter component of Algol. It took several more decades before Algol was recognized to be interacting. The dominant investigator of eclipsing binaries in the early decades of this century was Henry Norris Russell, and it is instructive to try to see that period through his eyes. This we can do by reading "The Royal Road of Eclipses" – the text of the first Russell lecture, which he gave himself in 1946 (Russell 1948). The emphasis of this lecture is on the "classical" motives for observing eclipsing binaries – the determination of masses, radii, luminosities, limb-darkening and density concentrations. Only in the last few pages (about ten per cent of the whole) are the phenomena mentioned that we now call interactions. No doubt this reflected Russell's own interests, but he could have justified his emphasis by pointing to the relative importance of his chosen topics, as he and his contemporaries perceived it, throughout his active career. Only in the last few years before he gave that lecture were astronomers beginning to recognize the reality of interactive phenomena as we have defined them. Russell's close colleague Dugan (1920) tried to explain the asymmetry of the light-curve of U Cephei by tidal distortion of one of the components, while Carpenter's (1930) velocity-curve for the same system, now seen as a vital clue to our understanding, was ignored. At about the same time as Carpenter was studying U Cephei, Walter (1931) saw that, in Algol systems, one component was barely stable, but no-one (including Walter himself) then knew enough about stellar evolution to understand the full significance of this fact. Joy (1942), following up an earlier discovery by Wyse (1934) of emission lines in the eclipse spectrum of RW Tauri, provided the first evidence of interaction that had to be taken seriously and which stimulated Otto Struve's surveys of interacting binaries – on which he reported (Struve 1948) at the same symposium during which Russell gave his lecture. Also just a few years before that lecture was the famous frontal attack on β Lyrae, led by Struve (1941), in which the leading close-binary specialists of the day combined forces to produce five major papers on the system in one issue of the Astrophysical Journal. Their interpretations read quaintly today, but Kuiper's (1941) paper contained many ideas of importance to our present understanding of interactions.

3. Into the Wilderness

Russell's lecture was a watershed: after thus summing up progress in the field as it appeared to him in 1946, he published very little more. Younger workers turned their attention to what we now call interacting binaries and the symposium at which both Russell and Struve spoke can even be seen as a symbolic falling of the former's mantle on the latter's shoulders. If Russell could have chosen who would follow him as the leading student of close binaries, one could imagine he might have chosen almost anyone else. Russell, primarily a theoretician, with his interest directed towards the eclipses of binary systems and his temperament leading him to search for the elegant graphical approach to computation, showed by his attitude to Carpenter's work that he underestimated the role of spectroscopy. Struve, primarily an observer, thoroughly trained in spectroscopy by Frost and with a temperament that led him to make inspired guesses without working them out, seldom if ever made photometric observations and pushed spectroscopy – and spectrographs – to their limits. His "reign" as the leading student of close binaries was destined to be short, however. Born twenty years after Russell, he was to die only six years after.

Russell had good reasons for caution about spectroscopy. His own career began just about the same time that useful determinations of radial-velocity became possible. During his life-time, most spectrographs were prism spectrographs attached to the Cassegrain foci of reflectors we would now call (at best) "middle-sized", or even to refractors – such as Struve's own instrument at Yerkes. They were slow and limited in precision – especially when applied to the measurement of velocities of B-type stars such as Struve was interested in. Photometry could give more information more precisely and more quickly, even before the development of photoelectric photometry. The first attempts at the "electrical measurement of starlight" were (according to Minchen 1895) made in Dublin in 1892 by Monck and Fitzgerald. The great pioneer was Russell's contemporary and compatriot, who published an important monograph (Stebbins 1928).

Before photoelectric photometry came into common use, and after visual or photographic estimates (as opposed to measures) had been found to be of little value for the purpose of finding orbital elements from light-curves (even in "well-behaved" systems such estimates led only to rough answers), most photometric observations were made with the visual polarizing photometer. In particular, the work of R.S. Dugan and his students, with such an instrument, gave light-curves from which relative sizes, orbital inclinations, etc. could be deduced with some confidence. In a polarizing photometer, the light of the comparison star was varied until it and the variable, seen side by side in the eyepiece, appeared equal in brightness. The eye is at its best in such circumstances, and the results could usually be trusted. Most of these remarks could also be applied to observations made with the wedge photometer, but the calibration of the wedge – which could change – added to the uncertainty. In fact, wedge photometers were little used compared with the polarizing photometer in the hands of Dugan's school (see , e.g. Wood 1946). The most accurate photographic work was done by taking plates deliberately out of focus, to provide a series of images of the telescope objective as seen in the light of the star. These images could be measured photoelectrically and compared with a series of images previously put on the plates in the darkroom. The method was extremely laborious and also little used, but its results were comparable with the best visual work. In all these methods, however, the uncertainties were such that variations from the standard model were rarely blamed on the system itself – which was assumed to show the same variation, cycle after cycle. Sometimes "night errors" were added to get a better light-curve – and frequently this was not reported, so a good deal of information about intrinsic changes has probably been lost.

Photoelectric photometry remained a difficult procedure and other astronomers were slow to follow Stebbins' lead, not because they failed to appreciate the value son 1991). Observations in different colours may now be treated simultaneously and even spectroscopic observations can be included in the one solution (Hill, 1979, Wilson, 1979, van Hamme and Wilson, 1984). For example, the mass-ratio of the system can now be determined photometrically, either in conjunction with spectroscopic observations or in their absence. The difficulties of old-fashioned rectification gave a bad name to such "photometric" mass-ratios (derived from the observed effects of distortion of the stars). Now, photometric mass-ratios are reliable enough to be used for systems in which only one spectrum is visible; they are particularly valuable for W Ursae Majoris systems, the spectrographic mass-ratios of which were, until recently, difficult to determine. Even the rate of rotation of the primary components of some systems can now be determined from the light-curve (Wilson 1989) which promises to be of great value for the understanding of these systems. It has become possible to specify more certainly what parts of the light changes observed between eclipses arise from causes other than ellipticity and reflection. The most obvious other cause is the presence of gas-streams in a system. No generally accepted method of analysis of the effects of gas-streams has yet been devised, although a start has been made for U Cephei (Olson 1980). Particular problems were experienced with the interpretation of light-curves (and indeed, of the velocity-curves) of W Ursae Majoris systems. Not until the seminal papers by Lucy (1967, 1968) was progress made with the light-curves, while the velocity-curves had to wait still longer (McLean 1981, McLean and Hilditch 1983).

Thus, 1970 can indeed be seen as the year in which students of binary stars stood on a threshold: they were armed with new analytical tools, new observational techniques were at hand, and the interest of the wider astronomical community in at least some classes of interacting binary had been aroused. How well we took advantage of that situation, the reader must judge when he or she has read the more detailed chapters that follow.

Plavec (1983) quoted Beer (1958) who, in turn, cited a forgotten source: "By crossing the borderline from the definite realm of geometry into the wilderness of astrophysics, we may run the risk of losing our way – but there is also the happy prospect of converting the wilderness gradually into a paradise". Plavec added the comment, perhaps sparked by the fact that Beer's paper was about W Serpentis, "Apparently the conversion of the wilderness has not yet advanced too far. At best we are in purgatory". Can we, nine years later, take a more optimistic view? Russell's Royal Road was certainly in the definite realm of geometry. The transfer of leadership in binary-star research from Russell to Struve symbolized the leaving of that realm and entry into what has sometimes looked like wilderness. Perhaps, even, some of us have, at times, lost our way. But there are signs of hope that we have all learned some important lessons and the detours of the decades before 1970 were not time wasted nor perhaps even unnecessary. Maybe we are at the gates of Paradise, but such recent arrivals that we are verily strangers there. On the other hand, many of our ideas may yet be turned on their heads several times over, and we shall have to be content to say with the poet

...Wilderness is Paradise enow.

The poet in question (Edward FitzGerald) claimed to be translating Omar Khayyam. Perhaps, after all, that eleventh-century Persian astronomer knew of the variability of Algol!

THE INTERACTING BINARY ZOO

J. SAHADE, G. E. MCCLUSKEY, JR. and Y. KONDO

When we look at the close binary zoo, we find a number of definite groups among them, most of which represent an interacting stage in their evolution. The groups we find among close binaries are:
- RS Canum Venaticorum stars
- ζ Aurigae stars
- Algols
- W Ursae Majoris systems
- cataclysmic variables*
- barium stars*
- symbiotic stars*
- X-ray binaries*

These groups are not listed in an order which implies any sequence in the evolution of close binary systems. In fact, there is no evidence whatsoever that one certain group evolves into any other, except perhaps in the case of the W Ursae Majoris stars which may become dwarf novae (Sahade 1959; Kraft 1962, 1963).

Those systems that are interacting must be undergoing slow mass loss and some of them, those that have been singled out with an asterisk, appear to be undergoing a second episode of mass loss.

Besides the groups we have listed, we also find a number of very peculiar systems, like β Lyrae (W Serpentis *et al.*) that do not seem to belong to any of them. Building on a suggestion by Kondo (1988) that "R Arae may be in the earliest stage of supercritical mass flow", Sahade (1986; 1987; 1988a,b,c) proposed that "the very peculiar systems are in the elusive stage of rapid mass loss" near the mass-ratio reversal phase, and that one could tentatively arrange them in a sequence that might describe the evolution of the gaseous structure in interacting binaries throughout the rapid mass loss episode. Such a sequence would be defined, before mass- ratio reversal, by R Arae, W Ser and GG Carinae, and, after mass-ratio reversal, by β Lyr, V453 Scorpii, U Cephei and the typical Algols, the sequence being given in the sense of the possibly advancing evolution of the gaseous structure. Sahade's suggestion is only a working hypothesis in an effort to make sense out of the peculiar interacting binary systems and is not to be taken to mean that one system would necessarily evolve into the next one and so on. Moreover, the suggestion may only apply to the evolutionary stages related to the first episode of mass loss; an understanding of the evolution of a peculiar system like SS 433 probably requires a different approach.

Most of the groups that we have listed will be discussed further in different chapters of this book.

Other groups of objects have, at one time or another, been considered to be binaries, but later work did not confirm such a belief. For instance, in the middle 50's it was thought that all Wolf-Rayet stars were close binaries and that their bizarre spectrum arose because of their binary nature. Later it became clear that this was not the case. Although a number of Wolf-Rayet stars were indeed members of binary systems, the large majority of them appear to be single objects. Actually, of the 159 Wolf-Rayet objects in van der Hucht *et al.*'s (1981) Catalogue, only 20 are definitely members of physical pairs. The same is true of the β Canis Majoris and δ Scuti objects and of the Be stars.

We should also mention the case of the central stars of planetary nebulae. In the late 70's the evidence seemed to suggest that they were all binaries but "at present it is too early", according to Méndez (1989), "to suggest a number for the percentage of close binary CSPN". The search for the answer in this case continues.

From what we have said so far it is clear that our zoo includes many different types of objects displaying a wide range of parameters and characteristics.

Ever since Ludendorff (1903) determined the orbital period of ϵ Aurigae, 27.1 years, this peculiar and still puzzling eclipsing interacting binary has represented the longest period found among eclipsing pairs. With regard to the shortest orbital periodicity, shorter and shorter periods have been found as better time resolution techniques became available. Until 1964, the shortest periods known were those of the nova-like variable UX Ursae Majoris (0.1967 days) and the old nova DQ Herculis (0.1936 days). Then Krzeminski and Kraft (1964) discovered that the recurrent nova WZ Sagittae is a binary with an orbital period of 0.05669 days, less than one and a half hours. This remained the smallest known value for an orbital period among close binaries until very recently when Abbles *et al.* (1989) announced the discovery of the millisecond pulsar PSR 0021–72 A in the southern globular cluster 47 Tucanae (NGC 104) as a binary with an orbital period of 32.4 minutes, which brings the range of periods in the close binary zoo to almost a factor of 440,000.

It is interesting to note that as we go to more evolved systems, which, of course, implies that we go to systems with more compact components, we find the shortest periods. And since the progenitor systems must have had larger, perhaps more massive components – except when their genesis follows a mechanism such as that discussed in a recent paper by Nomoto and Kondo (1990) –, and, consequently, longer periods, the conclusion is that mass loss does not necessarily mean period increase and that there should exist other mechanisms at work which bring the stars closer together as evolution proceeds. This question was brought up by Sahade (1959) some three decades ago, and recently, Eggleton (1986) has discussed processes that may shorten the period of a binary. Essentially the same problem has been dealt with by De Cuiper (1985) when considering possible progenitors of two neutron-star binaries, GX 301–2 and PSR 1913 + 16.

The conclusion from what we have said so far is that among the close binary zoo members there is a very large range in object sizes, going from the red supergiants that we find in the ζ Aur systems and in ϵ Aur, with radii as large as some 200 solar radii, that is, almost 1.48×10^8 kilometers, to the neutron stars which are supposed to be no larger than 10–20 kilometers in radius. These figures imply a range in the radii of the component stars, in the extreme case, of a factor of the order of 10^7.

As for the masses, if we exclude the case of 5 Lacertae, the elements of which are not only "difficult to assess" but also unreliable (Batten, Fletcher and MacCarthy 1989), we can state that the most massive system known is still Plaskett's star, HD 47129. The total mass of this pair is somewhere around 110–120 solar masses, each component having about 50 or 60 solar masses.

At the other extreme, we find objects that are substantially less massive than the Sun. Extreme cases among the Algols are provided by R Canis Majoris, S Cancri, and XZ Sagittarii, the fainter components having masses of $0.17 \pm 0.02 M_\odot$ (Tomkin 1985), $0.175 \pm 0.020 M_\odot$ (Popper and Tomkin 1984) and $0.26 M_\odot$ (Smak 1965), respectively. These values imply a range of masses of a factor of about 300. Among wide binaries, a dark component has been found with a mass of the order of twice that of Jupiter, that is, about 0.002 solar mass, and, among close binaries, there are probably masses much smaller than those just quoted for S Cnc, R CMa, and XZ Sgr, although at the moment there is no clear-cut case to mention here.

Regarding mass-ratios, we have come a long way since it was believed that the components of close binary systems were characterized by mass-ratios very close to unity. The first shock came when it was found that the mass-function in the case of R CMa, was very small, of the order of a few thousandths of the mass of the Sun, which indicated that the two masses were very small or that the mass-ratios departed drastically from unity, which is actually the case. Our understanding of binary evolution and our knowledge of what our zoo really contains make it predictable that we will find a wide range in mass-function values, a parameter which tells us much about the characteristics of the objects we are studying.

The interacting binary zoo is, therefore, varied and exciting and poses interesting problems and queries. We lack, however, a unified picture. Actually, we are dealing with a real puzzle and we are trying to put in place its different pieces. Progress has been enormous since Goodricke's time, particulary since space technology and computer facilites became available to us.

MASS LOSS IN INTERACTING BINARY SYSTEMS

J. SAHADE

1. Introduction

Evidence for mass loss in interacting binary stars was first found in the system of β Lyrae when Struve (1941) and Kuiper (1941) attempted to explain the peculiar spectral features that characterize this object (cf. Sahade 1980) and Kuiper (1941) applied the concept of zero-velocity surfaces. Then the question of gas streaming within the system arose. In the early picture of the system, β Lyrae was depicted as displaying gaseous streams from the two components, but this was later shown to be incorrect. Actually, in β Lyrae there is only one stream arising in the B8 II component and giving rise to an optically thick disk around the companion (Batten and Sahade 1973).

Further evidence for the existence of gaseous streams in interacting binary stars was provided by U Cephei (Struve 1944a), SX Cassiopeiae (Struve 1944b), and AU Monocerotis (Sahade and Cesco 1945) as well as by other systems. In every case the stream arose in the larger, less massive component.

Additional work led to the concept of the Algol paradox (cf. Struve 1948; Sahade and Wood 1978a) and it then became clear that in the course of evolution of these interacting binary stars, the more evolved, and now the less massive component, must have lost a considerable amount of mass. Further evolution may later produce a second episode of mass loss, as seems to be the case in cataclysmic binary stars, barium stars, and X-ray binary stars.

We find a relatively large number of Algol systems while we do not seem to find mass losing systems before reversal of the mass-ratio has taken place. This, in turn, suggests that in the evolutionary mass loss process there ought to be first an episode of rapid mass loss followed by an episode of slow mass loss. It is in this latter stage that we are able to detect and observe evolved systems.

Mass loss in a close binary system also gives rise to period changes, as pointed out by Kuiper (1941) and by Wood (1950) in an early attempt to understand the period changes in β Lyr and the erratic and irregular period changes found in a number of systems by Dugan and Wright (1937), respectively. Formulae for the prediction of the times of minima would then contain second power and third power terms, as in the case of β Lyr (cf. Sahade 1980).

Huang (1963) was first to compute the effect of mass loss upon the orbital period of a close binary; the increase in period of β Lyr, *in the conservative case,* implying a value of $\Delta M/(M_1 + M_2)$ of the order of $2 \times 10^{-4} M_0$ per year.

Bolton (1989) has discussed the effect of variations in magnetic field strength of the cool component, upon the orbital period in Algol binary stars, due to changes in this star's structure, and gives references to previous related discussions. The existence of magnetic fields, at least in some types of objects like Algol, the RS Canum Venaticorum stars and X-ray binary stars, is suggested by synchrotron-type processes giving rise to radio emission (cf. Sahade and Wood 1978). Recent references to work that implies detection of magnetic fields in several interacting binary stars are the papers by Stewart et al. (1989) and by Elias (1990).

Early on, it was thought that the mass that was lost by one of the components of an interacting binary was totally accreted by the companion. However, this could not be so because in many systems we observe features that suggest the existence of expanding, circumbinary envelopes. Moreover, the computation of binary evolution in the conservative case, i.e., assuming no mass loss to the system, could not reproduce observed Algol systems (cf. Plavec 1973). Other arguments are given, for instance, by DeLoore (1984). In consequence, mass loss from the system is an accepted fact at present, and important questions that are open relate to how much mass is actually lost to the system and how much mass is actually accreted by the component towards which the stream is directed. The loss and exchange of angular momentum is also critical to the evolution of an interacting binary.

Mass loss in interacting binary stars gives rise to a gaseous structure in which the stellar components are embedded. As is well known (cf. Batten 1970, 1973a, 1973b; Sahade 1973; Sahade and Wood 1978b), in such a structure we distinguish, besides the *gaseous stream* from the less massive component, a *circumstellar* envelope around the companion, and a *circumbinary envelope*.

2. Gas Streaming

The presence of gaseous streams in interacting binary stars could be inferred from the effects they produce, namely,

a) a distortion in the velocity curve of the component towards which the stream is directed, that is present immediately after second quadrature and results from the fact that the stream which projects upon the star, has larger relative velocities. Such a distortion explains the peculiar distribution of the longitudes of periastron in spectroscope binary stars, found by Barr (1908) and coined the Barr effect by Struve;

b) a (variable) weakening (or veiling) of the spectrum of the component towards which the stream is directed–because of its opacity (Sahade 1959)–when it is seen projected upon the trailing hemisphere of the star (cf. Sahade and Wood 1978);

c) emission lines at quadratures, as shown by Algol (Struve and Sahade 1957; Sahade 1958; see also Gillet et al. 1989 and references therein) and by UX Monocerotis (Struve 1947);

d) abnormalities in the light curve of eclipsing systems, such as depressions starting about 1/4 P before primary eclipse, etc.

All these effects can, of course, be variable.

Batten (1973a) has estimated that the electron density of the streams is of the order of 10^{13} electrons cm^{-3}, while, according to Warner and Nather (1971), this figure may be 10^3 times larger in the case of cataclysmic variables.

3. Circumstellar Envelopes

The existence of circumstellar envelopes in mass-losing close binary systems was first detected by Joy (1942) in the system RW Tauri, where he observed double H emission, that underwent eclipses at the time the large subgiant component passed in front of the brighter, early type companion.

The circumstellar envelopes are normally disk-shaped and their densities show quite a range in values depending on whether we consider a system like U Cep or SX Cassiopeiae, β Lyr, or a cataclysmic variable.

In the first case, we are dealing with thin, transient formations; in the second place, with relatively transparent, more permanent features (earlier called gaseous rings) while in the two last cases we have optically thick, flat envelopes that have always been known as accretion disks, particularly in the case of cataclysmic variables.

Lubow and Shu (1975) have modelled, within the context of the Roche model, the gas dynamics in semi-detached systems. They concluded that "if the detached component is smaller than a certain specified size, the stream results in the formation of a disk of material of prescribed size orbiting the detached component in a direct sense."

Thus, in the case of Algol, (Gillet, Mouchet and North 1989) the size of the detached component appears to be too large relative to the separation of the stars in the system and there is no circumstellar envelope formation; the emission observed comes from matter in the stream between Algol A and Algol B.

A survey for gaseous disks made by Kaitchuck and Honeycutt (1982a) and Kaitchuck, Honeycutt and Schlegel (1985) led to the conclusion that for orbital periods larger than five days "the gas stream moves around the trailing hemisphere of the primary and forms a permanent disk." For periods shorter than five days, "the gas stream strikes the primary star and disks are bound to be transient and unstable" and are rare. Peters (1989) finds further that the most variable systems are those in the five-six day range.

In the case of cataclysmic variables, where one of the components is highly evolved and small–a white dwarf in the novae–the circumstellar envelope is always thick. But in binary stars with no compact components, with periods longer than five days, we find cases with thin envelopes (the typical Algol systems) and cases with thick envelopes (β Lyr, V453 Scorpii) and this difference may be related to the mass loss rate in the system and then perhaps to whether or not the system is just evolving out of the rapid mass loss stage. In the case of thin envelopes we can still see the stellar spectral features, while in the case of the thick envelopes we only observe emission arising therein.

As for the characteristics of these stellar envelopes, we can say that they rotate in the same direction as the orbital motion (cf. Sahade 1960) but that the motions of the particles are non-Keplerian (Crawford 1981; Kaitchuck, Honeycutt and Faulkner 1989). There are normally differences between the leading and the trailing sides of the envelope which show in the line strengths, as already pointed out by Struve (1946b) and as was nicely illustrated in a recent paper by Peters (1989), and also in the velocities.

The emission lines are most probably broadened by turbulence. In a summary of our knowledge of β Lyr, Sahade and Wood (1978c) pointed out the fact that "the observations and several lines of argument suggest that turbulent mass motions on a massive scale are present in the envelope of β Lyrae", and recently, Kaitchuck and Park (1988) have stated, coming back to statements by Crawford (1981) and by Kaitchuck and Honeycutt (1982) that "there is growing evidence that this broadening mechanism [for the emission of TZ Eridani] is supersonic turbulence".

Olson (1989) has pointed out that the presence of a disk in a non-degenerate eclipsing binary is expected to produce "two distinct signatures in the light curves: 1) before and after primary stellar eclipses the disk produces 'dips' bordering the stellar eclipse; 2) around the secondary eclipse, the disk itself partially occults the cool star". The disk is also supposed to be responsible for the strong ultraviolet continuum found by Popper (1964) in KU Cygni and RZ Ophiuchi.

In cataclysmic variables, the disk provides most of the light of these systems and conditions in the the light curve are related to the interaction between the stream and the disk (cf. Sahade and Wood 1978d), a 'hot spot' being formed in the disk because of such an interaction, which in appropriate cases, gives rise to soft X-rays. Olson (1980) has found evidence for a 'hot spot' in U Cep, the relative importance of this formation depending on envelope thickness.

The questions of variable polarization and electron scattering in the disks is discussed in another chapter of the present book and, as a consequence, we will not consider them here.

We have been discussing the circumstellar envelopes that form around one of the components of a close binary in the process of mass loss that these objects undergo during their evolution. Such circumstellar envelopes have been termed accretion disks, and it might be in order to point out that one of the signatures that indicate that accretion actually does take place is found in systems, like U Cephei, RY Persei, RZ Scuti, S Velorum, etc., where the rotational velocities are much higher than the orbital velocities. Matter is accreted together with angular momentum and the systems with these characteristics are probably in an evolutionary stage immediately following the rapid mass loss stage, insufficient time having elapsed for synchronization between rotation and orbital motion to have been attained.

Another piece of evidence indicating that accretion processes take place in interacting binary stars is provided by the X-ray binary stars, which we understand in terms of the thermal energy that is released when the matter in the stream from the 'optical' component is accreted by the compact companion.

4. Circumbinary Envelopes

The existence of expanding circumbinary envelopes in interacting systems provides evidence for mass loss to the system. Such circumbinary envelopes are suggested by the presence in the spectrum of lines affected by dilution which yield velocities of approach in the range of -1300 to -170 kms^{-1} for different types of binary stars (cf. Sahade and Wood 1978b).

Sometimes evidence is provided by the presence in the spectrum of rather broad H emissions with a steep Balmer decrement and/or by the presence of emission lines that correspond to forbidden or semi-forbidden transitions, e.g., in W Serpentis, RY Scuti, v Sagittarii, γ_2 Velorum, etc. In these cases the velocities that are derived are rather small.

In the early picture of the circumbinary envelope of β Lyr, Kuiper (1941) concluded that it ought to spiral around the system and it is interesting to point out that the well distributed, phasewise, IUE observations of U Cep by Kondo, McCluskey and Harvel (1981) suggest a trend in the stream motion that may lead precisely in that kind of formation. However, a part of the gas stream was observed to be leaving the binary system; Kondo *et al.* reasoned, therefore, that the observations indicated "the presence of sources of kinetic energy for the gqs stream in addition to that arising from the conversion of the gravitational potentials".

5. Non-thermal Sources of Energy

During one of the discussion periods at the IAU Symposium No. 51, Smak (1973) and Popper (1973) brought up evidence of non-thermal effects in interacting binary stars by pointing out that the degree of excitation of the circumstellar envelopes seems to be similar, no matter whether we are dealing with early-type or intermediate- type primaries.

Actually, satellite observations have disclosed the presence of the high temperature resonance lines of N V, C IV, and Si IV in the spectra of interacting binary stars and the fact that non-degenerate binary stars like Algol and β Lyr are X-ray emitters.

Normally, one finds up to three regions where the resonance lines of NV, CIV and, SiIV are formed: one around the component of the system that has the circumstellar envelope, one at about the maximum velocity of approach of the circumbinary envelope and a third at its outermost edges, where the envelope merges with the interstellar medium. In the case of β Lyr, this has been clearly illustrated in a discussion of the line profiles (Aydin *et al.* 1988).

This picture which repeats itself in the other systems (cf. Sahade, Kondo and

McCluskey 1984) and in objects like λ Pavonis (Chen et al. 1989) suggests that the circumbinary envelopes first expand, reach a maximum value near the layer with maximum temperature, where the high temperature resonance lines in the ultraviolet are formed, and then the envelope decelerates. In turn, this picture poses the problem as to how this takes place and what are the mechanisms by which mass is actually lost to the system.

6. Mass Loss Rates

The determination of mass-loss rates in interacting binary stars is not an easy task. Recently, Olson and Bell (1989) attempted to derive mass-loss rate values for three systems by considering non-LTE and LTE predictions of the stream-produced light losses. The results obtained are as follows:

TABLE I
Mass-loss rate in $M_\odot yr^{-1}$

	non LTE	LTE
U Sge	4×10^{-8}	3×10^{-8}
AO Cas	5×10^{-7}	6×10^{-7}
RZ Sct	6×10^{-7}	6×10^{-7}

Earlier, Olson (1987) had obtained for U Sge:

$$\dot{M} \lesssim 5.2 \times 10^{-7} M_\odot yr^{-1}$$

by making use of Lubow and Shu's (1975, 1976) theory of stream structure. Olson and Bell's figures are smaller by a factor of 10^3 to 10^4 if we compare them to the figure we believe to roughly represent a case such as that of β Lyr, which would be closer to the rapid mass loss stage of evolution.

7. Wind and Wind Collisions

So far, we have considered mass loss through gaseous streams from one of the components of a close binary system, but, as we know, loss of mass in stars, also takes place through the mechanism of the so-called *stellar wind*. This process, as well as radiation pressure, must be particularly important in early type systems. HD 47129 (Struve, Sahade and Huang 1958) provides an excellent example of an interacting binary that displays a stream, and a circumbinary envelope which shows the effect of the deflection of the stream by radiation pressure.

In early type binary stars we should expect to find evidence for the effect of wind collision and recent efforts seem to have been successful in this regard.

Brandi, Ferrer and Sahade (1989) and Sahade and Brandi (1990) have found observational evidence, in the IUE spectra of γ_2 Vel, HD 47129, and AO Cas, respectively, for the kind of nebular formation between the two components that was predicted by Wallerstein *et al.* (1984) and by Willson et al. (1984) as a wind collision effect in two eruptive symbiotic systems. In addition, Shore and Brown (1988) and Gies and Wiggs (1990) have reported on detecting wind collision effects in the Wolf-Rayet eclipsing binary V444 Cygni and in the early type system AO Cas, respectively.

The effect of wind collision is an important element in the gaseous structure of at least the early type interacting binary stars and the pursuance of these studies will help in reaching a better understanding of the significance of the different processes taking place.

8. Other Mass Loss Mechanisms

The discovery of large erratic fluctuations in radial velocity and line intensity in systems such as UX Monocerotis (Struve 1947), and S Velorum (Sahade 1952) and the behavior of HD 47129 (Struve, Sahade and Huang 1958) and, in fact, in nearly all systems with gaseous streams indicates that a mechanism such as *prominence activity* might also contribute to mass loss in interacting binary stars (cf. Sahade 1960). Now with the extension of the range of energy in which we can perform observations, such a possibility has become an observational fact. Actually, flare activity is a common phenomenon in at least the most active kind of objects we are dealing with. Examples of such flare activity are found in the radio and X-ray regions of the spectrum in Algol and among the RS CVn stars. In 1983, a sudden, strong outburst in the EXOSAT 1-10 keV energy range was displayed by Algol (cf. White *et al.* 1986) with a peak luminosity of 1.4×10^{31} ergs^{-1}. Such activity implies mass transfer instabilities. X-ray outbursts are, of course, normally found among X-ray binary stars.

In a rather recent paper, Stickland *et al.* (1986) suggested that tidally-induced mass loss could take place in ι Orionis, an O9 III +B1 III system with a highly eccentric orbit. A theoretical discussion of such a possible source of stellar wind enhancement is given by Tout and Eggleton (1988).

9. Are the Very Peculiar Systems in the Neighborhood of the Mass-ratio Reversal Stage?

One of the facts that is clear from the observations is that the elements of the gaseous structure in interacting binary stars have different relative importance in different systems. Sometimes the gas in which the system is embedded dominates the spectrum and may be thick enough to obscure completely or almost completely, one or both stars. Examples are given by well-known, peculiar systems like W Ser, where the spectrum of the gaseous envelope strongly superimposes upon the stellar spectrum of the brighter component, or like V453 Sco and β Lyr, where the most massive component is hidden by a thick, circumstellar envelope.

It is very likely that the peculiar systems which are characterized by thick circumstellar (excluding the cataclysmic variables) and/or thick circumbinary envelopes are in the stage or very close to the stage of rapid mass loss and, therefore, very close to the mass-ratio reversal stage. Having this in mind and building on Kondo's (1988) suggestion that R Arae and HD 207739 are probably in or near the rapid stage of evolutionary mass loss, Sahade (1986, 1987, 1988a,b,c) has proposed a working hypothesis of the evolution of the gaseous structure in the neighborhood of the mass-ratio reversal and suggested that the very peculiar systems that puzzle us are precisely in such a stage.

The mass loss process in the stage that precedes the episode of rapid loss could be characterized by the presence of a rather variable, thick plasma in the system (R Ara), then by a thicker and thicker nebulosity in which the system becomes embedded (W Ser and then GG Car?). After mass-ratio reversal, a gaseous stream from the mass-losing component becomes distinctly observed and a thick circumstellar envelope surrounds the accreting companion (β Lyr). Such an envelope becomes thinner (V453 Sco) and thinner (U Cep) as time goes by. The less active Algol would be at the end of the sequence. If this suggestion happens to be correct, then we would perhaps have before us a set of objects that store the information we need to better understand the physics and the mechanism of mass loss in interacting binary stars.

SOME ISSUES IN THE THEORY OF MASS TRANSFER

S. H. LUBOW

1. Introduction

It is now widely recognized that mass transfer in close binaries has important implications to binary structure and evolution on a wide range of length and time scales. Many of the earliest and also subsequent studies concentrated on the evolutionary significance of mass transfer (see e.g., reviews by Paczynski 1971, Plavec 1973, De Greve 1986). Observations indicate that close binaries with degenerate or nondegenerate component stars demonstrate a variety of phenomena (see reviews by Plavec 1990; White 1989). A more detailed analysis of various processes in such systems requires considerations of gas dynamics. Accordingly, this paper is largely devoted to such processes. No attempt is made to provide a comprehensive review of recent theoretical developments. Instead, attention is given to a few major problems which are of broad interest. Issues requiring further analysis are described.

First, mass transfer through a gas stream is discussed. This discussion follows the order of mass flow. Namely, the dynamics of the mass losing star is discussed, followed by the dynamics of the gas stream and the interaction of the gas stream with the disk. Finally, some striking recent results on mass transfer through winds are briefly described.

2. Dynamics of the Mass Loosing Star

Any true understanding of mass transfer must include the dynamics of the mass losing star. Most of the important dynamical processes on the mass loosing star are not directly observable. Some difficult theoretical problems remain.

2.1. SYNCHRONIZATION AND CIRCULARIZATION

Synchronization and circularization processes require the action of both a tidal potential and a frictional mechanism. There are however important differences between the synchronization and circularization processes. For a fixed orbital angular momentum, circular binary orbits have the least energy. The energy in excess of the minimum amount for the binary's angular momentum is then responsible for the eccentricity. Any process that damps orbital energy can lead to circularization, even processes that are radial about one of the stars. Synchronization requires that stellar spin angular momentum transfer to orbital angular momentum. Such a process requires nonaxisymmetric forces together with friction to produce tidal torques. Synchronization occurs for each component separately; circularization is a mutual process that can involve the dissipation from both stars. Stars can appear synchronous in their photospheres even though their interiors are nonsynchronous.

The work of Zahn (1966a,b,c, 1975, 1977) formed the basis for understanding the observed high levels of synchronization and circularization in close binaries. For late type stars, the frictional mechanism is readily identified to be due to convective dissipation in the stars' outer envelopes. This dissipation mechanism can very effectively interact with the relatively strong tidal field in the outer parts of a star. Synchronization is achieved as a star experiences a tidal lag in its outer envelope that then causes an angular momentum transfer, as do the tides of the Earth slow down the Earth's rotation. Strong effective tidal dissipation due to convection can bring stars in close binaries into synchronous rotation on timescales that are very short compared with stellar evolutionary times.

There are some modifications to this picture. This convective synchronization mechanism is reduced in effectiveness when the typical convective cell overturn time becomes longer than inverse tidal driving frequency. Goldreich and Keeley (1977), in an analysis of solar oscillations, provided a model for determining the loss of effectiveness of convective damping in such situations. Their model uses a Kolmogoroff spectrum for turbulent cascade to determine the scaling of the effective viscosity with turbulent eddy overturn time. In the case of a binary, the driving frequency is the tidal forcing frequency experienced by material in the star, which is approximately twice difference between the rotation frequency of the star and the binary orbit frequency. Since convective cell overturn times inside a star are typically $\sim 10^7 s$, even fairly small levels of nonsynchronism can lead to substantial increases in synchronization times, due to the effects of reduced viscosity. As a result, very close binaries, such as dwarf novae, may experience synchronization times that are relatively long $\sim 10^7$ years for the mass losing star whose angular rotation rate departs from synchronism by 10% of the binary angular speed (Campbell and Papaloizou 1983). Furthermore, the rapid rotation in close binary stars modifies the convective cell flow pattern. Such a modification will likely also reduce the effectiveness of the convective friction, as motions of convective elements are restricted by angular momentum conservation (see discussion in Tassoul 1978).

Some recent work has concentrated on the orbital evolution of close binaries during the pre-mainsequence phase (Zahn and Bouchet 1989). One question here is whether the longest periods of binaries that are circularized can be used as a clock to determine cluster ages (Mathieu and Mazeh 1988). If circularization is largely rapidly achieved during the pre-mainsequence phase with little subsequent change, then the answer is no. Stellar models of single stars during the phases of cloud collapse and early stellar evolution are now available (Stahler, Shu, and Taam 1980; Stahler 1988). In their early stages of evolution, stars of one solar mass or less start on the Hayashi track with radii that about 5 times larger than their main sequence radii and are highly convective. This large radius and large convective mass makes tidal dissipation potentially very efficient. Even though the pre-mainsequence phase of evolution is short-lived, under some assumptions binaries are circularized during this phase. The major assumption, has to do with the prescription for the effectiveness of convective damping in the presence of nonsynchronous rotation, as discussed above. During the contraction phase along the Hayshi track, significant departures from synchronism may be expected. Under the Goldreich and Keeley (1977) prescription described above, the pre-mainsequence tidal evolution is small. With a prescription advocated by Zahn (1966b, 1989), nonsynchronous rotation plays a weaker role in limiting the effectiveness of convective friction. With this latter model, pre-mainsequence evolution sets a binary period cut-off for circularization at about 8 days, which is somewhat supported by observations (Zahn and Bouchet 1989). Subsequent eccentricity evolution is unimportant. However, the physical basis for this latter prescription is less clear and the observational situation is controversial (see e.g., Mathieu 1992).

Furthermore, some three-dimensional simulations of solar convection suggest that new effects may need to be investigated. These simulations indicate that solar convection displays large-scale gentle flows that extend over large regions of the convection zone (e.g., Stein and Nordlund 1989; see review by Spruit, Nordlund, and Title 1990). Rather than the usual model of turbulent cascade of convection to smaller scales, an inverse cascade occurs as smaller scale downdrafts merge into larger scale flows. The consequences of this picture of convection on circularization and synchronization of binaries has yet to be determined.

Also challenging is understanding the mechanism of synchronization for binary stars of early spectral type. Although such stars have convective cores, their cores lie too deeply within the central potential of the star to be strongly affected by the tidal field of the companion. Instead, Zahn (1975) provided a theory based on resonantly excited gravity waves. Observations by Giuricin, Mardirossian, and Mezzetti (1984 a,b,c) indicate that the eccentricity damping times are consistent with Zahn's theory, yet the synchronization times appear shorter than expected if the entire star is despun.

Savonije and Papaloizou (1983 a,b) and later Goldreich and Nicholson (1989a,b) further clarified Zahn's (1975) original model. Consider the case of a faster than synchronous or oversynchronous star. Gas elements in such a star experience time-periodic forcing by a nonaxisymmetric potential. In analogy with processes in planetary rings and other disk systems, such forcing leads to a net torque exerted by the perturber (e.g., Goldreich and Tremaine 1978). This torque occurs at locations where the local natural oscillation frequency in the star is a (small) multiple of the tidal driving frequency, as seen by gas in the rotating star. This frequency is small compared with the orbit frequency for nearly synchronous stars.

How then can a frequency match always be found as the driving frequency changes continuously in time? The answer lies in the fact that gravity mode frequencies (Brunt-Vaisala buoyancy oscillation frequencies) take on a very large range of values within such a star. Such frequencies depend on the degree to which the structure of the star departs from an adiabatic law. Well within the convective core, the buoyancy frequency is very low, since the core is nearly adiabatic. Within the radiative envelope, the frequency is very high, shorter than the binary orbital frequency, since the matter is very buoyantly stable. However, at the core-envelope interface, a rapid transition is made in gravity-mode frequency that can then match nearly any frequency posed by the star's level of nonsynchronism. At this location, the resonant effects are strongly felt.

The resolution of the resonant driving is to launch a gravity-wave that propagates outward, since inward of the core-envelope interface, i.e., inside the convective core, gravity-modes are evanescent. At this location, such waves have a very long wavelength, sufficiently long that the waves can interact with the tidal field of the binary and obtain a net angular momentum flux in the stress they carry. The flux strength is determined by the magnitude of the stellar density and tidal field where it is generated, i.e. near the core edge only. Outside the core, the wavelength of the propagating wave is very short compared with scale of the tidal field, resulting in small, rapid variations in the flux with little further accumulation. The wave transports a flux of negative angular momentum outwards, towards the stellar surface. What this means is that some binary orbital angular momentum is gained, due to the negative angular momentum flux in the wave. But until the wave damps, no torque is felt by the oversynchronous star. These gravity-waves can propagate large distances before damping. Accordingly, these waves travel all the way to the surface of the star where they must quickly break by radiative damping processes. The wave deposits some of its negative angular momentum in the outer layers of the star and the star is despun there. This remarkable process then behaves like a nonlocal tide – it is exerted deep inside the star, but detorques stellar material preferentially at the surface.

Since the moment of inertia of the matter in the thin wave damping region is likely very small, then very short synchronization times can be achieved on the stellar surface. Therefore, the outer layers are preferentially despun. This picture provides an explanation for the lack of agreement of Zahn's (1975) synchronization timescale for the entire star with that implied by observations of Giuricin, Mardirossian, and Mezzetti (1984a,b,c). However, the outer layers of a star cannot despin without some effect on the underlying layers. Circulation currents, caused by Ekman pumping and various instabilities, could transport this low angular momentum inward and despin the interior as well. Unfortunately, the efficiency of these transport processes is poorly known. Therefore, firm predictions cannot be made of the exact rate of stellar surface synchronism. Understanding these large-scale transport processes remains a challenge for the future.

If the stellar surface becomes synchronous, does this process stop or can the surface be subsynchronously despun by the waves launched at the still-nonsynchronous core? At the point the stellar surface reaches synchronism, the so-called corotation radius lies at the stellar surface. The gravity waves steepen and damp as they approach the corotation resonance, despinning the layers just inside corotation and inwardly migrating the corotation radius. For oversynchronous stars, the synchronization process proceeds from outside-in, until the core becomes synchronous (Goldreich and Nicholson 1989b).

2.2. MASS FLOW IN THE SYNCHRONOUS CASE

The mass transfer model by Lubow and Shu (1975, 1976) (see also Shu and Lubow 1981) dealt with the case of synchronous mass transfer. Some issues discussed in that study remain unresolved. Somehow, gas in a star must make its way to the inner Lagrange point (L1 point) as the binary system evolves. Pressure forces are responsible for driving gas to this region. The process is somewhat analogous to the escape of air from a leaky tire. The question remains as to what the flow pattern is and what the timescale is for matter well off the stellar equator to reach the L1 point. This question is important if one is to understand, for example, how a temporary nonequatorial irradiation of a stellar surface (e.g., generated by a companion star but shaded by a disk) influences that star's mass transfer rate (e.g. Kovetz, Prialnik, and Shara 1988).

In the upper, essentially photospheric, layers of the mass losing star, the horizontal pressure forces attempt to drive matter towards the L1 point. Material at high latitudes on the stellar surface must then gain angular momentum to reach the equator. Such angular momentum problems are also faced by the Earth's global atmospheric circulation. The atmosphere's angular momentum is transported by friction to the Earth's surface (Ekman Effect), and also by turbulent transport (e.g., Houghton 1977). The transport processes are often quite complex. Some of the turbulent transport of angular momentum in the Earth's atmosphere gives unexpected results. The angular momentum is transported against its gradient, resulting in an effective negative viscosity coefficient for the process (Starr 1968). Similarly, angular momentum transport mechanisms in stars then limit the flow speeds from high latitudes towards the equator. Again, we are limited by our knowledge of transport processes and instabilities. Such processes are likely to limit these equatorward speeds to very subsonic velocities.

On the equator, there is no angular momentum barrier. Material there can then freely flow towards the L1 point at velocities that are mildly subsonic. The flow is nearly parallel to the Roche equipotentials. Near the L1 point, the material begins to rapidly cross equipotentials as it gets pushed out of the star. The flow then rapidly accelerates to supersonic speeds and forms the gas stream. The sonic transition occurs within a small region about the L1 point of radius $\sim c/\Omega$, where c is the local gas sound speed and Ω is the binary angular speed. In this so-called L1 region, the Coriolis, inertial, pressure, and gravitational forces are all comparable and are of order $c\Omega$. All these forces are important in steering the flow through the L1 region.

2.3. MASS FLOW IN NONSYNCHRONOUS BINARIES

Much of the early theoretical work on nonsynchronous binaries was motivated by the desire to compute stellar parameters in binaries that transfer mass through a gas stream and therefore fill a Roche lobe, or something like it. The observed constraints on the size of the mass losing star are then used to constrain the binary star mass ratio, by assuming the mass losing star fills its Roche lobe. Together with other data, masses of neutron stars in low mass X-ray binaries are determined. Unfortunately, the Roche model only applies to exactly synchronous stars, and this fact leads to some uncertainty in the determined neutron star masses (e.g., Bahcall 1978).

The Roche model has the elegant property that surfaces of constant potential, density, temperature, and pressure coincide. Because of this property, the last closed equipotential that surrounds only the mass losing star defines an actual stellar surface, the Roche lobe. The star cannot substantially expand beyond this surface without losing mass. Early work invoked effective potentials to replace the Roche potential in various ways so that the above properties would be retained. The limiting surface of a nonsynchronous star would then be given by the outermost effective equipotential surface that encloses only the mass losing star. Some studies attempted to include an extra centrifugal term about the center of the mass-losing star to account for the nonsynchronism. Such attempts led to a relocation of the new effective Roche lobe in a manner that was roughly linear in the nonsynchronous velocity. One major problem with this approach is that the assumed velocity field is not parallel to the star at its surface. This difficulty is caused by the introduction of an axisymmetric velocity field about a nonaxisymmetric star, resulting in a violation of mass conservation.

In the so-called tidal lobe model (Davidson and Ostriker 1973), the mass losing star is not rotating in the inertial frame, and an effective potential was derived. However, the actual flow field on the mass losing star must be steady in the frame of the binary, where the tidal field is static. Therefore, there is an implicit circular velocity field in the binary frame with the same problems as just described.

More rigorous studies of slight nonsynchronism in stars with radiative envelopes (Lubow 1979) and later in stars with convective envelopes (Scharlemann 1981, 1982; Campbell and Papaloizou 1983) led to a different picture. Such studies were based on linearizing fluid equations about the synchronous case. The velocity field is calculated from first principles. Both the radiative and convective cases obtained similar results. The velocity is such that no effective potential can describe its effects. The elegant property of the Roche potential that surfaces of constant density, pressure, and temperature coincide is destroyed. In the radiative case, stellar surface temperature variations occur due to this effect.

On the equator of a star with either a radiative or convective envelope, the local velocity scales with the local effective gravity given by the Roche potential. This means that gas is decelerated on its approach towards L1. As a result, in linear order, the location of the effective L1 point for nonsynchronous flow does not change from the synchronous case and so the limiting surface of the star does not change from the Roche lobe. The reason is simply that the mass flow here is nearly along Roche equipotentials. As the L1 point is approached, the flow diverges because the equipotentials separate. To maintain a constant mass flux between equipotential surfaces, the flow must decelerate. At some level, the departures must become important, but no detailed results are available that predict departures from the Roche lobe in higher than linear order in nonsynchronism. For levels of nonsynchronism as high as the tidal lobe model described above, no results are available.

These linearized studies were conducted by perturbing about an equilibrium state that is in hydrostatic balance. Since the L1 region is not hydrostatic, these results are not strictly valid there.

Let us consider what new effects might occur in higher order of nonsynchronism. As is well-known in the context of galactic gas flows, transsonic flows cause shocks to occur. These shocks arise as matter along a streamline attempts to make a transition from supersonic to subsonic speeds, as viewed in the frame of the perturbing potential (i.e., stellar wave) (see e.g., Shu, Milione, and Roberts 1973). Can an analogous shocked flow occur here? Deep inside a star, the nonsynchronous flow along streamlines is likely to be purely subsonic in the frame of the binary, due to the high sound speed. Close to the stellar surface, even fairly mild nonsynchronous rotation rates of say 10% the binary orbit speed are supersonic. However, along streamlines, this rotational velocity is modulated by the tidal field. The possibility exists that such a modulation in the vicinity of the L1 point could cause streamlines to undergo a sonic transition and hence shock.

Such shocks could play a role in despinning the nonsynchronous outer equatorial layers in the star. In addition, they could play an important role in the dynamics of mass transfer through the L1 region. The shocks may be critical in allowing supersonic oversynchronous flow to make a fairly sharp turn in the L1 region (about 80 degrees) and emerge as a gas stream consistent with a ballistic trajectory. The most observationally significant aspect of nonsynchronism might even be the shock emission.

The main missing ingredient in performing such a calculation is a detailed model of the flow in the L1 region, numerical or otherwise. The previously obtained nonsynchronous flow solutions form an outer boundary condition on that flow.

2.4. ECCENTRIC MASS TRANSFER

Mass transfer in eccentric double-star systems is even more complex than in the nonsynchronous case. Changes in the effective gravity over the stellar surface occur periodically, leading to time-dependent effects with period equal to that of the binary. Several studies have invoked eccentric binaries as a means of modulating the mass transfer rate (e.g., Papaloizou and Pringle 1979, Hut and Paczynski

1984). The main question here is the magnitude of the minimum eccentricity that is required to modulate the mass flux in the gas stream by order unity amounts. Consider a noneccentric binary that transfers mass in a steady manner. As we discussed earlier, the Roche lobe filling star transfers mass as pressure forces drive the flow over the equatorial region at mildly subsonic speeds to the L1 region. The eccentricity in the binary imposes a periodic change in the binary separation.

To modulate the mass transfer rate by order unity amounts, two possible eccentricity values come into consideration. An eccentricity of order $\epsilon^2 = c^2/(\Omega d)^2$ with photospheric gas sound speed c, binary angular speed Ω, and binary separation d, causes the instantaneous Roche lobe over the stellar surface to oscillate by one density scale height, away from the L1 region. In doing so, the binary periodically modulates the amount of mass that is carried in the equatorial flow heading for the L1 region by order unity amounts. Notice however that an eccentricity of this magnitude causes the L1 point to shift by a very small amount $\sim \epsilon^2 d$, compared with the density scale height in the L1 region $\sim \epsilon d$. An eccentricity of order ϵ is required to cause the instantaneous L1 point to move substantially across the L1 region. This latter case is a likely upper limit to the eccentricity required to produce order unity mass flow modulations, since it would directly upset the steady-state dynamics of the L1 region. There, the local forces would be changed by order unity amounts within each orbit period. Such changes in forcing over a timescale of an orbit period would directly lead to order unity velocity and density modulations within the L1 region.

However, the first possibility, an eccentricity $\sim \epsilon^2$, must represent a lower limit, since anything smaller would not change the instantaneous mass flow anywhere on the stellar equator by order unity amounts. The main question is the extent to which this flow away from the L1 point can be driven through the L1 region within say half an orbit period. One way to understand this process is to consider the following idealized case. Suppose that the Roche lobe changes instantaneously from an equilibrium value for noneccentric mass transfer to a position one density scale height deeper in the star, taken away from L1, as a result of a binary separation change $\sim \epsilon^2 d$. Then the density of matter on the new Roche equipotential is increased by order unity amounts away from L1, while the density on that equipotential in the L1 region has hardly changed. The main issue is the extent to which these order unity density differences can be propagated into the L1 region in say half a binary orbit period. The time for flow at speed $\sim c$ away from L1 to cross the L1 region of radius $\sim c/\Omega$ is on the order of an orbit period. Therefore, this crude reasoning suggests that in fact the smaller eccentricity $\sim \epsilon^2$ may be adequate in modulating the mass transfer rate in the stream by order unity amounts. Detailed calculations are required.

Detailed 2d numerical calculations of this process were carried out by Edwards and Pringle (1987). They adapted a hydrodynamics code to follow the mass flow within a polytropic star. The potential contained only the gravitational contributions from the two stars. In doing so, they were able to take advantage of the axial symmetry about the line joining the two stars. The mass flow region analyzed included the effective L1 point of their potential and regions deeper in the mass losing star. The minimum required eccentricity to produce an order unity mass transfer modulation was indeed as argued above, $\sim \epsilon^2$. In fact, the instantaneous mass transfer rate they obtained was well approximated by the rate predicted by Paczynski and Sienkiewicz (1972) for a steady noneccentric flow at the instantaneous binary separation.

For a fixed eccentricity, they found that the percentage of mass flux modulation decreased with increasing mass flux for nonisothermal flow. The reason is due to the fact that with higher mass transfer rate, material from deeper inside the star is being transferred. This hotter material has a larger density scale height which is less sensitive to eccentricity changes.

Angular momentum restrictions on the flow need to be considered through the inclusion of Coriolis and centrifugal forces in future studies. Of course, 3d studies would be desirable.

3. Varying the Gas Stream

The narrow gas stream quickly develops as the gas becomes supersonic in an c/Ω neighborhood of the L1 point. The flow then achieves enormous Mach numbers $\sim \Omega d/c$. We consider here effects on the mass losing star or L1 region that could give rise to changes in properties of the gas stream.

The velocity field in the gas stream is largely determined by conservation laws. Its properties are surprising insensitive to starting conditions on the star, as can be understood from the behavior of ballistic trajectories that begin near the L1 point. Trajectories that start at L1 with velocities small compared to the binary orbit speed evolve to the so-called straight-line solutions as they proceed from the L1 point. These paths have a direction that depends only on binary mass ratio.

A more interesting possibility is that the density in the gas stream varies, due to a small disturbance on the surface of the mass-losing star. Such a density fluctuation may be due to a convective cell or magnetic fields. Consider a density fluctuation in the L1 region with characteristic spherical radius l. To be easily visible, its size must be $l \sim c/\Omega$, in order that the fluctuation occupy the full width of the stream. As the flow proceeds to the gas stream, this fluctuation is greatly stretched streamwise due to the large accelerations experienced by the stream. The streamwise length scale of the fluctuation becomes $l\Omega/c \sim d$, of order the entire stream length. Therefore, it is very difficult to introduce short pulses of matter in the stream. This condition is even more true of fluctuations that begin elsewhere on the mass-losing star, since even greater expansion occurs. Since the stream undergoes continuous acceleration, acoustic disturbances tend to be suppressed in the rarefractive flow. Furthermore, the shear in the flow is quite modest. Changing this shear is quite difficult. Its magnitude is determined by vortex-flux conservation (Kelvin's Circulation Theorem) which pins the vorticity in the stream to a value of -2Ω. Changing that value by even order unity amounts by processes in the L1 region would require a very extreme condition, a local vorticity of order $\Omega^2 d/c$. This would imply hypersonic differences in flow speeds across the L1 region.

A far more likely site of fluctuations in the flow is the point of impact of the gas stream with the disk. There, the large differences in flow speed between the stream and disk will give rise to the Kelvin-Helmholz instability. Such interactions must have important observational consequences.

4. Interaction of Gas Stream With Disk

As is well-known, the gas stream impact with the disk edge results in the often striking appearance of the so-called hot-spot at the disk edge. The exact nature of that interaction is quite complex and involves considerations of shocks with Mach numbers of $\sim 20 - 100$, well beyond what can be studied in a laboratory.

Substantial observational evidence has accumulated that indicates there are nonaxisymmetric features about the accretion disk. Evidence comes from studies of X-ray absorption dips, optical emission and absorption lines, and UV absorption (e.g., Parmar et al. 1986, Stover 1981, Naylor 1989). These features generally occur at about the same binary orbital phase (~ 0.7). There is substantial variation in the detailed structure of these features between successive binary orbits. These features are generally found at later orbital phases of the binary (i.e., the trailing side of the disk) but earlier in phase than the hot-spot, the point of intersection of the gas stream with the disk. Some attempts were made to explain the origin of X-ray absorption dips as being due to a warping or flaring of the disk edge (White and Mason 1985). From dynamical considerations, it appears that the gas stream is the likely explanation.

As the gas stream approaches the disk edge, its vertical scale height is 2-3 times larger than the disk scale height. Furthermore, a bow shock wraps around the disk edge and acts to deflect gas stream material away from the orbit plane. Material close to the orbit plane is stopped very effectively by the disk and is quickly entrained in it. However, at several scale heights off the orbit plane, the stream can pass over the disk edge. The dynamics of the gas stream flow beyond the bow shock is quite complex. The material in the stream and disk are subject to a strong Kelvin-Helmolz instability due to the large velocity differences. A turbulent wake may then form near the hot-spot region which could be responsible for the dips (Lubow and Shu 1976).

Some extensions to this model have been considered. Frank, King, and Lasota (1987) argued that the absorption dips in the low mass X-ray binary EXO 0748-676 come from a region well inside the disk outer edge. In their model, the X-ray absorbing material comes from a hot region where the gas has undergone a transition into a 2-phase medium of hot gas that surrounds the cool X-ray absorbing blobs. The inclination of this binary has been determined to be about 75 degrees. An additional benefit of this picture is that the material in the absorbing region need not be very far off the orbit plane because it is close to the central neutron star. According to the model of Frank, King, and Lasota (1987) some of the gas stream passes over this outer disk to form an inner disk. The inner disk radius is determined by a no-slip condition between the gas stream and outer edge of this postulated inner disk. Once formed, the inner disk provides a second point of impact. The gas stream material that flows over the disk collides with this inner disk. At this collision point, matter overflowing the inner disk is exposed to X-rays and forms the absorbing matter as blobs in the two-phase medium.

The model of Lubow (1989) concentrated on the vertical dynamics of gas several scale heights off the plane. The model demonstrated that once the stream passes over the disk outer edge, it interacts less with the disk because of an increasing mismatch in scale heights between the disk and stream. The reason is that this high stream material tends to continue almost ballistically at constant height off the orbit plane. Below the stream is the disk material, which is hydrostatic, and has decreasing height as the stream approaches the central star. As it progresses over the disk, the stream becomes affected by the vertical gravity of the binary. Near the periastron point for this flow, the stream strongly impacts vertically with the disk. The flow basically follows an inclined elliptical orbit. At impact, there is a large slip in horizontal velocity between the stream and disk. Near this point of impact, strong interactions may again occur between the stream and disk, e.g., shocks can form. The very large velocity differences between the stream and disk can again lead to strong effects of the Kelvin-Helmholz instability. The region of impact was associated with the location of the deepest portion of the dip observed in some systems. This location marks the point of maximum penetration of a gas stream over an accretion disk.

These latter two models obtain about the same orbit phase for the dip location, about 0.7, which is close to the observed phase in several systems.

Recently, Hellier and Mason (1989) have argued that in at least the case of low mass X-ray binary X 1822-371, the absorbing region instead comes from the disk outer edge. One possibility is that dips can be observed from both the disk edge and inner regions of the disk. The disk outer edge provides the site for the initial stream-disk interaction with some associated turbulence. Deeper within the disk, the above described process occurs. The actual site of observational importance depends on disk inclination. Another explanation is that is that the stream in some systems is better able to penetrate over the disk than in others, due to some details of the flow. Recall that the material responsible for dips is on the tail of the vertical density distributions. Therefore, small changes to the upper stream, disk, or bow shock structure could lead to large changes of the mass flow rate beyond the disk edge.

Other issues need to be explored. The material responsible for the dips is subject to the Kelvin-Helmholz instability at very high Mach numbers. We know very little about the effectiveness of this instability in this regime.

Sometimes weaker dips are also observed at phase 0.1-0.2 (e.g., Parmar et al. 1986). These models offer no explanation for such features.

5. Wind Driven Accretion

In massive X-ray binaries the dominant mode of mass transfer occurs through a wind, rather than Roche lobe overflow. Accretion onto the companion neutron star occurs through a process similar to that of Bondi-Hoyle accretion. The surprising recent result is that such flows have been found to be unstable at high Mach number, Mach numbers characteristic of X-ray binary winds. Since there are several excellent reviews on this topic, the discussion here need only be brief (e.g., Livio 1990).

The equilibrium state consists of flow through a bow shock that wraps around the neutron star (Matsuda et al. 1991). The flow in the postshock region accretes with little or no angular momentum onto the neutron star. The reason is that the angular momentum of upstream flow that approaches from smaller radii annihilates the angular momentum of the material that approaches from larger radii. (Davies and Pringle 1980, Livio et al. 1986a,b). As a result, no disk will form. What has been found numerically in two dimensions is that this flow becomes unstable as the bow shock oscillates about its equilibrium position. The shock cone appears to undergo strong oscillations without settling back to the equilibrium state (Matsuda, Inoue, and Sawada 1987, Taam and Fryxell 1988, Matsuda et al. 1991).

Due to these oscillations, the angular momentum of the material crossing the bow shock no longer cancels, and a disk now forms. As the bow shock oscillates, a disk with opposite angular momentum attempts to form. In the process, some of the overall disk angular momentum is cancelled and a burst of accretion occurs. This process repeats quasiperiodically on a characteristic timescale of about a few times R_a/V, where R_a is the classical Bondi-Hoyle accretion radius and V is the upstream wind speed.

In a time-averaged sense, little net angular momentum is accreted. There are just excursions in angular momentum about the small equilibrium value of zero. However, these excursions can lead to many observationally interesting processes. In particular, changes in the observed spin rate of the neutron star can be attributed to this process. Correlations are expected between the "waiting-time" for an X-ray flare and flare strength.

Much work remains to be done on this problem. The exact nature of the instability needs to be further explored. It is unclear what range of Mach numbers is required to produce the instability. Recent analytic studies suggest an instability of the accretion line, as well as the shock cone as the source of the "flip-flop" motions of the shock cone (Livio et al. 1992). A possible relation between vortex shedding also been suggested. The main question to be answered is whether this instability appears in three-dimensions, since preliminary work suggests that the instability is much less violent in three-dimensions (Matsuda et al. 1992).

6. Discussion

Considerable progress has been made in understanding the dynamics of mass transfer in close binaries, but fundamental problems remain. Some of these problems occur in broader contexts of astrophysical theory. Issues of angular momentum transport and the behavior of instabilities at high Mach number appear to be among the most important unresolved issues. Progress on such difficult problems will probably require the further use of numerical studies and simulations.

THE ALGOL-TYPE INTERACTING BINARIES

G. E. MCCLUSKEY, JR.

1. Introduction

The Algol-type systems were originally defined in terms of the nature of their light curves. Typically, the outside of eclipse brightness is fairly constant with the ingress and egress phases well defined for sufficiently deep eclipses. It was believed that these light curves were relatively uncomplicated implying that the stars were approximately spherical and "simple". The prototype, β Per = Algol, is a low activity system in modern terminology and fits these criteria reasonably well.

Kopal (1955) adopted the Roche model which he used to classify close binaries as detached, semi-detached, or contact systems. The term Algol system gradually came to mean a sub-group of the semi-detached binaries. This sub-group, occasionally referred to as classical Algol systems, is characterized by a B- or early A-type primary star which is on the main sequence and a G- or K-type sub-giant or giant star at its critical Roche lobe.

More recently, the meaning of Algol-type has been expanded in a somewhat vague manner to include most interacting binaries with a non-degenerate, moderate mass ($\sim$ to 10 solar masses) primary star and a non-degenerate secondary with no more than approximately 10 solar masses. It should be noted that the last decade or so has seen the almost complete replacement of 'close binary' by 'interacting binary'.

Initially, semi-detached and contact systems made up all or nearly all of the interacting binaries. It is gradually becoming apparent that contact with a critical Roche lobe is not the only way in which the evolution of stars in binary systems can differ from the evolution of single stars. Stellar wind enhancement, tidal forces, radiative heating and other poorly understood effects, particularly for evolved stars, can significantly alter the evolutionary history of stars in interacting binaries.

A physically meaningful upper limit to the initial mass of the primary star in an Algol-system might be defined by the condition that this star will avoid becoming a Type II supernova. For single stars, this condition gives an upper limit to its mass which is believed to be approximately 10-11 solar masses. The situation in interacting binaries is more complex but 10–11 solar masses is probably a reasonable upper limit. The lowest initial mass for the primary star in a system which can evolve into an Algol type binary is uncertain but a round number would be 1 solar mass.

Crawford (1955) and Kopal (1955) were the first to recognize that the Algol systems could only have formed by means of large scale mass exchange and mass loss. In the classical Algols the primary is presumed to be in the core hydrogen burning stage of its evolution and the secondary star has already undergone extensive mass loss via critical Roche lobe overflow. Very often the cool star has less than a solar mass and is almost always considerably less massive than its hot companion which is apparently of normal mass for a main sequence star of its spectral type. The cool star is almost invariably at its Roche critical lobe while the hot star is well detached from its lobe. The cool star is still losing mass at a rate which can vary from 10^{-5} to 10^{-7} solar mass per year for very active systems to less than 10^{-9} or 10^{-10} solar mass per year for the least active systems.

Variations on this basic scenario do occur. In some cases both stars are hot, e.g., V 356 Sgr, RZ Sct, or λ Tau or cool, e.g., RX Cnc, KU Cyg or AR Mon. It is important to remember that for O-type stars or supergiants with winds, the Roche critical lobes do *not* exist and the term semi-detached is meaningless.

Much of the recent progress in understanding the formation, evolution, and fate of Algol type binaries has been the result of improved observational techniques and the application of computers to data analysis and modelling. The advances in observational methods include improvements of standard methods, e.g., photometric and spectroscopic techniques; the application of new methods, e.g., high speed photometry and charge coupled devices; and the opening up of unexplored or poorly explored regions of the electromagnetic spectrum, e.g., space astronomy. The discovery of the existence of exotic interacting binaries such as the novae, X-ray binaries, SS 433, and binary pulsars has greatly stimulated interest in interacting binaries. In addition to innovation in technique and instrumentation, persistence of the observer is important. Since the photometric and spectroscopic signatures of mass flow in Algol systems are often highly variable on all time scales, an understanding of these phenomena requires a dense and long baseline of observations.

It will be convenient to divide the Algol systems into three subgroups and to discuss a few archetypes of each group. The most active Algol systems will be called dynamic Algol systems. These are systems in which the contact companion is losing mass at such a high rate that the system or at least the accreting star, is enveloped in a gaseous envelope. The light curves and radial velocity curves will show considerable distortion and variability. Consequently, it is precisely for these very interesting systems that even the basic photometric and spectroscopic elements will be extremely uncertain. This in turn makes it difficult to determine whether or not one star is at its Roche critical lobe. Moreover, the Roche critical lobe of the detached star may fill with gas so that the system mimics a contact binary. In a few cases, the accreting star is rotating so rapidly that it is more disk-like than spherical and may fill its rotational critical Roche lobe. There is weak evidence that these systems are semi-detached but often this is simply assumed to be the case. Examples are β Lyr, SX Cas, and V367 Cyg. It is not certain that β Lyr belongs in this subgroup since the mass of the current secondary star may be greater than 10 solar masses. In fact, β Lyr may be the most strongly interacting binary known and is still unique in many ways. The present secondary star which is accreting mass has not yet been definitively detected and the masses of both stars are quite uncertain. It is possible that eventually β Lyr will become a more "conventional" Algol system. Sahade (1980) has reviewed our understanding of β Lyr.

A second subgroup is the active Algol binaries. In these systems, photometric and spectroscopic observations clearly show the effects of mass flow but the amount of matter present in the flow does not make the detection of either component nearly or completely impossible and does not alter the appearance of either component beyond recognition. Good examples are U Cep, TT Hya, and SW Cyg.

Finally there are the low activity systems in which mass flow is normally quite sparse and intermittent. In the optical spectrum these systems often appear inactive but ultraviolet observations will almost always detect signs of activity sooner or later. Examples are U Sge, δ Lib, and S Cnc.

There are no clear cut boundaries between these subgroups and activity levels, for active and low activity Algols can change dramatically. When U Cep undergoes one of its outbursts, it is close to behaving like a dynamic Algol system. The spectrum of TX UMa in the far-ultraviolet puts it in the active category while its optical spectrum is usually quiescent.

In the following we will discuss recent observations of Algol systems from ground-based observatories, including the optical, infra-red, and radio regions of the spectrum and from space observatories. A brief discussion of chemical abundance anomalies will also be included.

A few recent reviews of Algol binaries are McCluskey (1982), De Loore (1984), Kopal (1984), Budding (1986), De Greve (1986), and McCluskey and Sahade (1987).

2. Ground-based Observations

The decade of the eighties saw much photometric and spectroscopic research devoted to the interacting binaries in general and to the Algol-type systems in particular. Slowly but surely our understanding of the dynamics of mass flow in Algols is becoming less obscure but is not yet clear.

2.1. OPTICAL STUDIES

The orbital periods of dynamic Algol systems are weeks or months and the light and velocity curves are rarely sufficiently observed for definitive studies. Spurred by dramatic discoveries in the ultraviolet to be discussed later, this situation is slowly being remedied.

Young and Snyder (1982) surmised the existence of disk-like structures, highly variable accretion activity, and as many as four thermal regions associated with RX Cas, SX Cas, V367 Cyg, RW Per and W Ser, with W Ser being the most active of the five. Andersen et al. (1988) studied SX Cas and found secondary eclipse to be much broader than primary eclipse due to a luminous disk around the primary star. Andersen, Pavlovski, and Piirola (1989) found the primary star in RX Cas to be completely obscured by a geometrically and optically thick disk with an equatorial temperature near 5500 K. A rapid period change of $\dot{P}/P \approx 6.3 \times 10^{-7}$/cycle is occurring. A mass loss of 10^{-6} to 10^{-5} solar mass per year from the cool star is probable. The systems SX Cas and RX Cas are quite similar in terms of orbital period, masses and radii. Both systems consist of a B-type star of 5–6 solar masses which is optically masked by a disk. This star does not appear to have evolved beyond the main sequence. The cooler star is a giant of approximately 20–25 solar radii and 1–2 solar masses. The orbital period is decreasing for SX Cas but increasing for RX Cas.

Besides β Lyr, RX Cas, and SX Cas, only W Cru, V367 Cyg, and W Ser are included in the dynamic Algol class. The systems V367 Cyg and W Ser have similar orbital periods but W Ser is more active and in terms of variability of activity W Ser is the most prominent dynamic Algol system. The V367 Cyg system probably consists of B8 III and A1 III components but the masses are very uncertain. It is possible (Li and Leung 1987) that the stars are fairly massive and that V367 Cyg is not an Algol binary as defined earlier. Plavec (1989) finds W Ser to consist of a B7 V star in a disk which gives it the optical appearance of a F6 II star and a secondary star later than F5. Guinan (1989) finds a period increase of 14 seconds/year for W Ser and no detectable secondary eclipse or spectrum. Whereas the orbital periods of RX Cas, SX Cas, V367 Cyg, β Lyr, and W Ser are between 12 and 37 days, W Cru has a period of 198 days. The visible star appears to be a G-type supergiant while spectrum of the companion star has not been detected.

It must be noted that the basic physical parameters, e.g., masses and radii, are uncertain to essentially unknown for the dynamic Algols.

Elias (1990a) has made detailed polarization studies of SX Cas. He finds that the hot star is surrounded by an optically thick disk of approximately 7 solar radii and optically thin material extends out to 24 solar radii. The disk thickness is at least 2 solar radii and its mass is at least 4×10^{-10} solar masses. Flares occur on the accretion disk and the gas streaming from the cool star is heterogeneous. A less detailed study of V367 Cyg indicates that the disk surrounding the hotter star has a total radius of about 25 solar radii. Elias (1990b) detected V367 Cyg, and W Ser at centimeter wavelengths, attributing this emission to thermalized gyrosynchrotron radiation.

The Algol system RW Per does not belong to the dynamic systems but it is peculiar. Wilson and Plavec (1988) found no evidence of a disk but the primary star is in such rapid rotation than it may be highly flattened. Olson (1989) found this star to be rotating at 325 km s^{-1} which is thirty times the synchronous rate.

He finds uncertain evidence for the existence of an optically thin accretion disk.

The best example of an active Algol system is the much studied U Cephei. Batten (1974) has thoroughly reviewed early work on this system. A long series of observations of U Cep has been undertaken by E.C. Olson. Olson (1978, 1980) analyzed data obtained during and after the activity outburst which lasted at various levels from late 1974 to late 1977. He proposed that an optically thick disk of matter accreted from the cool companion surrounds the primary star. The disk temperature is about 12,000 K as compared to 13,500 K for the primary star. A stream of gas ($\approx$ 10,000 K) flows from the cool star, leaving the L_1 point and impacting the disk. A hot spot ($\approx$ 20,000 K or more) is created by the impact. During high rates of mass transfer large continuum light loss in the U band, up to $0.^m7$, is observed on a time scale of several days. Smaller losses are seen at longer wavelengths. Olson (1981) estimates mass transfer rates as high as 4×10^{-6} solar mass/year during activity outbursts.

Olson (1985) summarized studies of U Cep. Due to accretion the primary star in the U Cep system has an equatorial bulge which is probably asymmetric in longitude and has a mass of about 3×10^{-12} solar mass and a temperature of 12000 K. The mass flow phenomena are highly variable in strength on time scales of a few to 20 days. During a mass flow outburst the outer hemisphere of the cool component brightens somewhat and about 70-80% of the projected hemisphere of the hot component is covered with a pseudo-photosphere with temperature of about 10000 K which causes a dramatic decrease in brightness outside of eclipse. The cool star is variable on a time scale of years. In 11 years three abrupt period changes of the order of $\dot P /P = 10^{-5}$ with alternating signs occurred. Kahn and Budding (1986) discuss the highly variable asymmetric 'W'-shaped variation of $H\beta$ through primary eclipse which was detected by Olson (1976). They conclude that Balmer emitting gas is concentrated around the primary star with a ring or disk-like distribution. This material extends from the photosphere of the star to a distance of about 0.6 of its radius above the equatorial region. Olson and Stoehr (1986) found a number of anomalies in RS Cep. The most unusual were large, variable color changes from eclipse to eclipse and during partial eclipse phases of the primary minimum. No simple explanation of these color variations was found. Olson (1986) detected occasional brightenings of RS Cep at all optical wavelengths which he attributed to accretion episodes. The disk around the hot star of RS Cep is asymmetric with most of matter concentrated on the trailing hemisphere of the primary star. The disk temperature was estimated to be approximately 6000 K. Kaitchuck, Honeycutt, and Faulkner (1989) observed six eclipses of U Cep in three months. The disk contributes to the continuum and to the emission lines. The emission line region of the disk is about 1.2 times the radius of the primary while the continuum contributing region extends out to 1.6 times the radius of the primary. Both dimensions vary. The emission lines are much broader than rotation would imply and the motion is non-Keplerian. During one eclipse a third set of emission lines was detected and interpreted as due to gas escaping from the system.

Among the short period Algol systems, U Cep is the most active with RW Tau next. Kaitchuk and Honeycutt (1982) found that a highly variable disk-like structure surrounds the primary star of RW Tau. The disk has a maximum radial extension of approximately one-half of the radius of this star. The disk displays highly non-Keplerian motions with the outer edge having the highest velocities. Olson (1985) finds an asymmetrical bulge around the equator of the primary star of RW Tau. It is concentrated on the following hemisphere near the region of stream impact. Brightness variations of 0.05–0.10 magnitude are observed for the cool star.

The low activity Algol binaries are perhaps characterized by U Sge. Khan and Budding (1986) observed a "W" feature variation in U Sge similar to that of U Cep but weaker. They conclude that the disks in U Sge and U Cep (when not highly active) are of similar size but due to the larger size of the U Sge system its disk is of lower density. There is more circumstellar and circumbinary gas in the U Cep system. Olson (1987) found brightness variations of the cool star and several abrupt period variations of different sign in U Sge. A significant brightness decrease appears at all wavelengths near phase 0.80, disappearing near phase 0.95. The gas stream from the cool star strikes the hot star's photosphere at about 500 kms^{-1}. Variations of 0.01–0.10 magnitude in the cool component's brightness occur on all time scales from days to years. Cool and/or hot spots may occur on this star. Olson and Bell (1989) find a mass transfer rate of $3 - 4 \times 10^{-8}$ solar mass/year and a gas stream temperature of about 8000 K. This mass flow rate is at least an order of magnitude below that of U Cep. Kaitchuck and Park (1988) detected transient accretion disk phenomena in TZ Eri. Emission lines were highly variable from eclipse to eclipse. The trailing side of the disk around the primary star extends from 1.2–1.6 or more times its radius while the leading side extends from 1.0–1.6 or more radii.

2.2. INFRA-RED STUDIES

Infrared observation is particularly useful for observing the cool components in Algol binaries, particularly for totally eclipsing systems. Tomkin (1978, 1979, 1981, 1985) and Popper and Tomkin (1984) obtained infra-red spectra of δ Lib, U Sge, U Cep, R CMa, and S Cnc and have derived relatively accurate masses. The cool components of R CMa and S Cnc have masses of 0.17 solar masses, making them the lowest mass stars yet found in Algol systems. Richards (1990) has obtained infra-red light curves of β Per. Time dependent changes in the depths and phases of both eclipses, in the asymmetric shape of secondary minimum and cyclic variations in the mean temperature of the secondary component are present. The overall variability is similar to that seen in the optical light curves of the RS CVn systems where starspots play a role. The application of infrared observing techniques to Algol systems is still in its infancy but promises to yield very interesting information in the future, particulary for longer period systems where dust may play a significant role.

2.3. RADIO STUDIES

This technology is just beginning to be applied to Algol systems. Slee et al (1987) observed a number of Algol systems at 5.8 and 8.4 GHz. Six systems ($\approx 30\%$) were detected. Four were detected once in 7 to 13 observations, one 3 out of 9 times and the most luminous source, δ Lib, was detected 13 of 22 times observed. A strong flare occurred at 8.4 GHz on δ Lib. Many others were probably not observed due to their relatively large distance. The mean observed power is 2.5×10^{10} watts per Hz and the mean brightness temperature is 1.1×10^{10} K. This emission is probably due to optically thick gyrosynchrotron emission. The high brightness temperature could imply the existence of a coherent process. Alternatively, the gyrosynchrotron source might be much larger than the cool star. Lestrade *et al.* (1988) have observed β Per at 2.3 and 8.4 GHz with the Very Large Array. The brightness temperature of the radio source varies from $3 - 50 \times 10^8$K and is indicative of gyrosynchrotron emission of mildly relativistic electrons in an active coronal region with magnetic fields of about 30 Gauss at unit optical depth. The size of the source is estimated to be about three cool star radii. Two outbursts occurred. One was a high brightness ($\approx 1.5 \times 10^{10}$K), broad band burst and the other was a short (≈ 15 minutes) outburst at 1.66 GHz with a brightness temperature of about 3×10^{10} K and a magnetic field of approximately 300 Gauss. Both of these are very likely associated with coronal loops. Stewart *et al.* (1989) detected six Algol systems at 8.4 GHz. The emission is highly variable for all of them and is indicative of magnetic fields of approximately 100 Gauss between the stars and 10^{3-4} Gauss on at least part of the surface of the cool star. Elias (1990b) has detected RZ Sct and W Ser at 3.6 cm and V367 Cyg at both 3.6 and 6 cm. This emission is probably thermalized gyrosynchrotron radiation.

3. Space-based Observations

Photometric and spectroscopic observations of Algol systems in the ultraviolet (912-3000Å) by a number of space observatories have provided a wealth of information. For binary systems containing a degenerate star the same is true in the X-ray region of the electromagnetic spectrum since such systems can emit powerful X-rays. With the successful launching of the EINSTEIN and EXOSAT X-ray observatories the detection of X-ray emission from a number of Algol binaries has been realized.

3.1. X-RAY STUDIES

White and Marshall (1983) detected six of nine Algol systems yielding X-ray luminosities of < 1 to 20×10^{30} erg s^{-1}. RY Gem was detected only once out of four observations at which time it had brightened by a factor of at least 20 via an X-ray flare. McCluskey and Kondo (1984) detected X-rays from R Ara, RZ Cas, δ Lib, and U Sge with energies similar to those found by White and Marshall

(1983). White *et al.* (1986) continuously observed Algol with EXOSAT for 35 hours centered on secondary eclipse. Quiescent emission at a temperature of about 25×10^6 K accounts for at least 80% of the X-ray continuum. No X-ray eclipse was detected so that the X-ray emitting region must extend over a projected area equal to or greater than the projected area of the cool subgiant. These X-rays are believed to originate in the corona of the cool star. An X-ray flare lasting about eight hours with a peak luminosity about three times the quiescent luminosity occurred. The peak temperature was near 60×10^6 K and the iron K-line was seen in emission. This flare probably originated in a magnetic loop located 0.1–0.2 K-star radii above its photosphere.

3.2. Ultraviolet Studies

A very large amount of unique observational material related to interacting binaries is now available, most of it provided by the International Ultraviolet Explorer Observatory (IUE). Recent reviews include Kondo, Boggess, and Maran (1989), Shore (1988), McCluskey and Sahade (1987), Sahade (1986), Plavec (1985), Rahe (1984), Plavec (1983), and McCluskey (1982). This paper will concentrate on research since 1985. High resolution observations of Algol systems with IUE have detected resonance absorption lines of high temperature ions of C IV and Si IV in essentially every system, at least at some phases and/or times and NV is commonly found in the hotter and/or more active systems (Kondo *et al.* 1979, Peters and Polidan (1984), McCluskey and Sahade (1987)). Emission lines of CIV, Si IV, and NV were first detected in β Lyr (Hack *et al.* 1975) and later in dynamic and active Algols (Plavec *et al.* (1982), Plavec (1983), Plavec (1989)). The existence of high temperature circumstellar and circumbinary plasma implied by these high ionization lines requires a nonthermal energy source which in Algols has invariably been attributed to accretional effects. It is quite clear from a large number of studies that mass flow is often far from conservative in Algol systems.

Polidan (1988) has observed the active Algol V356 Sgr with IUE and Voyager. The system consists of a B 3/4 V primary and a larger A2 II secondary. At wavelengths from 912-1500 Å an uneclipsed ultraviolet continuum is present. The emission lines are of invariant strength during the total eclipse of the B-star. No optical emission has been detected. The ultraviolet emission lines are very broad, almost symmetrical and have weak slightly shortward-shifted absorption components. It is proposed that gas from the A-star forms a disk around the B-star which has a wind, possibly magnetically assisted, which drives about 25% of this gas out of the system, mainly out of the orbital plane, while the rest is accreted. The mass transfer rate is about 4×10^{-7} solar mass per year. The B-star may be rotating near the centrifugal limit.

Polidan (1989) observed β Lyr with Voyager and found that no eclipses are detectable below from 912-1200 Å. A strong ultraviolet continuum is present. A rapid brightening by 50% in 40 hours occurred in this spectral region. He suggests that the ultraviolet spectrum from 912-1200 Å resembles that of a cataclysmic variable disk and that β Lyr behaves similarly to a large, massive CV-like accretion disk. If accretion is the ultimate source of the ultraviolet continuum, the radius of the invisible star in β Lyr would have to be no more than one solar radius.

McCluskey, Kondo, and Olson (1988) obtained ultraviolet spectra of U Cephei during and after its activity outburst of 1986 June with IUE. The ultraviolet continuum decreased by factors of 1.2 to 3.0 during the time of greatest activity. Narrow absorption components shortward-shifted by as much as 500 km s^{-1}, appeared as extra components in a number of C, Si, Al, Fe, Zn, and Mg lines. The Mg II line developed a flat-bottomed appearance with a 69% light loss and a half-width of at least 800 km s^{-1}.

McCluskey, McCluskey, and Kondo (1988) studied the low activity (optically) Algol TX UMa with IUE and found that absorption due to Si IV, C IV, and N V are always present. The resonance lines of Al III, Fe II, Mg II, and Si IV show strong phase dependent and secular variations indicative of both gas streaming and circumstellar/circumbinary material. Gas velocities as high as 500–600 kms^{-1} are present. TX UMa is nearly as active as U Cep when the latter is in its normal state.

McCluskey, McCluskey, and Kondo (1991) observed the low activity (optically) Algol system U Sge with IUE. The resonance lines of Si IV and C IV are present in absorption. They showed little phase variation over one orbit in 1983 but were weaker in 1980. It appears that most or all of the ultraviolet absorption spectrum arises in a pseudo-photosphere close to the surface of the primary star. Dobias and Plavec (1985) detected ultraviolet emission lines in U Sge during totality.

4. Chemical Abundance Anomalies

It is of great import to be able to determine the elemental abundances of both components of Algol systems in order to test and improve our theory of their evolution as well as that of their progenitors and successors.

Parthasarathy, Lambert, and Tomkin (1983) found [C/Fe]≈-0.5 and [N/Fe]≈ + 0.5 for the secondaries of S Cnc, U Cep, and U Sge. The iron abundance is solar. The secondaries in these systems have lost much mass, particulary S Cnc. Polidan (1988) detected no carbon lines at all in the ultraviolet spectrum of V356 Sgr during the total eclipse of the B-type primary by the A2 II rapidly rotating secondary. The B-star appears to have a normal carbon abundance.

Tomkin and Lambert (1989) determined abundances of the primary component of R CMa which has a very low mass secondary. They found [C/H] = -0.2, [N/H] = 0.4, [O/H] = 0.3, while S and Fe have solar abundances within the errors of determination. Cugier and Hardorp (1988) examined the primaries in eight Algol systems. Four of these: δ Lib, RS Vul, U Sge, and u Her, showed no carbon anomalies, while TX UMa, U CrB, β Per, and λ Tau showed carbon deficiencies by factors of about 1.7 to 2.5. Balachandran *et al.* (1986) found dramatic helium, carbon, nitrogen, and oxygen abundance anomalies in the B8 visible component of

β Lyr. Compared to the Sun, helium is 2.4 times overabundant, nitrogen is 20 times overabundant, carbon is at least 25 times underabundant and oxygen more than 16 times underabundant. Chemical abundance determinations for dynamic Algol systems are possibly subject to very large errors and should be regarded with great caution.

5. Discussion

The Algol systems present a wide variety of activity ranging from the very dynamic β Lyr (is it related to the Algols at all?) to the low activity systems such as U Sge. Nonetheless, there is a common thread running through this complex gamut. The amount of circumstellar/circumbinary matter, hereafter referred to as CS/CB matter, which is present in a given system is a major factor in determining how active a given system will be. This CS/CB matter makes itself known to us through photometric and spectroscopic perturbations of the ideal light and radial velocity curves. We then apply terms such as accretion disk, equatorial bulge, or pseudophotosphere in an attempt to "picture" what is actually happening.

The amount of CS/CB gas present depends on the physical characteristics and evolutionary status of the binary system and its components. Ultimately, we want to have a theory of evolution of interacting binaries which will provide us with an explanation of how the systems we observe arrived at their present status. A very great amount of effort has already been applied to this aspect of the problem. What we observe are systems in various stages of evolution, not necessarily uniquely related. Observers try to model details of CS/CB gas dynamics at given but different times in the evolution of these binaries. In order to arrive at a close interaction between theory and observation, we must understand gas dynamical (actually magnetogasdynamical) processes in interacting binaries. This is the primary and immediate goal of our observational programs. What have we learned?

The wide array of absorption and/or emission lines due to CS/CB gas in the Algol systems is indicative of a wide variety of thermal regimes. It is now becoming apparent that magnetic activity of the cool star in some Algol systems, detected via x-ray and radio emission, is adding fuel to the fire. Hall (1989) points out that the late type secondaries in most Algol binaries are rapidly rotating (for subgiants) convective stars and should have active chromospheres. The secondaries of U Cep, AR Mon, and S Vel have properties very similar to the chromospherically active stars.

As noted earlier, these cool stars may also be similar to the components of the RS CVn binaries. Clearly, no real progress in understanding gas dynamics in interacting binaries can be made as long as magnetic effects are neglected.

As discussed by Kaitchuck, Honeycutt, and Schlegel (1985), and Peters (1989), for Algols with periods less that 4–5 days, the stream of gas from the cool star strikes the primary star and only transient, unstable disk-like structures exist. For periods longer than 6–7 days, this stream circles the primary and a "permanent" disk forms. For periods of 5-6 days, the gas structure is highly variable. When disks form they apparently fill much of the critial Roche lobe and if the material becomes optically thick, the continuum of the primary can be considerably depressed or even completely blanketed. This occurs even in short period systems, e.g., U Cep, if the flow rate is sufficiently high. In the most dynamic systems, e.g., β Lyr or R Ara (McCluskey and Kondo 1983) an optically thick envelope can surround the entire system. In longer period systems, full sized subgiants or giants can exist, mass loss rates and cool star activity will be higher and a large amount of CS/CB gas can accumulate. These systems are generally the most complex.

One important phenomenon requiring study is the interaction of the stream with the photosphere of the accreting star. Koch (1989) notes the need to model the "splash" which occurs. This interaction is related to the development of a shear-dominated boundary layer (Shore and King 1986) which will generally be rather asymmetrically located around the equatorial region of the accreting star. This region is directly related to the equatorial bulge referred to earlier.

The situation with regard to chemical composition is rather obscure. The theory of CNO nucleosynthesis of He from H predicts order of magnitude changes of carbon and nitrogen abundances. Observations generally indicate factor of 2 or 3. Various scenarios involving mixing of envelope and core material are invoked to dilute the predicted abundance changes and fit the observations. Much more effort is required here.

As Guinan (1989) urges, coordinated, simultaneous, multi-wavelength observations are the best way to attack the difficult problem of understanding interacting binaries. The Algol systems provide us with a wide range of problems of crucial importance to astrophysics. In attacking these problems our motto should be: observation is king, theory is the servant, and modelling is the fool.

MAGNETIC ACTIVITY IN CLOSE BINARIES

Its Manifestations and Consequences

E.F. GUINAN and A. GIMÉNEZ

1. Introduction

Solar-like magnetic activity indicators such as photospheric spots, chromospheric emission, coronal X-ray and radio emission, and flare activity are commonplace in many cool stars with convective envelopes (spectral types of about F5 and later). Generally, the strength of stellar activity increases with more rapid rotation and later spectral types which corresponds to the increasing depth of the star's convective envelope. The origin of this solar-like activity is believed to be a magnetic dynamo in which magnetic fields are generated in the differentially rotating convective zones of the Sun and solar-type stars. Unfortunately, although the magnetic dynamo model is generally accepted to explain both solar and stellar activity, it is only poorly understood and inadequately tested to date (see Parker 1981; 1986).

For single solar-type stars, Skumanich (1972), Soderblom (1983; 1988) and others have found a clear relation between chromospheric activity and the age and rotation rate, in the sense that young solar-type stars rotate rapidly and have high levels of activity. Older stars like the Sun rotate more slowly and have correspondingly lower levels of magnetic activity. The simplest representation for solar-like stars is a decay of activity with time which is roughly proportional to $(\text{time})^{-1/2}$ (e.g., Skumanich 1972) or (a more recent suggestion) fit by an exponential decay law (e.g., Simon *et al.* 1985). The spindown of the Sun and other cool single stars is caused by angular momentum loss (AML) via magnetic braking by magnetized stellar winds. Whether the AML affects chiefly the outer layers or extends to the stellar core is a matter of current debate and study (cf. Pinsonneault *et al.* 1989).

As shown in the studies of Noyes *et al.* (1984) and Baliunas and Vaughan (1985), the majority of solar-like stars in the Sun's neighborhood are relatively old (t $\geq$ 10^9 yrs) and have activity levels comparable to our Sun's. Apart from very young stellar objects such as T Tauri stars and related pre-main sequence stars, the stars with the highest levels of activity are close binary systems composed of cool (G- to M-type) stars, i.e., the chromospherically active binaries such as RS CVn, BY Dra, W UMa, and related systems. In these binaries, the stars are usually forced to rotate relatively rapidly from tidal interaction and typically have rotational periods that are synchronized with their orbital periods. These chromospherically active binaries are strong coronal X-ray sources, having X-ray luminosities of typically

10^2 to 10^4 that of the quiet Sun. They have chromospheric and transition region line surface fluxes that are up to hundreds of times stronger than the Sun. Also, they quite often show migrating waves in their light curves which indicate that they have huge, cool starspots (or groups of starspots) covering up to 30–40% of their surfaces. In contrast, the Sun, even during the maximum of its Sunspot cycle, has $\leq 0.2\%$ of its surface covered with sunspots. In addition, large energetic flare-like events have been reported for several chromospherically active binaries (see Byrne 1989; Pettersen 1989). These flares, however, are much more powerful than even the largest solar flares and have energies of $\simeq 10^{34} - 10^{36}$ ergs.

Magnetic related activity is also probably present in Algol-type binaries and cataclysmic variable binaries (CVs) but it is more difficult to study because of the dominance of the hotter component at UV and optical wavelengths. Also, these systems have the cooler (= active) component in contact with its Roche lobe. Mass loss and mass transfer from the cooler, Roche-lobe filling star frequently produce light variations and spectroscopic features. Since the cooler components of Algol systems have spectral types and rotational characteristics similar to RS CVn stars, it is probable that they have similar levels of activity. This contention is supported by X-ray and radio studies in which several Algol-type systems, near enough to detect, appear to have X-ray and non-thermal radio properties similar to the RS CVn stars (e.g. , see White and Marshall 1983; Lestrade 1988; and Drake *et al.* 1989).

The cooler components of CVs are M-type stars with assumed deep convective envelopes and tidally-enforced rapid rotation periods of a few hours. These stars might be expected to have extremely high levels of surface activity that could have a profound influence on the physical structure and evolution of the binary itself, but they are difficult to study because of the large fractional brightness of the white dwarf and accretion disk at nearly all wavelengths. Although symbiotic binaries have M-giant components with presumably extensive convective envelopes, the long orbital periods ($P \gtrsim$ 1yr) and the complexity of these systems make it uncertain that they are chromospherically active.

This chapter reviews the evidence of dynamo driven magnetic activity in close binaries with cool components. The membership of a star in a binary system – in particular an eclipsing binary – provides a bonanza of information about the physical properties of the stars such as mass, radius luminosity, etc., not usually available for a solitary star. In addition to this, the membership of a cool star in an eclipsing system can provide invaluable information about the distribution and extent of the chromosphere and coronae of active stars through eclipse mapping techniques. Eclipse mapping techniques permit spatial resolution of magnetically active regions that are not presently possible with even the most sophisticated ground-based or orbiting telescopes. The role played by binaries in understanding and defining manifestations of stellar activity, possibly leading to a clearer understanding of the solar dynamo, is also discussed.

The inferred high levels of magnetic activity in close binaries with cool components should also have an important effect on their dynamics and evolution through

AML, which in a tidally-coupled system will cause a decrease in the orbital period and the semi-major axis of the system. AML can cause an initially short period (P < 5d) detached binary with solar-type stars to evolve into a W UMa-type contact system. Further, AML could cause the contact binary to coalesce, ultimately resulting in a rapidly rotating single star. The latter case has been suggested as a cause of the anomalous rapidly rotating G giant – FK Comae (Bopp and Stencel 1981).

2. The Importance of Binary Systems with Cool Components to the Study of Magnetic Activity

There are a number of reasons why close binary systems play an important role in the study of magnetic activity and dynamo models. Some of the reasons are discussed below:

1. If a star with a significant convective atmosphere is a member of a close binary system, tidal forces can cause the star to spin rapidly (equal to the binary's orbital period) and maintain a high level of magnetic activity throughout most of its lifetime. In the majority of cases this is why close binaries with cool, synchronously rotating components are so active and fairly numerous. Thus, close binary systems with cool components provide important testing grounds for studying high levels of presumably dynamo-generated, magnetic activity such as strong chromospheric and transition region (TR) line emission, coronal X-ray and radio emissions, starspots, and flares.

2. The same tidal forces that cause the star to spin rapidly in a close binary should alter the star's internal and surface rotation from that of a single star. As is well known, differential rotation (both latitudinal and radial) is supposed to be an important factor in the generation of magnetic fields and the development of activity cycles in most modern stellar dynamo theories (e.g. , Parker 1986; Gilman and DeLuca 1986 and refs. therein). Differential rotation should be significantly reduced for stars that are members of close binaries. Because of this, stars that are members of close binary systems in which tidal forces are strong can become laboratories for testing stellar dynamo models and the importance of differential rotation to the dynamo.

3. The membership of an active star in a binary system, in particular an eclipsing binary, permits its physical properties (mass, radius, luminosity, etc.) to be determined from the analysis of its light and radial velocity curves. This cannot typically be done for a solitary star. Also, the inclination of the star's rotation axis can be known from the orbital inclination on the reasonable assumption that the star's rotational axis is perpendicular to its orbital plane. An independent determination of the inclination of the star's rotation axis is very useful in interpreting and modeling the starspots frequently present on chromospherically active stars. For single stars with spots, the inclination of the rotational axis is usually a free parameter (or assumed) in the spot modeling analysis of

its light variation (see e.g. , Dorren and Guinan 1982a).
4. For a number of eclipsing binary systems with chromospherically active stars, it is possible to use the mutual eclipses of the components to map out surface and atmospheric features that otherwise could not be resolved. The UV, optical, and X-ray studies of AR Lac (see Walter *et al.* 1987; Neff *et al.* 1989; White *et al.* 1990), are good examples of the potential of the eclipse mapping technique.
5. For a small group of eclipsing binaries with an active, cool star component and a much smaller, hot, white dwarf or subdwarf companion, the hot star can be used as a beamlike probe of the cool star's atmosphere. The structure of the cool star's atmosphere (inner corona, transition region, and chromosphere) can be studied as the hot star passes behind the cool star's atmosphere prior to and after primary eclipse. This method has been exploited in the case of V471 Tau (wd + K2 V; P = 0.52d) by Guinan *et al.* 1986 and for FF Aqr (sdOB +G8 III-IV; P = 9.2d) by Baliunas *et al.* (1986). The results from these studies will be discussed later in the paper.

3. Activity in Detached Binaries with Cool Components: The RS CVn and BY Dra Binaries

3.1. Properties of RS CVn and BY Dra Binaries

3.1.1. RS CVn Binaries

Hall (1972; 1976) first proposed the term *RS CVn binary* be applied to systems which have the following properties:
1. Strong Ca II H+K emission lines are present, observable even outside of eclipses.
2. These binaries contain a hotter component with a spectral type of F- to G- and a luminosity class V or IV; the cooler component is usually a subgiant or giant K-type star which is the more massive component. Both stars typically lie well inside their Roche limiting surfaces.
3. RS CVn binaries have been subdivided into three groups according to orbital period: short period RS CVns ($P \leq 1$ day); *classical* RS CVns ($1d \leq P \leq 14d$); and long period systems usually containing at least one cool giant component with $P \geq 14$ days). In most cases, when the orbital period is less than about 20 days, the stars rotate synchronously so that the upper limit for the period of classical RS CVn stars should be extended to $P \approx 20$ days (see Linsky 1988). In current terminology, the W UMa-type contact binaries keep their identity as a class and are not included with short period RS CVn stars.
4. RS CVn stars have low amplitude quasi-sinusoidal light variations due to the presence of large, dark starspots. The photometric wave frequently is observed to *migrate* in phase relative to the orbital ephemeris.

Other important characteristics of RS CVn binaries are variable $H\alpha$ emission, strong UV chromospheric and transition-region (TR) line emissions such as

MgII h+k λ2800 and CIV λ1550, strong, probably coronal, X-ray and radio emission, flare activity, orbital period variations (which may be due to activity cycles), and color excesses. However, the infrared excesses reported for RS CVn stars are still controversial (see Scaltriti 1990; Reglero *et al.* 1990) but the ultraviolet excesses are quite well established. Also, some RS CVn stars display small variable amounts of intrinsic polarization at optical wavelengths ($P \leq 0.1\%$) but the source of this polarization is most likely circumstellar gas or dust (e.g., Pfeiffer 1979).

3.1.2. BY Draconis Systems

These stars have similar properties to the RS CVn binaries but consist of late dG, dK, or dMe- type stars with orbital or rotational periods typically from ≈ 0.5 to ≈ 20 days. The rotational period of the star may or may not be equal to the orbital period. Sometimes the rapid rotation of binary members results from youth rather than binary tidal effects. BY Dra stars, as originally defined by Bopp and Fekel (1977), include active (= young) single, main-sequence stars as well as members of detached binary systems.

As a group, BY Dra stars tend to have stronger $H\alpha$ emission and may have more frequent flaring activity than RS CVn systems. This could, in part, be due to the generally later spectral types – such as dMe stars included in the BY Dra group.

For the purpose of this chapter, we will treat the RS CVn systems (short period, classical, and long period varieties) and the BY Dra stars that are close binaries as a common group of chromospherically active, detached binaries.

Much work has been done at all wavelengths on RS CVn and BY Dra systems and many excellent review papers have been written (see e.g., Hall 1976, 1987; Bopp 1983; Linsky 1984, 1988, 1990; Rodonó 1988; Rodonó *et al.* 1986; Demircan 1990; Ibanoğlu 1990; Ramsey 1990; Strassmeier 1991 and refs. therein). Because of the extensive amount of work already published on RS CVn and BY Dra stars, we will not attempt a comprehensive review of these stars, but will instead discuss some new results and present several examples of magnetic related activity in these systems.

The characteristics of RS CVn and BY Dra binaries are summarized in Table I along with those of Algol, W UMa, and CV systems. In addition, the corresponding properties of the Sun are included for comparison. Symbiotic stars are not included in this table because of the paucity of evidence supporting magnetic activity.

3.2. STARSPOTS: STELLAR RUBELLA

Most of what is known about starspots on RS CVn, BY Dra, and other chromospherically active stars comes from the study of low amplitude quasi-sinusoidal light variations commonly seen in their broad-band light curves. Although the initial suggestion that these light variations were caused by starspots was made by Kron (1947) nearly 50 years ago, it was not until Hall's (1976) study that the

TABLE I
Summary of activity indications of chromospherically active binaries

Property	BY Dra Binaries	RS CVn Binaries	Algols	W UMa Systems	CV Binaries	Sun
Spectral Type (pri./sec.)	dG-dM / dK-dM	G-K IV-III / G-K V-IV	B5-F2 V / G-K IV-III	F-K V / F-K V	wd+disk / dMe	G2 V / —
Binary Type	detached	detached	semi-detached	contact	semi-detached	—
Orbital Period	$\approx 0.5 - 15^d$	$1-20^d$	$\approx 1 - 20^d$	$0.22-0.80^d$	$0.06-0.40^d$	—
Synchronized ($P_{rot} = P_{orb}$)	$P_{orb} < 3^d$ = yes $P_{orb} > 3^d$ = no (usually)	$P < 20^d$ = yes $P > 20^d$ = no (usually)	yes	yes	yes	$P_{rot} = 25.6^d$ (equator)
PHOTOSPHERE						
Starspots (% surf. cov.)	$\approx 5-20$	$\approx 5-30$	$\approx 5-10$ (Algol)	$\approx 5-15$	?	$\leq 0.2\%$
$<\Delta T> K$ (photo.–spot)	$\approx 300-1000$	$\approx 500-1200$	$\approx 500-800$ (Algol)	$\approx 500-1000$	?	1200-1600 (umbra)
Spot Distrib.	2 spots, 1 at high lat.	2 spots, oft. high lat.	?	2 spots	?	spot groups $40° - 10°$ lat.
CHROMOSPHERE						
F_{MgII} (ergs cm^{-2}s^{-1})	$\approx 10^6 - 10^7$	$\simeq 2 - 10 \times 10^6$	$\simeq 10^7$ (U Cep)	$\approx 5 - 10 \times 10^6$	?	$\approx 2 - 4 \times 10^6$
TRANSITION REGION						
F_{CIV} (ergs cm^{-2}s^{-1})	$\approx 1 - 5 \times 10^6$	$\approx 10^5 - 10^6$	$1-2\times 10^6$ (U Cep)	$\simeq 10^6$	?	$0.7 - 1.0 \times 10^4$ (global) $4 - 11 \times 10^4$ (plage)
CORONA						
F_x (ergs cm^{-2}s^{-1})	$\approx 10^6$	$\approx 10^6 - 10^7$	$\approx 10^6 - 10^7$	$\approx 10^7$	?	$\approx 0.9 - 4.0 \times 10^4$
L_x (erg s^{-1})	$\approx 10^{29} - 10^{30}$	$\approx 10^{29} - 10^{31}$	$\approx 10^{29} - 10^{31}$	$\approx 10^{29} - 10^{30}$	$10^{31} - 10^{32}$ (from disk)	$\simeq 10^{27}$
$T_1(K)$	$\sim 10^6 - 10^7$	$\approx 3 - 7 \times 10^6$	$\approx 3 - 7 \times 10^6$	$10^6 - 10^7$	—	$1-2 \times 10^6$
$T_2(K)$	?	$15 - 50 \times 10^6$	$15 - 50 \times 10^6$	$2 - 5 \times 10^7$	—	—
CORONA (radio)						
L_{6cm} (ergs s^{-1}Hz^{-1})	$\leq 4 \times 10^{14}$	$\approx 10^{15} - 10^{17}$	$\approx 10^{15} - 10^{17}$	$\approx 10^{14} - 10^{15}$	—	$10^{10} - 10^{11}$
f_{radio}/f_{bol}	$10^{-11} - 10^{-9}$	$10^{-7} - 10^{-8}$	$10^{-7} - 10^{-8}$	$\approx 4 \times 10^{-10}$	—	10^{-13}
$T_B(K)$	$\approx 10^7$	$\approx 10^8 - 10^9$	$\simeq 10^8 - 10^9$	$\simeq 10^7$:	—	$\sim 10^5$

starspot hypothesis, linked to chromospheric activity, was taken more seriously. Since that time, numerous light curves of RS CVn and BY Dra stars have been obtained and many studies have been conducted using different modeling techniques to analyze the light curves. In general, these studies verify the presence of large, dark starspots, typically $\approx$ 500-1000K cooler than the star's photosphere, and covering up to 30–40% of the star's surface (see e.g., Eaton and Hall 1979; Dorren et al. 1981; Dorren and Guinan 1982b; Rodonó 1986a,b; Rodonó et al. 1986; Strassmeier 1991). However, there are limitations to what can be extracted from modeling light curves because this technique suffers from ambiguity problems because of too many free parameters. In spite of this limitation, reasonable estimates can be made of the spot properties, including the number, areas, temperature of the spots, and their locations on the star's surface. Although the longitudes of the spots are well determined, the latitudes of the spots are usually uncertain due to trade-offs with spot area, temperature, and inclination of the star's rotation pole relative to the line of sight.

The existence of starspot regions has also been verified and studied spectro-

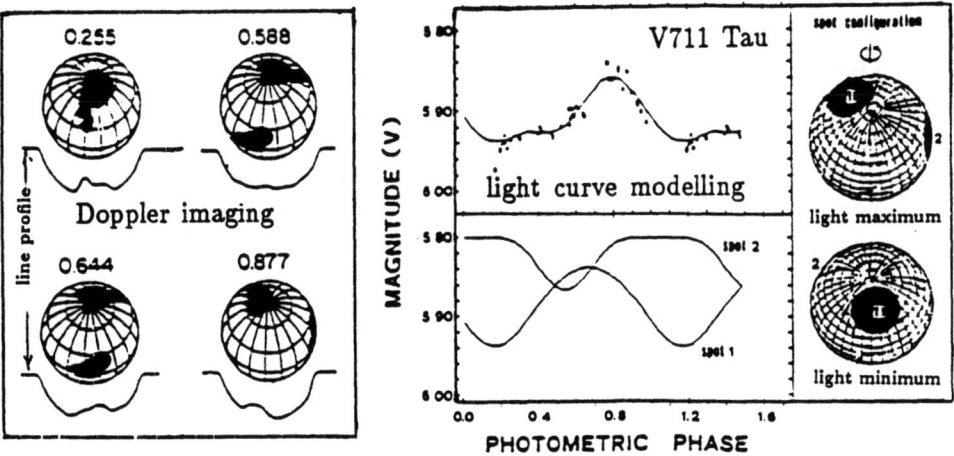

Fig. 1. A comparison of the distribution of starspots on the active K2 IV component of V711 Tau, independently determined from Doppler imaging method (Vogt 1988) and from light curve modeling (Rodonó et al. 1986). The spectroscopic and photoelectric observations on which the modeling is based were obtained over nearly the same time interval (Fall 1981).

scopically using the Doppler imaging technique developed by Vogt and Penrod (1983). With Doppler imaging the presence of dark regions on the star's surface is determined from the small distortions the spot produces on the rotationally broadened line profiles. This method permits the reconstruction of the surface brightness distribution of the rotating star. The Doppler imaging method requires high signal-to-noise and high resolution spectroscopy along with good phase coverage in order to extract the spot information. Doppler imaging is particularly sensitive to the latitude of the spot and thus is complementary to spot modeling of the light curves. Because of the changes in spot sizes and distributions, often on a timescale of weeks, it is necessary to have simultaneous (or nearly so) spectroscopy and photometry (see Rodonó 1988 and Strassmeier 1990, 1991 for a discussion of recent results). Unfortunately, Doppler imaging has had comparatively few applications, chiefly because of the quality and quantity of the spectroscopic observations needed, as well as the intensive computer data analysis that is required. As discussed by Linsky (1988) and Strassmeier (1990), analysis of near-simultaneous multi-wavelength light curves and Doppler imaging of spectroscopy yield qualitatively similar results on the surface distribution of starspots in the case of RS CVn stars, V711 Tau and EI Eri, respectively. A comparison of a Doppler imaging map (Vogt and Penrod 1983; Vogt 1988) and a starspot map from light curve modeling (Rodonó et al. 1986) for observations made of V711 Tau in 1981 is shown in Figure 1. The agreement between the results from these two different methods is encouraging.

In addition to Doppler imaging and light curve modeling, the properties of spots can be inferred from high signal-to-noise TiO band spectroscopy at $\lambda 8860$Å. As discussed by Ramsey (1990), most RS CVn stars have excess TiO absorption for their spectral type. This excess TiO absorption is ascribed to the presence of cool starspot regions on the these stars. Ramsey and Nations (1980) first employed TiO band spectroscopy and spectrum synthesis techniques to determine a temperature difference between the photosphere and spot region of V711 Tau of $\Delta T \approx 1000K$. This is in good agreement with the spot temperatures inferred from other techniques. Subsequent work has been carried out on several other bright RS CVn and related systems (including II Peg, UX Ari, V833 Tau, and HD 82558) that have yielded spot temperatures and coverages in satisfactory agreement with photometric results (see e.g., Huenemoerder and Ramsey 1987; Huenemoerder et al. 1989; Saar and Neff 1990). Where comparisons are possible, the three techniques yield qualitatively similar results. Generally, the modeling indicates at least two large spot regions, usually located at high stellar latitudes (sometimes even at the rotation pole) which cover typically 5 to 25% of the visible surface of the star. The spot temperatures generally are between $\Delta T \approx 500 - 1200K$ cooler than the star's photosphere. A brief summary of starspot properties is given in Table I. A recent star by star listing of starspot temperatures and area coverages is given by Strassmeier (1991).

3.2.1. Spot Cycles

A growing number of RS CVn and BY Dra stars have been systemically observed with photometry for over a decade or more. Many of these stars show evidence of long-term variation in the light amplitudes and in the mean light levels of their light curves that could be due to solar-like activity cycles. Evidence of activity cycles in RS CVn and BY Dra stars is given, for example, by Rodonó (1981); Dorren and Guinan (1982b); Evren (1990); Ibanoğlu (1990); Maceroni et al. (1990) and Olah (1990). Most cycles are between 6 and 14 yrs.

Also, Hall and Kreiner (1980) and Hall (1990) find evidence for cyclical decreases and increases in the orbital periods of chromospherically active binaries Hall (1990) suggests that these cyclical alternating period changes are a consequence of solar-type magnetic cycles.

An interesting example of evidence for an activity cycle in an active RS CVn star is given by Dorren and Guinan (1990) for V711 Tau (G5 V + K0 IV; P = 2.84d). Figure 2 from this paper shows a plot of the seasonal mean values of $H\alpha$ index, V-magnitude, color index, and C IV $\lambda 1550$ emission line flux over time. The light variation appears to show a systematic rise and fall suggesting the presence of an activity cycle with a period of 11-14 yrs. However, as shown in the figure, the $H\alpha$ index and color index measured for this binary do not seem to correlate well with the observed brightness changes. Moreover, the CIV $\lambda 1550$ emission flux, which arises from the transition-region, appears to rise and fall a few years out of step with the star's luminosity. This study indicates that the star was brightest when the activity levels inferred from the transition region lines was greatest. This effect is opposite to what would be expected if the starspots were the *only* contribution to the long-term brightness variations.

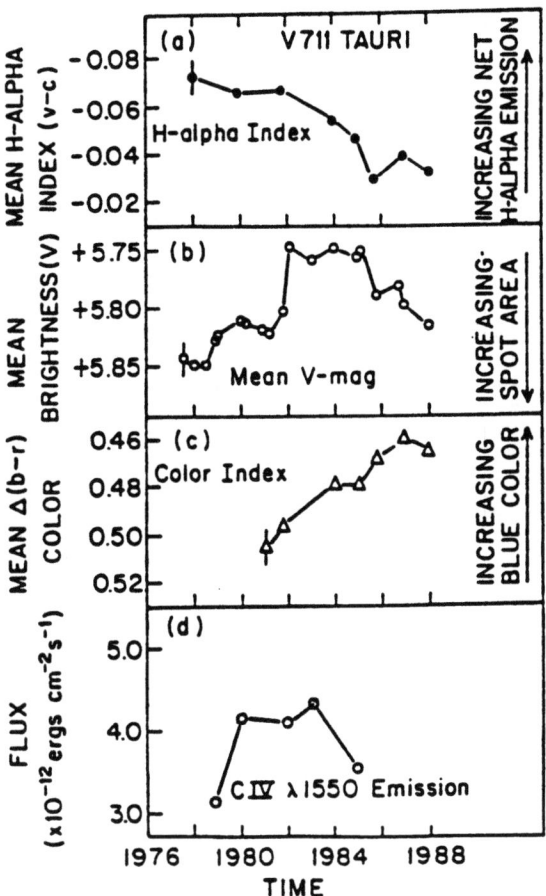

Fig. 2. The mean seasonal values of Hα index, V-magnitude, color index (b-r), and C IV λ 1550 emission line flux for the RS CVn binary, V711 Tau, are plotted versus time (from Dorren and Guinan 1990).

These results are interpreted as evidence of a significant facular (white light faculae) contribution to the star's luminosity. They suggest that the CIV λ1550 line emission yields a more direct measure of magnetic activity while the long-term light variation is produced by competition between the blocking effect of the dark starspots and enhancements in luminosity from white light faculae and a facular network similar to that observed in the Sun (see Foukal and Lean 1986, 1988). More work on other active stars is needed before we can try to understand the long-term behavior of these stars.

TABLE II
Differential rotation measures

Star	Spec. Type	Prot. Days	$\Delta\Omega/\Delta\phi$ (rad s^{-1}deg^{-1})	Reference/Comments
		Single Solar-Type Stars		
Sun	G2 V	25.6	6.0×10^{-9}	Direct measures of sunspot motions (Allen 1981)
HD190406	G1 V	13.6	$\approx 3.0 \times 10^{-8}$:	Estimated from Mt. Wilson Ca II H&K data from Baliunas et al. 1985. $\Delta\phi_{1-2} = 20°$ assumed.
HD149661	K0 V	~11.0	$\approx 2 - 3 \times 10^{-8}$	Estimated from photometry of Dorren and Guinan (1982a). $\Delta\phi_{1-2} = 20°$ assumed.
		RS CVn Stars		
SV Cam	G3 V + K4 V	0.59	2.6×10^{-10}	Busso et al. (1985)
AR Lac	G2 IV + K2 IV	1.98	2.1×10^{-10}	Rodono et al. (1986)
V711 Tau	G5 V + K2 IV	2.83	$-2 \times 10^{-12*}$	Rodono et al. (1986)
VV Mon	G2 V + K0 III	6.05	5.6×10^{-10}	Busso et al. (1985)
SS Boo	G0 V + K1 IV	7.61	1.5×10^{-9}	Busso et al. (1985)
RU Cnc	F5 IV + K1 IV	10.16	1.5×10^{-10}	Busso et al. (1985)
CQ Aur	dG2 + K0(IV)	10.49	$\approx 1.1 \times 10^{-9}$	Busso et al. (1985)
λ And	G8 IV-III + ?	54.5	3.0×10^{-9}	Dorren & Guinan (1984)

*The negative value found for delta $\Delta\Omega/\Delta\phi$, if real, indicates that the rotation rate increases with increasing stellar latitude.

3.2.2. Differential Rotation

Sunspots are important tracers of differential (latitudinal) rotation on the Sun (cf. Newton and Nunn 1951; Tang 1981). There is a decrease of the Sun's rotation rate (Ω) with latitude (ϕ) that can be fit by the formula from Tang (1981) of Ω (deg/days) $\approx 14.37 - 2.60 \sin^2\phi$. Most solar dynamo models incorporate differential rotation as a key element in explaining the ≈ 11 year activity cycle (and ≈ 22 year magnetic cycle) of the Sun. According to Parker (1955, 1986), Brandenburg et al. (1990), and others, the physical basis of the magnetic dynamo in the Sun and solar-type stars is the cyclic conversion of the poloidal to toroidal magnetic fields through differential rotation and the conversion of toroidal to poloidal fields by helicity in the convection zone. For example, the Babcock-Leighton model of the sunspot cycle requires differential rotation to intensify and amplify the magnetic fields and drive the cycle (cf. Babcock 1961). Thus, differential rotation appears to be a critical element in magnetic dynamo theory. Except for the Sun, where the tracers can be spatially resolved, it is difficult to measure differential rotation in most stars (see Gray 1977; Bruning 1981).

However, with the availability of long-term photometry of stars with large spotted regions, such as RS CVn stars, it has now become possible to estimate differential rotation for several active stars. Busso et al. (1984, 1985) and Rodonó et al. (1986) have determined differential rotation rates for several RS CVn stars from a cross-correlation analysis of the light curves of these systems obtained over a few years. The differential rotation rates are determined from modeling the observed phase shifts and period variation derived from the light curves. These

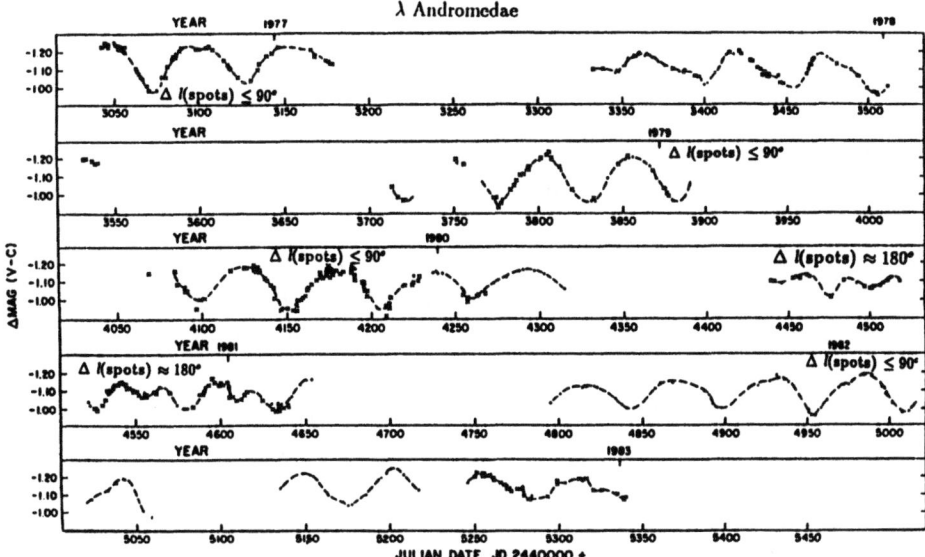

Fig. 3. The light variations of λ Andromedae from 1978–1983 plotted versus time. The evolution of the light curve with time is produced by the effects of differential rotation of two starspot regions on the surface of the G8 III–IV component. When the longitudinal separation of the starspots is less than ≈ 90° the light curve has a large light amplitude and is nearly sinusoidal; when the longitudinal separation of the spots is ≈ 180°, the light curve displays two maxima and minima during the ≈ 54d rotation period of the star (from Dorren and Guinan 1984).

results are included in Table II.

Probably the most definitive determination of differential rotation comes from the analyses of the light curves of the long-period RS CVn system λ And (G8 IV-III + ?; P_{rot} ≈ 54 days) by Dorren and Guinan (1984). Photoelectric photometry of λ And has been carried out nearly continuously since 1977. The light curve undergoes systematic variations in shape and light amplitude with time and sometimes has two minima and maxima during a single rotation. Representative light curves of λ And are shown in Figure 3. They analyzed the changing light curves using a non-static starspot model and found that the spot region at the higher stellar latitude (ϕ_1 ≈ 50° − 70°) lags behind the rotation of the faster moving, larger spot region located at ϕ_2 ≈ 30° − 50° by about ≈16° per rotation. When the longitudes of the spot regions are ≈ 180° apart, the light curves showed two unequal minima and maxima during one ≃ 54 day rotation cycle as shown during 1981 in the figure. On the other extreme, the light curve has the largest amplitude and a nearly sinusoidal form when the two spots are within 90° of each other in longitude. This occurred during 1976, late 1978, late 1979 and during early 1982. In addition to light curve changes, the apparent photometric period varies between 52 days and 56 days. This may be caused by differential rotation as the spots drift in latitude.

The measured values of the differential rotation rates for the RS CVn stars are given in Table II. In addition, the differential rotation rates for the Sun and two other single solar-type stars are also given for comparison. In general, the differential rotation rates inferred for the shorter period RS CVn systems are about $\approx$ 10-30% of the solar value while the non-synchronously rotating G8 IV-III component of λ And and the two single, active main-sequence stars appear to have values of $\Delta\Omega/\Delta\phi \approx 5$ times the solar value. The differential rotation rates only approximately determined because of uncertainties in the spot latitudes. Even smaller differential rotation rates ($\approx$ 0.1-0.01 times the solar value) are found for several RS CVn stars by equating changes in the photometric period or migrating wave period with the effects of spots migrating to different latitudes on a differentially rotating star (cf. Rodonó 1986a,b).

It is not surprising to find small differential rotation rates for RS CVn binaries with synchronously rotating components because strong tidal coupling should diminish differential rotation. What is most puzzling, if not surprising, is the high activity levels of these stars in spite of the fact that differential rotation, an assumed crucial element to dynamo generated magnetic fields, is significantly reduced. The fundamental question arises: *How important is differential rotation in solar and stellar dynamo theory?* More observations are needed, together with a more satisfactory magnetic dynamo theory.

3.3. ENHANCED CHROMOSPHERIC AND TRANSITION REGION (TR) LINE EMISSION

One of the hallmarks of RS CVn, BY Dra, and related stars is strong CaII H+K emission and sometimes Balmer Hα emission. Following the solar analog, these emissions, which are many times stronger than observed in the Sun, are identified with enhanced chromospheric emission from plage-like regions and the chromospheric network. Solar-type stars showing enhancements in these lines are said to be *chromospherically active* stars. Numerous observations of these stars with the *International Ultraviolet Explorer (IUE)* satellite since 1978 has led to an explosion of information on the chromospheric and transition region line emissions that fall conveniently within the $\lambda\lambda$1150-3200 wavelength range of the satellite. For G-, K-, and M- type stars, the ultraviolet continuum is very weak or absent which allows the UV chromospheric emission features such as MgII h+k λ2800 and TR line emissions such as N V λ1240, Si IV λ1400, and CIV λ1550, to be easily measured. These UV emission lines are important diagnostics for studying stellar activity and yield vital information on the physical conditions of the chromospheres and transition regions of these stars (see Jordan 1986). Moreover, the study of the HeII λ1640 yields some information on the physical conditions of stellar corona. There is now an almost overwhelming amount of information that has been gleaned about chromospheric and TR activity from the IUE and ground-based spectroscopy. We refer the reader to a number of recent reviews: Jordan 1986; Rodonó 1986a; Zwaan 1986; Jordan and Linsky 1987; Linsky 1988, 1990; Ramsey 1990; and Guinan 1990.

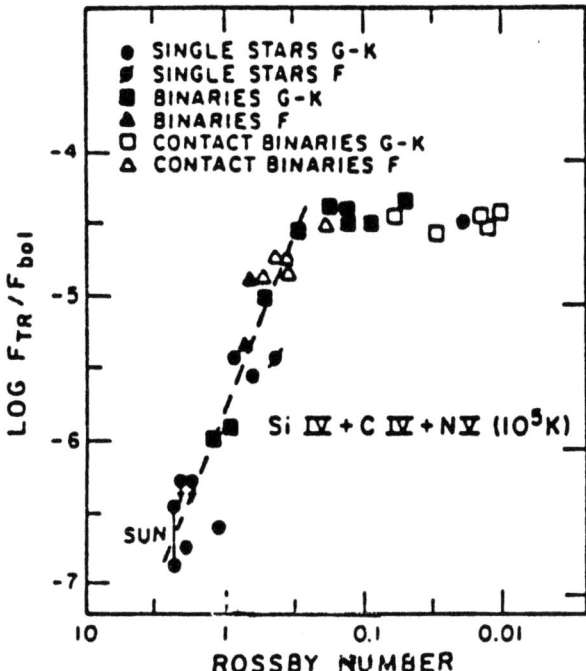

Fig. 4. The ratio of the total transition-region (TR) emission flux relative to the bolometric flux (F_{TR}/F_{bol}) is plotted against the Rossby number, R (from Vilhu 1984).

One of the most important results concerning chromospheric and TR activity of RS CVn and related stars is the tight relationship of these line emissions with rotation and convection. The latter two quantities – rotation and convection – can be parameterized by the Rossby number $R = P_{rot}/\tau_c$, where τ_c is the turnover time of convection. The values of τ_c can be determined empirically for main-sequence and subgiant stars from spectral type or (B-V) colors (cf. Noyes et al. 1984; Stepien 1990 and refs. therein). Two of the most important results are presented in Figures 4 and 5 from Vilhu (1984, 1987).

In Figure 4, the ratio of the total TR emission flux relative to the bolometric flux (F_{TR}/F_{bol}) is plotted versus Rossby number R. The different types of stars and binaries are plotted with different symbols in the figure. As shown, there is a very tight relation between F_{TR}/F_{bol} and Rossby number for the more slowly rotating single stars (including the Sun) and RS CVn-type binaries for values of $R \geq 0.4$. For the more rapidly rotating active stars (mostly shorter period RS CVn and W UMa systems with $R \leq 0.4$) the strength of the net TR line flux levels off and remains nearly constant at $\log(F_{TR}/F_{bol}) \approx -4.4$. This upper limit has been interpreted by Vilhu (1984; 1987) as a saturation limit in which the star's surface is completely filled with emitting structures and active regions. A similar relation is found between X-ray emission and Rossby number. Both of these relations are in accord with general predictions of dynamo theory.

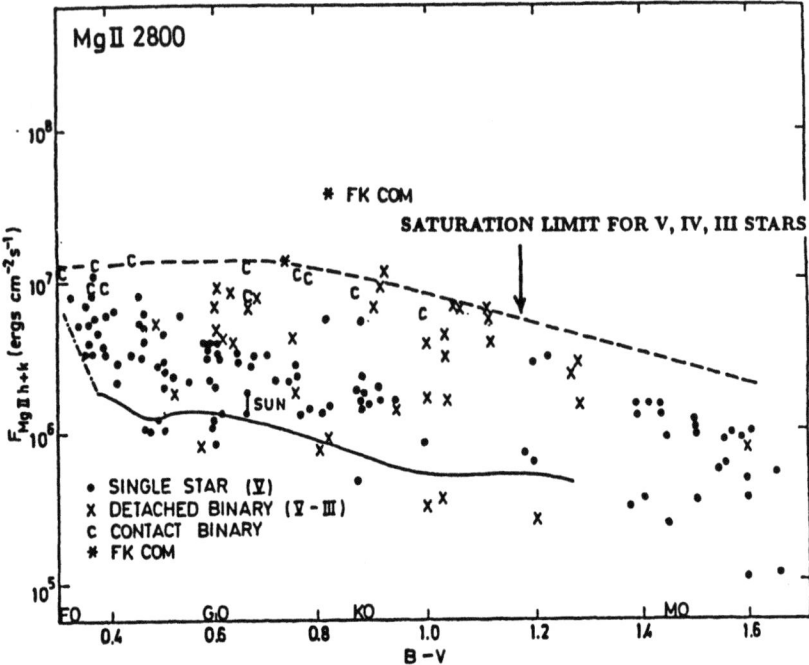

Fig. 5. The surface fluxes of the chromospheric Mg II h+k (λ 2800) emission line for different types of chromospherically active stars are plotted versus $B-V$ (from Vilhu 1987). The upper limit of Mg II emission for each value of $B-V$ arises from the saturation, the complete filling-in, of the available surface of the active star with emitting regions. The shorter period binaries are located near the saturation limit.

Figure 5 shows a plot from Vilhu (1987) of the chromospheric emission line fluxes (F_{MgII}) for single stars, RS CVn, BY Dra, and W UMa stars plotted against $B-V$. The most rapidly rotating stars for each value of $B-V$ are mostly the RS CVn, BY Dra and W UMa type binaries. These stars define an apparent upper envelope of activity for values of $B-V$. Again, the upper limit of chromospheric activity is likely due to the saturation effect, i.e., the complete filling in of the available surface with magnetically controlled structures. Zwaan (1991) has noted that the subgiant and giant components of RS CVn systems have chromospheric, TR and coronal emissions that are often 10-100 times greater than would be expected from single stars with comparable spectral types and rotation periods. He suggests that this *hyper-activity* is somehow produced by enhancements in the dynamo efficiency by the tidal interactions between the components of the binary system. For the most part, however, comparisons are being made between rapidly rotating subgiants and giants of binaries and single main-sequence stars. Perhaps, the lower surface gravities and, possibly, more extensive convective zones of the subgiant and giant stars also play an important role in the apparent enhancements of magnetic activity in these stars. More observations are definitely needed to resolve this question.

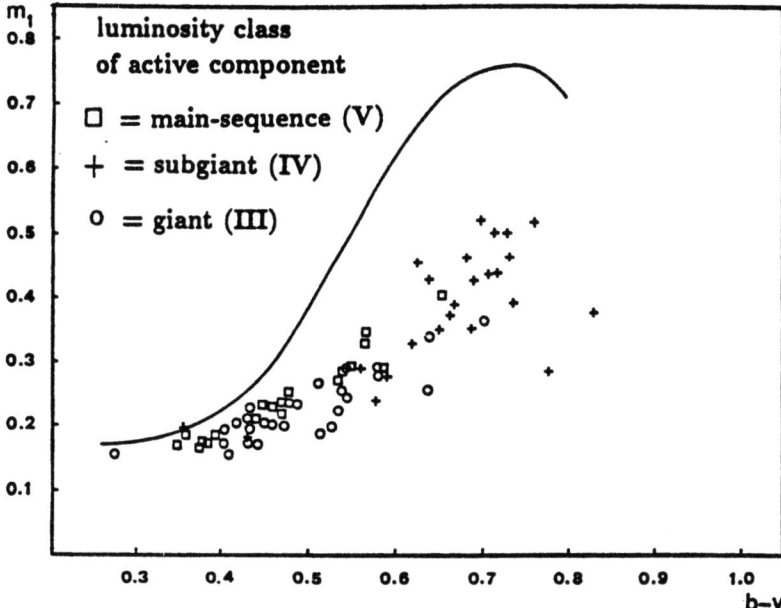

Fig. 6. The m_1-$(b-y)$ diagram with the RS CVn and BY Dra binaries plotted. Binaries with two main-sequence components are plotted with the □ symbol. The binaries with subgiant and giant components (RS CVn stars) are plotted with (o) and (+) symbols, respectively. The continuous curve denotes the standard relation for unevolved stars of solar abundance (from Giménez et al. 1991).

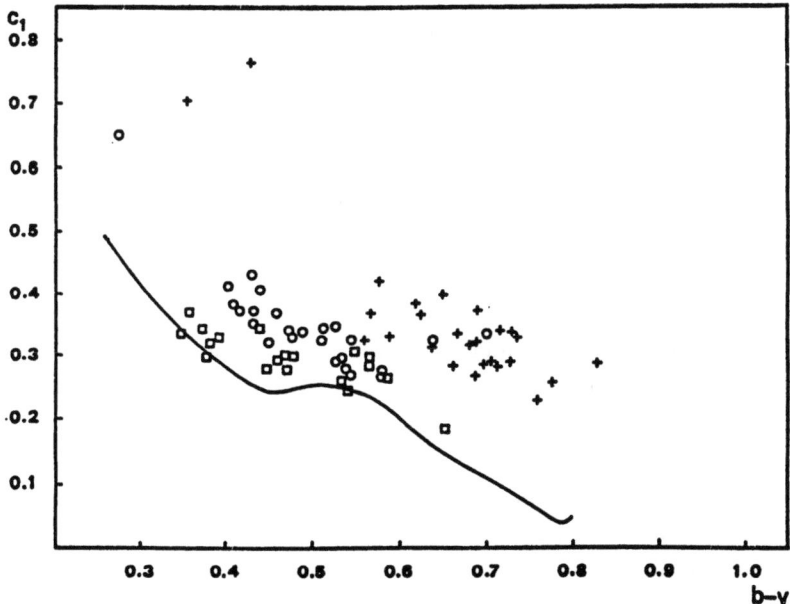

Fig. 7. The c_1-$(b-y)$ diagram, showing the locations of the chromospherically active binaries. As shown, the chromospherically active binaries tend to have larger values (≈ 0.10 mag) of the c_1 index compared to normal main-sequence stars. A numerically larger value in the c_1 index corresponds to an ultraviolet excess. Symbols are as defined in Figure 6.

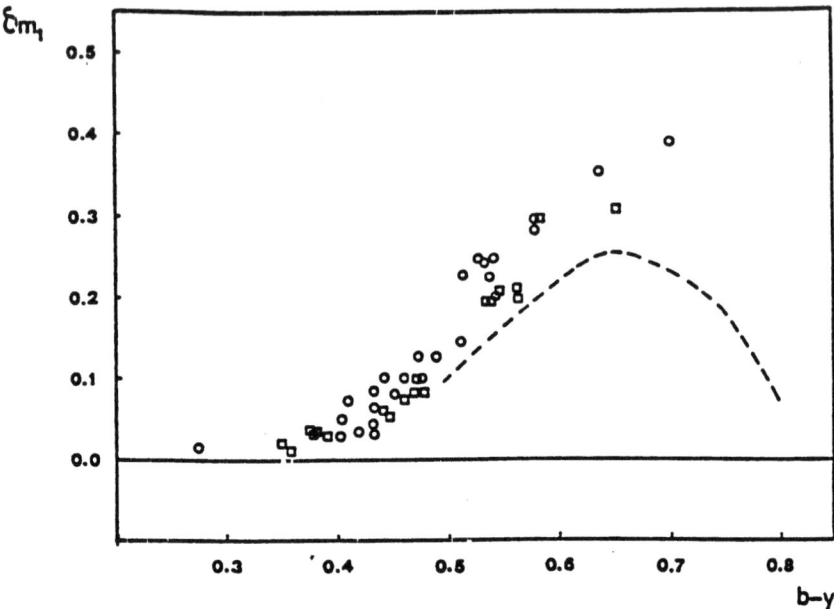

Fig. 8. The δm_1-$(b-y)$ diagram, showing the positions of the chromospherically active binaries. As discussed in the text, the apparent deficiency in the m_1 index (δm_1) is too large to be explained by abundance anomalies. The most likely explanation is the filling-in of the cores of some absorption features with emission.

3.3.1. Effects of Activity on Photometric Color Indices

Giménez et al. (1986; 1991) and Reglero et al. (1987) have conducted uvby photometry of 72 northern hemisphere, chromospherically active binaries – mostly RS CVn and BY Dra systems. The uvby photometry was made to investigate the effect that chromospheric activity may have on the Strömgren m_1, c_1 and $(b-y)$ indices. This was accomplished by plotting the measured Strömgren indices in m_1 and c_1 versus $(b-y)$ diagrams with respect to the corresponding calibrations for normal main-sequence stars (see Figures 6, 7, and 8). The sample of RS CVn and BY Dra systems was divided into three distinct groups according to their evolutionary status. Group 1 contains systems with both components within the main-sequence (mostly BY Dra binaries). Group 2 includes those binaries with at least one of the components classified as subgiant (mostly classical RS CVn systems); and Group 3 corresponds to those with at least one giant component (mostly classical and long period RS CVn binaries).

In Figures 6 and 7, the corresponding positions in the m_1-$(b-y)$ and c_1-$(b-y)$ planes, respectively, are plotted with different symbols for each group: ($\square$) for Group 1, (0) for Group 2 and (+) for Group 3. The continuous lines denote the standard relations for unevolved normal stars as given by Crawford (1975) and Olsen (1988) for F-type stars and Olsen (1984) for the later types. As shown in Figure 6, the BY Dra and RS CVn binaries have systematically smaller values of

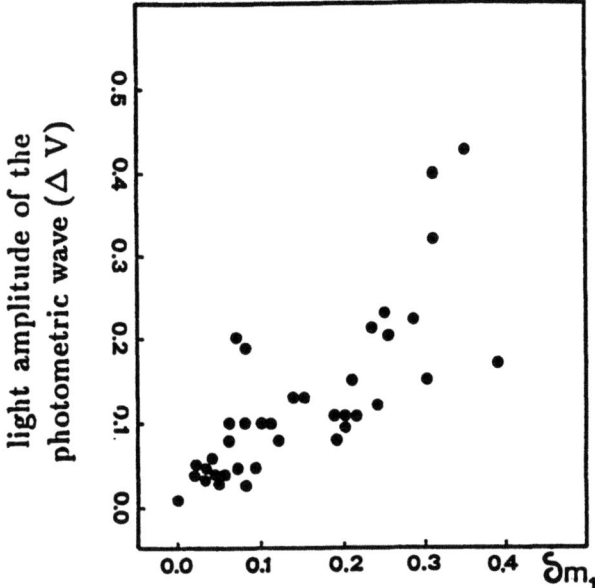

Fig. 9. A plot of the deficiency in the m_1 index (δm_1) versus light amplitude of the photometric wave (ΔV). The value of δm_1 appears to increase with increasing light amplitude. The light amplitude is related to the spot coverage and therefore to chromospheric activity (from Giménez et al. 1991).

m_1 compared to the standard relation for normal stars. The difference between the observed value of m_1 index (δm_1) for the binaries becomes greater for increasing values of *(b-y)* which corresponds to decreasing temperature (i.e., increasing convective zone depth) of the stars. It is evident from Figure 7 that the c_1 index permits the separation of luminosity classes, and the behavior of the surveyed active binaries is comparable to that of normal individual stars with respect to this index. On the other hand, Figure 8 shows a systematic deficiency in the m_1 index, too large to be explained in terms of metal underabundance or evolution. In terms of standard calibrations, the observed values of δm_1 would represent metal deficiencies up to ≈ 6 dex in Fe/H with respect to the solar value (using the calibration by Olsen, 1984). Furthermore, a distinct linear relation is found, independent of evolution, between δm_1 and *b-y*. Chromospheric activity is expected to fill-in the core of metallic lines in such a way that the observed indices would correspond to metal deficient dwarfs. The effect would obviously be strongly dependent on the relative importance of the continuum and thus the detected dependence on temperature or, for that matter, on *b-y*.

The importance of stellar activity on the observed deficiency in the m_1 index can be investigated by plotting m_1 versus the light amplitude of the photometric wave for the stars in Groups 1 and 2 of the sample (see Figure 9). The light amplitude should be roughly proportional to the spot coverage and the spot coverage should depend on the degree of solar-type activity. For the most part, the light amplitudes were adopted from the maximum values given by Strassmeier *et al.* (1988). Though the scatter in Figure 9 is certainly large, probably due to the number of parameters involved (e.g., the comparison of possibly different epochs of activity and the use of combined light rather than that of the more active component of the binary), a definite correlation is seen. From this plot, it appears that the more active stars (those with the largest light amplitudes = more spots) have the greatest anomalies in their observed m_1 indices. This is in good agreement with the study of Giampapa *et al.* (1979) in which measurements of the quiet solar photosphere and active solar regions in the *uvby* system show that the effect of the active regions was to decrease the apparent metal abundance derived from the m_1 index.

In another study of RS CVn stars, Pettersen (1983) suggests that chromospheric activity may affect the photometric indices by filling in the cores of strong non-magnetic lines as a result of non-thermal heating in the lower chromosphere. Finally, in a recent paper, Basri *et al.* (1989) have shown that stellar activity in late-type, main-sequence stars produces perturbations to the spectral line profiles through pseudoemission features and broader wings but shallower cores.

As discussed by Reglero *et al.* (1990) the presence of infrared excesses in RS CVn-type binaries is still an open question and different investigators reach contradictory conclusions. The origin of the IR excesses have been attributed to the presence of circumstellar material or the presence of cool starspots on the stars, or both effects. Independent indications of circumstellar material in some RS CVn systems have been given by Verma *et al.* (1987) from IRAS data and by Morris and Mutel (1988) from radio observations. Also, several RS CVn and BY Dra binaries have small, variable amounts of linear polarization ($P \leq 0.1\%$) at optical wavelengths that are best interpreted as the result of scattering from cool, transient, circumstellar material (e.g. Pfeiffer 1979; Scaltriti *et al.* 1992). However, in the recent study of RS CVn itself, Reglero *et al.* (1990) find that the IR excess of the cooler, more active K2 IV component is clearly reduced once the effect of starspots is accounted for. This result certainly does not rule out the presence of circumstellar matter in the system, however. Simultaneous *UBVRI* and *HJKLM* photometry might help resolve this issue.

3.4. CORONAL X-RAY EMISSION

Ever since the discovery of strong, soft X-ray emission (0.1-4.5 keV) from RS CVn and BY Dra systems by HEAO-1 (Walter *et al.* 1978), these stars have been frequently observed by subsequent X-ray satellites *Einstein, EXOSAT, Ginga,* and most recently *ROSAT*. The results of the X-ray surveys indicate that the X-ray luminosity is correlated with the binary period and hence stellar rotation. There are also tight correlations between X-ray emission and other stellar activity indicators such as chromospheric and TR line emissions and radio emission (see e.g., Zwaan 1986). The X-ray luminosities of RS CVn systems generally are $L_x \approx 10^{29} - 10^{31}$ ergs/s while BY Dra binaries tend to have $L_x \approx 10^{29} - 10^{30}$ ergs/s. In either

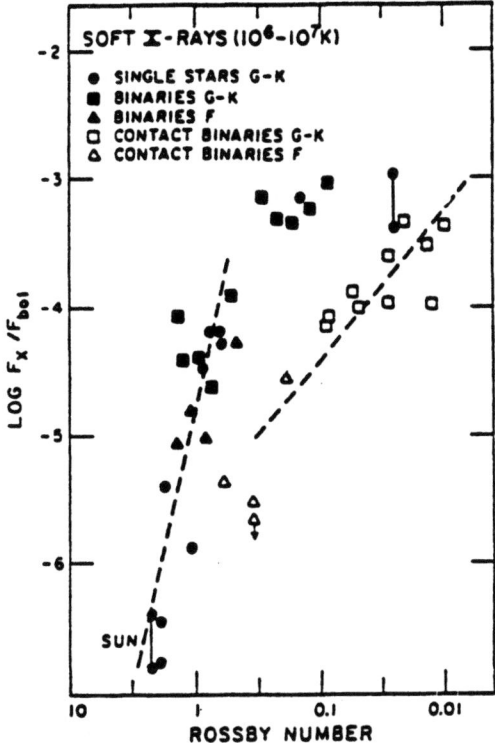

Fig. 10. The log of the normalized x-ray flux (F_x/F_{bol}) is plotted versus Rossby number R. Single stars as well as binary stars are plotted with different symbols (from Vilhu 1984).

case the values of L_x for these stars are $\approx$ 100-10,000 times the X-ray luminosity of the quiet Sun ($L_x \approx 2 \times 10^{27}$ ergs/s). There is now little doubt that this X-ray emission comes from enhanced coronal emission from these stars and other binaries containing a rapidly rotating cool member such as Algol systems and W UMa-type binaries. By analogy to solar coronal X-ray emission, the X-ray emission originates from hot (T $\approx 10^7$ K) plasma confined by closed magnetic loop structures in the star's corona (cf. Rosner et al. 1978). To attain the observed high values of X-ray emission, however, the loop structures must be hotter, more numerous, and occupy a larger volume than those observed on the Sun.

Figure 10 from Vilhu (1984) shows a plot of the normalized X-ray flux (F_x/F_{bol}) versus Rossby number. Single stars as well as binary systems are plotted. As shown in the plot, there is a tight correlation between the normalized X-ray emission and Rossby number for the less active, slowly rotating stars from the Sun up to R $\approx$ 0.5. At fast rotation rates (i.e., R $\leq$ 0.5), the X-ray emission reaches a maximum level value near $\log(F_x/F_{bol}) \simeq -3.0$. The contact W UMa binaries appear to follow a different relation and tend to have X-ray fluxes less than their RS CVn and BY Dra star counterparts with the similar Rossby numbers. Also, as discussed later, the W UMa systems have significantly smaller radio luminosities than RS CVn,

BY Dra, and Algol systems. As discussed by Vilhu (1984, 1987), the leveling off of X-ray emission for the most active RS CVn and BY Dra stars (with R ≤ 0.5) suggests that saturation occurs when the loops fill the available coronal volume.

The temperature of the corona can be derived from the X-ray energy distributions. The first important measures of coronal temperatures came from the *Solid State Spectrometer (SSS)* on the Einstein satellite (see Swank *et al.* 1981). Later determinations of coronal temperatures of Capella and σ^2 CrB were made with the Transmission Grating Spectra with EXOSAT (Lemen *et al.* 1989). These studies and others of Algol systems (White *et al.* 1986) indicate a bimodal X-ray energy distribution fit by a two temperature plasma with $T_1 \approx 4 - 7 \times 10^6$K and $T_2 \approx 2 - 8 \times 10^7$K. The two-temperature X-ray result has been challenged, however, by Majer *et al.* (1986), who find evidence of a continuous X-ray energy distribution.

Swank *et al.* (1981) propose that each of the two thermal components of the X-ray plasma originate from loops with different characteristics. As discussed by Linsky (1990), without additional information this question cannot be answered now. Perhaps the X-ray observations now being made by *ROSAT* will help resolve this problem.

3.4.1. X-ray Mapping of Stellar Coronae: The Case of AR Lacertae

Recently, White *et al.* (1990) have applied this X-ray eclipse/rotational modulation technique to the bright (nearby ≈ 50 pc), totally eclipsing RS CVn binary AR Lac. This binary has an orbital period of 1.98 days and consists of a G2 IV primary ($M_h = 1.35 M_\odot$; $R_h = 1.54 R_\odot$) and a K0 III-IV secondary ($M_c = 1.36 M_\odot$; $R_c = 2.82 R_\odot$). The system is detached and has an orbital inclination of i ≈ 88° and semi-major axis of a = 9.2 $R_\odot$. Because of its relatively short orbital period and tidally forced rapid rotation of its solar-like components, AR Lac is a strong X-ray source with $L_x \approx 10^{31}$ ergs s^{-1} (White *et al.* 1990). The only drawback with interpreting the X-ray observations of AR Lac (not present in Algol) is the possible confusion in locating active regions because both stars of this binary have high levels of activity.

These X-ray observations of AR Lac were made during July 1984 with EXOSAT. The binary was observed continuously over its orbit for 2 days. They report observing a <u>definite</u> X-ray eclipse at low energies (< 1 keV) during primary minimum and a smaller decrease preceding the secondary minimum (Figure 11). At higher X-ray energies of > 1 keV no significant orbital modulation or X-ray eclipses are evident in the data. The low energy component can be fit with a temperature ≈ 7×10^6K. The failure to observe an orbital modulation or eclipse at high energies indicates that the high temperature plasma envelopes the entire binary. They suggest that the high temperature corona may be confined by closed magnetic structures with dimensions comparable to the binary separation (≈ 10^8 cm) and that it may be flare heated plasma, ejected from flaring loops.

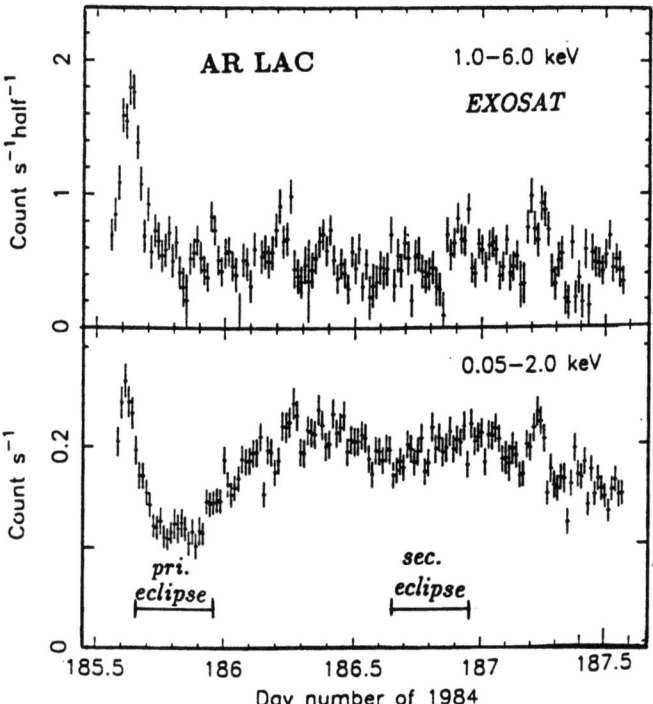

Fig. 11. X-ray observations of AR Lac, made during July 1984 with *EXOSAT* by White et al. (1990). An x-ray eclipse is seen at low energies (<1 keV) during primary minimum and a smaller decrease in x-ray flux is apparent preceding the secondary minimum. At higher x-ray energies (>1 keV) no significant orbital modulation or x-ray eclipses are evident.

Modeling of the low energy X-ray modulation was carried out using maximum entropy techniques. There are multiple solutions due to uncertainties in the relative heights of the different emitting regions and on identifying from which star the emission originates. They find that the 7×10^6K emission originates from one or two regions covering to 15% of the K star and 25% of the G star. The modeling gives a peak coronal pressure of ≈ 100 dyne cm^{-2} which is over $\approx 10^2$ times larger than the Sun's quiescent corona, but comparable to pressures determined for solar flare regions. One possible solution is illustrated in Figure 12. The X-ray map was made under the assumption that both stars are X-ray sources and that the height of the corona above the G star is ≈ 1.5 R$_\odot$ and ≈ 0.05 R$_\odot$ on the K star.

These studies illustrate the power of using the eclipse mapping techniques to study the structure of coronae of active stars. This X-ray study of AR Lac complements the work done by Walter et al. (1987) and Neff et al. (1989) using IUE data. It would be *extremely* valuable if simultaneous Doppler imaging of ground-based and UV spectroscopy could be coordinated with X-ray mapping by the ROSAT X-ray observatory. The increased count rate expected from ROSAT (about 10 × EXOSAT) would provide much better signal to noise in the data and also permit better time and spatial resolution and identification of coronal structures in AR Lac and other eclipsing chromospherically active binaries.

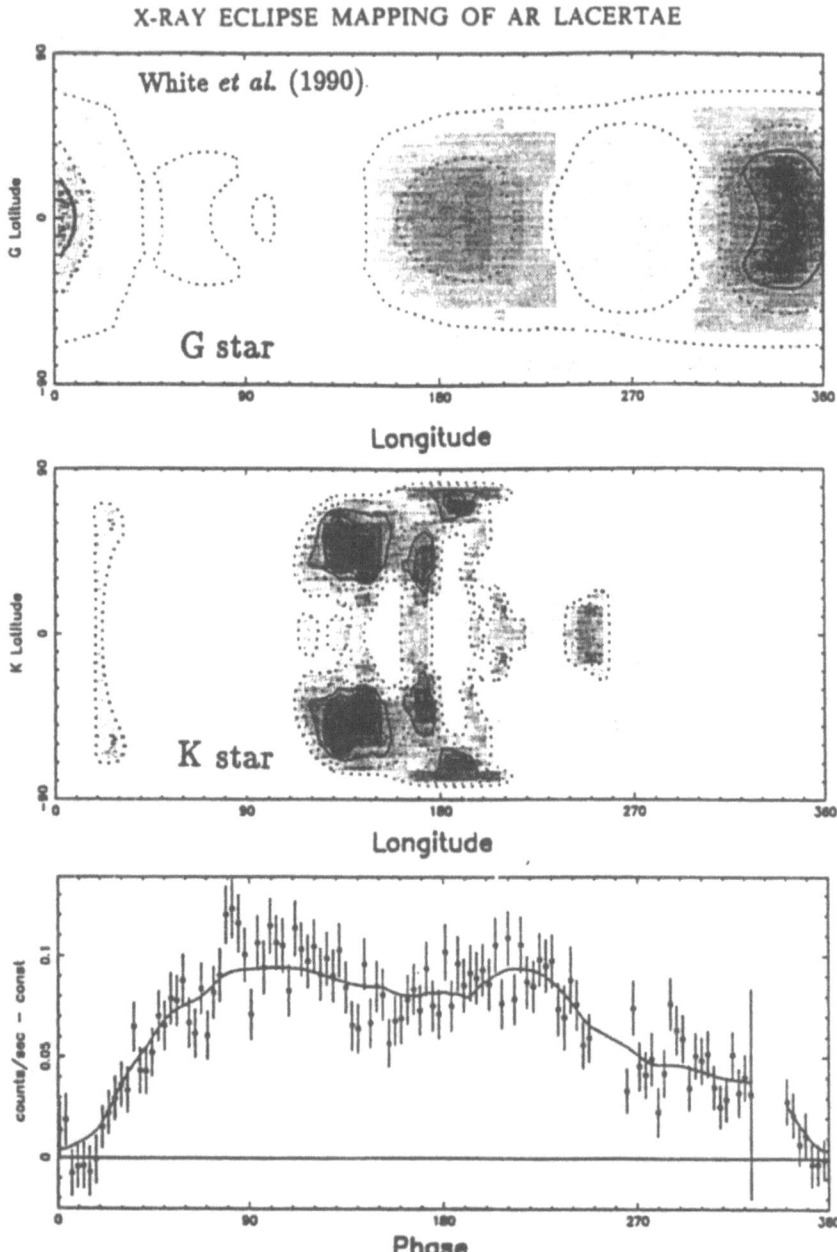

Fig. 12. X-ray map of AR Lac from White *et al.* (1990). This map assumes that both stars are x-ray sources and that the height of the corona above the G star is $\approx 1.5\ R_\odot$ and $\approx 0.05\ R_\odot$ on the K component. Other x-rays maps, computed with different assumptions are given by White *et al.* (1990).

3.5. RADIO EMISSION OF RS CVN AND BY DRA BINARIES

The Sun is a well known source of radio emission. The Sun's radio emissions, both the slowly varying *quiescent* emission and radio outburst activity, are generally believed to be intimately associated with magnetic activity within the Sun's corona and magnetosphere. For the most part, the bulk of the centimeter radio emission originates from gyrosynchrotron and thermal bremstrahlung processes in thermal and relativistic plasma in the corona (see Dulk 1985; Hjellming 1988; and Morris and Mutel 1988 for reviews).

Because most (if not all) of the Sun's radio emission is related to magnetic activity, it is not surprising that stars with higher levels of magnetic activity (inferred from enhanced chromospheric, X-ray coronal, and starspot activity) should show correspondingly high levels of radio activity. Radio investigations of RS CVn and BY Dra binaries show this to be true. In a recent study of RS CVn and related chromospherically active stars by Drake *et al.* (1989), about half of the over 120 active binaries studied at 6-cm are radio sources. They have radio emission levels of up to several orders of magnitude stronger than the Sun and other single, slowly rotating, solar-like stars. In addition, large radio flares have been detected for several of the most active and better studied systems like V711 Tau, UX Ari, and II Peg (see Byrne 1989 and refs. therein). These radio flares have similar characteristics to the X-ray flares observed for the same stars.

Drake *et al.* (1989) find no significant correlation of the 6-cm radio luminosity (L_{6cm}) with either orbital or rotational period. However, they do find a statistically significant (although weak) correlation of the 6-cm radio luminosity with the rotational velocity of the active component of these binaries which is $L_{6cm} \approx V_{rot}^{1.0\pm0.3}$. The lack of a significant dependence of 6-cm radio emission on orbital period may indicate that the radio emission is due to their rapid rotation and probably *not* to the binary nature of the emitting star. This would imply that the star's membership in a binary seems to have no direct consequence on its radio emission, except that the membership in a close binary keeps the star spinning rapidly.

It should be noted, however, that young, single main-sequence solar-type stars with moderate rotation velocities ($V_{rot} \approx 10\text{-}20 \text{ kms}^{-1}$) have not yet been detected as cm-radio sources even though they are strong X-ray sources (Drake *et al.* 1991; Drake 1991). For example, Elias and Dorren (1990) have failed to detect any 3-6 cm radio emission from the young, rapidly rotating ($P_{rot} \approx 2.80$ days) and chromospherically-active, single G0 V star HD 129333. So, perhaps binarity may play some role in enhancing radio emission. This point should be more thoroughly investigated.

The study of Drake *et al.* (1989) does indicate a strong correlation between normalized radio luminosity (L_{6cm}/L_{bol}) and the normalized soft X-ray luminosity (L_x/L_{bol}) in the sense of $L_{6cm}/L_{bol} = (L_x/L_{bol})^{1.4\pm0.1}$. A similarly strong correlation is also found between L_{6cm}/L_{bol} and the normalized CIV λ1550 line emission luminosity L_{CIV}/L_{bol} in the form $L_{6cm}/L_{bol} \approx (L_{CIV}/L_{bol})^{2.2\pm0.3}$. These relation-

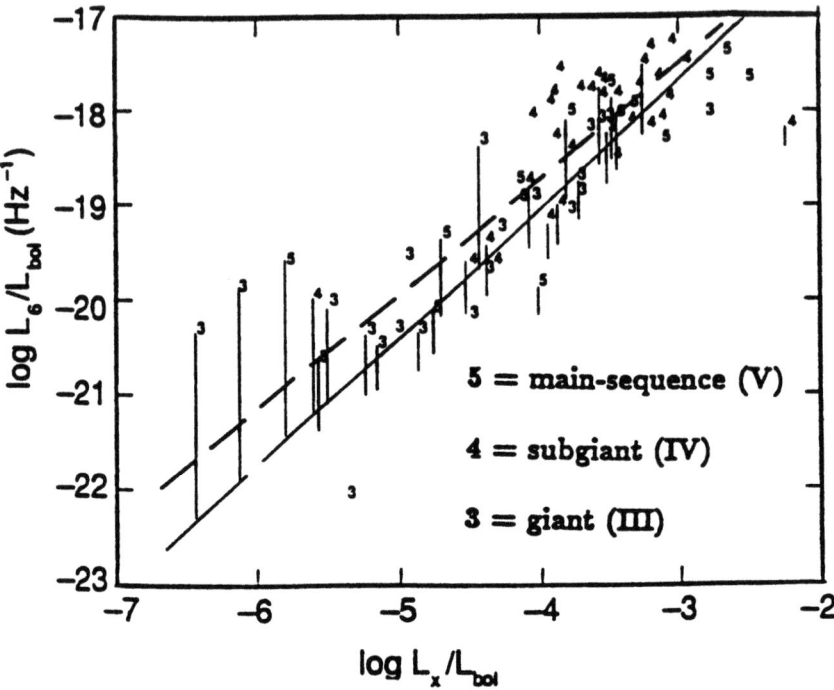

Fig. 13. The normalized 6 cm radio luminosities L_{6cm}/L_{bol} are plotted against the normalized x-ray luminosities L_x/L_{bol} for chromospherically active binaries (from Drake et al. 1990).

ships are clearly shown in Figures 13 and 14 from Drake et al. (1989). This study unambiguously demonstrates the close relationship among radio emission, X-ray and chromospheric-transition region activity of these active binaries.

They also find strong evidence that physically connects the observed 6-cm quiescent radio emission of the stars in their sample with the hot component ($T_h \approx 25 - 70 \times 10^6$K) of the X-ray emission found from *Einstein* observations. If this physical association is true, then the 6-cm radio emission and the high temperature component of the X-ray emission are both produced by the same $\approx 50 \times 10^6$K thermal electrons. This would indicate that the radio emission arises from thermal gyrosynchrotron radiation from an extended halo region. They estimate that the electron density is Ne $\approx 2 \times 10^8 \text{cm}^{-3}$ and the average magnetic field strength is $B \approx 200$ Gauss.

3.6. FLARES

Powerful, often long-lasting flare events have been observed at all wavelengths for RS CVn and BY Dra systems. Summaries of the flare characteristics and the stars

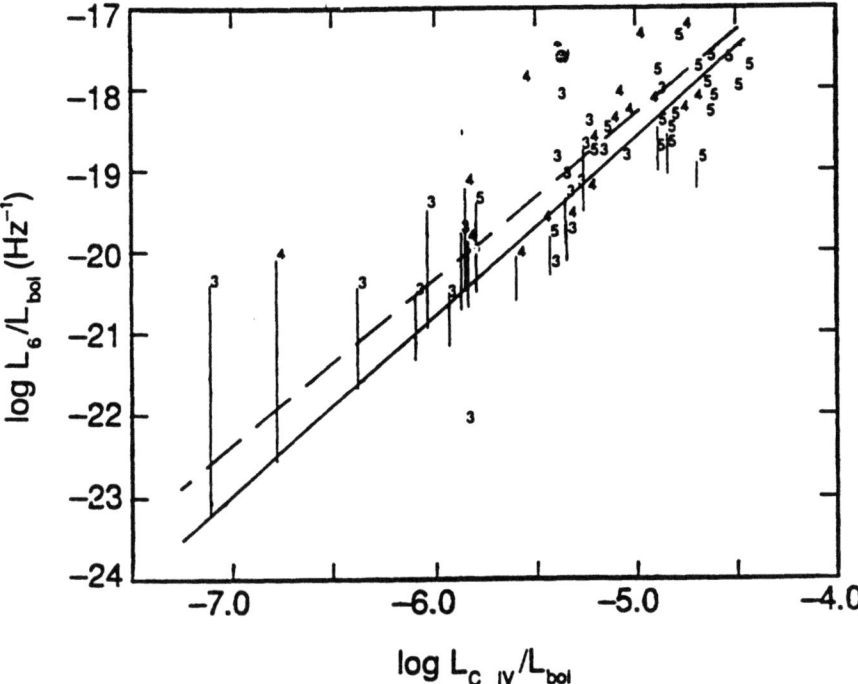

Fig. 14. A plot showing the relationship between L_{6cm}/L_{bol} versus $L_{C\ IV}/L_{bol}$ for chromospherically active BY Dra and RS CVn systems (from Drake et al. 1990).

from which they were observed are given by Bryne (1989), Kuijpers (1989) and Pettersen (1989). Table III gives a summary of flare characteristics of the RS CVn stars.

As discussed by Bryne (1989), the basic solar flare model of a localized injection of energy at the base of a magnetic loop structure, followed by radiative cooling, appears to fit many of the properties of the stellar flares. It is interesting that there are significant differences between flares on RS CVn and dMe stars, including some BY Dra stars. As discussed by Bryne (1989), Haisch (1989) and Kuijpers (1989), the flare events at X-ray, UV, and radio wavelengths on RS CVn systems are over $10^2 - 10^3$ times more energetic than the flares of dMe stars and other active single, main-sequence stars. In addition, the duration of the flare events on RS CVn systems are long and can last up to 7 hours, as in the case of the UV flare observed by Baliunas et al. (1984) for λ And with IUE.

It has been suggested by a number of investigators (e.g., Mutel et al. 1987; Kuijpers 1989; and Bryne 1989) that the long duration, highly energetic flares observed for RS CVn systems may be directly related to their binary nature rather than just their tidally enforced rapid rotations. It is possible that the magnetic fields of both components of the binary could interact and produce violent relaxations between the individual stellar magnetic fields (see e.g, Uchida and Sakurai 1983).

TABLE III
Summary of the flare characteristics of CR CVn systems[1].

X-ray	
Integrated X-ray Energy (erg)	$10^{34} - 10^{36}$
X-ray temperature (K)	$6 - 10 \times 10^{7}$
X-ray electron density (cm^{-3})	$3 - 9 \times 10^{11}$
Soft X-ray (0.1-2 keV) Emission measure (cm^{-3})	$10^{53} - 10^{54}$
UV Transition Region	
Integrated transition region energy (ergs) (from *IUE*)	$10^{35} - 10^{37}$
C IV λ 1550 Å Emission measure (cm^{-3})	$10^{51} - 10^{54}$
Transition region electron density-N_e(cm^{-3})	$5 - 10 \times 10^{10}$
Radio	
L_{radio}(5GHz) erg s^{-1} Hz^{-1}	$10^{14} - 10^{18}$
L_{radio}/L_{bol}	$\approx 10^{-7}$
All Wavelengths	
Total Flare energy (all wavelengths)(ergs)	$10^{34} - 10^{36}$
Characteristic Dimension of Flare Emitting Region (cm)	$\approx 10^{11}$
Average Flare Duration (sec)	$\approx 10^{2} - 10^{4}$

(1) Adopted from Bryne(1989) Table V and Kuijpers(1989) Table I.

This could explain, in part, the extended radio and X-ray emission observed from some RS CVn systems which have dimensions comparable to the size of the binary orbit. The disconnection and reconnection of magnetic field lines between the stars could lead to powerful releases of energy that could explain the energetic flares. This effect is known as *duplicity-enhanced activity* from Kuijpers.

As discussed later in this chapter, the radio observations of Algol systems, which contain only one active component, do not seem to support this idea.

3.7. DIRECT DETECTION OF MAGNETIC FIELDS FROM SPECTROPOLARIMETRY

The Zeeman-broadening technique used by Saar (1991) and others to measure magnetic field strengths on mostly single, solar-type stars, cannot usually be applied to RS CVn and BY Dra binaries. This is chiefly because the line profiles of these binaries are dominated by rotational broadening and also complicated by the visibility of the lines from both components of the binary. Recently, however, Donati *et al.* (1990) have reported directly detecting a localized magnetic field from the more active K-subgiant of the RS CVn star, V711 Tau. They used the Zeeman-Doppler Imaging method described in their paper, in which the magnetic field is determined from circular polarization measurements made from high resolution, high signal-to-noise spectropolarimetry. Their observations, obtained at 0.82-0.87

phase, indicated a localized magnetic region with $B = 985 \pm 270$ Gauss located near the equator of the star and covering about 18% of its total surface area. Observations made of the same star one night earlier at 0.48-0.52 phase did not indicate the presence of a significant magnetic field.

The magnetic field strength they determined for V711 Tau is smaller than that found in sunspot umbrae ($B \approx 1200 - 1600$ Gauss), but covers a much greater surface area. Their result is in good agreement with the results reported by Saar (1991) for about 30 single solar-type stars using the Zeemann broadening technique.

The Zeeman-Doppler imaging method is a powerful tool for gaining access to spatial information of magnetic fields on V711 Tau and other chromospherically active binaries.

3.8. PHYSICAL PROPERTIES AND EVOLUTIONARY STATUS

An important step toward understanding the physical mechanisms involved in RS CVn and BY Dra-type binaries has been facilitated by Popper's study of spectroscopic orbits and the determination of absolute dimensions of 19 eclipsing binaries with CaII H+K emission (see Popper 1988a,b; 1990; 1991). These studies include RS CVn itself and the interesting case of the detached binary Z Her (F4 IV-V + K0 IV; P = 3.993d) in which, unlike the vast majority of RS CVn-type binaries, the hotter F4 IV-V component is more massive than its cooler companion. This is difficult to explain since both components of Z Her appear to lie well inside their Roche lobes and do not appear to be exchanging mass. One possible explanation is that the cooler star has lost mass, perhaps through enhanced stellar winds (cf. Tout and Eggleton 1988).

The cooler components of AR Mon (G8 III + K2-3 III; P = 21.21d) and RZ Cnc (K1 III + K3-4 III; 21.64d) definitely fill their Roche lobes and probably are transferring and losing mass (see Demircan 1990). Moreover, the cooler components of UX Ari (G5 V + K0 IV-III; P = 6.44d), SS Cam (F5 V-IV + K0 IV-III; P = 4.82d), AR Lac (G2 IV + K0 IV; P = 1.98d), SZ Psc (F8 IV + K1 IV; P = 3.97d) and RT Lac (G9 IV + K1 IV; P = 5.07d) nearly fill their respective Roche lobes. Demircan (1990), furthermore, suggests that RZ Cnc and probably RT Lac are in the later stages of mass transfer. All of the above mentioned stars could soon become Algol systems once rapid Roche lobe overflow (RLOF) ensues.

A comparison of the physical properties of the components of classical RS CVn binaries with evolutionary models has been carried out by Montesionos *et al.* (1988) and by Demircan (1990). From a sample of 31 systems, Demircan (1990) finds strong evidence that the classical RS CVn binaries are divided in two distinct groups according to their X-ray luminosities. For the strong X-ray emitting group (log $L_x > 30.85$), the orbital period, mass ratio, and orbital separation (= semi-major axis) are relatively smaller and the the rotational velocity and Roche lobe filling percentage (RL%) are relatively larger. The X-ray surface flux F_x first increases with increasing RL%, until RL% ≈ 0.6 and then tends to decrease with increasing

RL%. The short period systems in this sample with rapidly rotating G dwarfs with RL% ≈ 0.6 – such as ER Vul (G1 V + G2 V; P = 0.698d) have the highest values of F_x. According to Demircan, the important parameters in determining the two distinct groups of RS CVn binaries in L_x are Mass (M), mass ratio (q), semi-major axis (a), period (P), stellar radius (R), rotation velocity (v), and surface gravity (g). The effect that the membership of the star in a binary may play with activity is dependent not only on P and v through tidal synchronization, but also on q and the size of the orbit, which may play a role in the possible interaction of the magnetic loops. Studies of a larger sample of stars with well-determined physical and orbital properties are needed to answer this important question.

4. Magnetic Activity in Semi-detached Binaries with Cool Components: The Algols

Algol-type binaries are interacting semi-detached systems composed of a hot (spectral types ≈ B5-F2), nearly spherical primary and a cooler, less massive companion in contact with its Roche equipotential surface. Algols have apparently evolved from formerly detached binary systems in which the originally more massive star has evolved off the main-sequence and is transferring mass to its now more massive companion. Spectroscopic studies of Algol systems show that many Algols have gas streams and accretion disks surrounding the hotter component. From light curve and spectroscopic analyses, the cooler components of Algols appear to have spectral types, radii, and rotation rates similar to the active components of RS CVn stars. The major differences between the cool components of Algol systems and RS CVn stars are that cool components of Algols are evolved stars that have lost considerable mass, and, in addition, have their surfaces in contact with their Roche lobe. However, because of the overwhelming relative brightness of the hotter B-to-A primaries of Algols at UV to near infrared wavelengths, it is difficult in most cases to ascertain the activity of their cooler components. Because of this, magnetic activity in Algol systems is inferred chiefly from X-ray and radio observations in which these emissions presumably arise from the cooler, secondary components of the systems (cf. Hall 1990).

4.1. STARSPOTS ON ALGOLS

There are basically two methods for studying the characteristics of starspots on the cool components of Algols: (1) Observation at infrared wavelengths where the cool star and its spots contribute a larger portion of the system's light, and (2) Study of *totally* eclipsing Algols in which the light contribution of the hot star is eliminated by the occultation of the cooler star.

Recently, Richards (1990) has analyzed the 1.2 μm IR light curves of Algol (B8 V + K2 IV; P = 2.87d) itself, obtained during the 1960's, and found strong evidence of a variable RS CVn-type photometric wave. Figure 15 from Richards

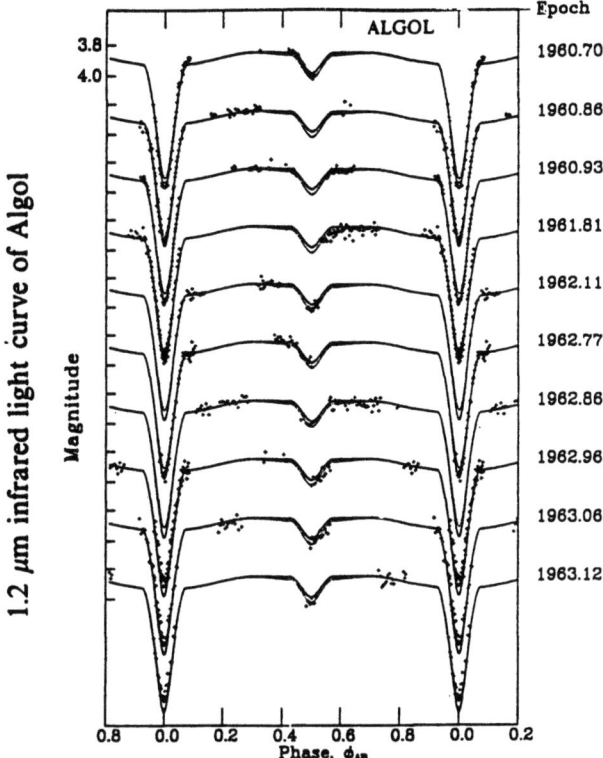

Fig. 15. The 1.2 μm light curves of Algol obtained during the early 1960s by Chen and Reuning as given by Richards (1990). The distortions and changes in the infrared light curves are produced by variable amounts of starspots on the K IV component of the system. The observations are compared with theoretical curves computed by Richards from a set of fixed parameters. The solid curve represents the case where the average temperature of the cool star is $T_K = 4500$K, and the dotted curve is for $T_K = 5000$K.

shows the variability of the 1.2 μm light curve of Algol with time. The theoretical light curves for the binary are shown together with the data. Modeling of the epochal light curves indicates temperature changes of the cool star of a few hundred degrees. She explains the time-dependent changes in the depths and phases of both eclipses, as well as in the asymmetric shape of secondary minimum, as arising from starspots on the K2 subgiant secondary. An important follow-up study would be to obtain modern high precision IR light curves of Algol and several other Algol-type binaries to determine the spot coverage and temperatures. High Precision *HJKL* light curves would be very valuable in studying this problem.

The totally eclipsing Algol system U Cep (B7 V + G8 III-IV; P = 2.49d) offers another opportunity to study activity on the cool components of Algols. During primary eclipse, the B7 V star is totally occulted, producing a deep, nearly flat-bottomed eclipse. At primary eclipse only the outer hemisphere of the cool star is visible, making it possible to measure directly its surface properties. Olson (1985) has studied nearly 9 years of photometry of U Cep and found evidence of cyclic variability in brightness of the G8 III star from changes in the depth of primary eclipse. It appears that the depth of mid-primary eclipse varies periodically with time. These low amplitude (≈ 0.05 mag) light variations could arise from changes in the surface temperature of the G-star. If we carry this suggestion a bit further, the brightness variations could be caused by a ≈ 4-6 yr spot cycle. In addition to the cyclic light variation, there are ≈ 0.1 to 0.2 magnitude light increases. These rapid variations in brightness could be due to flare-like events or white light faculae on the G-star. Moreover, Olson (1985; 1987) finds evidence for a correlation between the large brightness increase of the cool star with times of enhanced gas flows and accretion in the system. Further observations are needed to make more definitive conclusions.

4.2. CHROMOSPHERIC ACTIVITY ON ALGOLS

The total eclipse of U Cep also provides an important opportunity to investigate the chromospheric and transition region of the G8 III-IV component in the ultraviolet with the IUE satellite. Many important chromospheric and TR line emissions are found in the $\lambda\lambda 1150$-3200Å wavelength band of the IUE. Numerous UV spectra of U Cep have been obtained with IUE, chiefly to study gas accretion (see Plavec 1985; Giménez et al. 1990). Several spectra were obtained during totality and these reveal the rich emission line spectrum, chiefly of the cool star (see Figure 16). As shown in the figure, the UV spectrum is similar to the corresponding emission line spectra typical for RS CVn stars. Except for the absence of the HeII $\lambda 1640$ line, the emission line fluxes are similar to those measured for RS CVn stars of comparable periods and spectral types. Plavec (1985) has suggested that these emission lines arise from the accretion disk that surrounds the hot star. However, recent high dispersion IUE spectra of the MgII h+k emission feature taken during primary eclipse by us indicate that most of the emission originates from the cool star.

4.3. X-RAY CORONAL EMISSION IN ALGOLS

Probably the best evidence for solar-like magnetic activity in Algol binaries comes from X-ray studies. Algol itself was first detected as an X-ray source over 15 years ago by Schnopper et al. (1976). Subsequent X-ray studies of Algol and several nearby Algol-type systems (see White and Marshall 1983) indicate that they have quiescent X-ray emission in the range $L_x \approx 1 - 7 \times 10^{30}$ ergs/s. This is similar to the X-ray luminosities of RS CVn stars. Moreover, the X-ray spectrum of Algols can be modeled with two temperature components ($T_1 \approx 1 - 6 \times 10^6$K; $T_2 \approx 20 - 40 \times 10^6$K), also similar to the RS CVn stars (see Swank et al. 1981; White et al. 1986). X-ray surveys of field stars indicate that the mid-B to late-A main-sequence primaries of Algols should have X-ray fluxes several orders of magnitude lower than the typical X-ray luminosity ($L_x \approx 10^{30}$ ergs/s) of an active cool star (cf. Pallavicini et al. 1981). Thus, the X-ray emission observed in Algols

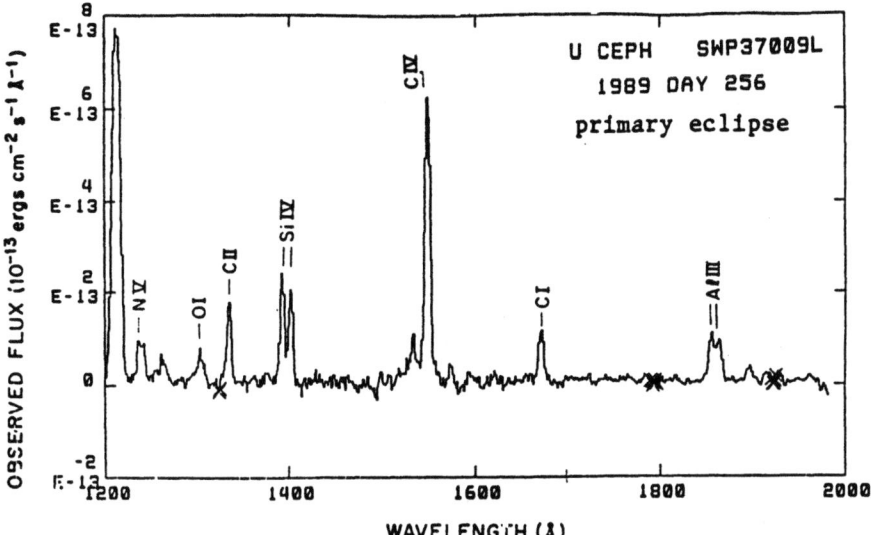

Fig. 16. The *IUE*/SWP spectrum of U Cep obtained during the total eclipse of the hotter B7 V component of the system. The emission lines are similar in strength and kind to those found in chromospherically active binaries (from Giménez et al. 1990).

has been reasonably associated with the coronal emission of the cooler star of the system.

White *et al.* (1986) used the EXOSAT observatory to make 35 hours of continuous X-ray observations of Algol over about half of its orbit. These observations were centered on the eclipse of the K-star by the B8 V primary star to investigate the spatial structure of the cool star's corona. As pointed out by White *et al.*, Algol is an almost ideal candidate for an eclipse measurement because the eclipse is nearly total and the occulting B-star is X-ray dark so that the complication of two coronal X-ray sources in one system (as in the case of RS CVn stars) is eliminated. However, no obvious X-ray eclipse was detected, indicating that the cool star's corona extends out to at least one stellar radius ($\geq 3R_\odot$). X-ray observations of Algol made with the *Ginga* X-ray satellite for about ≈ 2 days in January 1989 also found no evidence of significant X-ray eclipses (see Stern *et al.* 1990).

4.4. X-Ray Flares

More dramatically, both White *et al.* (1986) and Stern *et al.* (1990) detected powerful, long-duration X-ray flares in Algol during these observation intervals. The X-ray flare observed by EXOSAT lasted about 2.4 hours and had a 0.1-10 keV maximum luminosity of $L_x \approx 1.4 \times 10^{31}$ ergs/s and a peak temperature of $T_{max} \approx 6 \times 10^7$K. The detection of FeXXV emission in the X-ray spectrum of the flare indicates that the flare is produced by thermal heating of the plasma, and, by analogy to solar flares, plasma is confined by a magnetic loop or loops

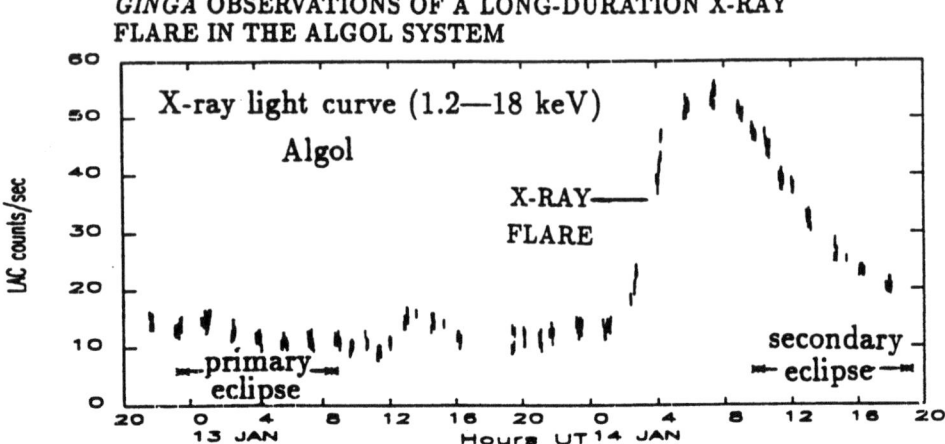

Fig. 17. X-ray flare from Algol observed with the Ginga X-ray satellite in 1989 (from Stern et al. 1990).

extending from the surface of the cooler star (White et al. 1986). As shown in Figure 17 from Stern et al. (1990), the X-ray flare detected during 1989 by *Ginga* was even more energetic than the 1983 flare, lasting over 12 hours. The peak temperature of this flare was $T_{max} \approx 70 \times 10^6$K and its peak luminosity was $L_x \approx 10^{31}$ ergs/s. This T_{max} is about 3 times hotter than that of the most energetic solar flares, but comparable to the peak temperature of the EXOSAT flare event. However, the characteristics of the X-ray flares observed for Algol are not unusual in intensity or duration when compared to the X-ray flares observed in some of the more active RS CVn stars – such as V711 Tau and UX Ari (see Linsky 1990). It may be that X-ray flares on Algol are frequent because two major X-ray flares were detected in about 3-1/2 days of monitoring.

4.5. RADIO STUDIES OF ALGOLS

As in the case of X-ray studies, only a small number of Algol systems have been studied at radio wavelengths (chiefly at 3 cm and 6 cm) with the VLA. As discussed by Umana et al. (1991), the radio properties of Algols appear strikingly similar to the more common RS CVn binary. For example, the average radio luminosity at 6 cm of $<L_{6cm}> \approx 1.6 \times 10^{16}$ ergs s^{-1}Hz^{-1} found for the Algol systems compares very closely to the median radio luminosity for the larger sample of 53 RS CVn binaries of $<L_{6cm}> \approx 2.5 \times 10^{16}$ ergs s^{-1}Hz^{-1} reported by Morris and Mutel (1988). In addition, the Algols observed appear to have high brightness temperatures of the order of $T_B \simeq 10^8 - 10^9$K which are similar to the values of T_B observed for RS CVn stars and which indicate a non-thermal origin for the radio emission (Umana et al. 1991). As in the case of RS CVn stars, the radio observations of Algols appear consistent with gyrosynchrotron emission from mildly relativistic electrons interacting with the magnetic field of the cooler,

TABLE IV
Radio properties of Algol-type binaries.

Star	Spectral Class.	Period (days)	Obs. Date	Radio Flux Density (mJy)	Distance (pc)	Radio Luminosity (6cm) (10^{15}ergs^{-1}Hz^{-1})
YZ Cas	A2 IV + F4	4.467	Nov 84	≤ 0.25	81	≤ 2
RZ Cas	A2 V + G1 IV	1.195	Nov 84	3.25	75	21.9
			Feb 89	1.25	-	8.40
AS Eri	A0 V + G6 IV	2.664	Nov 84	≤ 0.09	200	≤ 4.30
RZ Eri	A5 m + G8 IV	12.240	Nov 84	0.97	143	23.7
			Nov 84	2.31	-	56.7
R CMa	F1 V + K1 IV	1.135	Nov 84	0.36	43	0.79
UX Mon	A6 ep + G3p IV	5.904	Nov 84	0.19	500	56.8
TT Hya	A3 + G7 IV	6.953	Nov 84	≤ 0.06	182	≤ 2.38
δ Lib	A0 IV + G2 III-IV	2.327	Nov 84	≤ 0.48	100	≤ 5.74
α Cr B	A0 V + G3 V	17.359	Nov 84	0.63	25	0.47
TW Dra	A5 V + K0 III	2.806	Nov 84	3.9	190	168.4
			Feb 89	0.30	-	12.9
AI Dra	A0 V + F9 IV	1.19888	Nov 84	≤ 0.09	182	≤ 3.56
RY Aqr	A3 + G8 IV	1.966	Feb 89	0.22	230	13.91
V505 Sgr	A5 V + K2 IV	1.182	Jun 89	3.05	120	52.5
DL Vir	A3 + K1 IV	1.320	Nov 84	≤ 0.17	128	≤ 3

* The negative value found for delta $\Delta\Omega/\Delta\phi$, if real, indicates that the rotation rate increases with increasing stellar latitude.

assumed-active component of the systems. Table IV, adopted from the study of Umana et al. 1991, lists the radio properties of the Algols studied so far with the VLA.

The radio observations of Algols provide a means of testing the model advanced some time ago to explain the strong radio emission of RS CVn binaries (see Uchida and Sakurai 1983). In this model the strong activity observed in RS CVn systems arises from the interactions between the magnetic fields of both components. In Algol systems, however, only the cool secondary star is expected to have magnetic activity and no magnetic interactions between the components appear possible. The similarity between the radio emission in Algols and RS CVn stars indicates that the strong radio emissions of RS CVn stars originates from the individual stars rather than from the magnetic interactions between them.

4.6. THE RADIO PROPERTIES OF ALGOL

Because of its nearness to us (d ≈ 27 pc) and its favorable position in the northern sky, Algol is the best studied, chromospherically active star at radio wavelengths. Algol was one of the first *radio* stars detected; it was initially detected in 1971 by Wade and Hjellming (1972). Algol was the first radio star observed by the *Very Long Baseline Interferometer* (VLBI) in 1975 and 1976 (see Clark et al. 1975; 1976).

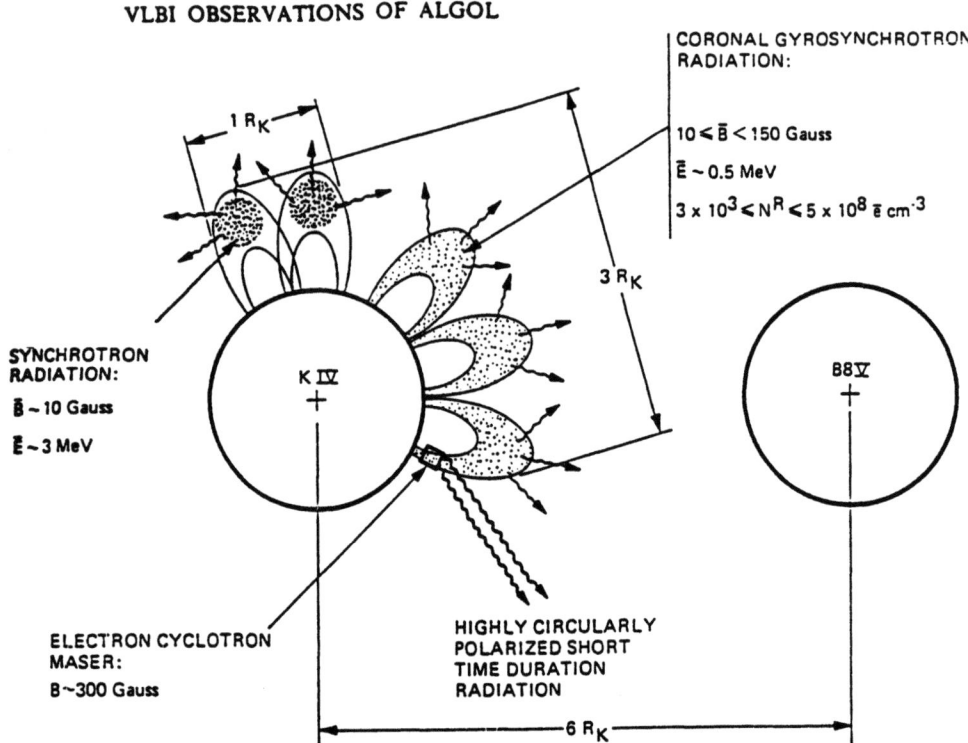

Fig. 18. A schematic diagram of Algol showing the assumed sources and mechanisms of coronal radio emission in the system. As shown, the sources of radio emission in the binary are assumed to be associated with the K-subgiant star (from Lestrade *et al.* 1988).

The angular size ($\approx$ 5 milliarcseconds (mas) one time and $\approx$ 2 mas on another occasion) of the radio emitting region and a brightness temperature $T_B > 10^8$K were determined from these first VBLI observations. Subsequent multifrequency and polarization VLBI observations were carried out using more arrays by Lestrade *et al.* (1988). The results of this study indicate that the brightness temperature of the radio source in Algol is 3×10^8K $< T_B < 5 \times 10^9$K and that it is consistent with gyrosynchrotron emission from an active coronal region. They determined the size of the radio emitting region as $\approx$ 3 times the radius of the K IV star. The inferred magnetic field strength is $B \approx 30$ Gauss. In addition, two remarkable radio events occurred indicating the presence of synchrotron radiation with $E \approx 3$ MeV and electron cyclotron maser radiation with $B \approx 300$ Gauss. Figure 18 from Lestrade *et al.* (1988) gives a schematic geometric representation of the possible sources of coronal radio emission observed for Algol.

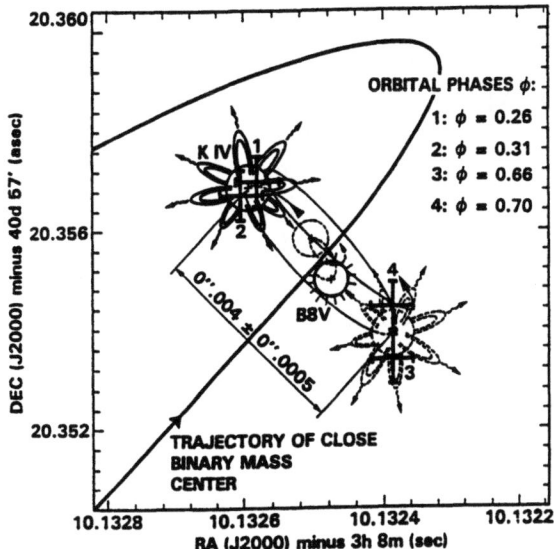

Fig. 19. *VLBI* astrometry of Algol carried out by Lestrade *et al.* (1992). As shown, the source of radio emission is definitely the K-subgiant component of the system.

4.7. ASTROMETRY OF ALGOL WITH VLBI

Ultra-precise differential astrometry with positional accuracies of $\leq \pm 1$ mas are now possible with the VLBI (see e.g., Lestrade 1988). This has led to astrometric identification of the radio source in the Algol system by Lestrade *et al.* (1992). As shown in Figure 19 (from Lestrade *et al.* 1992), highly accurate positions of the radio source of Algol were measured with the VLBI over two consecutive orbits when the binary stars were near the opposite quadratures of its 2.87 day orbit (i.e., at $\simeq 0.25P$ and $\approx 0.75P$). These observations *unambiguously* identify the K2 IV secondary of the binary as the star responsible for the radio emission in this system. The orbital motions and positions of the components of Algol were computed from the well known orbital elements of the system. This remarkable observation leaves no doubt that the cool component is the radio source.

5. Magnetic Activity in W Ursae Majoris Contact Binaries

Except for the cool dM secondary components of CVs, the F- to K- components of W UMa-type contact binaries have the fastest rotation rates of solar-type main-sequence stars. With orbital periods between $0.2d \leq P \leq 0.8d$, strong tidal forces cause the components of W UMa systems to rotate synchronously (i.e., $P_{rot} = P_{orb}$), producing rotational velocities typically $\approx 80\text{-}160$ km/s. Thus, these systems can provide information on stellar activity and magnetism of solar-like stars at the extreme limits of rotation. Excellent summaries of the properties of W UMa stars are given by Rucinski (1985; and in this volume). A summary of the magnetic-related properties of W UMa systems is given in Table I.

The first indications of high levels of solar-like activity in W UMa systems can be traced to the early report of CaII H+K emission in several W UMa systems by Struve (1950) and the suggestions by Binnendijk (1965, 1970) that the asymmetries and changes in the light curves of some W UMa systems were caused by the presence of starspots. In addition, Kuhi (1964) reported a large flare in one W UMa system. The formal connection of W UMa binaries to RS CVn stars and to solar-type magnetic phenomenon, was made by Hall (1976), who included W UMa stars in his class of short period (P < 1 day) RS CVn stars.

5.1. ENHANCED CHROMOSPHERIC, TRANSITION REGION AND CORONAL X-RAY ACTIVITY

The recognition of the extremely high levels of solar-type activity in W UMa stars was revealed in the ultraviolet and X-ray regions with the IUE satellite and the Einstein-, EXOSAT-, and ROSAT X-ray satellites. Observations of W UMa stars with the IUE have shown that these stars are rich sources of ultraviolet line emissions. They have among the largest observed surface fluxes of high temperature (10^5K), transition region emission lines such as C IV $\lambda1550$, NV $\lambda1240$, SiIV $\lambda1400$, and He II $\lambda1640$ as well as moderate temperature (10^4K) chromospheric emission lines such as MgII h+k $\lambda2800$, Lyα $\lambda1216$, and OI $\lambda1300$. Moreover, most W UMa systems are strong X-ray sources with typical X-ray luminosities of $L_x \approx 10^{29} - 10^{31}$ ergs/s (e.g., see Cruddace and Dupree 1984). This strong, generally soft X-ray emission is similar in strength and energy distribution to that observed in the most active RS CVn and Algol systems as well as in very young stars. It is believed to originate in the hot ($\approx 10^6 - 10^7$K) extended coronae associated with these systems. Summaries of the ultraviolet and X-ray properties of W UMa stars are given by Eaton (1983); Rucinski and Vilhu (1983); Cruddace and Dupree (1984), Vilhu *et al.* (1989) and references therein. In analogy with solar-type activity observed in RS CVn binaries, these line emissions most likely arise from dynamo generated magnetic fields because these stars possess (tidally enforced) rapid rotation and deep convective zones, at least in the cooler, shorter period systems. However, because of their strong tidal interactions, the components of W UMa systems should not possess significant differential rotation (radial or latitudinal) which is believed to be an important (if not necessary) component of stellar dynamo theory.

Although W UMa stars have among the highest levels of solar-type activity of main-sequence stars, their activity levels (as defined by chromospheric, transition-region, and coronal X-ray emissions) are not as high as expected from a simple extrapolation of the rotation-activity relations defined chiefly by the more slowly rotating RS CVn, BY Dra, and young single stars (see Figure 10). As discussed previously for RS CVn stars, this leveling off of activity at high rotation rates has been interpreted by Vilhu (1984, 1987), Vilhu and Walter (1987) and Vilhu *et al.*

(1989) as being due to a *saturation* effect in which the entire surface of the stars are totally filled-in by magnetic structures. As shown in Figure 5, W UMa stars are close to the upper limit of observed MgII surface fluxes for a particular value of B-V.

5.2. STARSPOTS

As first noted by Binnendijk (1965) and Kwee (1966) the light curves of W UMa systems frequently show asymmetries and undergo changes in shape, height, and depth with time. These changes appear to be similar to those observed in RS CVn and BY Dra stars but are more difficult to study because of the usually larger light variations produced by the eclipses and tidally distorted stars. The best explanation for the observed asymmetries and changes in the light curves of W UMa stars is the presence of dark starspots on the component stars. Modeling of the light curves of several W UMa systems has been carried out, for example, by Yamasaki (1982), Bradstreet (1985), and Bradstreet and Guinan (1990). The spots generally are found on the larger, cooler members of the systems and cover about 5 to 15% of the star's visible surface. The spots have temperatures about 500-800 K cooler than the stellar photospheres. To a first approximation, the spots on W UMa stars appear to have similar properties (temperatures and sizes) to those found on RS CVn and BY Dra stars. Moreover there is evidence in at least one W UMa system, VW Cep, that there is a systematic variation in the spot coverage with time which could indicate the presence of a spot cycle (see Bradstreet and Guinan 1990).

5.3. CORONAL RADIO STUDIES

As discussed by Rucinski (1992) the radio emission from W UMa systems appears to be significantly lower than that of RS CVn and Algol binaries with comparable X-ray emission. Although the radio data available for W UMa stars is still incomplete, it appears that the ratios of their radio (3-6 cm) luminosities relative to their total bolometric luminosities of $L_{radio}/L_{bol} \simeq 10^{-12}$ to 10^{-11} are much smaller relative to the values of $L_{radio}/L_{bol} \simeq 10^{-8}$ to 10^{-7} observed for active RS CVn and Algol binaries (Drake *et al.* 1986; Rucinski and Seaquist 1988; Vilhu *et al.* 1988). At the present time with only fragmentary data, it is difficult to interpret this apparent *deficiency* in the radio emission of W UMa systems relative to RS CVn stars.

5.4. A CASE STUDY: VW CEPHEI

In the next section we will discuss the results of coordinated ground-based photometry and ultraviolet spectroscopy of one of the most best studied W UMa systems: VW Cep. This star may serve as a guide to the behavior of other, less well-studied members of its class.

VW Cep consists of $\simeq$ G5 V and G8 V components in contact with their Roche limiting surfaces; the orbital period is P = 0.27 days. Analysis of the changing, and

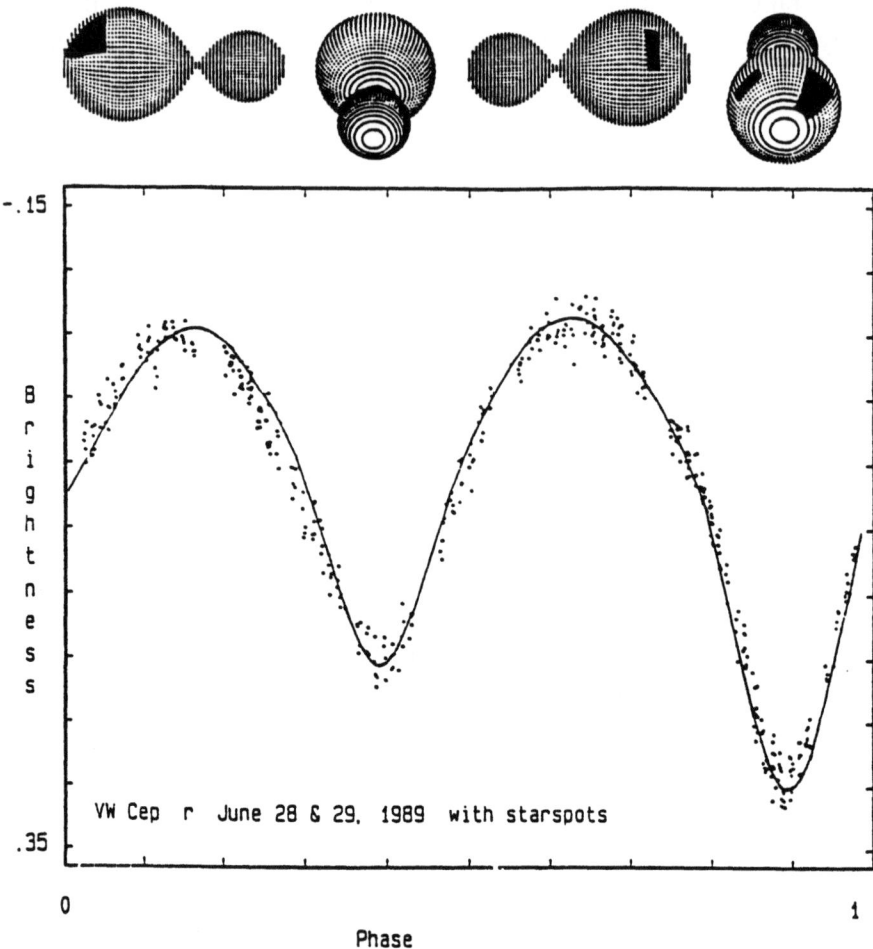

Fig. 20. The red (λ 6600) light curve of VW Cep obtained in June 1989 is plotted. The asymmetries in the light curve and the unusually deep primary eclipse have been modelled with cool starspots as shown in the three dimensional representation displayed at the top of the figure. The solid curve drawn through the data represents the theoretical light curve generated by the starspot program (from Bradstreet and Guinan 1990).

often asymmetric, light curves of VW Cep indicates the presence of cool starspots regions, chiefly on the larger, cooler star. This is illustrated in Figure 20 which shows the red light curve of VW Cep and the fit to the data with the starspot model of Bradstreet (1985). Also shown are three dimensional representations of VW Cep displayed at different orbital aspects corresponding to the phases directly below them. At that time, the modeling indicates that spots about 500 K cooler than the star's photosphere cover about 6-10% of the surface of the larger, cooler component of the system.

IUE observations were obtained around its orbit during 1987, 1988 and 1989. Figure 21 (Bradstreet and Guinan 1990) shows the SWP low dispersion spectra

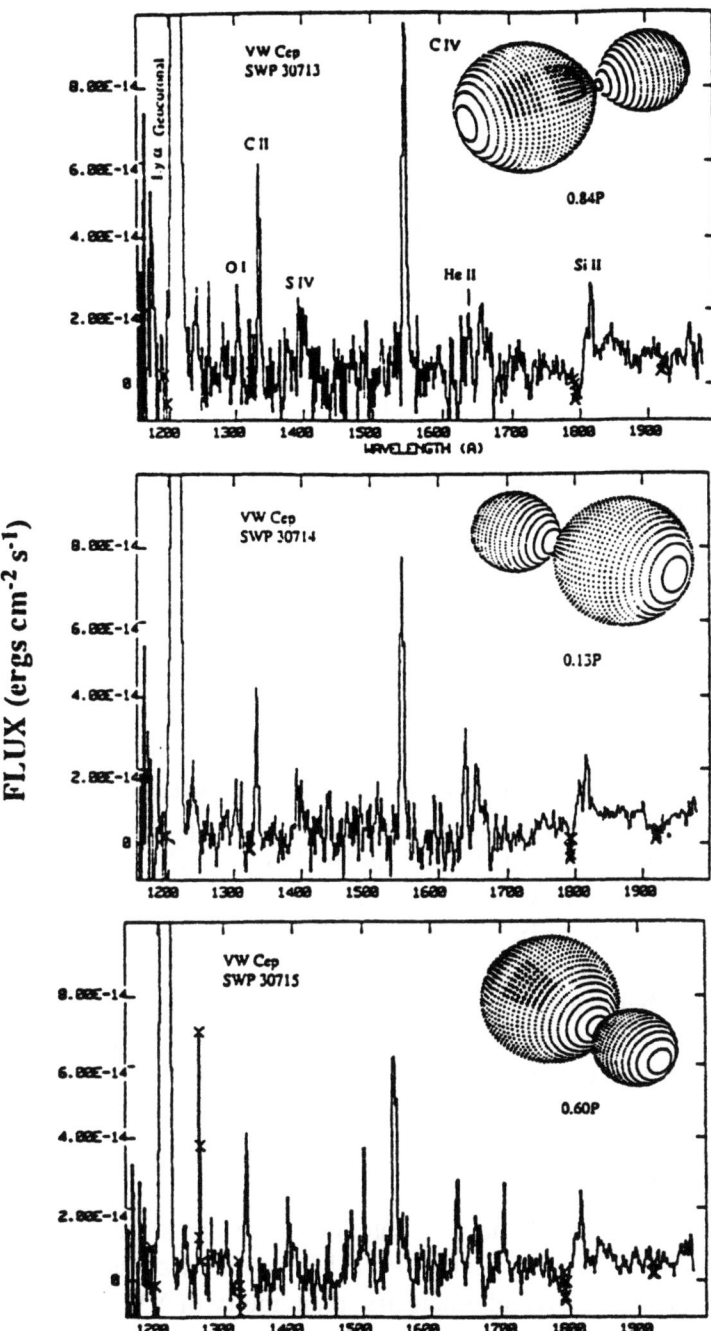

Fig. 21. SWP spectra of VW Cep obtained during April 1987 with the IUE satellite. The top spectrum (SWP 30713) was centered at 0.84P when the starspots were most visible. The middle spectrum (SWP 30714) was centered at 0.13P, and the bottom spectrum (SWP 30715) was centered at 0.60P. The insets in the upper righthand corners show representations of VW Cep and its spot distributions at the appropriate phase of mid-exposure (from Bradstreet and Guinan 1990).

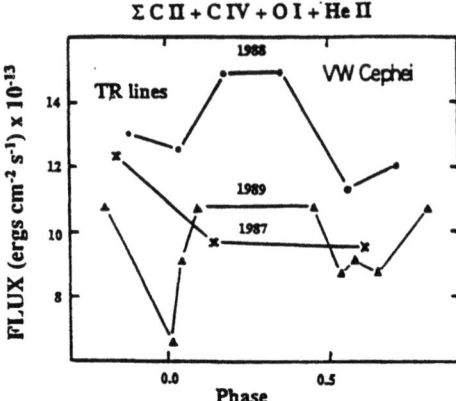

Fig. 22. The integrated emission line fluxes for C II λ 1335, C IV λ 1550, O I λ 1305, and He II λ 1640 versus phase for VW Cep. Data from 1987, 1988 and 1989 are plotted (Bradstreet and Guinan 1990).

of VW Cep obtained in April 1987. The strongest emission features are indicated in the figure. Three-dimensional representations of the binary are shown at the corresponding phase of the spectrum. The sizes and locations of the spots shown on the stars were derived from an analysis of contemporaneous light curves. As shown in the figure, the top spectrum (SWP 30713) was secured when the starspots were most visible and displayed the strongest emission features. The other two spectra were obtained when the active regions were only partially visible.

Figure 22 shows the integrated emission line fluxes for CII λ1335, C IV λ1550, OI λ1305, and HeII λ1640 versus phases of VW Cep for 1987, 1988, and 1989. As shown, there appears to be a phase dependent variation seen in the 1988 and 1989 data. The mean level of the 1988 emission line fluxes is about 40% higher compared to the other years. Thus, the observed phase dependency of the emission line fluxes appear to be correlated with the visibility of spot regions inferred from contemporaneous light curves.

5.4.1. *Evidence for an Activity Cycle in VW Cep*

Because of the longevity of the IUE satellite, there are a great many archival observations of VW Cep going back to 1978. Moreover, by coincidence, this is the same year in which extensive photoelectric photometry of VW Cep was initiated at Villanova University. The long term photometric and UV spectrophotometric behavior of VW Cep from 1978 to 1991 is shown in Figure 23. The top panel represents the changing of the levels of maximum light from the ground-based photometry. The difference in brightness in the sense [mag(max II)-mag(maxI)] is plotted against time in which max II = 0.75P and max I = 0.25P. The variations in the relative heights of the maxima of the optical light curves appear to arise from the migration of spot regions on the surface of the larger, cooler star of the binary.

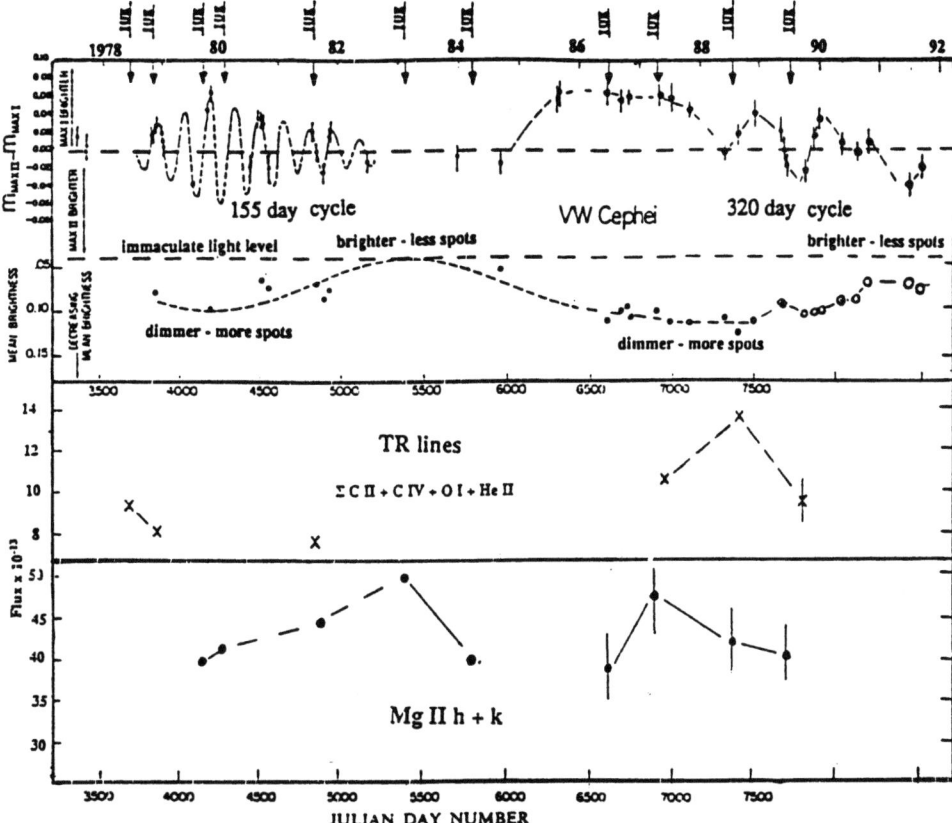

Fig. 23. The long term photometric and spectroscopic behavior of VW Cep is shown from 1978–1990. The top panel shows the changing of the levels of maximum light in the sense [Δ m (MAX II) − Δ m (MAX I)] where MAX I and MAX II are 0.25P and 0.75P, respectively. The second panel from the top shows the mean light of the system. The third panel from the top plots the sum of the integrated emission fluxes of the transition-region (TR) lines. The bottom panel displays the integrated emission fluxes of the Mg II h+k emission lines.

There appear to be transient ≈155 day and ≈320 day cyclic patterns during 1978-1981 and 1988-1989, respectively. These cyclic variations may arise from small differences between the rotational periods of the starspot regions and the orbital period of the binary. The wave migration periods are apparently the beat period between the orbital period and the rotational period of the photosphere region where the spots were located at that time. For example, the 320 day migration period indicates that the region on the star where the spots were located has a period 21 seconds longer (0.1%) than the assumed synchronous rotation period of the stars. The migration of the spotted region around the surface of the cooler star produces the varying light levels of the maxima and minima of the light curve. At times, when the spots are small, such as in 1982-84, the asymmetries are smallest and the mean brightness of the system is greatest and vice versa, as in 1979-80,

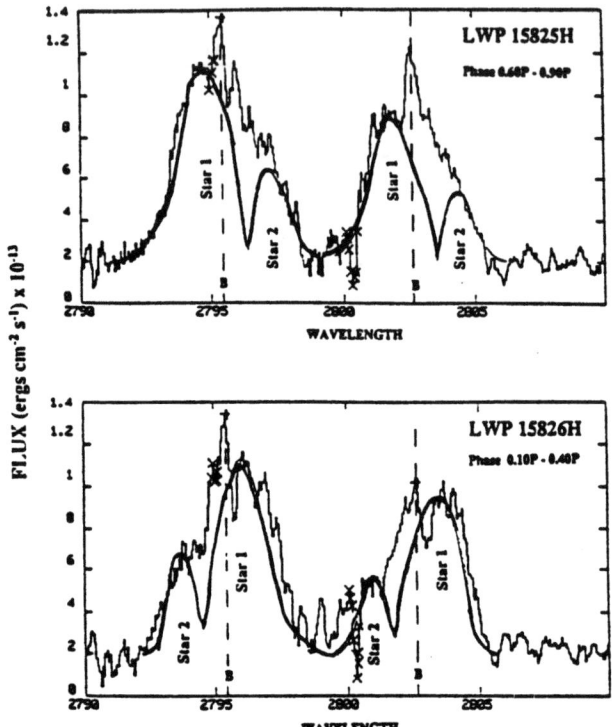

Fig. 24. Two high dispersion *IUE* LWP spectra showing the Mg II h+k emission features. The top spectrum (LWP 15825H) was centered at 0.75P; the bottom spectrum (LWP 15826H) was centered at 0.25P. The smooth curves designated Star 1 and Star 2 represent the synthetic line profiles calculated from the light and radial velocity curve solutions. The broken lines marked B are the wavelengths of the barycenter of the binary. The excess emission near the wavelength of the barycenter is easily seen. This emission could arise from the neck region joining the two components of the binary (from Bradstreet and Guinan 1990).

1986-88. A spot cycle of about 8-10 years may be present in which the spots attained their maximum areal extent during 1979-80 and 1987-88. This possible spot cycle produces the long-term variation in the mean light level of the system as shown in the second panel from the top of the figure. The recent photometry obtained during 1991 indicates a possible increase in the average brightness of the system.

The bottom two panels in Figure 23 show the mean values of the TR lines and the mean integrated fluxes of the MgII h+k lines. These data cover the years 1978-1989. The early (1978-1981) TR line fluxes are significantly less than the more recent data (1987-89), corresponding to about a 70% increase in the range of integrated line fluxes.

There is no conclusive evidence for an activity cycle in the high temperature TR emission lines. In part, this may be due to less IUE coverage of VW Cep with the short wavelength (SW) camera and the longer exposure times ($\approx$ 70-120 min) needed for a satisfactory, but still underexposed spectrum. However, it is also important to realize that the TR lines are more sensitive to flare activity so that any activity cycle could be hidden by the larger scatter in the line fluxes.

However, there is a suggestion of an activity cycle in the chromospheric UV lines. VW Cep was observed more frequently with IUE using the LWR/P cameras chiefly to measure the MgII h+k lines. The bottom panel of Figure 23 shows a plot of the mean integrated MgII h+k fluxes of all the usable IUE data. The MgII lines appear to vary in a systematic way on a time scale of a few years. The total range of variation is 25-30%; the maximum chromospheric emission occurs near the beginning of 1983 and again at the beginning of 1987. This variation could be part of an activity cycle operating in the system. More observations over the next few years with IUE will be crucial in defining long term variation, with the possibility of confirming an activity cycle in the chromosphere.

There does not appear to be, at least with available data, a clear correlation between the light curve variability and the chromospheric and TR line fluxes as may have been found for II Peg (Marstad *et al.* 1982). This is both perplexing and interesting in light of the expectations of dynamo-related magnetic activity. Perhaps, as indicated in the recent study of the Sun (Foukal and Lean 1988) and of the very active RS CVn system V711 Tau (Dorren and Guinan 1990), the light variations of chromospherically active stars over their activity cycles could be dominated more by white light faculae than dark spots. However, the gaps in the IUE data, especially for SWP spectra and the TR line emissions, make it difficult to discern correlations even if they are present.

5.4.2. Evidence for Enhanced Extra-Chromospheric Emission from VW Cep.

High dispersion IUE spectra of VW Cep were obtained during 1989 near the opposite quadratures of its orbit by Bradstreet and Guinan (1990). Observing VW Cep near the orbital quadratures (i.e., at 0.25P and 0.75P) permitted the resolution of the individual MgII h+k $\lambda 2800$ emission feature of each star. These spectra are shown in Figure 24. The synthetic emission line profiles of each star are indicated in the figure along with the position of the system's barycenter. These Doppler-broadened synthetic emission line profiles were computed by Bradstreet from the published orbital elements and physical parameters of the system (Hill 1989). As shown in the figure, the larger, more luminous star contributes nearly 3.5 times as much MgII emission as its smaller, less luminous companion.

However, the most interesting aspect of the MgII h+k emission line profile is the sharp, nearly stationary emission feature located near the rest wavelength of barycenter of the binary. This feature is also present on the four additional high dispersion LWP spectra of VW Cep available in the IUE archives. An examination of the available high dispersion IUE spectra of two other W UMa systems – W UMa and 44i Boo – also indicated the presence of this emission feature in some of the MgII h+k line profiles. Bradstreet and Guinan (1990) discuss several possible explanations for this stationary emission feature: (1) The emission could originate from the neck region of the binary where the two components interconnect and exchange gas and energy; (2) It could arise from circumbinary gas or (3) from an extended region of shocked-heated gas located between the two stars where their stellar winds interact. Additional high dispersion spectra with IUE, or other satellites will be needed to verify the presence of the extra emission and shed some light on its nature. At the present time, the neck region seems to be the leading candidate for this emission.

6. Magnetic Activity from the Cool Components of Cataclysmic Variable(CV) Binaries

The cool secondary components of CVs are rapidly rotating dM stars that are in contact with Roche lobes and are transferring matter onto a white dwarf primary. The orbital periods of CVs generally range from $\approx$ 2-10 hours and because of strong tidal coupling the cool components are forced to rotate synchronously with their orbital periods. In the shorter period CVs this causes the M-stars to have the fastest rotational periods of any main-sequence star. These stars generally have rotational periods of $\approx$ 10 times faster than most BY Dra stars. Because of their tidally induced rapid rotations and late spectral types (implying deep convective zones), the cool components should have extremely high levels of dynamo-related magnetic activity. However, most of the luminosity of a CV is produced from the release of gravitational energy of matter transferred from the cooler star to a viscous accretion disk. Because of this, it is difficult to study directly the properties of the cool star.

However, according to Warner (1988) and Bianchini (1990), the magnetic activity of the cool star can be inferred from the effect that the magnetic activity can have on the system's mass transfer rate. They propose that small changes in the mass transfer rate of the secondary star arising from the presence of an activity cycle could be observed from changes in the luminosity of the accretion disk and possibly also from changes in the system's orbital period. Bianchini (1990) has suggested that the $\approx$ 3-14 year periodic, or quasi-periodic, light variations apparent for some CVs result from the effect of solar-type activity cycles operating in the cool secondaries of these binaries. The changes in luminosity in these systems are thought to result from changes in the mass transfer rate from the cool star, which in turn, results from magnetically induced small changes in the cool star's radius with respect to its Roche lobe. Analysis of the long-term light changes of 19 CVs by Bianchini (1990) indicates evidence for cyclical variations in brightness with an average period of $\approx$ 6 yrs. As shown in Figure 25, the distribution of periods found for CVs by Bianchini is similar to period distributions of activity cycles found for single main-sequence stars and for RS CVn systems.

These luminosity variations are believed to arise from changes in the mass-transfer rate which can be ascribed to small fractional changes of the radii of the

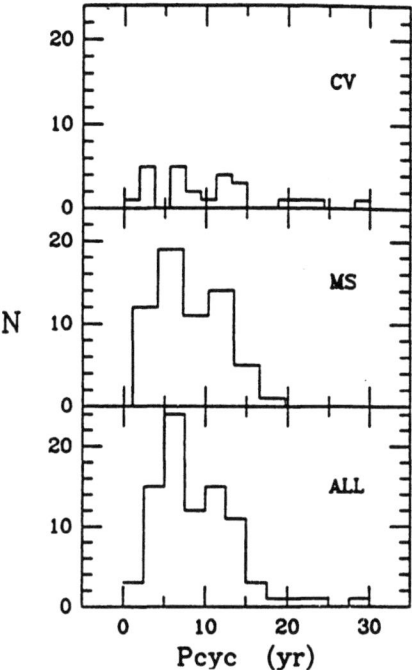

Fig. 25. Histograms for cycle-periods indicated for CVs and main-sequence stars (MS) from the study of Bianchini (1990). As discussed by Bianchini, the Kolmogorov-Smirnov test indicates that the two period distributions are identical, suggesting a peak near 6 yr.

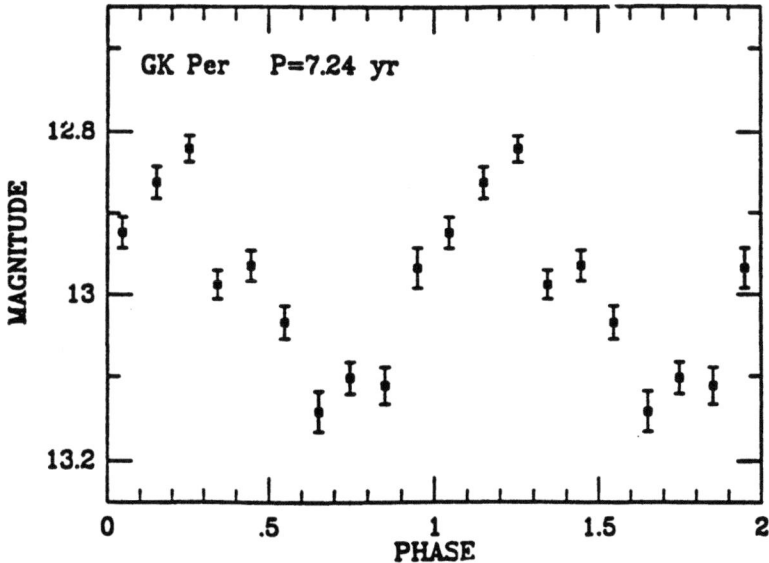

Fig. 26. The brightness variations during quiescence of the old nova GK Per, folded with a period of P=7.24 yr (Bianchini 1990). The data analyzed was obtained from 1920–1986.)

dM secondaries of $\Delta R/R \approx 0.6 - 3 \times 10^{-4}$ as expected for a typical solar cycle. For example, in Figure 26 from Bianchini (1990) the brightness measures of the old nova GK Per from 1920 to 1986 are plotted with a period P = 7.24 yr., obtained by a Fourier analysis of the data. However, as pointed out by Cannizzo and Kenyon (1986) the long term cyclic light variations of GK Per and other related systems could be caused by disk instabilities rather than magnetic activity cycles.

Further evidence in support of magnetic activity cycles in CVs is provided by Warner (1988). He has investigated the changes in orbital periods of some eclipsing CV systems and finds that the timescales and magnitudes of the observed period changes are consistent with magnetic activity cycle hypotheses. These orbital period changes are attributed to small changes in the cool star's radius and quadrupole moment caused by variations of the magnetic field density in its convective layers over an activity cycle (see Applegate and Patterson 1987 for a fuller discussion of this mechanism).

Although far from conclusive, the cyclic behavior of the brightnesses and the period variations of CVs appear to be consistent with the presence of activity cycles in these systems. Perhaps long-term monitoring of these systems in the IR, where the cool component contributes more light, may lead to the unambiguous detection of spots and spot cycles on the cool star. This is a promising area of research which needs to be further explored.

Before leaving the discussion of magnetic activity in CVs, we note the recent evidence of Horne and Saar (1991) that Balmer emission lines observed from the accretion disks of some CVs show similar behavior between emission and period as chromospherically active stars. They suggest that these emission lines arise from a magnetic dynamo operating in the accretion disk. This is an interesting suggestion and deserves further study.

7. Activity in Long Period Binary Systems with Cool Components: Symbiotic and ζ Aurigae Binaries

Symbiotic binaries typically are detached binaries which consist of a cooler M giant star and an accreting, hotter white dwarf, subdwarf, or sometimes a main-sequence companion. The orbital periods generally range from ≈ 1 to several years. A full description of the properties of symbiotic binaries is given by Kenyon (1986). At the present time, there is little compelling evidence of magnetic activity in the cool components of symbiotic systems. The lack of activity is not very surprising because of the relatively long orbital and rotational periods of the cool stars in most of these binaries. Moreover, the usual indicators of activity such as starspots, chromospheric line emission, etc., would be difficult to identify, even if present, with magnetic activity of the cool component because of the complex light and spectral variations arising from pulsations of the cool star and accretion processes assumed to be taking place in these stars. X-ray emission is sometimes present but it appears to originate from the accreting (usually white dwarf) companion.

The evidence for activity in symbiotic binaries is weak and is given by Bianchini (1990) for the symbiotic nova RR Tel. This system is composed of a Mira-type variable with a 387 day pulsation period and a $\approx 0.7 M_\odot$ white dwarf companion. Bianchini finds some evidence for a ≈ 11.4 yr periodicity in the system's light curve that is attributed to a possible activity cycle in the cool star. He suggests that these cyclic light variations are produced by cyclical mass-transfer rate variation caused by a putative magnetic activity operating in the M III component. The evidence for magnetic activity in this system is, however, far from convincing.

This is not to say that the cool components of all symbiotic stars are chromospherically inactive. For example, Altamore *et al.* (1981) argue that the high ionization emission lines observed in the spectrum of the symbiotic star Z And (M3 III + hot source (hs); P = 757d) may be formed in a solar-type transition region around the cool star and not in the nebular region near the hot subdwarf companion. We also point out that dynamo-generated magnetic fields might be expected for the shorter period symbiotic stars in which the M giant fills (or nearly fills) its Roche lobe. In this case, tidal effects will force the cool star to rotate synchronously with its orbital period. It would not be surprising that a red giant with a rotation period of a few hundred days might be magnetically active because of its extensive convective envelope. From Kenyon (1986), three symbiotic stars believed to have lobe filling M giants are T CrB (M4 III + hs; P = 227d), AX Per (M5 III + hs; P = 681d), and CI Cyg (M4 III + hs; P = 855d). Because of the complexity of these systems, it will be difficult to assess the presence of magnetic activity.

ζ Aurigae-type binaries generally have orbital periods of years. They consist of a supergiant G or K star and an accreting (wind accreting) main-sequence $\approx$ B companion. The supergiant components are slow rotators and most likely because of that show no evidence of significant magnetic activity. They do, however, appear to have warm, extended chromospheres (see Guinan 1990 and refs. therein).

8. Eclipse Mapping: Vertical Cross-sections of the Chromospheres and Transition Regions of Active Cool Stars

Several chromospherically active stars are members of eclipsing binaries containing hot white dwarf or subdwarf companions. These eclipsing binaries are especially important because they provide a rare opportunity to study the structure of the chromospheres and transition-regions of active cool stars. In these systems the hotter, smaller white dwarf or subdwarf star can be used as a probe of the cooler star's atmosphere near primary eclipse when it passes behind its atmosphere. The best studied examples of this special group of eclipsing binaries are V471 Tau (wd + K2 V; P = 0.525d) and FF Aqr (sdOB + G8 III; P = 9.2d). These stars are the short period, chromospherically active diminutive cousins of the long period ζ Aurigae binaries.

8.1. ATMOSPHERIC STRUCTURE OF AN ACTIVE K DWARF: THE CASE OF V471 TAURI

Few stars are such rich sources of important information in stellar astrophysics as the 9th magnitude Hyades eclipsing binary V471 Tauri. It consists of a hot DA2 white dwarf and a detached K2 main-sequence star which move in a circular orbit with a period of $\simeq$ 12.5 hours. The system is a member of the Hyades cluster and has a presumed age of 0.6 Gyr. Since its discovery as an eclipsing binary in 1969 by Nelson and Young (1970), V471 Tauri has been recognized as an important astrophysical laboratory for many reasons and has been observed extensively with IUE and in the X-ray with EXOSAT and more recently ROSAT.

Similar to the case of the ζ Aurigae stars, V471 Tauri offers a unique opportunity to study directly the vertical structure of the atmosphere of an active K-dwarf by using the white dwarf as an almost point-like probe of its outer atmosphere. Observations of the system made with the IUE satellite from 1979 to 1986 (see Figure 27) reveal the spectrum of an hot extended region of the K star's atmosphere superimposed on the nearly featureless continuum of the white dwarf when the hot star is near the limb of the cool component (Guinan et al. 1986a). Absorption features of CIII (λ1175), SiIII + OI (λ1300), CII (λ1335), SiIV (λ1400), CIV (λ1550) have been detected with equivalent widths of 0.5-3.0Å. Although these features appear only when the white dwarf is near the limb of the cool star and are not seen at other orbital aspects, sometimes the absorption lines are weak or absent even though the white dwarf is close to the cool star's limb. Ground-based photometry of the star reveals the presence of a variable photometric wave superimposed on the light variations expected from the eclipse and binary proximity effects. Analysis of the light curves reveals the presence of large starspot regions on the cool star which vary in extent and distribution with time (cf. Ibanoğlu 1990). From contemporary light curves obtained with the IUE observations, it appears that the strength of the UV absorption features are strongly correlated with the longitudinal distribution of the spots in the sense that when the spots were located near the limb of the K star at primary eclipse, the lines were detected, and vice versa. This indicates that the atmosphere of the K star is inhomogeneous and that the FUV spectral features arise when the line of sight to the hot star intercepts gaseous material lying above (or flowing from) the active regions of the cool star. Analysis of the line strengths indicates a temperature of $T \simeq 10^5 K$, an electron density of $N_e \simeq 10^{10} - 10^{11}$ cm^{-3} and a limiting vertical height above the cool star's surface for the gas of 300,000 - 400,000 km. The inferred properties of the plasma, along with its apparent location above active sites, indicate that these atmospheric structures may be related to the *cool* loops or active region plumes observed on the Sun but are much more extensive in size. Figure 28 shows a schematic diagram of the binary system in which the stars and the assumed overlying loop structure of the K-star are depicted to approximate scale.

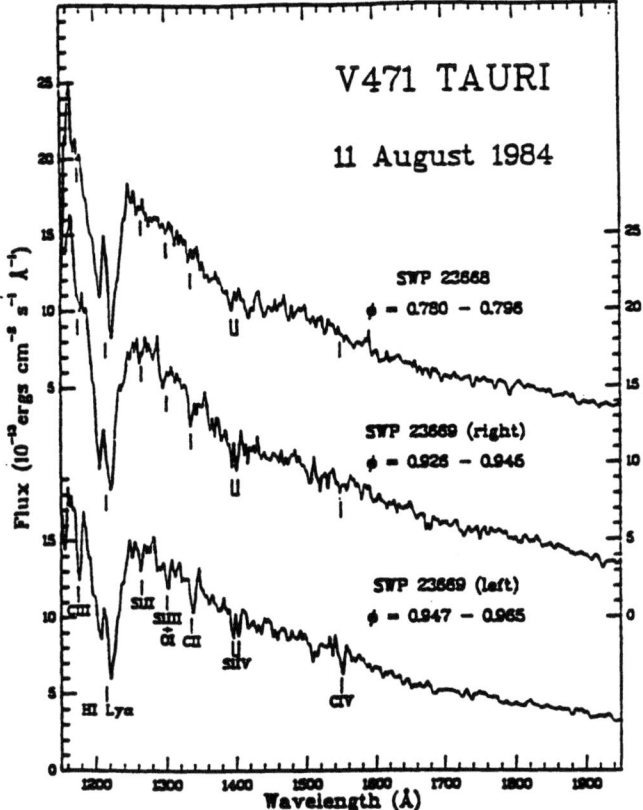

Fig. 27. Representative UV spectra of V471 Tau obtained with the SWP Camera of the *IUE* satellite. These spectra are mainly those of the hot white dwarf star because it overwhelms the light of the K-dwarf companion at UV wavelengths. The top spectrum, obtained near the quadrature of the orbit ($\phi = 0.79P$) shows a broad Lyα absorption feature from the white dwarf's photosphere. The lower two spectra were obtained prior to the star of primary eclipse at $\phi = 0.935P$ and $\phi = 0.955P$, respectively. (First contact occurs at 0.967 phase.) Absorption features of C II, C III, C IV, O I, Si III, Si II, and Si IV are clearly seen in the spectra prior to the eclipse of the white dwarf. These absorption features appear to originate from large-scale structures in the atmosphere of the K dwarf as the light from the white dwarf shines through them.

8.2. PHYSICS OF THE CHOROMOSPHERE OF AN ACTIVE K-DWARF

The nearness of the synchronously rotating K2 V component to a hot white dwarf with well-determined physical properties provides an *unique* opportunity for learning about the structure and physical properties of the transition region (TR) and chromosphere of a main-sequence K star. V471 Tauri provides an opportunity in this case to probe the cool star's chromosphere by studying the effect that the incoming radiation from the white dwarf has on the cool star's atmosphere. The radiation field and incoming flux of the white dwarf are well determined by EX-

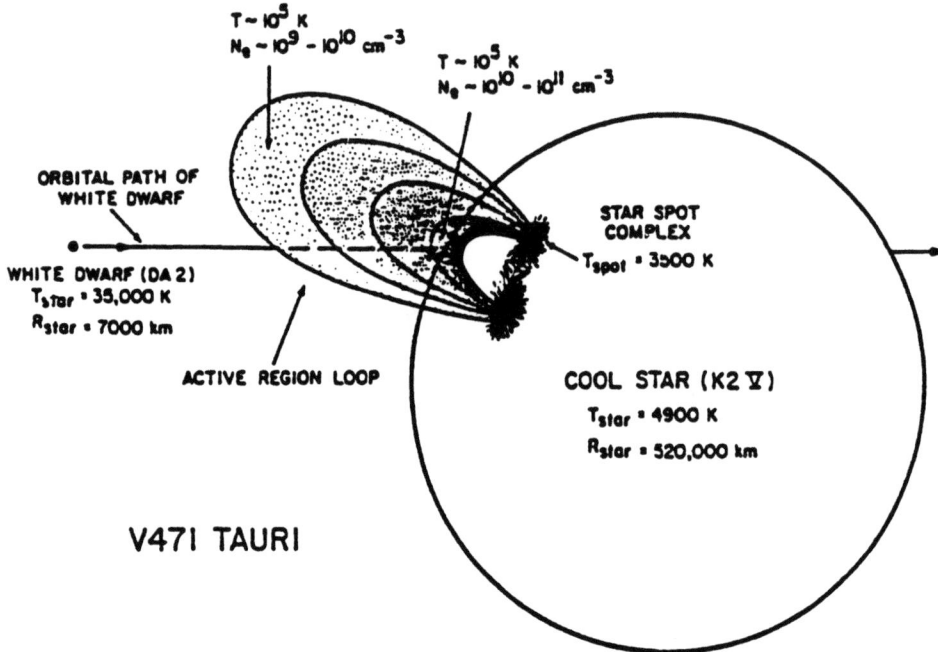

Fig. 28. A schematic diagram of V471 Tauri, showing the surface and atmospheric features of the K2 V star, inferred from optical photometry and from ultraviolet spectroscopy. The physical properties of the stars and the atmospheric "loop" feature are given (Guinan *et al.* 1986).

OSAT, IUE, and ground-based photometry. IUE observations of the V471 Tauri over its entire orbit reveal that the net MgII h+k emission flux is strongly phase dependent. The MgII emission is strongest near 0.50 P when the irradiated hemisphere of the cool star is facing the Earth and smallest when the non-irradiated hemisphere is in view, near primary eclipse (i.e., 0.9 P-0.1 P). The enhancement of the MgII h+k emission between the irradiated and non-irradiated hemispheres is $\approx$ 2.5 times. Ground based studies by Bois *et al.* (1988) and Young *et al.* (1988) also reveal that the Hα emission is phased locked in the same way as the MgII emission. However, the study of the continuum at ultraviolet to visible wavelengths indicates no significant heating of the K-star's photosphere – i.e., there is no appreciable *reflection effect* in the star's continuum. From the IUE data, the MgII h+k surface flux due to the irradiation by the white dwarf is $\approx 5.6 \times 10^6$ ergs s^{-1}cm^{-2}. This value was obtained by subtracting out the MgII emission of the non-irradiated side of the star, assuming that to be the background chromospheric contribution from the K-star.

Preliminary calculations show that the white dwarf FUV radiation is sufficient to produce the observed enhancements in the MgII h+k, and Hα line emissions. However, detailed modeling of the data has not been completed. With the existing IUE data, the modeling should yield good estimates of the physical conditions in an active star's atmosphere. Figure 29 shows a schematic diagram of the effect of

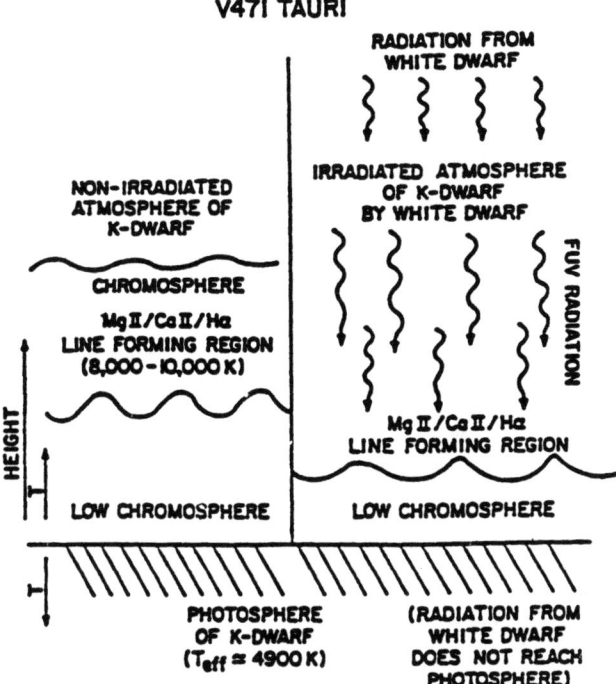

Fig. 29. The comparison of the properties of the non-irradiated and irradiated atmosphere of the K-dwarf component of V471 Tauri. The incoming radiation from the hot white dwarf alters the structure of the coll star's chromosphere. In the irradiated case, stronger chromospheric emission is produced in a lower, but denser region of the star's chromosphere. Almost none of the white dwarf's radiation penetrates to the K star's photosphere.

the irradiation of the cool star's atmosphere by the hot white dwarf. Young *et al.* (1988) have followed a similar approach in the interpretation of the variations of Hα emission with orbital phase.

8.3. ATMOSPHERIC STRUCTURE OF AN ACTIVE G GIANT: THE CASE OF FF AQR

FF Aqr (=BD-03°5357) is an upscale analog of V471 Tau. FF Aqr is a 9.21d totally eclipsing binary which consists of a chromospherically-active, heavily spotted G8 III-IV star and less massive, hot helium-hydrogen sdOB subdwarf. The binary nature of FF Aqr was discovered by Dworetsky *et al.* (1977) when *UBV* photometry revealed a sharp occultation eclipse lasting about 13 hrs with an ingress/egress time of only $\approx$ 24 min. The light curve of FF Aqr is unusual and is dominated by a large photometric wave with an amplitude of up to $\approx$ 0.35 mag at yellow wavelengths. The phasing of the maximum and minimum light as well as the large amplitude of the photometric wave cannot be explained by the usual binary star interaction effects. However, Dorren *et al.* (1983) were able to successfully model the light curve by assuming that chiefly one hemisphere of the G8 III-IV star is extensively covered with cool starspots about $\approx$ 800°K cooler than the star's photosphere. This

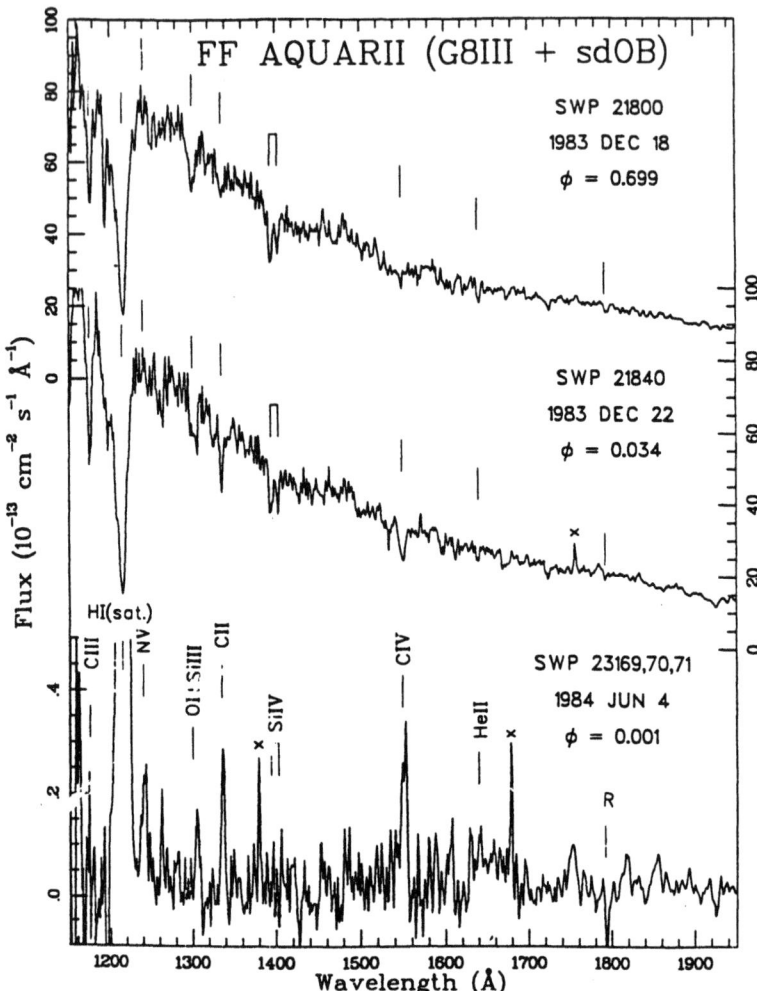

Fig. 30. Ultraviolet spectra of FF Aqr taken at different orbital phases. During the total occultation eclipse of the sdOB component (bottom spectrum) the emission lines of the G8 III star are visible. The subdwarf's spectrum is shown in the figure's top spectrum. Just past 4th contact (middle spectrum), the absorption strengths of C IV and C II are enhanced relative to the spectrum obtained near the orbital quadrature (Baliunas et al. 1986).

analysis indicates that at least ≈ 43% of the cool star's surface is spotted! Thus, the cool star of FF Aqr has one of the largest spot contrasts of any chromospherically active star. Subsequent photometry indicates substantial seasonal changes in the amplitude and phase of the photometric wave (Etzel 1991; Guinan and McCook 1991).

Observations of FF Aqr have been carried out in the UV with IUE (see Dorren et al. 1982; Baliunas et al. 1986). Ultraviolet spectra of the system during primary eclipse (when only the G8 III-IV star is visible) reveal the presence of strong chromospheric and TR line emission from the cool star (see Figure 30). Moreover, IUE observations made prior to and after primary eclipse indicate the enhancements in the absorption line strengths of the subdwarf – most noteworthy are CII λ1335 and CIV λ1550 lines. Because of the crowded absorption line spectrum of the hot sdOB star, the analysis is not as straightforward as in the case of the nearly line-free spectrum of a DA white dwarf. However, subtraction of spectra of FF Aqr obtained near the quadratures of the orbit from those obtained when the subdwarf was near the limb of the G-star, indicates residuals in the absorption line strengths. Increases in absorption line strengths of chiefly CII, OI, SiIV, and CIV are apparent when the sdOB star is near the cool star's limb prior to and after primary eclipse. This is similar to the behavior of V471 Tau and indicates that the absorption line enhancements originate in the extended chromosphere and TR of the active G8 giant.

9. Consequences of Magnetic Activity in Binary Systems: Angular Momentum Loss and Magnetic Braking

The spindown of single solar-type dwarfs with time is now well documented (see Skumanich 1972; Soderblom 1983; Simon *et al.* 1985). The angular momentum loss (AML) is assumed to be caused by magnetic torques in a stellar wind (Schatzman 1962; Mestel 1968; Mestel 1984). This mechanism is referred to as magnetic braking. The torques produced by the magnetic field in the wind depend on the strength of the magnetic field, the Alfvén radius, and the extent of the stellar wind. Also, according to Mestel (1984), AML depends on the stiffness of the magnetic field at large distances from the star. Except for the Sun, this information is not known and the theory is very complicated and still uncertain (Kawaler 1988).

As first suggested by Huang (1966), AML via magnetic braking could be very important in close binaries with cool stars, causing the orbital separation and period to decrease with time. In close binary systems where tidal forces are strong enough to cause synchronization of the components, the total angular momentum of the orbit is tidally coupled to the rotational angular momentum of the star, i.e., spin-orbit coupling. Thus, the angular momentum of the binary bleeds away with time from magnetic braking, produced by torques in the stellar winds. Quite opposite to the case of a single solar-type star, the binary components *spin up* with time as the orbital period decreases. Presumably the magnetic activity of the stars also increase with the more rapid rotation. As discussed by Van't Veer (1979), Vilhu (1982), Guinan and Bradstreet (1988) and others, this process can cause a detached binary composed of solar-type stars (i.e., stars having convective envelopes) to become a short period W UMa-type contact system. In addition, once the binary becomes a contact system with a short period ($P < 0.5d$), general relativistic gravitational radiation also can become a significant source of AML. AML from both mechanisms can ultimately lead to the coalescence of the binary into a single, rapidly rotating star such has been suggested for FK Com stars and blue stragglers.

Guinan and Bradstreet (1988) have studied the space motions of a sample of 34 W UMa systems. They found that the average space motion of $<S> = 61$ km/s (relative to the Local Standard of Rest) and the relatively large U, V, W velocity dispersions of these systems are consistent with old disk population stars. Although there are several low velocity, young disk stars in the sample, the average kinematic properties of the W UMa systems are similar to those of the evolved, old disk stars such as white dwarfs, planetary nebula, and shorter period RR Lyr variables. They estimate a mean dynamical age for the group of $t \approx 5 - 10$ Gyr (1 Gyr = 10^9 yrs). The old age of most W UMa stars seems also to be supported by the presence of 7 W UMa binaries in the old ($\tau \simeq 5 - 10$Gyr) open cluster NGC 188 (Kaluzny and Shara 1987).

Guinan and Bradstreet (1988) have derived a relation for the angular momentum loss rate for binary systems with tidally-coupled solar-type stars from magnetic braking. They adopted the spin-down rate law from single solar-type stars in the form:

$$V_{rot} \approx 1.4 \times 10^5 t^{-1/2} \tag{1}$$

where V_{rot} is expressed in kms^{-1} and time (t) in years.

The expression for the change in orbital period with time due to spinorbit coupled angular momentum loss is:

$$dP/dt \approx 1.1 \times 10^{-8} q^{-1}(1+q)^2 (M_1 + M_2)^{-5/3} k^2 (M_1 R_1^4 + M_2 R_2^4) P^{-7/3} \tag{2}$$

In this relation M_1, M_2 and R_1, R_2 are the masses and radii of the component stars expressed in solar units, q is the mass ratio (q = M_2/M_1), k^2 is the gyration constant (typically ranging from 0.07 to 0.15 for solar-type stars), and P is the orbital period expressed in days. The change in period with time (dP/dt) is in units of days/yr. Since the star's rotational and orbital periods are assumed to be synchronized, P = P_{orb} = P_{rot}. In addition the time needed for the period to change from an initial long period to a shorter value through AML is obtained by integrating eq. (2) to obtain:

$$t(yrs) \approx 0.30 b (P_o - P_f)^{10/3} \tag{3}$$

where b includes all the constant terms on the right side of eq. (2). In eq. (3), P_o and P_f are the initial and final orbital periods of the binary system expressed in days, and t is the time needed to go from P_o to P_f. To estimate the time needed for a binary system to reach its contact state, the final period is set equal to the critical period at which the two components reach contact. The critical period depends on the mass of the binary.

Using the above expressions and adopting a value of $k^2 = 0.10$, the rate of decrease of the orbital period from AML can be computed as a function of time and initial orbital period. Figure 31 shows the decrease in orbital period of originally detached binaries with solar-type ($M_{1,2} \approx 1.0 M_\odot$) stars having initial orbital periods of 1 to 5 days. For the example given, $M_1 = 1.11 M_\odot$ and $M_2 = 0.74 M_\odot$ and the critical (= contact) period = 0.315d. Systems having initial periods of $P_o > 5$d may not experience spin-orbit coupling as main sequence stars because the tidal effects are small due to their relatively large orbital separations. For tidal coupling

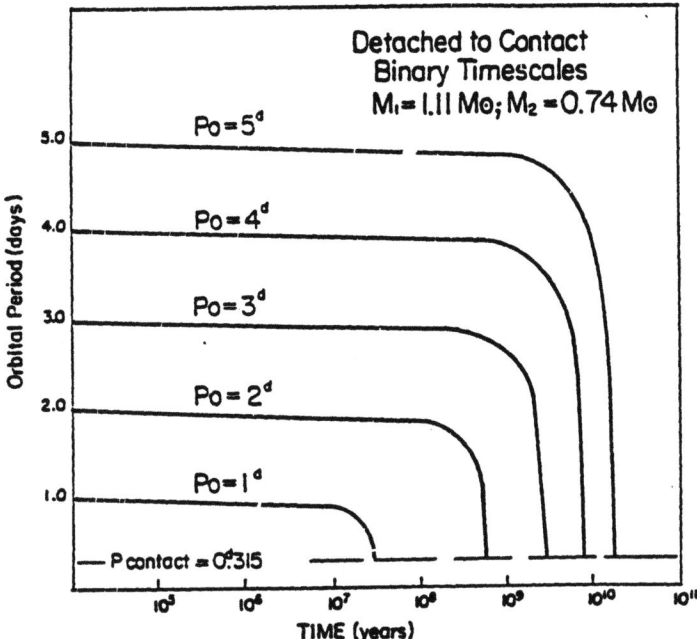

Fig. 31. The decrease in orbital periods of originally detached binaries with initial orbital periods of 1–5 days. The decrease in periods is due to angular momentum loss from magnetic braking, assuming spin-orbit coupling of both components. For the example given, the period of the system upon reaching contact is $P_c = 0.315^d$ (from Guinan and Bradstreet 1988).

to become effective, the fractional radius of the star (r = R/a) should be r > 0.1. For example, for a binary system with two solar-like stars ($M_{1,2} = 1 M_\odot$; $R_{1,2} = 1 R_\odot$), the fractional radius would be $r_{1,2} \approx 0.1$ at P ≈ 3d.

As shown in Figure 31, the time to reach contact depends strongly on the initial period of the system. Stars with initial orbital period of P_o = 5d (assuming spin-orbit coupling) would take t ≈ 17 Gyr to reach contact, whereas for a short period system with P_o = 1d, contact would occur in t ≈ 30 Myr (1Myr = 10^6yr). Once contact occurs, the system becomes a W UMa-type binary and its evolution will also strongly depend on mass exchange between component stars and mass loss from the system. Although AML from gravitational radiation becomes important for short period contact systems (see Webbink 1976;1985), it is likely that magnetic braking still plays the dominant role, based on the high levels of magnetic-related activity manifested by W UMa binaries. It is difficult to determine the amount of angular momentum lost by magnetic braking during the contact stage. Many assumptions have to be made to estimate it (see e.g., Mochnacki 1981; Van't Veer 1991). At the present time, insufficient information is available on the strengths

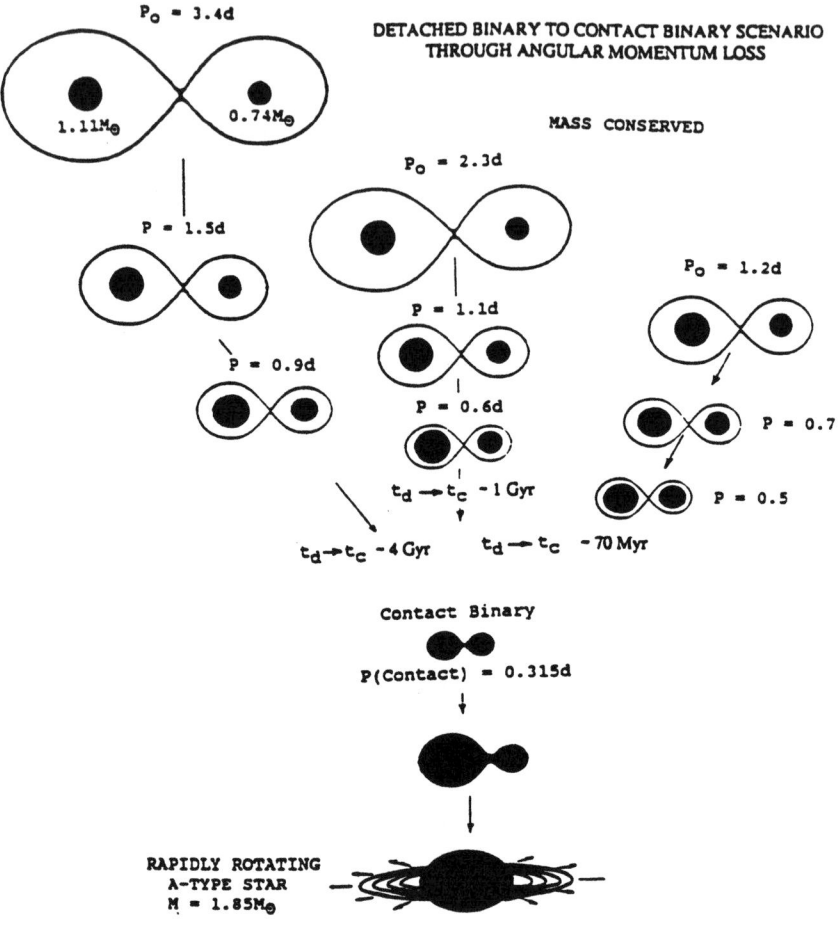

Fig. 32. A possible scenario for the origin and evolution of W UMa-type binaries (Guinan and Bradstreet 1990). Initially detached binaries with solar-type components lose angular momentum via magnetic braking and become contact binaries. As shown (drawn approximately to scale) detached binaries of diverse initial periods will funnel into contact systems. The evolution of the contact binary into a rapidly rotating star with a possible outflow of gas from its equator (Webbink 1976) is also shown. The end result is a rapidly rotating A-type star or a rapidly rotating F- or G-giant star.

and configurations of the magnetic fields and winds form these strongly interacting binaries. From radio studies, however, it does appear that magnetic field strengths may be weaker than for detached binaries (see Section 5). Following the arguments of Vilhu (1982), Mochnacki (1981) and Van't Veer (1991), it appears the contact stage is relatively brief, 0.5 Gyr $< t_c < 5$ Gyr.

A possible scenario for the origin and evolution of short period binaries (P < 4d) with solar-type components has been discussed by Guinan and Bradstreet (1988). Figure 32, from this paper, illustrates the evolution of initially detached binaries into contact systems and ultimately into single stars via magnetic braking. With

AML from magnetic braking following a $v_{rot} \approx t^{-1/2}$ law, the times necessary for these systems to reach contact (in the case presented of $P_f = 0.315d$) is 4 Gyr, 1 Gyr, and 70 Myr, for detached binaries having initial periods of 3.4d, 2.3d, and 1.2d, respectively. In calculating these models, it is assumed that the radii and masses of the stars remain constant. From this study it appears possible that most W UMa stars may be formed from low mass, detached binary progenitors.

Initially detached, short period binaries of low mass ($M_{1,2} < 1.3 M_\odot$) appear to funnel into the contact stage via AML at different rates depending on their initial periods. This funneling effect produces contact systems from longer period progenitor binaries. The apparent average old age of W UMa stars of $t \approx 5\text{-}10$ Gyrs is explained by the relatively long lifetimes of the stars as detached systems prior to reaching contact. However, to account for the relatively high numbers of W UMa stars relative to their assumed short period progenitors, it is necessary for the contact stage to last for at least ≈ 1 Gyr. Of course, the progenitors of W UMa binaries would be the shorter period BY Dra and RS CVn systems. However, some of the more massive RS CVn binaries could become Algol-type systems if the more massive component evolves to fill its Roch lobe before reaching the contact stage.

The evolution of a W UMa system depends on the age and masses of the components prior to reaching contact. If the primary should exhaust its hydrogen fuel just prior to reaching contact, it will fill its Roche lobe and form a short period semi-detached system. In this case, mass exchange will occur on a thermal timescale and the resulting binary might resemble the short period β Lyrae type systems, known as V1010 Oph stars (see Shaw 1990).

For the shortest period system in the figure, the stars reach the contact configuration in such a short time that the components are only slightly evolved from ZAMS. ER Vul (G0 V + G1 V; P = 0.698d) could represent a pre-contact binary which may have evolved through AML from a binary with a slightly longer period. Since the components of ER Vul are little evolved, it must be relatively young and its initial orbital period should be less than 2 days.

The evolution of W UMa binaries is complicated and is still not well understood, but it seems plausible that they ultimately coalesce into single stars (see Webbink 1976; Bopp and Stencel 1981; Rucinski 1992). As shown in Figure 32, the end result of W UMa evolution could be a rapidly rotating A-type star which could evolve into a rapidly rotating giant, becoming a FK Comae star. As pointed out by Van't Veer (1979) and Guinan and Bradstreet (1988), a significant number of single stars could have originated as short period binaries that have coalesced into single objects after passing through a contact binary stage. The presence of blue straggler stars in old clusters, high velocity (= old?) A- and F-type stars, and FK Comae stars, could be explained by this scenario (e.g., see Tutukov and Yungelson 1987; Eggen and Iben 1990; Van't Veer 1991). These coalesced single stars would in time evolve through AML into apparently *normal* slowly rotating main sequence or giant stars. These stars could be distinguished from Pop I objects possibly by their higher space motions as in the case of W UMa stars.

10. Conclusions

This review paper summarizes our progress in the study of magnetic activity in close binaries. Some of the more important conclusions are given below.

- Manifestations of enhanced solar-like magnetic activity are apparent in nearly every close binary containing at least one star with a convective envelope (spectral type $\approx$ F5 and later). It is even possible for the M-giant components of the shorter period symbiotic stars to have magnetic activity but it is difficult to assess in these complicated systems.
- The subgiant and giant G- and K- components of close binaries seem to be the most active stars for a given rotation period or rotational velocity. Maybe the presence of a deeper convective zone or lower surface gravity (or both) play a role in enhancing magnetic activity. Zwaan (1991) suggests that tidal effects may play an important role.
- In several RS CVn systems where there is sufficient data, differential (latitudinal) rotation appears significantly smaller, $\Delta\Omega/\Delta\phi \approx 0.1\text{-}0.01$ than that observed for the Sun. This result is not surprising because of the strong tidal interaction effects present in these systems. The differential rotation rates measured for single solar-type stars and the non-synchronously rotating G8 IV-III primary of λ And are similar to that of the Sun.
- Differential rotation appears to be very much diminished in chromospherically active binaries with tidally coupled components. However, differential rotation is an important element in most stellar dynamo theories. Nonetheless, chromospherically active binaries with synchronously rotating components display all of the characteristics of magnetic activity found in the Sun and in more active (= younger) solar-type stars. This could indicate a drastically different dynamo mechanism is operating in the these close binaries.
- Activity cycles, possibly resembling the Sun's ≈ 11 yr. cycle, appear present in a number of RS CVn, BY Dra, and W UMa systems in which adequate long-term observations are available. Also, as discussed by Hall (1990), cyclic increases and decreases in the orbital periods of many chromospherically active binaries including Algols with cool secondaries, have been attributed to the effects of activity cycles. Moreover, the apparent cyclic behavior of outbursts in some CVs and a few symbiotic binaries could be produced by magnetic activity in the cooler components of these systems.
- Analyses of light curves with geometric spot models and Doppler imaging techniques indicate the presence of large, dark starspots on the surfaces of chromospherically active binaries. These methods complement each other and yield results that indicate the presence of at least two large spots (or groups of spots). Moreover, there is a tendency for the spots to be located at high stellar latitudes, sometimes even at the rotational poles of the stars, and

to have long lifetimes of the order of years. In the case of the Sun, sunspots usually lie between 10-40 degrees above and below the equator and persist for a few months. However, the recent suggestion by Dorren and Guinan (1990) that white light faculae may be important in the brightness variations of chromospherically active stars could complicate the interpretation of the light curves of RS CVn and BY Dra stars with standard spot models.

- There is some evidence that there is an *enhancement* of magnetic-related activity for cool stars in close binaries compared to single stars with similar spectral types and rotation rates. This effect has been referred to as *hyperactivity* by Zwaan (1991) or *duplicity-enhanced activity* by Kuijpers (1989). This *duplicity-enhanced activity* is most evident at radio wavelengths where active, single main sequence stars with high levels of X-ray emission have radio luminosities at least 10 times less (may even be 100 times less) than RS CVn, BY Dra, and Algol systems with comparable X-ray emission. It has been suggested that the observed enhancements in the *radio* activity for these binary stars is produced by interactions of the magnetic fields between the binary components. However, this mechanism does not seem to explain the high activity of Algol systems which have only one active cool star. On the other hand, W UMa systems with two cool components are weak radio sources relative to the RS CVn, BY Dra, and Algol binaries. This is an interesting, but unsolved problem.
- Radio observations of chromospherically active binaries are playing increasingly important roles in understanding stellar magnetic activity in the coronae of these stars. As shown in the case of Algol, VLBI observations are now capable of submilli-arcsec astrometry of radio sources. High precision radio maps of nearby active binaries should prove valuable. Multifrequency, polarimetric radio observations are permitting the determinations of the magnetic field strengths and physical conditions in the coronae of active stars.
- Eclipse mapping techniques at all wavelengths are proving to be extremely valuable for isolating magnetic structures that otherwise could not be resolved. Examples include AR Lac, Algol, V471 Tau, and FF Aqr.
- Recent direct measures of the surface magnetic fields of the active RS CVn binary, V711 Tau, have been carried out from spectropolarimetry using the technique of Zeeman-Doppler Imagining. This technique can be applied to determine magnetic field densities and areas on rapidly rotating stars. The method of measuring magnetic fields using Zeeman-broadening (see Saar 1991) works well for stars with sharp-line profiles, but is not useful for most chromospherically active binaries where the spectral features are rotationally broadened. The Zeeman-Doppler imaging method looks very promising and can yield two dimensional maps of the magnetic fields on the surfaces of the stars.
- Binary systems serve as important laboratories for studying stellar dynamos and magnetism. However, the consequences of magnetic fields on the dynam-

ics and evolution of close binaries are proving to be just as important. AML through magnetic torques in the winds of the component stars can cause the orbit of the binary to decrease and the stars to rotate faster with time. AML can cause initially detached binaries with cool stars to become short-period contact systems and ultimately lead to the coalescence of the binary into a single star. A significant fraction of single stars could have originated as close binaries that merged; thus magnetic fields may play an important role in the formation of single stars from close binaries.

All of this progress points toward further important advances in both our understanding of magnetic dynamos in stars (and in the Sun) and the impact of magnetic dynamos on binary star dynamics and evolution. Although much has been accomplished so far in defining magnetic activity in these stars and its consequences, much more work is needed before we understand the processes involved. This summary points up the present inadequacies in stellar dynamo theory and the need for further theoretical and observational work. Some of this has already begun.

Even now, important binary systems too faint for study with existing telescopes and instrumentation are becoming accessible for study at nearly all wavelengths thanks to the newest technology: γ-ray (GRO), X-ray (ROSAT, AXAF, and Astro-D), extreme ultraviolet (EUVE or FUSE), ultraviolet (IUE, HST), optical (Automatic Photometric Telescopes, global networks of APTs = GNATs, CCD Spectroscopy and photometry, polarimetry, spectropolarimetry, and interferometry, infrared (ground-based, airborne (SOFIA), and satellite (SIRTF)), and radio (VLA, VLBI and RADIOASTRON). Moreover, multi-frequency and long-term studies should prove very valuable for understanding the important time-dependent manifestations of magnetic activity. In short, the future looks active and exciting for research in magnetic activity in close binaries.

Acknowledgements

We thank J.-F. Lestrade for sending us data prior to publication. We also thank Carol Ambruster, David Bradstreet, Steve Drake, Scott Kenyon, and Harold Nations for comments and valuable discussions about the manuscript. We also wish to thank Joan Feuer and Elizabeth Jewell for their help in the preparation and editing of the manuscript.

EFG wishes to acknowledge the hospitality of the Harvard-Smithsonian Center for Astrophysics, where some of this research was carried out with support from the *Visiting Scientist Program* of the Smithsonian Institution. This work was partially supported by grants from *NASA*: NAG 5-382 and NAG 5-1703.

CONTACT BINARIES OF THE W UMa TYPE

S. M. RUCINSKI

1. Introduction

W UMa-type systems are eclipsing binaries with orbital periods between about 5 and 18 hours showing continuous variations of light. They were defined as a class in general reviews of Binnendijk (1965, 1970, 1977) and Rucinski (1985a). The W UMa-type stars occupy a special position among binary stars for a number of reasons. We will enumerate those reasons below, stressing, if necessary, that some seem to be based on misconceptions or old, invalid data.

First of all, they have the least amount of angular momentum that binary stars can have. It is true that this amount is still many orders of magnitude more than single stars can store in rotational motion, even when rotating at break-up, but – among binaries made of normal, unevolved stars – W UMa systems are at the edge of the angular-momentum distribution. In simple terms: they are as close as binary stars can be. The only type of binaries with even closer components are various types of cataclysmic variables which must have lost a lot of angular momentum during their long evolution.

The other distinction is the supposed high frequency of W UMa systems. Shapley (1948) pointed out that they are very frequent among eclipsing systems and among variable stars in general. The first part of the statement is correct and simply says that easiest to discover eclipses in W UMa systems relative to majority of eclipsing binaries. But the second part is more difficult to evaluate because comparison is made between quite dissimilar objects. In any case, we suggest that the finding of Shapley should be interpreted exactly as it was formulated at the time of his studies and certainly should *not* be taken to mean that W UMa systems are very frequent in space. In fact, the contrary is probably true (with some very important exceptions, see Secs. 3 and 4). The best estimate of their frequency at present is one W UMa system per about one or two thousands of ordinary stars (as always to a factor of 2 both ways). This frequency is quite low in comparison with the canonical 50 to 100 percent of all stars being binary-star members. Thus, a W UMa is something rather special and unusual in the world of binary stars.

And, finally, W UMa-type variables are supposed to be "easy" objects, especially from the point of view of photometric observers: one light curve can be covered in one night! However, with the rather moderate information content of broad-band photometry, this aspect is not as attractive as it may seem. We have now many light curves for many systems but progress of research without spectroscopic support is slow. We should remember that for some time the "rectified" model of spherical stars by Russell-Merrill served quite well and light curves could be "modelled" quite successfully. It is a separate matter that the model was not physical – it could explain photometric observations! Scores of light curves were solved using the R-M method until Lucy (1968a, 1968b) presented his more physical, contact model[1]. But, is it the final model free of any systematic deficiency? We cannot decide on that matter using only photometry (take as an example the so-called W-type syndrome, Sec. 2.3). Spectroscopy is the only line of attack. But, to do this we must be able to resolve very heavily blended lines and – preferably – be able to analyse the eclipse effects within the spectral lines which requires observations of high spectral resolution and high signal-to-noise. For high-resolution spectroscopy, W UMa systems are quite faint and must compete for time on the largest telescopes against projects, such as cosmology, where they lose in general. From the theoretical point of view the W UMa systems are also not easy: in spite of intense research, they are the only type of binary stars that we don't know how they originate and into what they evolve!

The distinctions listed above clearly confine W UMa systems to a small "niche" which may seem to have few connections to other areas of stellar astrophysics. It is the intention of this review to point out that this "niche" is interesting, not only by its own intellectually stimulating, extreme properties but also because this niche has, in fact, much relevance to other areas of contemporary stellar research.

There have been many reviews about W UMa systems. The older work concentrating on light curve solutions was summarized by Binnendijk (1965, 1970). Another older paper, by Eggen (1967), is still very useful as the reference and place to organize one's thoughts. The area become much more active and acquired more astrophysical significance after the seminal papers by Lucy (1968a, 1968b) and the introduction of the contact model. A lot of activity took place in the 1970' and the field was really flourishing for some time. This work was summarized in a somewhat pedestrian book review by Rucinski (1985a) and in more specific reviews on the evolutionary state of these systems by Smith (1984) and by Mochnacki (1985). The solar-type magnetic activity was reviewed by Rucinski (1985b). There appear new reviews from time to time (Rucinski 1986) but the general impression is that since the early 1980's the field became somewhat stagnant. Few new studies seem to deal with W UMa systems and there has been no breakthrough research which would lead to resolution of major difficulties described in the aforementioned review papers. The present summary is written with the intention of pointing out again where the major difficulties are and what type of work is especially needed in the present situation.

---

[1] In fairness, we should state that Russell-Merrill solutions with *correctly performed rectification* can give amazingly similar results to those of the full contact model, cf. Mauder (1972). However, equality of temperatures is then unexplained (see below).

2. W UMa Systems as Contact Binaries

It is best to start with a somewhat unconventional statement: It was *not* the light curves which lead to the development of the contact model; the model was needed to reconcile the strange discrepancy between very dissimilar masses of components, as indicated by spectral studies, and practically identical surface temperatures, indicated by light curves. *Without the spectroscopic data, the contact model was not necessary.* Once the model was postulated, everything started to fall in place. Indeed, the three papers of Lucy (1967, 1968a, 1968b) really changed everything in our understanding of W UMa systems which, before, had quite unusual and unexplainable properties.

2.1. THE GEOMETRIC MODEL

There are three essential aspects of the contact model, as suggested by Lucy. First, his finding (Lucy 1967) that the gravity brightening law, $T_{\text{eff}} \propto g^\beta$ for cool, convective-envelope stars is much shallower ($\beta \simeq 0.08$) that the von Zeipel law ($\beta = 0.25$) admitted the possibility of much stronger distortions of stars than had ever been envisaged before. In particular, a common, optically and geometrically thick envelope could be considered as a means of equalizing the temperatures between components.

Second, stellar interior models built with an arbitrarily-strong energy transfer through the envelope had general properties not too disimilar from actual W UMa systems (Lucy 1968a). But the agreement was not perfect - the models were somewhat too restrictive – and much activity in the 1970' (summarized best by Smith (1984)) attempted to rectify this. But the most important property, the equality of temperatures, was no longer unexplainable.

Third, Lucy (1968b) showed that contact geometry does reproduce the observed light curves very well. The model had an appeal of simplicity and yet of a better physical support. In short: the original picture of two unrelated, strongly distorted stars which – for some unexplained reason – had equal temperatures, has been replaced by one simple configuration of strictly-known properties. This model is now the only one remaining as a basis to reproduce light and radial-velocity curves of W UMa systems.

The essential geometric properties of the contact model have been described in numerous papers, starting with that by Lucy (1968b). A detailed description of the same light-curve synthesis approach was given by Rucinski (1985a). Different integration schemes were described by Wilson and Devinney (1973), Wilson and Biermann (1976), by Hill (1979) and by Mochnacki and Doughty (1972a), the latter particularly convenient for contact binaries (see also Binnendijk 1977). All these methods differ only in quadrature schemes but have in common the same parametrization of the problem. A contact system is described by an equipotential $\psi = \psi(q; x, y, z)$ located between the inner and outer critical potentials of the

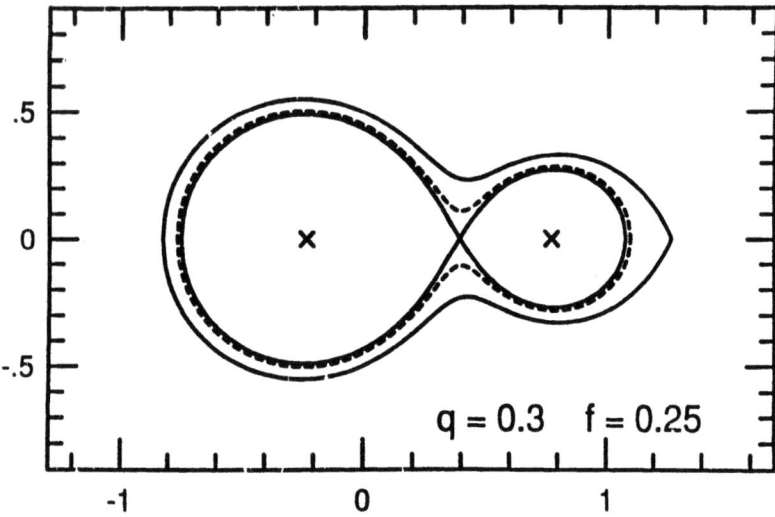

Fig. 1. The Roche geometry describes synchroneously orbiting stars with solid-body rotation in the combined potential of two point masses and of the centrifugal force. A contact binary is supposed to fill one equipotential whose relative "distance" from the inner and outer critical equipotentials is measured by the fill-out parameter, f. The mass-ratio is defined as $q = M_2/M_1 \leq 1$. The continuous lines in the figure show the two limiting equipotentials. The actual contact system fills an equipotential surface located between these two limits (the broken line).

restricted three-body problem (Figure 1). The mass-ratio, $q = M_2/M_1$, is the second physical parameter. These two parameters fully describe the geometry of the system. Even variation of gravity can be calculated simply as $g = |\nabla \psi|$. All other parameters are auxiliary and are related either to the local atmospheric parameters or to the way we observe the system. In particular, we see the orbit at an arbitrary inclination to the line of sight, i, so that the full geometric set of elements is (ψ, q, i).

Calculations of light or radial-velocity curves consist of the simple summations of contributions from local atmospheres at all visible points of the contact configuration, properly binning them in time and in radial velocity. One can use here atmospheres of arbitrary degree of sophistication, in arbitrary spectral bands, with simple or complicated limb-darkening properties, etc. The light curves so generated look amazingly close to the observed ones. And, indeed by adjustment of parameters one can usually find a good set of parameters (ψ, q, i) which best reproduces observations. Of course, something must be assumed about atmospheres before the fitting process (gravity-brightening exponent β, limb-darkening coefficient(s), etc.) but the idea is not much different from that used to calculate light-curves for other eclipsing systems.

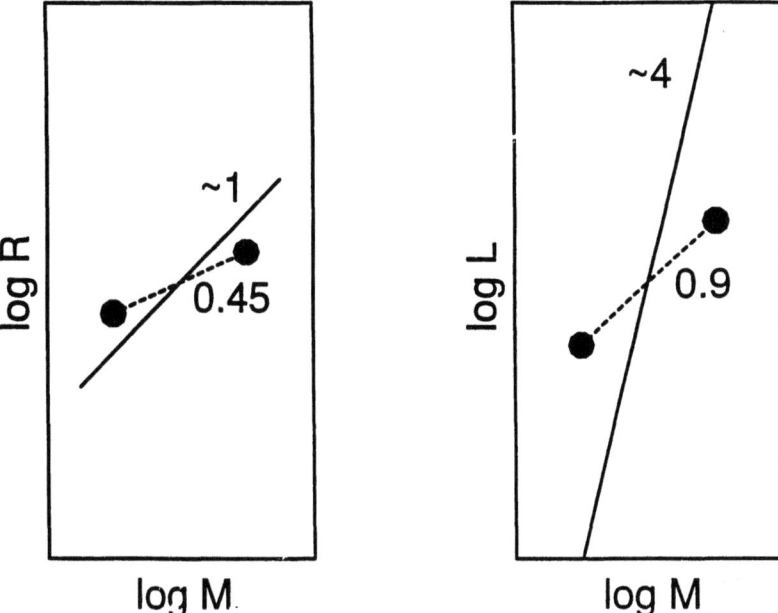

Fig. 2. Schematic mass-radius and mass-luminosity relations for contact binaries, compared with slopes of these relations for the lower Main Sequence stars. The internal mass-radius relation for the Roche geoemtry is: $R_2/R_1 \propto (M_2/M_1)^{0.45} = q^{0.45}$. Since temperatures are both components are almost identical, $L_2/L_1 \propto (R_2/R_1)^2$ which explains the shallow mass-luminosity relation. Note that this relation is absolutely decoupled from the internal luminosities generated by each star. In fact, it seems that the more massive components, which generate most of the energy, sit right on the Main Sequence relations (Maceroni et al. 1985) implying that the symbols should be appropriately shifted in the above figure.

2.2. LIMITATIONS OF THE MODEL

The contact model is still a simplification of a rather complicated physics. Let us enumerate the disputable points.

First of all, we know that there exists a huge energy transfer between components: one star produces almost all the energy radiated by the system, the other star is basically inert. W UMa systems seem to be made of "normal" stars with a "normal" relation between mass and nuclear luminosity: in simple terms – more massive stars are more luminous. Even on the Main Sequence this relation is as steep as $L \propto M^4$ (or even steeper). Differential evolution normally leads to an even stronger dependence. Thus, with a typical mass-ratio of about 0.3, this means that one component is several times more luminous in terms of its nuclear luminosity that the other component. This flux, which can amount to many solar luminosities, is transfered somehow through the narrow neck between stars. In cool stars with deep turbulent-convection envelopes such a transport must be convective. Can we assume therefore that there is no net mass transfer?

As was shown by Lucy(1976), Flannery (1976) and Robertson and Eggleton

(1977), contact systems are indeed unstable to the thermal time-scale mass exchange. This instablity is cyclic and the contact configuration is only one part of the cycle, the other one is the a semi-detached configuration (ses Sec. 6.2). True, the mass exchange is slow (10^7 years) when compared with the dynamical time-scale (fraction of day). But, in view of the energy and mass exchange processes, is the hydrostatic equilibrium a tenable assumption in the outermost layers that we actually observe?

The second simplification relates to the brightness distribution over the contact configuration. Normally it is assumed that the local surface brightness (equivalently the bolometric flux) depends on the local gravity according to the Lucy's modification (Lucy 1967) of the Von Zeipel law, $T_{\text{eff}} \propto g^\beta$, where the theoretically predicted $\beta \simeq 0.08$ (for a recent confirmation, see Sarna 1989). We can assume that this theoretical prediction is a solid one and, under this assumption, determine other parameters. But there are problems with this approach: there are observational indications that β is actually smaller, closer to $\beta \simeq 0.04$ (Eaton *et al.* 1980; Hilditch 1981), and there is a theoretical prediction (which can be questioned) that $\beta = 0$ (Anderson and Shu 1977). But is all this really valid in the situation of a strong energy flux between components? Why don't we see a gradient of surface brightness along the contact configuration, in the direction of the energy flow? It is possible that all these uncertainties are related to what we call the "W-type syndrome", which will be discussed in Sec. 2.3. The value of β for convective stars in general is a separate, very interesting matter (Rucinski 1989) which we will not discuss here.

And, finally, the geometry of the system. The shape of the contact configuration is usually assumed to be that of the binary Roche model. The equipotential surface is simply the one corresponding to the gravitational attraction of two point masses and of the centrifugal term in the rotating system of coordinates: $\psi = 2/(1 + q)/r_1 + 2q/(1 + q)/r_2 + d^2$ (Figure 1)[2]. Such a simple potential assumes rigid rotation of the whole configuration. We are not sure, whether this assumption is really fulfiled in real systems; there are arguments going both ways. On one hand, W UMa systems consist of cool stars with extensive turbulent convection in the envelopes. Such a convection may prevent establishing large-scale motions and break up convective cells into small-scale elements. This should produce large viscosity and lead to rigid rotation. On the other hand, numerous observations suggest existence of persistent asymmetries in the external layers of W UMa stars which manifest themselves as light maxima of unequal height (due to photospheric spots perhaps?) or velocity shifts of chromospheric lines (e.g. emission cores of CaII in CC Com, Rucinski *et al.* 1977). Indeed, deviations from the exact rigid-rotation regime would not be hard to imagine for such extended structures as contact binaries. But the outer layers are filled with chromospherically-active regions (cf. Sec. 5) which may produce the asymmetries so that what we see may be more

---

[2] The potential in physical units is: $V = \psi G(M_1 + M_2)/2A$ (r_i are the distances from the gravity centres and d is the distance from the axis of rotation).

relevant for understanding of the dynamo action in these conditions than to the rotation law. In summary: we assume the rigid-rotation law for want of anything better and still tractable, but is it really justified?

Each of the assumptions listed above is a real limitation of the model and each still requires further intensive research.

2.3. THE W-TYPE SYNDROME

The contact model has one systematic deficiency: for almost all cooler systems, the light curves are slightly wrong in that eclipses of more massive components are shallower instead of being deeper of the two. Although its magnitude is quite moderate, just a few percent, this systematic effect may be due to a serious inadequacy of the model. But, equally possibly, it may be due to an unaccounted but minor problem of the model. In this situation, let us have a brief look at what we do understand in the matter of relative depths of eclipses in a contact system.

Geometry of the Roche model tells us that the more massive component has a slightly larger mean gravity of the two components, $\overline{g_1} > \overline{g_2}$. For the standard gravity brightening law, $T_{\text{eff}} \propto g^\beta$, with non-zero β, this component should therefore have a slightly higher mean effective temperature, $\overline{T_{\text{eff},1}} > \overline{T_{\text{eff},2}}$. For this reason, eclipses of this star should be deeper. But there is also a second reason why this should be so: during such eclipses *two* limb-darkened disks are visible in place of one, as during the eclipse of the less massive star. Thus, eclipses of more-massive components should be deeper. And they are, but primarily among somewhat hotter W UMa systems, at spectral types earlier than about F5-F8. For almost all cooler systems, the unexplained reversal takes place.

The two types of contact systems, those that conform to the model, and those that do not, are called types A and W, respectively. From the geometric point of view, there is nothing exciting about type A – it simply follows the model. It is interesting, however, that these somewhat hotter A-type systems may be more evolved or may have entirely different origins from the rest of W UMa systems; we will return to this in Sec. 2.4. The real problem is with the W-type and we do not know how to solve it. Since the "W-type syndrome" is confined to the late-type systems, we have good reasons to suspect stellar activity as this mysterious underlying cause. A few speculative possibilities, all related to stellar activity were discussed before (Rucinski 1985b). It could be spots forming preferentially on the more-massive component, it could be belts of spots of the same relative extent on both components or, finally, it could be due to reduced limb-darkening. One can think of various reasons why the degree of limb-darkening could be reduced in W UMa stars: it could be chromospheric emission or bright plages preferentially visible at large angles. The possibility of altered limb darkening can be tested and the requirements are very unrealistic: a reduced limb-darkening is not sufficient to explain the effect, one needs considerable limb brightening! The infrared data do not support the idea of an abnormal limb darkening. The IR limb darkening coefficient is normally expected to be smaller than in the optical range and the eclipses should become correspondingly more shallow. This is exactly what is observed (Jameson and Akinci 1979; Shenavrin and Zhukov 1984) indicating a rather normal dependence on the wavelength.

The most popular way out of the W-type difficulty is easy but not satisfactory in the opinion of the author. By assuming that the less-massive component is slightly hotter (by about 5% typically, Rucinski 1974), one can remove the depth-of-minima discrepancy. This simple assumption seems to work well for accurate optical light curves (Linnell 1986) but encounters difficulties with passing a more stringent test: in the UV range the temperature difference should translate into a huge difference in eclipse depths with a transition to the β Lyrae-type light-curve and this is *not* observed (Rucinski 1976; Eaton, Wu and Rucinski 1980). Apparently, the depths preserve the discrepancy in all wavelengths as if something grey produced the W-type syndrome.

One possiblity, which could be checked is the preference for spot formation on the more massive component. It can be shown that this component should have a relatively deeper convective envelope than its companion (Rucinski 1992). This may lead to the more vigorous dynamo action in this component and preference for dark spots there. There are some indications that the more massive components are indeed preferentially more active (Barden 1985; Bradstreet and Guinan 1988).

The W-type syndrome is telling us something but we are unable to decipher the message. Maybe the problem is indeed a minor one and related to some simple modifications of the surface distribution, as we tend to interpret now. Or, may be it is related to the flow of energy between components (one such an idea of the hot matter from the more-massive component preferentially heating the super-adiabatic surface layers of the secondary was presented by Whelan (1972) and Mochnacki and Whelan (1973) but it has weak foundations). Or, maybe the reasons are much deeper beyond our current understanding of contact (or – blasphemy here! – maybe even non-contact) configurations.

2.4. THE PERIOD-COLOUR RELATION

The period-colour (PC) relation (Figure 4) was discovered by Eggen (1961, 1967). It is one of the best proofs that W UMa systems are really in contact. Indeed, for progressively shorter orbital periods, progressively smaller stars can be accomodated in the geometrically same Roche model. If these stars are similar – in this case close to the Main Sequence – then the size sequence will create related mass and effective temperature sequences. Thus, the observational P-C relation is, in fact, an expected period-effective temperature sequence.

Since its discovery, the PC relation has been used as an important tool for testing models of contact binaries and, in particular, their evolutionary state. Most models are quite restrictive in reproducing the whole range of periods and colours and it is relatively easy to spot prcblems with models when the period-colour is used as a testing tool.

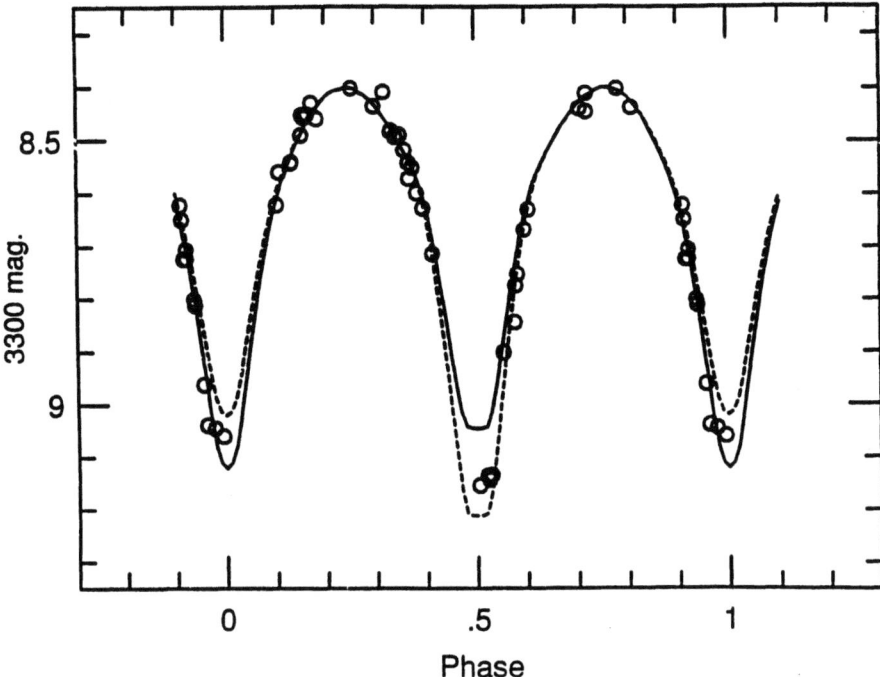

Fig. 3. The W-type syndrome in W UMa itself. The theoretical light curve for W UMa with phases counted from the eclipse of the more-massive component, as suggested by spectroscopic observations, is shown by the continuous line. Introduction of a 5% temperature excess of the less-massive component results in an interchange of the relative depths of the eclipses (the broken line). The observations shown here have been obtained with the ANS satellite working at $\lambda = 3300$Å (Eaton *et al.* 1980) where the small temperature excess is expected to produce a larger effect than in the visual range. In fact, the shape of the light curve is approximately the same indicating that something else may be needed to explain the W-type problem.

Let us have a closer look at the relation. It starts at the low temperature end close to B-V=1.24 and P=0.221 day of CC Com and B-V=1.07 and P=0.224 day of V523 Cas. It is quite significant that contact systems with shorter periods and lower temperatures do not exist. Obviously, it is not easy to discover such systems and we cannot exclude the possibility that not enough late-type dwarfs have been searched for the W UMa-system characteristics. The relation starts climbing from that point along, very roughly, (B-V) = 3.75 log P + 4. Higher up, systems with longer periods (for the colour) start to appear and the relation becomes broader. The width, at solar colour, reaches a factor of about 1.5–2 in the orbital period. Longer periods, for the same geometry, must correspond to larger, i.e. more evolved, stars and, indeed, the relation becomes even wider for early spectral types where evolution must be even more important. Across the whole diagram, the left edge of the relation is the locus

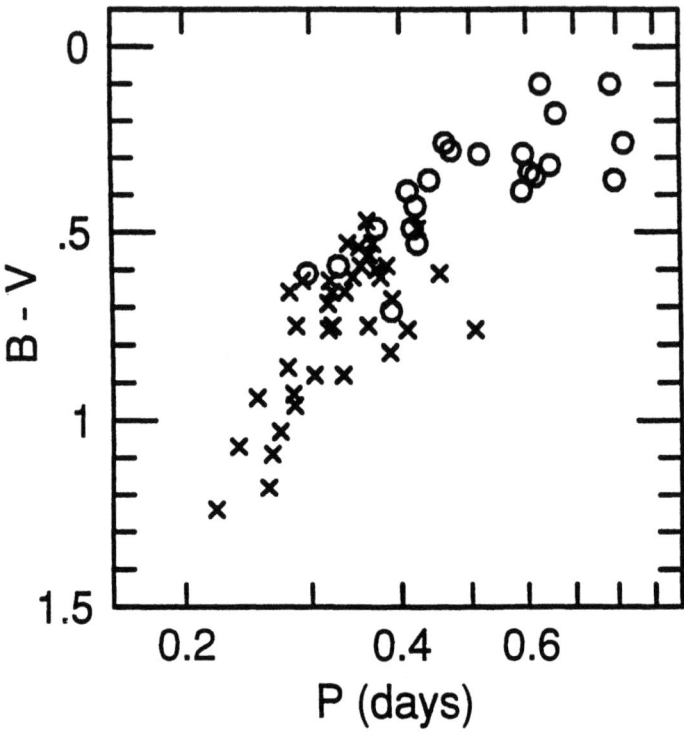

Fig. 4. The period-colour relation, based on the observational data, as listed by Mochnacki (1985). W-type systems are marked by crosses, A-type systems are marked by circles.

of the most compact, i.e. least evolved systems and this is where we find most of the "typical" W UMa systems.

Mochnacki (1981) used the PC relation as a tool to analyse the evolutionary status of contact binaries. He pointed out that the effective temperature must be always corrected for an effect related to the energy transfer between components. Namely, the temperature corresponds to the energy coming primarily from the more-massive component but redistributed over the whole contact configuration. Therefore, the temperature of the primary is always lower than a non-contact star of the same evolutionary state would have without the secondary attached to it. The correction is relatively easy to make because it depends only on the mass-ratio (as a measure of relative sizes and radiating areas of components). Thus, a relation which is particularly convenient for comparison with theoretical studies is the one between the orbital period and the *energy-transfer-corrected* colour of the primary[3].

In his investigation, Mochnacki has shown that some systems must consist of

---

[3] Moss (1985) pointed out that the mean density also requires a correction by a small amount of about $\Delta \log \bar{\rho} \simeq 0.1$.

components which have evolved within the Main Sequence band. The most common location of these systems is the upper part of the period-colour relation. The mass-ratios of such systems are usually quite small, somewhere around $q \simeq 0.1$. Bright systems, AW UMa with $q = 0.08$ and ε CrA with $q = 0.11$, and with spectral types around F0 are good examples of such evolved systems. And these are theoretically "unproblematic" systems as one can envisage ways converting a normal, ZAMS W UMa system into such a more extreme one by just re-arranging masses, i.e. an evolution within the Main Sequence with constant angular momentum (cf. Sec. 6). Such systems are commonly called A-type systems. We should remember that the "A-type" means that such systems simply show light curves which can be explained by the normal Roche-model contact configuration. We do not understand at present why the evolved systems conform with the geometric contact model so well when cooler, lower-mass systems almost always show the W-type syndrome.

The early-type, low mass-ratio binaries are not the only evolved systems. There also exists a group of oversized, but cooler (middle/late G-type) systems of sufficiently large orbits to indicate that the amount of angular momentum is too large for originally contact systems. As pointed out by Mochnacki, these systems must have come from even wider, non-contact binaries. Good examples are OO Aql with $q = 0.84$ (Hrivnak 1989) or AH Vir at spectral types as late as G8–K0. This large amount of angular momentum suggests that at least some contact binaries form late, from originally detached binaries. This very interesting result has open up a new avanue in studies of contact systems which may be thought of as one of the intermediate or end products of normal binary-star evolution.

We should remember, however, that in addition to the above-described systems showing definitive signatures of evolution, most systems seem to be evolved only slightly or entirely unevolved. Thus, the picture is not simple at all.

The particularly nice aspect of the results of Mochnacki is that they have been obtained almost directly from the P-C relation. Both of these two quantities are easily and accurately observable and the evolutionary effects are visible almost directly in the period-colour diagram.

2.5. THE MASS-RATIO AND DEGREE OF CONTACT

The mass-ratios of contact binaries are always different from unity. Contrary to the detached binaries, where equal-star combination may be the most common situation, components of contact binaries definitely do not like being identical. The largest mass-ratios known is that of SW Lac[4] ($q = 0.88$ with that of OO Aql, $q = 0.84$.) Since W UMa systems have freedom of exchanging matter between components, avoidance of $q = 1$ must be due to a some sort of instability. Much reseach has been devoted to why W UMa systems avoid this solution and two main instabilities at $q = 1$ have been identified. Thus both, the dynamical instability

---

[4] Zhai and Lu (1989) reduced this number recently to $q = 0.78$.

(Williams and Roxburg 1976) or the thermal instability (Lucy 1976; Flannery 1976; Robertson and Eggleton 1977) can drive solutions away from the equal-mass configuration.

If the system is bright enough, its mass-ratio can be measured using spectroscopic techniques. The measurements are not very easy because spectral lines are broad and heavily blended. The spectrum-deconvolution techniques work well in such cases providing values of q accurate to a few percent (McLean 1981, 1983; Hilditch and King 1986; Hill 1989; Hill et al. 1989a, 1989b). Such results are quite independent of photometric solutions. Another possibility is to use a photometric method of fitting theoretical light-curves to the observed ones and look for the best q. Results obtained that way are model-dependent and, in most cases, very poor because of the weak sensitivity of the light-curve shapes to changes of q. In only one situation the method can work: for total eclipses. As was shown by Mochnacki and Doughty (1972a, 1972b), knowledge of a duration of the total eclipse offers a unique way of determining a combination of (q, i). The total amplitude of light variations permits to separate these parameters. The model-dependency still remains but one can obtain a well defined set of parameters. But, we stress, this is possible only for totally-eclipsing systems; for partially-eclipsing systems, the mass-ratio should not be determined from photometry alone!

A parameter measuring the degree of contact can be defined in various ways. A definition based on the potential ψ, as in Figure 1, is most frequently used. Since we assume that ψ is confined between those that pass through the Lagrangian critical points L_1 and L_2, a convenient parameter is one that measures the distance, in potential, from the two critical equipotentials. One can therefore use f, as in the formula: $\psi = \psi_1 + f \times (\psi_2 - \psi_1)$. Here, ψ_1 and $f = 0$ is for the inner, and ψ_2 and $f = 1$ is for the outer critical equipotentials, respectively. Such a definition frees us from the dependence of ψ on q. But it is not perfect, because the same f means quite different degree of geometrical contact (both in absolute sense and relative to the size of the system) for different q (see Figure 5 below). But nobody used anything better than that. One should also be aware that there exist various definitions of f differing by a sign or a constant, etc. (see Mochnacki 1981).

All determinations of the degree of contact are model-dependent. And all are correlated strongly with the assumed value of the gravity-brightening exponent β (cf. Sec. 2.1). The smallest β produces the strongest contact and the largest permissible $\beta = 0.25$ (the von Zeipel law) may even lead to solutions *not* requiring contact (Hill 1979). The present data suggest that f is small (Rucinski 1985a), $f < 0.15$. Most of the solutions have been performed with the Lucy convective law with $\beta = 0.08$. There are some indications that this value may be still too large and that $0 < \beta < 0.04$ (Rucinski 1989). But this value would not make f much larger. Thus, a rather thin neck between stars seems to be what contact binaries prefer. We do not understand why contact binaries, on the average, prefer weak contact but, if indeed real, this may be the most important feature of these systems. Maybe somewhere here is buried a mysterious mechanism of self-regulation which

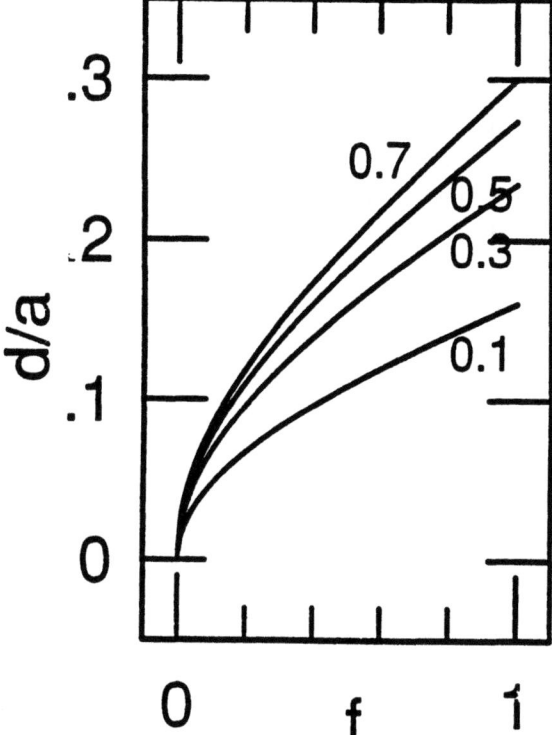

Fig. 5. The widths of the "neck" between stars, d, as function of the degree of contact, f, for a few values of the mass-ratio, q (labels on the figure). The width of the neck is expressed in units of the separation between components, a.

may keep stars always in contact, but in a very special one: never too strong and never too weak. We will return to this subject when discussing the evolution of contact systems (Sec. 6).

Before we conclude this section, we should point out that weak contact with $f < 0.15$ should not be interpreted that stars touch each other just barely. In fact, the geometrical contact is still quite substantial and the neck between stars has typical dimensions of about 5–10% of the stellar radii (Figure 5). Therefore, the outer parts of both stars can be in very good thermal contact even for very small f. It seems that use of this parameter should abandoned because it is not a useful one for weak (i.e. shallow) contact. The directly-defined, relative geometric extent of the neck, say d/a (where d is the thickness and a the binary separation), would serve us much better.

3. Space Density of W UMa Systems

We will consider now frequency of occurrence of W UMa systems in the field, among ordinary stars. The next section will concern this frequency in stellar clusters.

As was briefly described in the Introduction, the subject of the supposed high spatial frequency of W UMa systems has been surfacing since the time of Shapley (1948). As we know, this matter now looks quite different than 40 years ago. First, the comparison of the numbers of W UMa systems with the total number of variable stars is simply useless: one compares the proverbial apples and oranges! We recognize now so many classes of variable stars that there is no point to relate W UMa systems to all these numerous examples of stellar variability. The other comparison, the number of W UMa systems versus the number of eclipsing binaries, is more meaningful but is heavily biased by the observational selection for detection of W UMa systems. Chances of discovering a contact system are indeed quite high: The contact geometry increases probability of eclipses and even for a small inclination brightness variations can be quite large. The latter are due to changes in visible area, sometimes called also the "ellipticity variations". Chances of discovering radial velocity changes are also quite high for contact binaries because amplitudes can reach up to 500 km/s.

Probability for discovery of a contact system depends very strongly on the mass-ratio. When this quantity is very different from unity, one of the components becomes very small and its presence is difficult to detect: mutual eclipses become very shallow, spectral lines of the less-massive component are difficult to detect and the radial-velocity variations of the more-massive component have a reduced amplitude. This is a normal situation for binary stars consisting of Main Sequence stars, very well known in all studies of the binary frequency (Abt 1983 and refs. therein). Normally, it is treated by various incompleteness corrections of differing degrees of credibility. But, with W UMa systems, the situation is somewhat different. Because less-massive components have temperatures abnormally high (i.e. equal to those of more-massive components), they are also abnormally bright and hence still detectable. This explains why we know quite a few low mass-ratio systems like AW UMa with $q \simeq 0.08$. We could not imagine detecting such small mass-ratios among any other type of binaries (except for well separated visual binaries and cataclysmic variables where signatures of the presence of secondaries come in a very indirect way). Summing up: W UMa systems are very easy to detect, even if they have small mass-ratios.

There exist a few attempts to estimate the selection effects for discovery at the low mass-ratio (van't Veer 1975a; Budding 1982; Duerbeck 1984). Corrections to the frequency estimates which are postulated to take into account the invisible very low-mass secondaries are sometimes quite large (van't Veer 1978). They obviously depend on the assumed detection limits for light and radial-velocity variabilities. It is therefore not surprising that the current estimates for spatial frequency of W UMa systems differ considerably and cannot be confined to better than about 1/500 to 1/2000 of ordinary dwarfs being W UMa systems.

All estimates of the spatial frequency of W UMa systems start from the "General Catalogue of Variable Stars". Its most recent, the fourth edition of the Catalogue (Kholopov 1985) lists 563 stars as having EW-type light curves. However, many

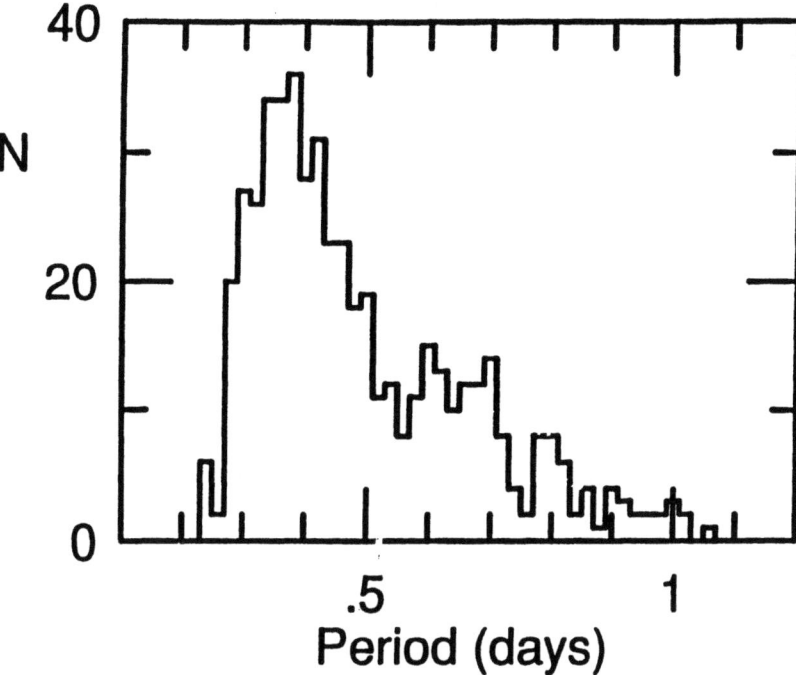

Fig. 6. Distribution of periods for systems classified as possessing EW-type light curves in the newest edition of the General Catalogue of Variable Stars (Kholopov 1985). Systems with periods more poorly known than 0.0001 day were eliminated before plotting. Subsequent distribution plots contain only systems with periods shorter than 0.75 day.

have poorly known periods. After eliminating those with periods entirely unknown or known to worse that 0.0001 day the sample consists of 514 systems. Their distribution is shown in Figure 6. As always in such distributions, one has a distinct feeling that what is seen there are different selection effects for discovery. But the sharp fall-off at very short periods seems to be real. The genuine, late-type W UMa systems form the main concentration in the distribution in the period range between $0.25 < P < 0.5$ day. Beyond $P = 0.5$ day extends the domain of contact binaries with spectral types earlier than A. Statistics of the number of contact binaries in successive magnitude intervals does not actually show any major difference if one limits himself to $P < 0.5$ day (327 systems) or uses the somewhat larger sample extending to $P < 0.75$ day (458 systems). The relevant data are listed in the table. Notice that for most of the systems the photometric data are photographic or in the B-band (also probably photographically derived). Since there exists a progression of colours with periods (the P-C relation discussed in Sec. 2.4.) and absolute magnitudes (on the Main Sequence), it was deemed safer to leave these data in their respective systems and not to mix the samples.

TABLE I

Statistics of systems with the EW-type light curves in the 4th Edition of the General Catalogue of Variable Stars. Systems with periods known to better than 0.0001 day. (Number of systems in bins $(m - 1/2, m + 1/2)$.

m	V	pg	B
5	1	0	0
6	1	0	0
7	2	1	0
8	10	0	0
9	14	2	1
10	19	3	0
11	42	13	1
12	19	26	4
13	3	51	4
14	6	92	7
15	1	77	9
16	4	30	2
17	2	3	2
18	0	4	1
19	0	1	0

Figure 7 gives the distribution of systems in one-magnitude bins, for both samples (visual and blue) independently. It is quite obvious that both samples suffer strong incompletness biases for $V > 10$ and for $B > 13$. Actually these biases may appear even for brighter systems and the cumulative distributions could help in resolving this possibility. Such distributions are graphically shown in the two parts of Figure 8. Somewhat surprisingly, the cumulative distributions do not confirm the existence of any obvious biases for magnitudes below the limits mentioned above. The visual sample seems to follow the standard $N_m(\text{vis})$ curve up to $V < 9$ and then starts to flatten out; the blue sample seems to follow the $N_m(\text{pg})$ curve up to pg< 13. Interpretation of Figure 8 is simple: the selection effects are quite uniform over the ranges quoted above. But the matter of the *absolute* selection effect remains open as can be seen by comparing the two panels of this figure: at any magnitude the numbers of systems with photometric data in B or pg are much smaller! To estimate this absolute effect, the dotted lines in the figure give the curves $N_m(\text{vis})$ and $N_m(\text{pg})$ from Allen (1973) but shifted in such a way that they would give numbers of W UMa systems as if their frequency was 1/1000 of all stars. Apparently, by displaying linear trends in the cumulative plots, the samples seem to be internally uniform and not much discriminating against fainter systems over a wide range of magnitudes. The blue sample is apparently based on searches confined to smaller areas of the sky (open clusters? the Milky Way?) and does not approximate the true frequency of the W UMa systems. The

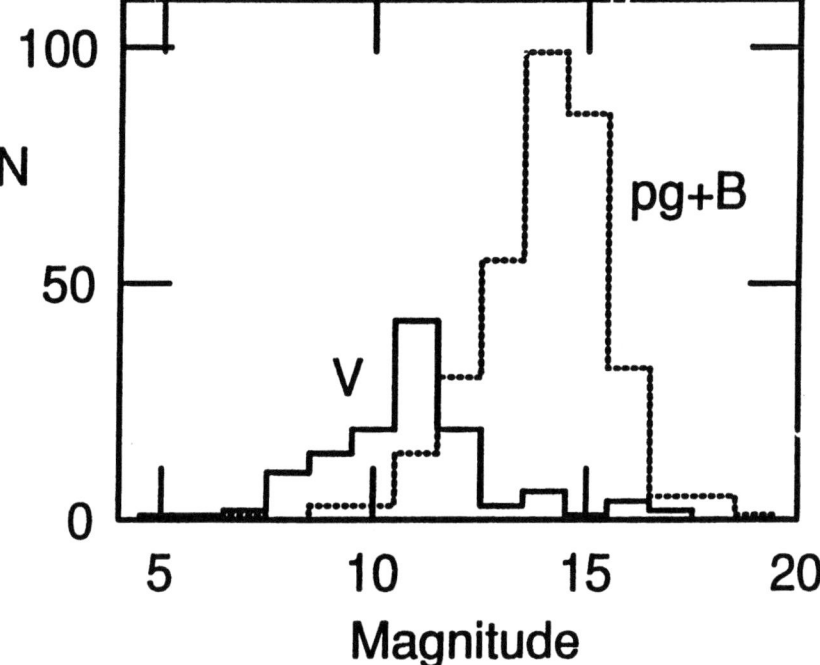

Fig. 7. The magnitude distribution of EW systems with $P < 0.75$ day. The data are for two independent samples with photometry in the visual system (V) and in the blue system (the combined B-band and photographic samples).

visual sample is apparently more complete (28 systems to $V < 9$) but may still be far from the actual frequency of occurence. The galactic latitude distributions shown in Figure 9 confirm the above interpretation.

We can use also the Bright Star Catalogue in an attempt at estimating the frequency of the W UMa-type systems. We know that stars listed in this Catalogue have undergone a much more thorough scrutiny than any other sample. Among these stars there are only 2 W UMa-type systems: ε CrA ($V_{max} = 4.74$) and 44i Boo B ($V_{max} = 5.8$). Taking into account that many bright stars are distant giants, presence of these systems gives us a reasonable guess on the frequency of W UMa systems in the field: It seems to be of the order of one such a system per a thousand of ordinary dwarfs. Obviously, two systems is a terribly small sample but, at least, we know that it is drawn from the best source list possible.

Summing up: the best current estimate for the spatial frequency of W UMa systems remains about 1/500–1/2000 of ordinary dwarfs. This frequency is not high especially when compared with the total frequency of all binary systems. Thus, the W UMa system is are, in fact, an oddity in the world of binary systems. They are such an oddity either because conditions which lead to their formation are very special or because the contact stage lasts for short time.

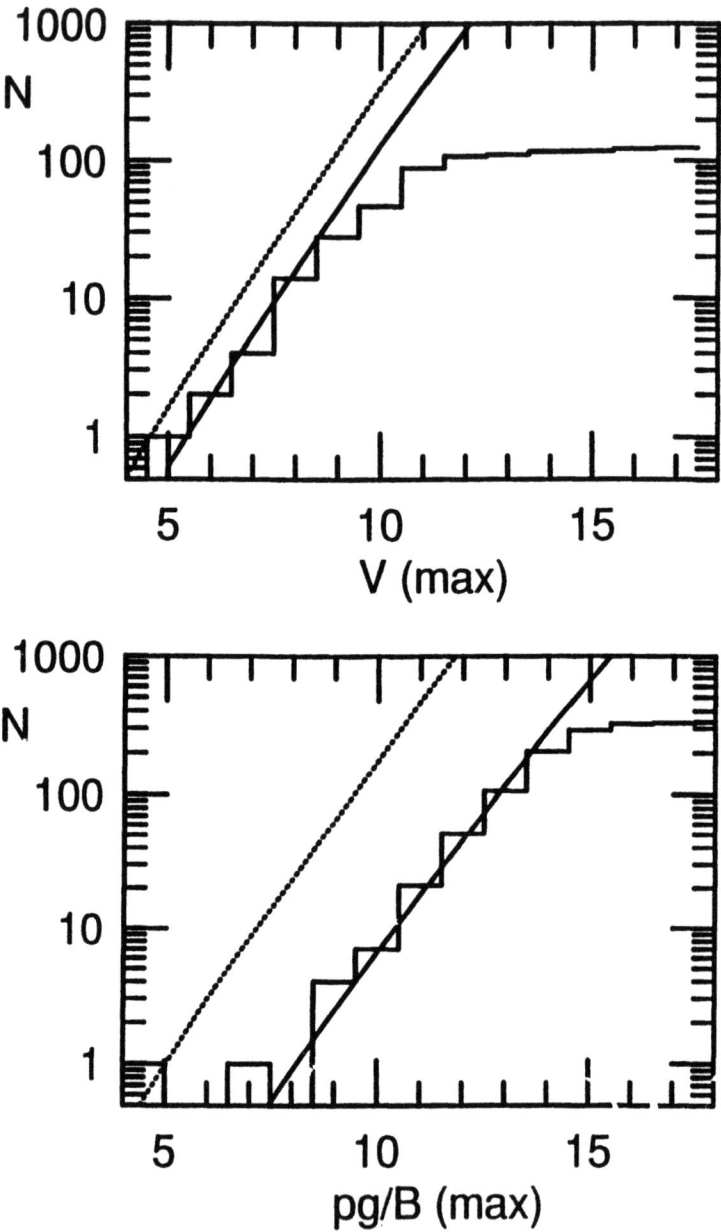

Fig. 8. The cumulative magnitude distributions for the visual sample (upper panel) and for the blue sample (lower panel). The continuous lines give fits of the functions $N_m(\text{vis})$ and $N_m(\text{pg})$ to the data over the ranges of reasonable agreement (see the text). The normalization factors are: 3×10^{-4} and 4×10^{-5}, respectively. The dotted lines give the expected distributions calculated on the basis of the N_m functions for the field stars with the assumed frequency of W UMa systems equal 1/1000 of the field stars.

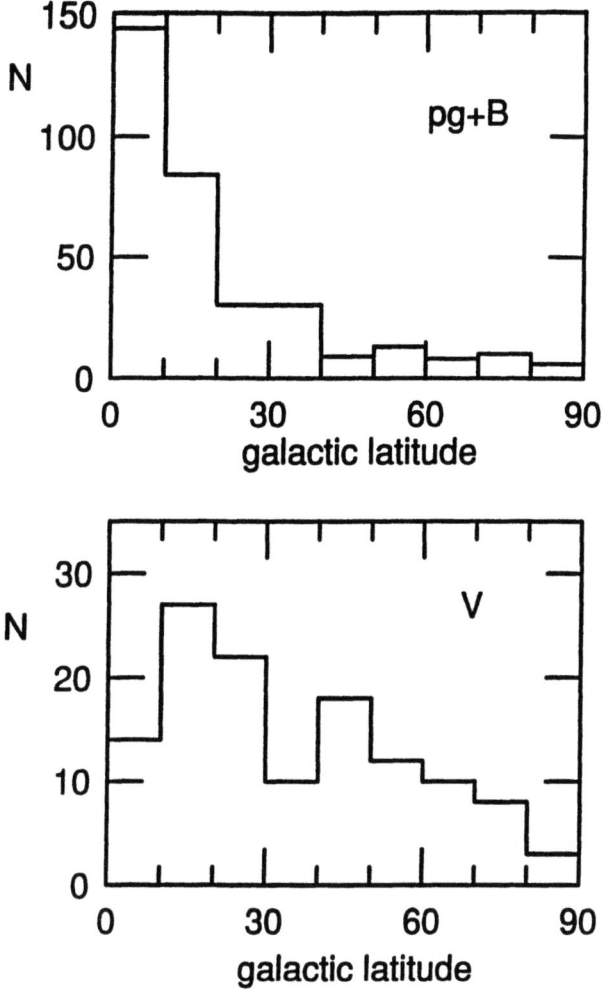

Fig. 9. Distributions of W UMa systems in the galactic latitude (both hemispheres together). Apparently, most systems with photographic or B-band data are located within the Milky Way band ($|b| < 20°$).

4. W UMa Systems in Stellar Clusters

Presence of W UMa systems in stellar clusters, as for any other type of stars, has a special significance as it permits an unambiguous check on the age and abundace characteristics. A number of W UMa systems were suggested to belong to various clusters on the basis of photometric data (e.g. Van't Veer 1975b). However, one must be careful with taking these data at face value. The main reason is the uncertainty in membership which will remain for most clusters, unless supported by proper-motion studies. We must remember that areas of open clusters are most carefully checked for presence of variable stars. Thus, variables situated in front and behind a cluster may be more easily detected than in other parts of the sky and then erroneously assigned the cluster membership. This actually the case of AH Tau, at one time considered to be a member of Pleiades and of AD Cnc, supposedly belonging to M 67 (Samec and Bookmyer 1987, Samec *et al.* 1989). Similarly, membership of all contact systems in the loose grouping Coll 359 (Rucinski 1980, Van't Veer 1980) is very highly questionable. Therefore, many independent lines of evidence must be brought up to be able to consider a contact system as indeed a member of the cluster. Good examples of such systems are TX Cnc (Whelan *et al.* 1973) in Preasepe and AH Cnc (Whelan *et al.* 1979) in M 67. A "mini-cluster" of two unquestionably related W UMa systems in one visual binary is formed by BV Dra and BW Dra (Batten and Lu 1985, Kaluzny and Rucinski 1986).

It is somewhat easier to find W UMa systems in distant clusters where "reasonable" photometric properties, such as spatial proximity supported by the period-colour data, give more confidence in assigning the cluster membership. The situation concerning the frequency of occurence of contact systems in open clusters has recently become quite confused due to a few very new discoveries. The best known case is the very old open cluster NGC 188. Four W UMa systems in this cluster have been known since the work of Efremov *et al.* (1964) (see also Baliunas and Guinan (1985)). In recent years Kaluzny and Shara (1987) using time-sequence CCD photometry discovered three more contact systems and three related stars, possibly of the FK Comae type (see Sec. 6). With seven W UMa systems, their relative frequency in NGC 188 reaches 3 percent of all stars with comparable spectral types and luminosities. For some time NGC 188 seemed to be an exception because all other investigated clusters, both younger than NGC 188 and slightly older (Mel 66), did not show any W UMa systems (Kaluzny and Shara 1988). Things received a new twist when the team Kaluzny, Krzeminski, Kubiak and Mateo (Mateo and Krzeminski 1990) discovered 4 W UMa systems, 2 Algols and one RS CVn system in the old cluster Tombaugh 2 and the other, even larger team (Gilliland *et al.* 1991) studied the central region of M 67 re-discovering AH Cnc and finding two more contact systems there.

Van't Veer (1982) suggested even before the recent work of Kaluzny and Shara that NGC 188 is special case in having orbital planes of its binaries oriented preferentially in space with us being close to this plane. Obviously, many astronomers would like to avoid such a picture, especially for such an old and well relaxed system as NGC 188. An intuitively simpler explanation was that only this particular cluster is so abundant in close binaries. Now, it looks like *some* old clusters are very abundant in contact systems, some are not and we do not know the reason for this dichotomy.

Frequency of contact systems in open clusters in general, with the exclusion of NGC 188, M 67 or Tom 2 is not inconsistent with the general-field frequency of about one or two such systems per a thousand ordinary dwarfs (Sec. 3). Too few clusters have been studied to state anything about any relationship with age or chemical composition. The three clusters very abundant in contact systems with their advanced age and somewhat reduced metallicities invite an interpretation that W UMa systems form at later stages of stellar evolution. There is also the question of the cluster richness. Eg. NGC 188 is a very rich cluster. It is rich even now but was much richer before, in its initial stages of dynamic evolution. Does this richness have something to do with the abundance of W UMa systems?

Let us set aside old clusters and turn to young clusters. W UMa systems are not observed there. But these clusters are normally quite sparse and contain typically no more than tens or hundreds of stars. If frequency of contact systems is similar to that in the field, we simply cannot expect to see them there. Thus, although many facts point to the formation of W UMa systems at later stages of binary-star evolution (Sec. 6), we cannot assume this as proven. Besides, it is so difficult to study lower Main Sequence stars in young clusters, that our inability to detect the W UMa systems there should not be terribly surprising.

One type of young analogues of W UMa systems are early-type contact binaries which frequently occur in open clusters. There are normally discussed separately but may be there is no real reason to do so. As pointed by Popper (1982), some among early-type contact systems are very similar to W UMa systems: the components have identical temperatures, mass-ratios are different from unity, etc. The problem with early-type systems is the difficulty of sorting them out of various other types of interacting binaries more akin to β Lyrae. It is, however, quite possible that a count of early-type contact systems would give us the frequency of their occurence not too different from that for genuine W UMa systems.

And finally the globular clusters. There are two tantalizing observational results. First, Irwin and Trimble (1984) found eight variable stars in the cluster M 55 (NGC 6809) which may be either W UMa systems or spotted active stars. The latter possibility would be also interesting as it would mean that these are also binaries (because single stars would have braked their rotation by now to very low rates). The variables in M 55 are faint, at $m \simeq 18\text{-}21$ so that accurate photometry requires a large telescope. And indeed, Mateo and Krzeminski (1990) very recently have discovered that at least two of the variables in M 55 are short-period binaries; one is certainly a W UMa system with $P = 0.259$ days. And the second result: two of the blue stragglers studied (by Mateo et al. 1988) in NGC 5466, NH 19 and NH 30, turned out to be W UMa-type systems (Mateo et al. 1990) (there is also a short-period Algol system there). Of particular importance are the photometric properties of NH 19 and NH 30: both systems are unusually blue for their periods, even after application of (large) corrections for blanketing. Equivalently, the orbital periods of both systems are too short for their colours when compared with Population I W UMa systems (see Sec. 2.4). This interesting property of the NGC 5466 systems urgently requires further studies.

The discoveries in M 55 and NGC 5466 are so exciting not only for the obvious advantages of knowing and being able to study contact systems in globular clusters but also because of the theoretical prediction by Webbink (1979) who gave arguments for very restrictive conditions on very old contact systems. These conditions, due primarily to gravitational radiation, would produce systems with small mass-ratios, shallow eclipses and short periods.

5. Magnetic Activity

For a long time, until the IUE and X-ray satellite era, there were few indications that W UMa systems are the most active among solar-type stars. With hopelessly rotationally-blended spectra, relatively faint magnitudes and short time-scales for variability, these systems were simply difficult to study. The data were quite fragmentary: a few studies suggested photospheric spots as an explanation for the frequent light-curve changes; there was one report of a large flare (Kuhi 1964); Struve (1950) noted presence of the Ca II emission. These older results were briefly summarized by Rucinski (1985b).

It is obvious that activity of W UMa systems must be discussed on the wider background of the solar-type activity in late-type stars. As we know now, late-type stars – when rotating rapidly – can display extremely diversified range of phenomena tracable to the magnetic dynamo. The dynamo operates through poorly understood processes related to insertion of vorticity into turbulent convective motions. But the end product, genaration of magnetic fields, is amazingly simple in its principal characteristics: the activity level seems to depend on only two parameters, the effective temparature and the rate of rotation. The over-all picture is so simple that reduction to one parameter similar to the Rossby number, $\rho = \tau/P$, may be actually possible. Many studies, starting with Noyes et al. (1984), have analysed the dependence on ρ for normal stars and the emerging picture seems to be quite convincing. In the formula for ρ, τ is the turn-over time for convective motion deep in the convective envelope (where magnetic field amplification has a chance to take place before field expulsion) and P is the period of rotation of the star. The turn-over time depends practically only on the effective temperature of the star[5] and ranges between fraction of a day for F-type stars and several days to weeks for K-M-type stars. The expected relation between ρ and activity is simple: When ρ is large, either because of long τ or short P, activity is expected to be large; the reverse should hold for small ρ. And this is precisely what is observed!

In spite of success in relating ρ to activity, much research is still needed to establish all details and extend the range of applicability of this dependence. Although not very bright, and not too numerous, the W UMa systems may play a very important role in this process, primarily because of their period-colour dependence (Sec. 2.4.). This dependence produces combination of effective temperatures and rotation periods quite unlike general trends observed in normal stars. In normal stars that we sample in the field, the most common combination is: late spectral

---

[5] In theoretical predictions, τ also depends on the convection-efficiency parameter, $\alpha = l/H_p$. This uncertainty leads to τ predictable to a multiplicative coefficient but with almost unchanged dependence on T_{eff}. Stars "know" what value of α is the right one so that this uncertainty should not appear for groups of similar stars.

type → slow rotation, early spectral type → rapid rotation. This is because of the magnetic breaking phenomenon which is more efficient in late-type stars. But with W UMa systems the relation is just reversed: cooler stars rotate more rapidly! A look at the definition of ρ above tells us directly that the range of this number can be extended very considerably that way.

Although W UMa systems may be indeed one of the best classes of objects to test the Rossby-number concept, we should be prepared for surprises. W UMa systems most probably have quite different inner structure and angular momentum distributions than normal stars and it would be really amazing if W UMa systems followed exactly the same universal relationships! In fact, it is easier to invent reasons why W UMa systems should *not* follow the same dependences as for single and detached-binary components than to find reasons for similarity. But, so far, everything suggests (Rucinski 1985c, Vilhu and Walter 1987, Vilhu *et al.* 1989) that W UMa systems have very high but "normal" activity levels ... Apparently, activity is a very robust phenomenon!

The following indicators of steady activity can be studied in W UMa systems (in order of wavelengths): the coronal X-ray emission, the UV line-emission originating in chromospheres and transition regions, and the coronal radio emission. The line emission in cores of optical-range resonance lines ($H\alpha$, H and K of Ca II) cannot be as easily used as in single stars because of the strong rotational broadening. Unsteady emission due to flare-like phenomena can be observed in all spectral ranges, including the optical where W UMa systems may reveal very infrequent but powerful brightenings, similar to that observed by Kuhi (1964).

As seen in the X-rays and in the chromospheric and transition-region line-emissions, W UMa systems seem to follow the "saturated" dependence on ρ similar to that observed for the most active among the BY Dra- and RS CVn-type stars (Vilhu and Walter 1987). Figure 10 adapted from the paper by Vilhu *et al.* (1989) shows that W UMa systems practically delineate the highest levels of Mg II activity. Most probably the "saturation" takes the form of the total filling-in of the stellar surfaces by magnetic structures.

The radio emission seems to be low, at least as compared with RS CVn-type stars. Only very few W UMa systems have been detected in the radio, even with the VLA system. The data are very fragmentary and their temporal variability precludes definite statements. We should remember that even the most active stars convert only very tiny fractions of their energy output into the radio emission. For RS CVn-type stars, which are known to be most active in the radio: $L_{\text{radio}}/L_{\text{bol}} \simeq 10^{-7}$ (Gibson 1985). But, for W UMa systems, this efficiency seems to be much lower, at about $10^{-12} - 10^{-11}$ or so (Hughes and McLean 1984, Drake *et al.* 1986, Rucinski and Seaquist 1988, Vilhu *et al.* 1988). It would be useful to establish a better statistics on the radio emission of W UMa systems but one could do it only by having much observing time on the largest radio telescopes in the world.

An interesting aspect of activity are photospheric spots. There is no question that they form on the W UMa stars and that they provide a ready explanation for

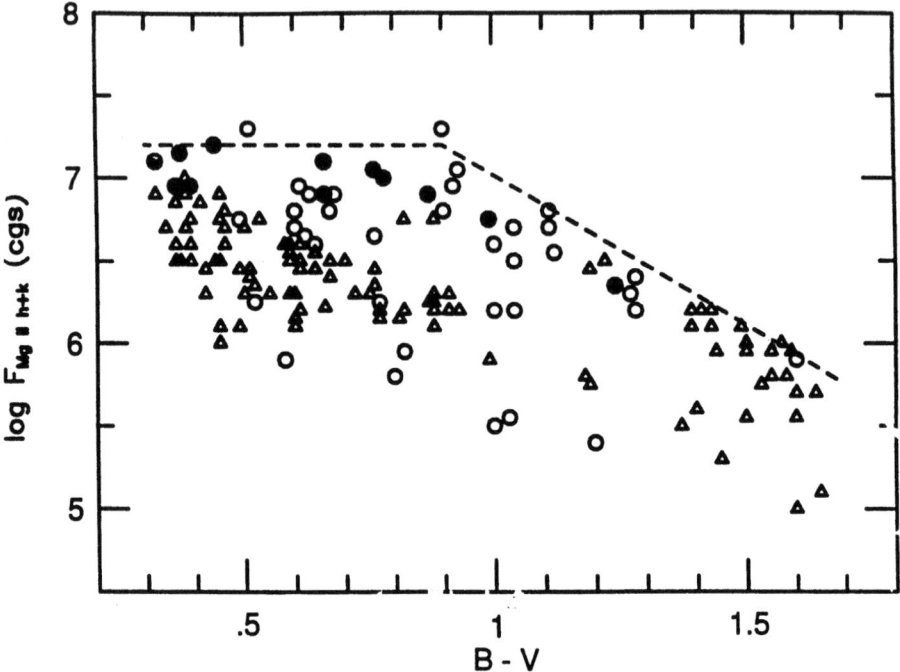

Fig. 10. The surface flux in the emission core of Mg II line at λ2800 after Vilhu et al. (1989). The symbols are as follows: tringles -single stars, open circles – binary components, filled circles – W UMa systems. Note that the W UMa systems are very close to the upper limit of observed fluxes and that the shortest-period W UMa system, CC Com, delineates the red end of known contact binaries.

changes and asymmetries of the light curves (Binnendijk 1965, Bell et al. 1984, Bradstreet and Guinan 1988). For some systems interesting additional periodicities have been discovered of differing degree of credibility: VW Cep: 11.8, 46, 647 and 718 days (Walter 1979, 1983; Leung and Jurkevich 1969), AE Phe: 56 and 100 days (Walter and Duerbeck 1988), W UMa: 500 days (Rigterink 1972). The spots can also offer the simplest explanation for the W-type light-curve phenomenon (Sec. 2.3). We know one case of a system (TZ Boo) which switched from an A-type curve to a W-type one in a relatively short time (Hoffman 1978). This requires to postulate that huge spots formed alternately on each of the components of TZ Boo and produced this large change.

The spots have some unpleasant properties. Not only do they spoil light curve solutions but are very difficult to describe in a quantitative way. Attempts to model them are also very questionable as solutions are usually quite far from being unique. For that reason we should concentrate on systems with particularly convenient geometry in order not to have too many free parameters to handle simultaneously. Fully eclipsing systems with $i \simeq 90°$ may be one such category. Other systems may be those with extreme mass-ratios. For example SW Lac with $q = 0.88$ is very close to being an equal-mass binary system, yet its less massive component seems to have a slightly higher surface brightness. In other words, SW Lac "knows" that it should be a W-type system in spite of the very small difference between components. Stepien (1980) pointed out the importance of SW Lac in this context. SW Lac is one of those systems which show the most violent and largest changes of their light curves (Rucinski 1968).

Period changes observed in almost all W UMa systems (Kreiner 1977) are frequently explained by activity-related phenomena, mostly because of the characteristic time-scale for sudden changes of the order of a few years which is reminiscent of the solar-cycle time-scale. The period changes have a typical size of about $\delta P/P \simeq 3 \times 10^{-5}$ and very much look like random-walk variability with the alternating sign of δP. Hall (1990) in his excellent review of period changes and magnetic cycles established a typical length of a cycle in the W UMa systems of the order of 50 years. The data that he presented suggest a very broad distribution ranging between 20 and 80 years with a few cycles as long as 180 years. The cyclic variability of the periods can be explained by many phenomena which may operate in contact systems. The simplest mechanism would be a simple re-arrangement of mass and an associated change in moments-of-inertia. Such mass flows may very well be controlled by magnetic fields.

6. Internal Structure and Evolution

6.1. GENERAL THEORY OF THE CONTACT MODEL

The first models of contact configurations were computed by Lucy (1968a) on the assumption that the turbulent convective motions are able to transfer any necessary luminosity to make entropy (or, simplifying very considerably, the surface temperature) equal in the whole common envelope[6]. Having this luminosity as an adjustable parameter, he was able to construct stable contact models. Two solutions were possible, one with two identical stars (which is not interesting and not observed), and one with quite dissimilar stars. The second, the interesting solution required that the binary components be different and *not* mutually related via a homology relationship. In simple terms, they could not be scaled versions of one another.

Data on the evolutionary state of W UMa systems were very sketchy at the time of Lucy's models and he elected to make them out of Zero-Age Main-Sequence stars. It was a very reasonable assumption, especially for the first attempt. And then a difficulty emerges almost immediately: such components cannot be easily made too different! Lucy solved this by assuming that the nuclear energy sources are somewhat strange. This forced the internal structures of both components to obey different homology relations but did not satisfy everybody. Also, the models

---

[6] This assumption would restrict contact solutions only to cool stars with convective envelopes. Existence of hot contact systems is then a problem: either the entropy is equal there for other reasons or efficient turbulent convection is somehow induced by the contact environment.

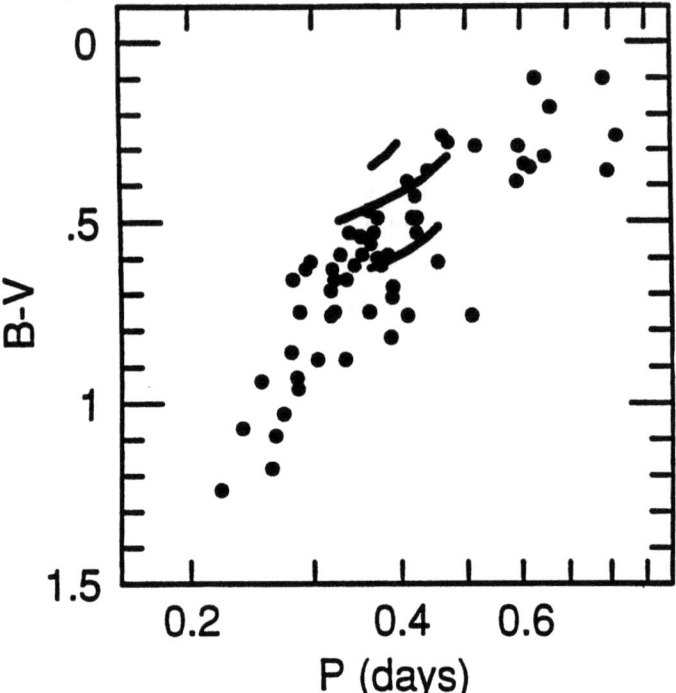

Fig. 11. Three sequences of models computed by Lucy (1968a) superimposed on the period-colour diagram. Note the very short extent of these sequences.

could reproduce only very short segments of the P-C relation (Figure 11). Soon afterwards models with different reasons for dissimilarity started emerging, among them the most obvious reason being the differential evolution. Much activity in the 1970' concentrated along this line of approach. Many points remained not clear, however: Was the more-massive component more evolved while all the time in the contact binary? Or, rather, did the contact system form from two differently evolved, but originally detached, stars? Or, perhaps still other reasons induced dissimilarity of components?

6.2. THE TRO MODEL

Almost 15 years ago Flannery (1976) and Lucy (1976) showed that by permitting normal, MS components in a system to exchange mass (in addition to energy) one can considerably increase the range of solutions. The dissimilarity between stars was in this case induced by the mass-transfer effects. If the mass-transfer between components were permitted, the matter would normally flow from the less-massive to the more-massive component. This flow generated large gravitational terms in the energy equation in the envelope (mass outflow and mass setttling, respectively) which made both stars very much unlike each other. The mass transfer within the contact configuration took place in the thermal time scale of the less-massive component. The interesting property of this flow was that - in spite of its direction - it would *not* lead to the eventual disappearence of the secondary component. Instead, an oscillation in the mass-ratio would set up and its pattern would be as follows: With the assumption of the angular-momentum conservation, the flow from less- to more-massive component leads first to enlargement of the orbit and the eventual breaking of contact. At this stage, the roles reverse. The less-massive component shrinks inside its Roche lobe because it is cut off from the energy of the primary. The primary, in turn has too much energy, expands and starts transfering mass to the secondary. The evolution enters a semi-detached stage, which is relatively rapid, in the thermal time-scale of the more-massive component. The orbit starts shrinking until contact is re-established and everything repeats again. Note that the two branches of the cycle – for typical systems with unequal components ($q \neq 1$) – have quite different durations.

The above described model of evolution through cycles of thermal instability has been given the name of Thermal Relaxation Oscillation (TRO) model. It seemed for a while to be the real solution to all difficulties. In particular, it permitted a very wide range of stars to form contact structures. There was even a reason why TRO would decay with time and stable contact systems possibly form: the nuclear evolution, quietly progressing inside the primary component, would make stars different enough to stabilize in one configuration. Maybe, it is the right model. It has only two uncertain or questionable points: (i) the semi-datached stage is not observed among shortest-period W UMa systems, (ii) the model assumes conservation of the angular momentum. Whereas the first objection can be defended by pointing out incompletnesses in statistics, short duration of the semi-detached stage, etc., the second objection is more substantial. We know that contact binaries are very magnetically active. They should therefore lose angular momentum. With a continuously shrinking orbit, the oscillations should simply die out. Besides, the mass-transfer in the semi-detached stage could also lead to the angular-momentum loss.

There is no question that some elements of the TRO model must be right. Two stars connected together in a contact system must evolve through exchange of both energy and mass. But, there is also no question that the angular-momentum loss must be addressed as a separate subject.

6.3. ANGULAR-MOMENTUM LOSS

One can look at the evolution of binary stars as an evolution of angular momentum. The binaries form because pre-stellar clouds have too much angular momentum. Then, both single stars and binaries, can only lose the momentum which leads to slowing down of rotation or to shrinking of orbits. Since W UMa systems contain least angular momentum among binary stars, they may be the most direct product of evolution of this quantity.

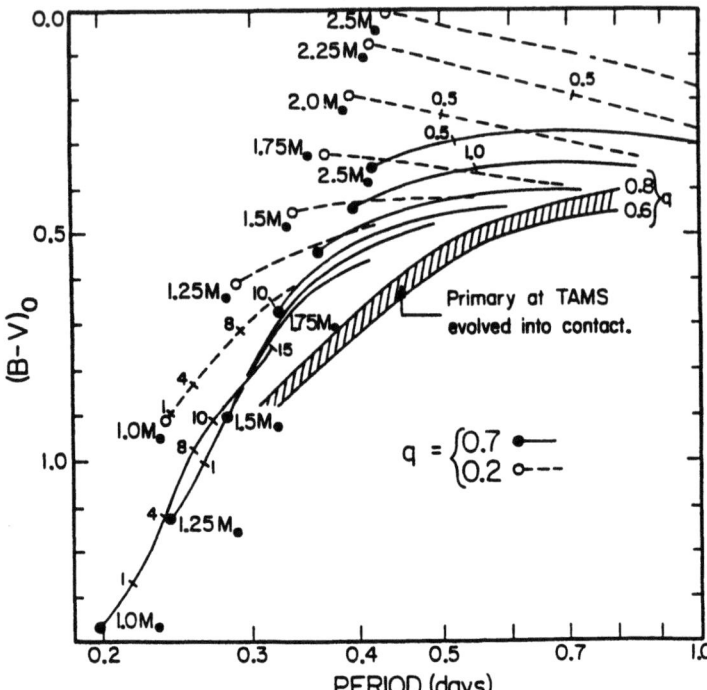

Fig. 12. The period-colour diagram obtained by Mochnacki for systems evolving under the influence of the gravitational radiation AML (Figure 9 in Mochnacki, 1981; by permission of the author). Note that the observed distribution of systems on this diagram (Figs. 4 and 11) can be easily explained that way even without making any recourse to the magnetically driven AML.

The angular momentum loss (AML) may play a role not only in formation of contact binaries, it may be very important also during their evolution. Thus, originally detached stars on wide orbits may be forced in contact. And, once in contact, they may be forced to shrink their orbits and transform *within* the somewhat restricted regime of conditions permissible for the contact systems. As was shown by Rahunen (1981), by an appropriate tuning of the AML rate one can always keep components of a system in good contact, without any trace of the TRO cycles. If the contact stage is to last long, the tuning of the AML must be very precise indicating that some sort of self-regulation operates here.

We should note that even quite modest amounts of the AML may produce relatively large effects on parameters of contact systems. For example Mochnacki (1981) was able to populate practically the whole observed period-colour diagram (Figure 12) by assuming that the gravitational radiation is the only source of the AML.

Eventually, the process of the continuing AML must obviously lead to merging of stars and production of rapidly-rotating giants. A consistent scenario of such evolution up to and including the giant stage, was first given by Webbink (1976). His scenario of evolution up to the giant stage found much support with identification of FK Com stars as possible results of contact-binary merges (Bopp and Rucinski 1981, Bopp and Stencel 1981). Such stars apparently exist also in NGC 188 (Harris and McClure 1985; Kaluzny and Shara 1987). The rapidly-rotating giants of the FK Com type are expected to quickly shed any traces of their origins by losing much of their inherited angular momentum through very strongly elevated activity.

A possible scenario for formation of contact binaries from initially detached binaries through AML was suggested by Vilhu (1982). It starts with a detached system of two solar-type (i.e. convective-envelope) stars, close enough for the tidal synchronisation of rotation and orbital revolution. The magnetic breaking of the rotation, acting via this synchronism, extracts the angular momentum from the orbital motion: the orbit shrinks. But this leads to an even better synchronism and faster rotation. This, in turn, increases activity of components and an even faster approach. At some point, a contact binary is formed. At this point, a re-arrangement of mass and settling on a particular mass-ratio must happen very quickly, in a dynamical time-scale. A contact system that will emerge will depend on properties of both components, on their evolution, the initial mass-ratio, etc. Most probably, the final mass-ratio will be more extreme so that the orbit may be even wider (in spite of the contact already established). A new life will start for the binary.

It is not clear if the AML mechanism is the only one that can produce contact binaries. Leonard (1991, private communication), building up on his earlier research (Leonard 1989), thinks that collisions of detached binary systems in cores of clusters could explain at least a sizeable fraction of W UMa systems there. Such binaries are known to settle toward the cluster cores and their collisions are quite likely because of the large cross-sections (effectively equal to orbital separations). After the violent process of interaction and exchange of components, the typical end product is a contact binary and two single stars. There should exist and obvious relation between efficiency of this process and the cluster richness; maybe we see some indications of this in the high numbers of W UMa systems in NGC 188 and Tom 2.

Both mechanisms, the AML and binary collision, form contact systems from mature binaries. This late formation of contact systems seems to agree with the data on their kinematics (Guinan and Bradstreet 1988) and their presence in old open clusters. While the kinematics suggest a rather old population, similar to that of the Sun and older (8-10 Gyr), the open cluster data are somewhat confussing. Presence of 7 W UMa systems among 300 brighter stars in the old cluster NGC 188 (Kaluzny and Shara 1987), and of one system in Preasepe (TX Cnc; Whelan et al. 1973) and of three systems in M 67 (Whelan et al. 1979; Gilliland et al. 1991), agrees with the notion of the late formation. But then there is the result of Kaluzny and Shara (1988) that six clusters as old as NGC 188 (or even older, Mel 66) do not have a single W UMa system. This result is not understood at present; apparently other factors than age produce the increadible abundance of contact systems in some clusters. An interesting point is similarity of the W UMa systems in NGC 188. Possibly NGC 188 originally had many detached binaries and many of them simultaneously transformed into W UMa systems. We should note that six (out of seven) systems in NGC 188 have similar periods in the range 0.29–0.34 day (the only exception, star "v-5", has a different, β Lyr-type light curve and $P = 0.586$ day).

The above observation suggests that the contact stage is relatively short one and that, in order to evolve in comparable time scales, the systems and their progenitors must have similar parameters. The visual binary system composed of two W UMa binaries, BV Dra ($P = 0.35$ day) and BW Dra ($P = 0.29$ day) consists of two systems which seem to be somewhat more different (Kaluzny and Rucinski 1986). This would suggest a longer duration of the contact stage.

6.4. TOWARDS A GLOBAL PICTURE

The pre-Main Sequence and Main Sequence evolution of W UMa systems remains a mystery. Actually, we do not know if it is possible for a W UMa system to form in the pre-Main Sequence stage and then always stay in contact. Since the stellar evolution has a distinct symmetry in the pre- and post-Main Sequence stages (in that pre-MS and post-MS stars are always larger than on the MS), a configuration with *stonger* contact must precede a contact MS configuration. The present data seem to suggest that Zero-Age-Main-Sequence contact binaries do not exist, at least among solar-mass (and lower-mass) systems. Possibly, strong over-contact in the pre-MS stage leads to coallescence. But, have we checked enough young clusters to be certain of the statistics?

The most likely scenario is formation from (slightly) evolved, detached systems, either via AML or via binary collisions. This would explain the increased density of W UMa systems in old clusters and kinematic characteristics of an old population. This point is the most evident for more-massive, A-type systems with spectral types A5-F5 (Guinan and Bradstreet 1988) which belong to the kinematic old population but whose primaries are simply too massive to be unevolved after 8-10 Gyears. These large masses most probably resulted from mass re-arrangement while in contact configuration. Possibly, they started as detached binaries with components of similar (about solar) mass and then, after establishing contact, re-configured into more extreme mass-ratio, early-type systems.

The life-time of a contact binary may be quite considerably prolonged by the ability to adjust the mass-ratio to the most convenient value. We do not know what is the principle controlling this "most convenient" value but the AML is almost certainly involved here. Note that some sort of self-regulation must apply to the AML: if it were too efficient, contact binaries would quickly merge into single stars; if it were not efficient enough, they should show the TRO cycles. Because of this self-regulation and general independence of the initial conditions, the contact configuration may be a common "graveyard" of many different types of binaries.

Guinan and Bradstreet (1988) call this a "funneling mechanism" driving all binaries into the same domain of short-period contact binaries with little "memory" of their formation.

A number of questions remain to be addressed. Among them, we consider the following as most interesting:

1. Absence of contact binaries with periods $P < 0.22$ day and temperatures lower than that corresponding to B-V=1.24 (CC Com). Definitely, gravitational radiation (Paczynski 1967) at such short periods does have a strong influence on their evolution providing an additional mechanism of the AML and such binaries may not last long. But, if their frequency of occurence is a constant fraction of all dwarfs and since late-type dwarfs are very numerous, there should be many very short-period W UMa systems. One cannot eliminate this problem by pointing out that at such low temperatures the contact systems will not form because of the homology restiction discussed in Sec. 6.1. (the stars being too similar to each other when they have extensive convective zones). At about K5 the radiative cores are still quite large and stars are definitely *not* fully convective (cf. eg. Table V in Rucinski (1988) where sizes of radiative core of late-type dwarfs are tabulated). It can be shown that the full-convection limit (the "Hayashi line" for contact binaries) is a hard one and systems cannot exist with temperatures lower than those corresponding to about $B - V \simeq 1.5$ (Rucinski 1992). But there is still a large gap between this limit and the currently known, reddest systems.

2. Preponderence of genuine contact systems for $P < 0.45$ day and a large number of almost contact binaries with periods close to this limit (Hilditch, King and McFarlane 1988; Hilditch 1989; series of paper by Kaluzny in Acta Astronomica in 1983-1986 with summary in Kaluzny (1985); series of papers by Yamasaki *et al*, mostly in Publ.A.S.Japan starting in 1984, see also Yamasaki *et al.* (1990)). Existence of many short-period ($P \simeq 0.45$), Algol-type systems with very tight orbits yet not in contact is especially puzzling. Why have they stopped short of going into contact when predictions suggest strongest loss of angular momentum at this stage and fastest evolution towards contact? Are these systems representations of the semi-detached branches of the TRO cycles? Further studies of the interesting binary V361 Lyr (Kaluzny 1990; see also discussion after Hilditch (1989)) which may be a contact system either in the formation stage or in the broken-contact TRO stage may be especially relevant for this problem.

3. A self-regulation of AML for contact systems. What keeps the AML rate at exactly necessary level so that W UMa binaries do not coalesce too soon and do not show the TRO oscillations (at least for $P < 0.45$ day)?

4. Presence of contact binaries in globular clusters. Previous estimates of Webbink (1979) suggested that, mainly because of the gravitationa-radiation decay, it should be exceedingly difficult to find W UMa systems in globular clusters. Yet, a new discovery of apparently quite normal W UMa systems in NGC 5466 may require a revision of this view.

7. Conclusions

The W UMa systems are of particular interest for studies of the angular momentum loss (AML) due to the magnetized wind. Because of the peculiarity of the AML from tidally-locked binaries, such stars revolve and rotate progressively faster as they lose angular momentum. This makes them even more active up to a high, probably 'saturated' level in the braking mechanism. It appears that the AML has the driving role in the formation and subsequent evolution of the W UMa systems. Because in their majority these systems consist of low-mass components, their AML-controlled evolution may be at least comparable in its rate – or even faster – than their nuclear evolution. Thus, our understanding of these systems crucially depends on the progress in the field of the "solar – stellar" connection. Unfortunately, the third factor – in addition to the AML and chemical evolution – complicates the picture. These are basically invisible (taking place inside the thick common envelope) but powerful energy and mass exchanges between components. These flows can modify moments-of-inertia and blur the simple relation between the angular momentum and the main observable, the orbital period.

In the galactic field, the W UMa-type systems are a relatively rare type of binary systems. Their frequency in the field is only one per about 1000 ordinary dwarfs. This scarcity can be understood as either due to the short duration of the W UMa stage or to difficulties of formation at the low end in the angular momentum distribution of binary systems. Of particular interest are the recent discoveries of numerous W UMa systems in some of the old open clusters and in globular clusters. Possibly, some (or maybe all) systems form by collisions of detached binaries in cores of clusters.

Acknowledgements

I would like to thank Dr. Nancy Evans for reading the manuscript and contributing interesting and useful comments. Special thanks are also due to Dr. Stefan Mochnacki for interesting discussions and for the permission to use one figure from his paper.

ACCRETION DISKS IN CLOSE BINARIES

E. MEYER-HOFMEISTER and H. RITTER

1. Introduction

With the improvement of observational techniques in astronomy the importance of accretion processes in objects of quite different nature, ranging from double stars in a close orbit to active nuclei of galaxies, was recognized. Wherever a surplus of angular momentum keeps mass from falling directly onto a central object an accretion disk might be formed. This can happen around single objects (for example in connection with star formation or probably in active galactic nuclei) as well as in binary systems. Mass transfer in close binaries is a classical example. There accretion disks can occur in systems of various combinations of main-sequence stars, evolved stars or compact stars as primary and secondary component in a close orbit. The accretion onto a compact star is especially important because the accretion luminosity is proportional to the inverse of the radius of the mass-gaining star. For a neutron star the release of gravitational energy from accreted matter per unit mass is much higher than the release of energy from nuclear burning of the same amount of hydrogen to helium and heavier elements. Thus, accretion disks can be very bright. They are, therefore, very important for our understanding of the physics displayed in binary systems.

In Section 2 we will give a short description of the geometry of the equipotential surfaces of close binary stars, the 'Roche model', consequences for mass overflow and the formation of accretion disks. The fact that probably at least half of all stars are in binary systems, of which many will undergo mass transfer during their evolution shows the importance of the investigation of these processes. As this chapter is somewhat complementary to others in this book which are devoted to different classes of close binaries, we wish first to evaluate those systems in which we would expect an accretion disk (Section 3). Although we know about the existence of accretion disks for some classes, such a systematic compilation might be helpful. The observational evidence for disks is then discussed in Section 4.

The properties of geometrically thin disks in close binaries are best known from investigations of cataclysmic variables and X-ray binaries. The vertical structure of stationary disks and their evolution with time as well as hydrodynamic models are discussed in Section 5. Disk instabilities are of special interest. The comparison of theoretical models with observed light variations allows us to check our knowledge of the physical processes which produce the time-dependent features (Section 6).

Recent opportunities for observations in many wavelengths (UV, optical, X-rays of different hardness) demand increasing effort to get good theoretical spectra for accretion disks (Section 7).

Theoretical models of disks which are not geometrically thin (due to a high rate of mass transfer), and self-gravitating disks are only discussed very briefly here. The same is true for the boundary layer between disk and central star.

2. Roche Geometry, Mass Transfer and Disk Formation

We give here only a brief description of basic concepts. These ideas are included in several review articles (references will be given later in connection with the problems discussed). Of special interest to the reader might be the very instructive books "Accretion power in astrophysics" by Frank *et al.* (1985) and "Interacting binary stars" (Pringle and Wade, 1985), where many of our points of interest are discussed in more detail.

In a binary, mass transfer from one star to the other star can occur in different ways: (1) If one component ejects mass in a stellar wind and a part of that material is captured gravitationally by the nearby companion, we have "wind accretion". (2) If the binary orbit is sufficiently close, matter from the outer layers of one star can flow directly to the companion ("Roche lobe overflow"). A consequence of the latter process is a rather high specific angular momentum of the transferred material, relative to the accreting star. The chances for disk formation are therefore much higher than in the case of wind accretion.

A basic tool for describing mass overflow in close binaries and the formation of accretion disks is the "Roche model" (Kopal 1959). In this model the total potential (gravitational and centrifugal forces) is approximated by the Roche potential Φ_R. For Φ_R it is assumed that the mass distribution of both stars is spherically symmetric (usually a good approximation due to the fact that the stars are centrally condensed), that the stars rotate synchronously and that the orbit is circular. Equipotentials are displayed e.g. in Frank *et al.* (1985, Figure 11). The innermost equipotential surface which encloses both stars defines the critical "Roche lobe" of each star. The corresponding value of Φ_R will be denoted by Φ_{crit}. The section of Φ_{crit} with the orbital plane looks as a figure-of-eight. In the following we denote the mass of primary (originally the more massive star and therefore more evolved) by M_1, the mass of secondary by M_2, the mass ratio by $q = M_1/M_2$ (note that q is sometimes defined in the inverse way, as e.g. in Frank *et al.* (1985)).

The Roche equipotential surface $\Phi_R = \Phi_{crit}$ provides a very useful tool for distinguishing three basic types of binaries.

A binary system is called 'detached' if none of the stars fills its critical Roche lobe or 'semi-detached' if one star fills it. If both stars fill or overfill their critical Roche lobe simultaneously the binary is called a 'contact system'. For more information on the Roche geometry see Kopal (1959) and Mochnacki (1984). To get an idea of the geometry in various double stars an approximative formula for the size of the critical Roche lobe is helpful (Paczyński 1971). Denoting by $R_{L,2}$ the radius of a spherical star (Roche radius) which has the same volume as the largest equipotential closed around star 2, we have approximately

$$\frac{R_{L,2}}{a} = \begin{cases} 0.38 - 0.20\log q & \text{for } 0.05 < q < 1.25 \\ 0.462(1+q)^{-1/3} & \text{for } q > 1.25 \end{cases} \quad (1)$$

where a is the orbital separation.

The value $R_{L,1}$ (for star 1) follows from the same formulae after replacement of q by q^{-1}.

As a consequence of mass overflow in a semi-detached binary the mass ratio and with it the Roche radii change. The change of the radius of the mass losing star is important for the question of whether the mass overflow is stable or not. Since a detailed discussion of stability of mass transfer is rather involved we refer the reader to Webbink (1985).

Solution of the equations of the restricted three-body problem shows (e.g. Flannery 1975) that a single test particle which flows over towards the accretor (into an otherwise empty Roche lobe) traces a non-closed, self intersecting and highly excentric trajectory in the orbital plane around a sufficiently small accretor. In the Roche model it is straightforward to compute the specific angular momentum of the overflowing matter with respect to the gainer. If this is a compact star (a white dwarf, a neutron star or a black hole), the specific angular momentum of the overflowing gas is too high for the gas to reach the star directly. Instead, because the single particle trajectory intersects itself, the gas flow undergoes self interaction as a consequence of which it circularizes. If, on the other hand, the accreting star is sufficiently extended, as are e.g. the gainers in some Algol systems, the single particle trajectory and thus also the gas stream from the donor will impact on the accreting star.

In cases where a more or less circularized flow around the accretor forms, viscosity will subsequently lead to a spreading of the flow in radial direction, in which there is a net outward flow of angular momentum at the expense of a net inward flow of matter. In this way an accretion disk is formed (e.g. Lynden-Bell and Pringle 1974). In such a disk matter spirals continuously inwards from the outer rim to the inner edge (close to the accretor's surface) and angular momentum is at the same time transported continuously outward to the outer edge (compare Pringle 1981), where tidal interaction with the secondary keeps the disk from overflowing the Roche lobe. More detailed investigations (Paczyński 1977; Papaloizou and Pringle 1977) show that the disk will remain inside about 80%–90% of the gainer's Roche radius. This is also confirmed by detailed two-dimensional hydrodynamic simulations (see Sect. 5.3 for references).

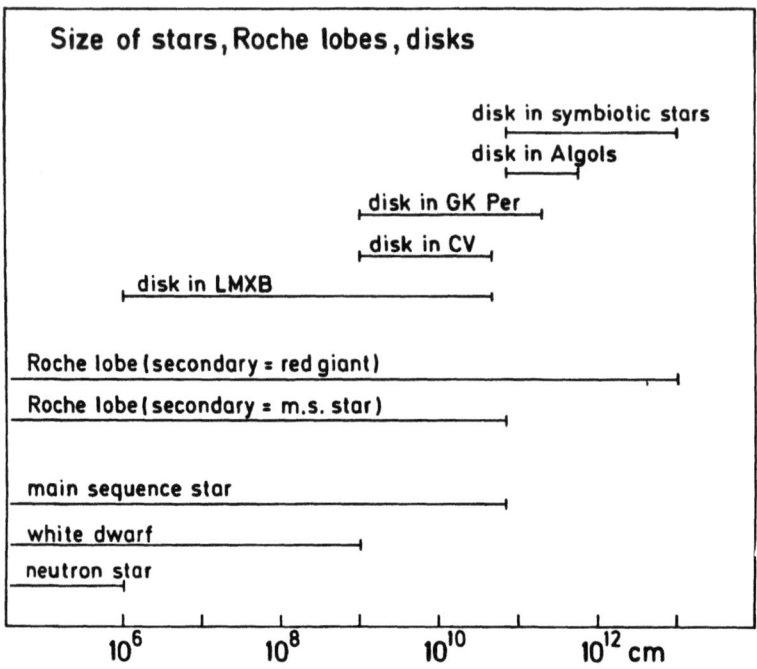

Fig. 1. Comparison of the typical sizes of possible accretors (m.s. star, white dwarf, neutron star), of the Roche radius of possible donors (m.s. star, giant) and of the accretion disks in related binaries.

3. Existence of Accretion Disks in Close Binary Systems

We expect the formation of an accretion disk wherever the specific angular momentum of material flowing towards a star is so high that it cannot directly fall on it. The more compact and therefore the smaller this star, the better is the chance for the formation and persistence of an accretion disk.

In Figure 1 we compare typical radii of possible accretors (compact stars or a main sequence star of $\sim 1 M_\odot$) and of possible donors, i.e. either a main sequence star or a giant with the space available to a disk around the accretor in different types of binaries. (Note that the maximum size of the accretion disk is, to a good approximation, of the same order as the critical Roche radius of the donor.) In Algol systems, disk formation around the non compact secondary is marginal. The occurrence of accretion disks in Algols was investigated and reviewed by Smak (1989). Observations are consistent with the theoretical view according to which disk formation is restricted to systems with a gainer of small radius. (For further information on accretion disks in Algol systems see Batten (1988)).

Members of close binary systems might undergo mass transfer at different stages in their evolution. In an evolution which starts with both components on the main sequence, it is always the more massive component which evolves faster and thus fills its critical Roche lobe first and starts transferring mass to its companion. In many cases, the primary's remnant after the first mass transfer will later become a compact star. If in such a system at a later time the secondary starts transferring mass back to the now compact primary, this will almost invariably result in the formation of an accretion disk around it. For a more detailed description of the evolution of binary stars see e.g. Webbink (1985).

We list in Table I examples of binaries in different evolutionary phases, with emphasis on systems with an accretion disk. Some classes are heterogeneous in the sense that, although the constituent objects are phenomenologically similar, the underlying binaries are not of a uniform type, as e.g. in the case of symbiotic stars, where the accretor may be a main sequence star or a white dwarf. The class about which we have the most detailed information are the cataclysmic variables (CVs), where disks are most thoroughly studied, observationally and theoretically (see Section 5). Disks in symbiotic stars are similar to those in CVs but on a larger scale (for a review of symbiotic stars see Kenyon (1986), for the role of disks in these systems, see Duschl (1988)). Apart from the frequent combination compact star/main sequence star we find also binaries with a larger separation containing a compact star and a star which has reached a more advanced evolutionary stage (e.g. GK Per or some long-period low-mass X-ray binaries). AM Her stars which probably do not contain a disk are discussed in the chapter by King and Lasota (this book). The closest systems are the ones which consist of two compact stars. Double white dwarfs, in particular, have recently become of interest as possible supernova type I precursors. But their intrinsic faintness as detached systems and their short lifetimes as semi-detached ones makes it extremely difficult to find them. Only semi-detached systems with a very low-mass white dwarf as a donor have been found so far (one LMXB and a few CVs, see e.g. Ritter 1990). These are the most compact binaries with an accretion disk.

Besides the accretion via Roche lobe overflow there is also the possibility of accretion from a stellar wind. This is particularly relevant for detached systems consisting of an O/B star and a neutron star (massive X-ray binaries). For a detailed description of wind accretion see Frank *et al.* (1985), for a review of X-ray binaries White (1989).

4. Primary Observational Evidence for Accretion Disks

Direct observational evidence for accretion disks in binary systems or elsewhere, say in form of a spatially resolved picture of such a disk, does not (yet) exist. This is because even in the nearest disk-containing systems, i.e. cataclysmic binaries (CBs), the angular size of the disk is far below available angular resolution. At a distance of $D \gtrsim 50$pc and with a radius of $r_D \lesssim 1 R_\odot$, the disk subtends an angle $\theta_D = 2r_D/D < 0.2$mas. Thus all evidence for accretion disks must necessarily be indirect. Consequently, we must now ask what is the primary (indirect) evidence for accretion disks (in binary systems). Basically, there are two main pieces of

TABLE I
Synopsis of close binaries with and without an accretion disk.

primary \ secondary	main-sequence star *)	evolved star**)	white dwarf	neutron star or black hole
main-sequence star *)	[binary T Tauri stars] [RS CVn stars] Algols (AD) (TAD) {W UMa stars = contact systems}	symbiotic stars Type I as e.g. CI Cyg, Z And, AR Pav (AD) Algols (AD), (TAD)	*)main-sequence star or slightly evolved **)evolved star, but not yet a compact star [ ] detached systems	
evolved star**)	[Wolf-Rayet binaries] [binary planetary nebulae]		(AD) evidence for an accretion disk (TAD) evidence for a transient accretion disk	
white dwarf	[pre-cataclysmic binaries] non-magnetic CVs: UX UMa stars (AD) dwarf novae (AD) DQ Her stars (AD) AM Her stars	long period CVs as GK Per (AD) recurrent nova (AD) symbiotic stars (AD) symbiotic novae (AD)	[double white dwarfs] AM CVn stars (AD)	
neutron star or black hole	massive X-ray binaries (AD) (wind accretion) low mass X-ray binaries (AD) HZ Her/Her X-1 (AD) SS 433 (AD)	long period low mass X-ray binaries (AD)	[binary pulsars] 4U1820-30 (AD)	[binary pulsars]

Comments: in semi-detached systems the mass gaining star is listed as the primary
in detached systems the more evolved star is listed as the primary

evidence which lend strong support to the existence of "geometrically thin, roughly axisymmetric luminous gas flows in the binary's orbital plane and in prograde rotation around one component" i.e. accretion disks, in systems such as CBs, Algol systems or low-mass X-ray binaries (LMXBs). These two pieces of evidence are 1) the results of light curve analysis of eclipsing systems and 2) the interpretation of the line profiles of the observed spectra. In fact, it was because of such observations that the entity 'accretion disk' became part of the now 'standard model' of a CB or a LMXB. In the following, we shall discuss briefly in more detail the observational evidence for accretion disks and its limitations.

4.1. ANALYSIS OF THE LIGHT CURVES OF ECLIPSING BINARIES

High-speed photometric observations of eclipsing CBs and LMXBs (e.g. Warner 1988) provide a unique opportunity for studying the relative arrangement, the angular extent (in orbital phase) and the relative brightness of various luminous (and dark) components, such as a disk, a hot spot, the accreting star and the mass donor, in the binary system. Given a suitable model, an observed light curve can then be decomposed into the various contributions. Recent examples for the application of this technique are OY Car (Schoembs et al. 1987, Wood et al.

1989a), Z Cha (Wood et al. 1986; Warner and O'Donoghue 1988; O'Donoghue 1990) and U Gem (Zhang and Robinson 1987). As an example, we show in Figure 2 the light curve of OY Car and its decomposition. Using the standard model of a CB (or a LMXB where appropriate) light curve decomposition yields the disk's relative brightness and its photometric radius (in units of the orbital separation). With repeated observations of dwarf novae over an outburst cycle (or longer) one can even monitor temporal variations of the photometric disk radius (derived from the hot spot's position), as e.g. in Z Cha (O'Donoghue 1986; Zola 1989), U Gem (Smak 1984a) and IP Peg (Wood et al. 1989b).

A potentially more powerful but at the same time also more computer demanding technique for light curves analysis is the eclipse mapping technique (or eclipse tomography) pioneered by Horne (1985). With a minimum of assumptions, this technique allows the reconstruction of a two-dimensional brightness distribution (i.e. temperature distribution) of the eclipsed object, i.e. here of the disk and the hot spot. The radial temperature profile of the disk which one can derive from this may then be compared with corresponding theoretical predictions. Systems to which this technique has been successfully applied are OY Car (Wood et al. 1989a); Z Cha (Horne and Cook 1985; Wood et al. 1986; Warner and O'Donoghue 1988); VZ Scl (O'Donoghue et al. 1987); RW Tri (Horne and Stiening 1985).

4.2. Spectroscopic Observations, Interpretation of Line Profiles

The visual spectrum of a typical dwarf nova in minimum is characterized by prominent emission lines, mostly the Balmer series and HeI lines. If the system in question has a high orbital inclination (inferred either from the occurrence of an orbital hump or eclipses), these emission lines are often double-peaked, with one peak blue-shifted and the other red-shifted from the line center. Very early in the study of CBs (see e.g. Kraft, 1963), these double-peaked emission lines have been associated with gas circling around the compact star, because the rotational motion (of the gas in the disk), via the corresponding Doppler-effect, accounts in a natural way, at least qualitatively, for the observed line profiles. If one accepts the disk-hypothesis, the one thing that one can learn immediately from the changes of the line profiles through an eclipse is that the gas flow is in *prograde* rotation. Apart from this, a detailed and quantitative interpretation of the observed spectra in terms of an accretion disk suffers from a number of serious problems: First, the observed spectra are not pure disk spectra, but rather a composite of the spectra of all light sources in the system. Second, because of our insufficient knowledge of how and where in the system the emission lines are formed it is virtually impossible, without making far-reaching assumptions in the first place, to extract quantitative information about the gas flow producing these lines. To make things even more complicated we note that there are a number of high-inclination systems (such as V1315 Aql, SW Sex or DW UMa, for references see e.g. Ritter 1990) in which the emission lines are not double-peaked or at least not around the whole orbit. On the

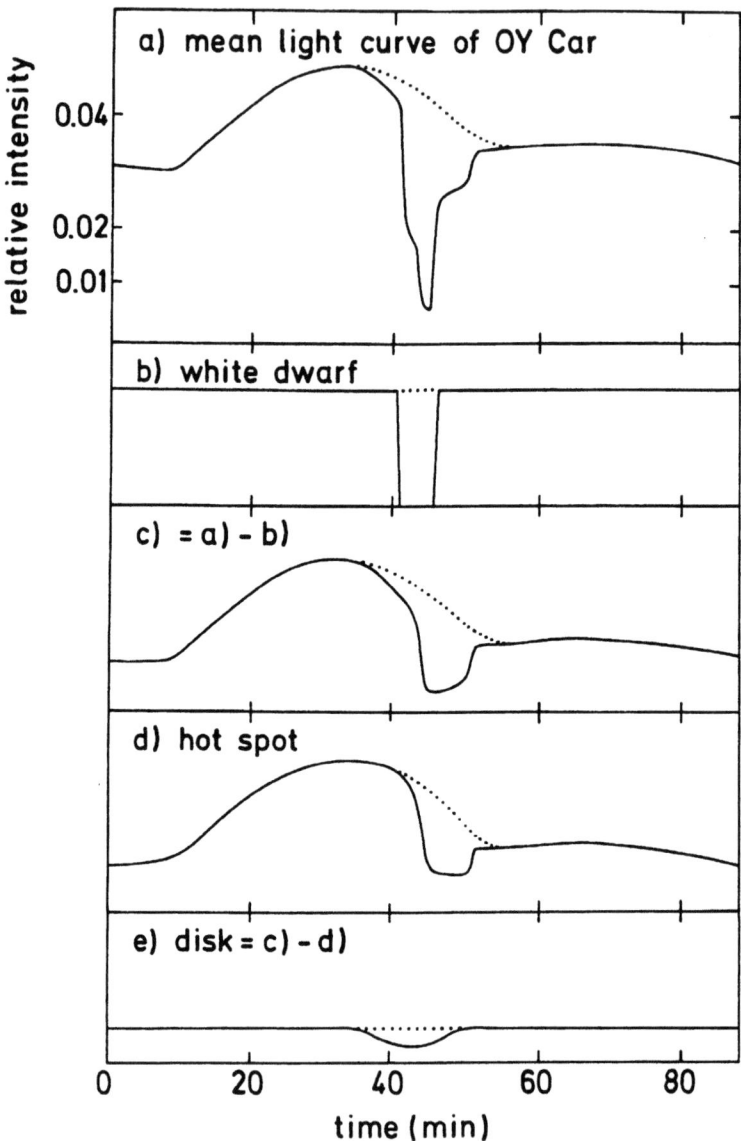

Fig. 2. Mean light curve in B of OY Car in quiescence (a) and its decomposition into contributions from the white dwarf (b), the hot spot (d) and the accretion disk (e). Dotted parts show the light curve ignoring the eclipses. After Schoembs et al. (1987).

other hand we observe double-peaked emission lines from systems which we are almost certain do not contain an accretion disk. A case in point is the AM Her star EF Eri (Bailey and Ward 1981). Thus the presence or absence of double-peaked emission lines by itself is not an unambiguous indicator for the presence or absence of a disk.

Despite all these caveats, the analysis of emission line profiles is a frequently used method to extract more detailed information about the physical properties of the accretion disk. A particularly promising method is the Doppler imaging technique (or Doppler tomography) which uses the changing of the line profiles through an eclipse to reconstruct a two-dimensional map of the disks emissivity in a given spectral line in velocity space (Marsh and Horne 1988). This technique has recently been applied to U Gem (Marsh et al. 1990) and IP Peg (Marsh and Horne 1990).

Let us return now to the question posed at the beginning, namely which is the evidence for accretion disks. At a first glance, from what has been said above, it appears not to be very compelling, particularly because every single observational feature could also be explained in different ways. However, the attractiveness of the disk model derives from the fact that an accretion disk can account for numerous different observations in a natural and physically plausible way. The disk hypothesis gets even more convincing in the context of explaining observational phenomena which we have not yet even mentioned, i.e. the different types of dwarf nova outbursts. We shall return to these matters in more detail in Section 6.

5. Basic Theory of Geometrically Thin Disks

5.1. STATIONARY DISKS

In Section 3 several examples of binaries were given where we expect the formation of an accretion disk. Because of viscosity, the material which flows over from a Roche lobe filling secondary star will quickly dissipate all relative motion perpendicular to the orbital plane and all highly non-circular motion. It settles to the orbital plane and circles around the primary with an angular frequency that is in a good approximation given by the Keplerian value $\Omega_K = (GMr^{-3})^{1/2}$, ($M$ mass of accreting star, r distance from the center of the star).

The viscosity which operates in disks in close binaries is clearly non-molecular (this has been inferred from the time scale on which dwarf novae decline from outbursts (Bath and Pringle 1981)). Different physical processes were studied as a possible mechanism for the transport of angular momentum. The most obvious mechanism appears to be some kind of turbulence as it occurs e.g. in convective regions. However, the mechanism which could drive such a turbulence in those parts of the disk which are convectively stable is still a matter of debate. Some recent investigations dealing with turbulence-induced viscosity are due to Cannizzo and Cameron (1988), Canuto et al. (1988), Scott (1990). Dynamo-induced small-scale magnetic fields were suggested alternatively as responsible for the friction (Eardley and Lightman, 1975). Hydrodynamic instabilities (Drury 1977; Papaloizou and Pringle 1984, 1985; Narayan et al. 1987) have also attracted considerable attention, but the associated growth rates are too modest for thin Keplerian disks. An alternative mechanism is angular momentum transfer by shock waves generated by the flow in the tidally deformed potential (Matsuda et al. 1989). But the effective viscosity determined from such numerical simulations of the two-dimensional adiabatic flow is also (too) small.

Since we do not know the origin of the viscosity, the frictional stress $\tau_{r\phi}$ (which has the dimension of a pressure) is usually parametrized following Shakura and Sunyaev (1973) by introducing a dimensionless number α

$$\tau_{r\phi} = \mu r d\Omega/dr = \alpha P \quad , \tag{2}$$

where P is the pressure and μ the effective viscosity. This parametrisation is roughly equivalent to the expression $\mu = \alpha \varrho H v_s$ (ϱ density, v_s sound velocity, H vertical pressure scale height). Eq. (2) does not specify whether P is the total pressure or the gas pressure. Furthermore, α need not be a constant but could, in principle, depend itself on other dimensionless quantities of the problem, such as H/r. (For a more detailed discussion of α prescriptions see e.g. Meyer and Meyer-Hofmeister (1982) or Duschl (1989)). As a result, within the framework of eq. (2), various viscosity parametrisations are possible and have also been used in different investigations.

Given the nature of the central star, its mass and radius, the remaining essential parameter which determines the disk structure is the mass flow rate $\dot{M}$. For typical rates of up to $10^{-7} M_\odot$ yr^{-1} the accretion disk is geometrically thin, $H/r \approx 1/30$, for example in cataclysmic variables. It can be shown that in the thin disk approximation, i.e. if $H/r \ll 1$, the radial pressure gradient is small compared to the radial gravitational force and the radial heat flow is small compared to the vertical heat flow. Under these circumstances, the radial and vertical structure is largely decoupled. For a more detailed description of the thin disk approximation the reader is referred to the reviews by Pringle (1981) and Frank et al. (1985).

The vertical structure of the accretion disk can therefore be treated at each distance r in the same way as the structure of a star. The three equations defining the vertical structure are those of hydrostatic equilibrium, energy balance (viscous heating/cooling) and energy transport (radiative or convective). The balance between heating and cooling includes the generation of heat from the viscous stresses, the only relation through which the radial coupling between adjacent rings enters explicitly. The solution of the differential equations yields pressure, temperature and energy flux density as functions of the height z above the midplane. The height of the photosphere (optical depth = 2/3) follows from computations. The boundary values are determined by a photospheric boundary condition (as in stellar structure computations). The effective temperature T_e at the photosphere is related to the mass flow rate $\dot{M}$ and the viscosity integral

$$f = \int_{-\infty}^{\infty} \mu dz \qquad (3)$$

$$\sigma T_e^4 = \frac{9}{8} \frac{GM}{r^3} f = \frac{3}{8\pi} \frac{GM\dot{M}}{r^3} \qquad (4)$$

The relation between f and $\dot{M}$ used in (4) holds only for stationary disks with zero net angular momentum flow. The vertical structure also yields the mass in the disk, described by the surface density

$$\Sigma = \int_{-\infty}^{\infty} \varrho \, dz \qquad (5)$$

Equation (4) shows how the temperature increases radially in the disk towards the inner edge. The boundary conditions in radial direction are determined at the outer edge of the disk by the mass overflow from the secondary star and at the inner edge by the specific angular momentum accreted on to the central star. In very cool regions, $T_e \lesssim 4000$K, the disk can become optically thin.

Vertical structure calculations have been performed by several authors, starting with the work of Cannizzo and Wheeler (1982), Faulkner et al. (1983), Meyer and Meyer-Hofmeister (1982, 1983b) and Smak (1982). The dependence of the resulting $f - \Sigma$ relation on α-prescriptions, opacity values, more sophisticated atmosphere calculations (including the viscosity in the optically thin regions or frequency-dependent opacities) was investigated by many authors. A recent investigation with references to other calculations is Mineshige and Wood (1990). Much of this work was done with the aim to improve models of dwarf nova outbursts. We show results for typical mass flow rates in Figure 3.

The main results of the vertical structure calculations are as follows. The hot disk has radiative energy transport and Σ decreases with decreasing $\dot{M}$. But for $T_e \lesssim 10^4$K partial ionization of hydrogen sets in. The corresponding increase of the opacity enforces energy transport by convection. This, in turn, results in a lower midplane temperature (note that the effective temperature at the photosphere is determined by $\dot{M}$), a lower viscosity and an increase of surface density. Finally for sufficiently low effective temperature the surface density decreases again with decreasing T_e. This non-monotonic change of $\Sigma(T_{\text{eff}})$ (compare Figure 4) has important consequences for the stability of the disk. For disks around neutron stars, and accretion rates, $\dot{M} \gtrsim 10^{-10} M_\odot \text{yr}^{-1}$ radiation pressure becomes important at small radii and leads to another non-monotonic change of the surface density with temperature. This is, however, not the case if the friction is coupled to the gas pressure alone (cf. Eq. (2)).

For relatively high mass overflow rates ($\dot{M} \approx 10^{-6} M_\odot/\text{yr}$) which seem to be relevant for the disks in symbiotic stars, H/r may no longer be small. In geometrically thick disks the radial gradients $\partial P/\partial r$, $\partial T/\partial r$ become important

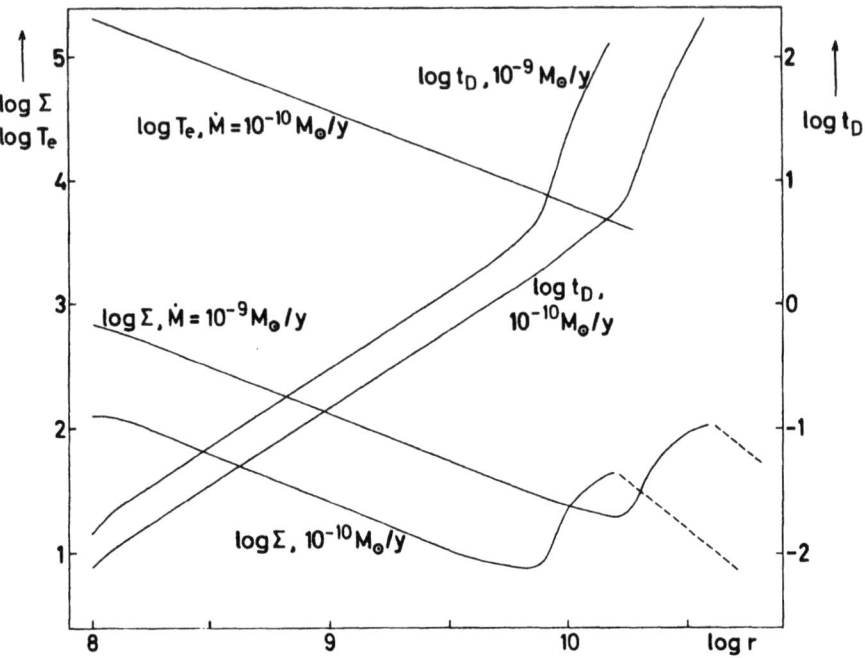

Fig. 3. Surface density Σ as function of radius r for different mass flow rates $\dot{M}$, effective temperature T_e and diffusion time $t_D = 2\pi r^2 \Sigma/\dot{M}$ in days. Dashed lines: optically thin regions (viscosity parameter $\alpha = 1$).

(compare Frank et al. 1985). The so-called "slim" disks are in between thin and thick disks. This approximation is relevant for disks close to a black hole. For "slim" disk models see Abramowicz et al. (1988) and Kato et al. (1989).

In disks around a neutron star there is an additional effect, which changes the vertical structure essentially. The irradiation from the central star leads to a heating of the disk's photosphere and reduces therefore the temperature difference between photosphere and midplane (Meyer and Meyer-Hofmeister 1990; Tuchmann et al. 1990).

For a modelling of disks in Algols see Hubeny (1989), and for disks in symbiotic stars Duschl (1988).

5.2. EVOLUTION OF NON-STATIONARY DISKS - DISK INSTABILITY MODEL FOR DWARF NOVA OUTBURSTS

Calculations of the vertical structure of an accretion disk yield for given values of radius r and mass overflow rate $\dot{M}$ (or equivalently f or T_e) the surface density Σ. The resulting relation between f and Σ, the viscosity-surface density relation $f(\Sigma)$ has an "S"-shape in the temperature region of partial hydrogen ionisation. The

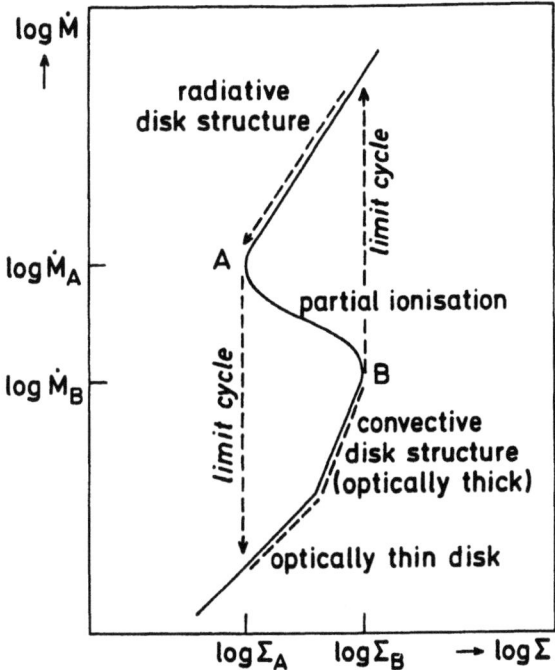

Fig. 4. Schematic drawing of the viscosity-surface density relation (for stationary disks equivalent to an $\dot{M} - \Sigma$ relation). The limit cycle between hot and cold state is indicated (see Section 5.2).

intermediate branch with a negative slope $\partial f/\partial\Sigma$ is unstable against disturbances of the stationary radial mass flow (Lightman 1974) and also unstable on the much shorter thermal timescale against disturbances of the vertical thermal structure (e.g. Meyer 1984). For values of surface density Σ in the interval $\Sigma_A < \Sigma < \Sigma_B$ (see Figure 4) we have two stable branches $f(\Sigma)$, one associated with the hot radiative structure and the other with the cool convective (or optically thin) structure, and an unstable branch where the gas is partially ionised. This means that for a given range of surface density Σ the disk structure can be in three different states, depending on the mass flow rate. A stationary disk can be stable for both, a high ($\dot{M} > \dot{M}_A$) or low ($\dot{M} < \dot{M}_B$) mass flow rate, but not for the intermediate rate ($\dot{M}_B < \dot{M} < \dot{M}_A$) which corresponds to partial ionisation. If the rate of mass overflow from the secondary falls in the latter range no stationary disk structure is possible. Instead the structure alternates in a limit cycle between the hot and the cool branch (as indicated in Figure 4). This limit cycle works coherently at different r. The coherence is mediated by transition waves which separate regions of hot and cold state. The matter within these waves ("heating front" and "cooling front") changes on a thermal timescale from one state to the other (Meyer 1984). The change between high and low mass flow rate results in a high and low luminosity of the disk, which, in turn, has been associated with outburst and quiescence in cataclysmic variables (Meyer and Meyer-Hofmeister 1984). We describe this time dependent evolution of the disk in the following. The values of α appropriate for this evolution are derived from observed time scales of the dwarf nova outburst cycles. These yield $0.1 \lesssim \alpha \lesssim 1$ (Bath and Pringle 1981); for special systems α can be even smaller than 0.01, as estimates for GK Per (Cannizzo and Kenyon 1986) or for symbiotic stars (Duschl, 1988) show.

The evolution of the accretion disk with time follows from the diffusion equation for the surface density Σ, derived from conservation of mass and angular momentum (Lightman 1974). In the regions of transition from the cool to the hot state or vice versa one needs, in addition, an equation which describes the relaxation to thermal equilibrium on the thermal timescale.

In time-dependent outburst calculations this complex process was modeled in different approaches (Cannizzo and Wheeler 1982; Meyer and Meyer-Hofmeister 1984; Papaloizou et al. 1983; Mineshige and Osaki 1983; Smak (review 1984b)). Such models give a good description of the observed light changes in dwarf novae (for a comparison with observations see Section 6). Furthermore, detailed models, including a code for computing the spectral evolution (see Section 7) were constructed to allow for a better comparison with the observations. In Figure 5 we show as an example a theoretical lightcurve for VW Hydri.

We may summarize the main results of the disk instability model as follows: outbursts will occur if the mass overflow rate $\dot{M}_{ov}$ from the secondary is such that the corresponding disc structure would have $\partial f/\partial \Sigma < 0$, i.e. lies on the intermediate temperature branch of the triple valued $f - \Sigma$ relation for any value r in the disk. To avoid this, i.e. to have $\partial f(r)/\partial \Sigma(r) > 0$ everywhere, the disk must be either in the hot or cool state everywhere. To find out whether a stationary disk can be stable or not for a given mass overflow rate one has to compute the effective temperature as a function of r.

As has already been mentioned above, irradiation can essentially influence the vertical structure. The "S" shape of the $f - \Sigma$ relation can be changed so much that the unstable branch disappears. The above described dwarf nova instability is then replaced by a diffusive instability which can lead to long period light changes (Meyer and Meyer-Hofmeister 1990, Mineshige et al. 1990).

5.3. HYDRODYNAMIC SIMULATIONS OF ACCRETION DISKS

Real accretion disks in binaries are non-axisymmetric vortex flows. Realistic models, therefore, are necessarily 3-dimensional. The models discussed above, however, are all axisymmetric and thus at most two-dimensional. Since a detailed modelling in three dimensions is not yet feasible, the study of non-axisymmetric disks requires sacrifice of one dimension, the obvious choice being the vertical coordinate. Basically two types of numerical approaches have been used for studying the hydrodynamics of a two-dimensional accretion flow in the Roche potential:

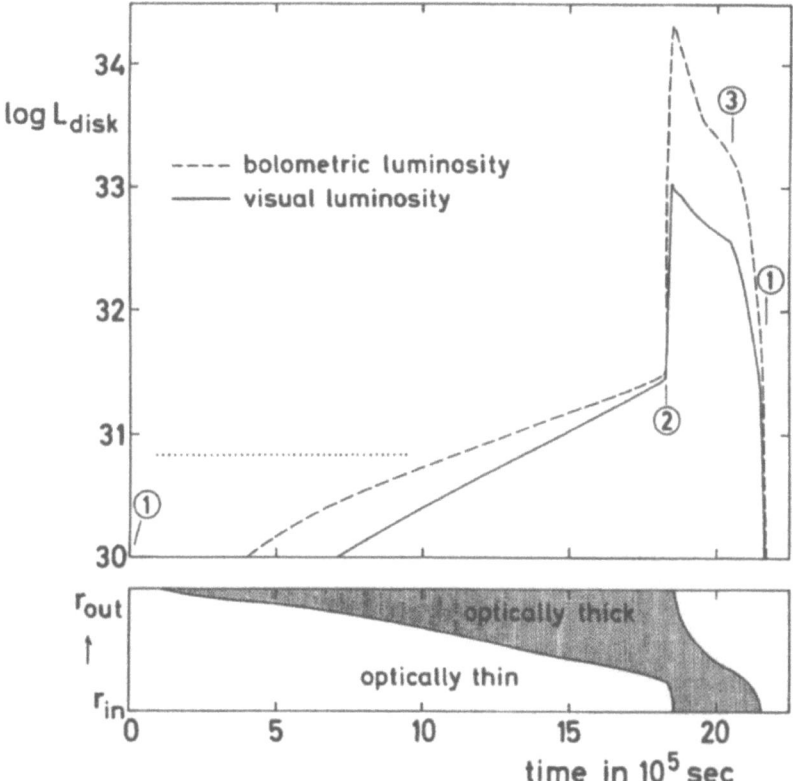

Fig. 5. Theoretical lightcurve for VW Hydri (from Meyer and Meyer-Hofmeister 1988) (1) state of lowest surface density, (2) and (3) start of heating and cooling wave. Dotted line: estimate for contributions from hot spot, white dwarf, secondary star. Below: extent of the optically thin and thick regions in the disk.

The first one is the "particle in cell" (PIC) method introduced by Lin and Pringle (1976) and subsequently used by a number of authors (most recently by Whitehurst (1988) and Hirose and Osaki (1990); see also references given therein). Since the disk is assumed to be geometrically thin, only the motion in the orbital plane has to be computed. The motion of the fluid is followed by solving the ordinary differential equations of the restricted three-body problem for many particles, representing fluid elements, in the Roche potential. The effects of viscosity are taken into account through averaging at each time step the velocities of the particles which are within a pre-specified cell. As an example of such a calculation, we show in Figure 6 the density distribution in the orbital plane of a disk at 6 equidistant times as computed by Hirose and Osaki (1990).

The other way of computing non-axisymmetric accretion flows consists of solving the partial differential equations of hydrodynamics in a given potential, as has e.g. been done by Różyczka (1985) and Sawada et al. (1986). In both papers

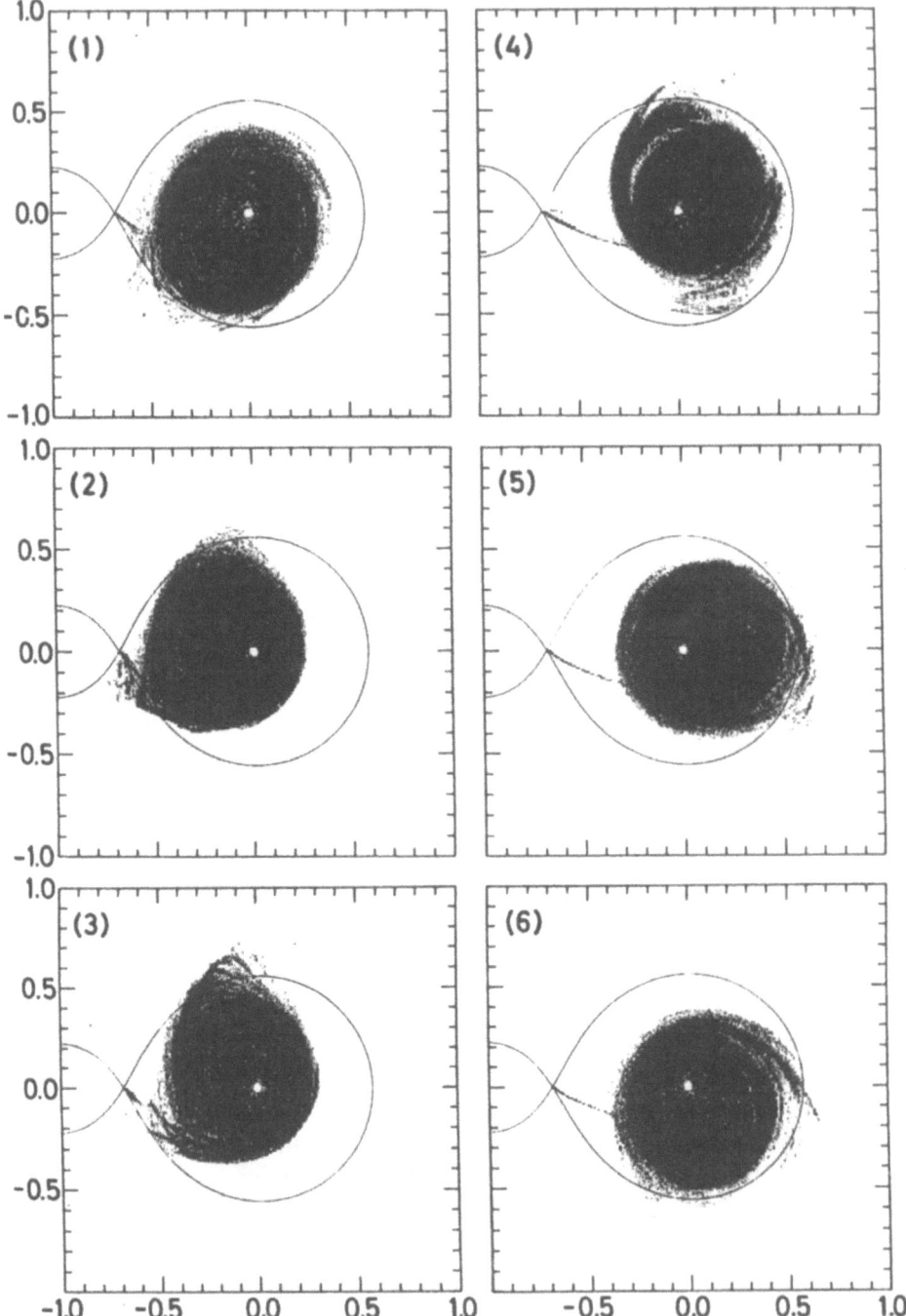

Fig. 6. Density distribution of matter in the accretion disk at six phases in the superhump period at equal time intervals from Osaki and Hirose (1990) (computed as the results in Hirose and Osaki (1990), Figure 7a). The mass ratio of the binary is $q = 6.67$, the scale in units of the orbital separation.

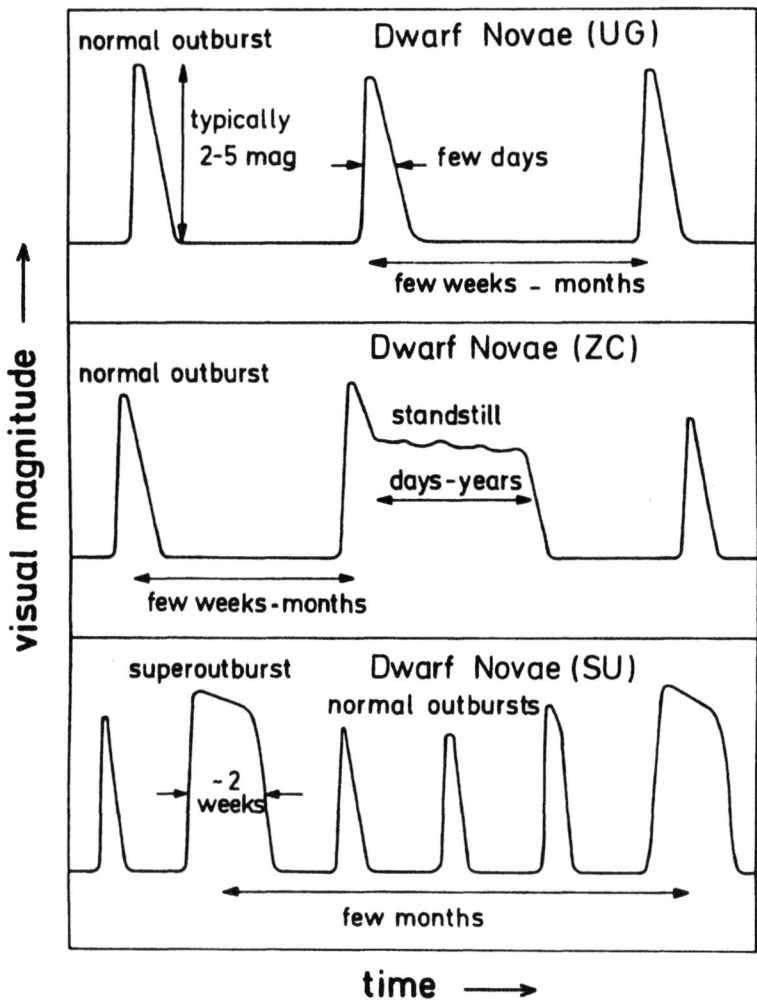

Fig. 7. Schematic outburst light curves of the three dwarf nova subtypes U Gem stars (a), Z Cam stars (b) and SU UMa stars (c).

the advantages of the hydrodynamic calculations over the PIC methods are also discussed. References to earlier work may also be found in these two papers. More recently, Schwarzenberg-Czerny and Rózyczka (1988) and Rózyczka and Spruit (1989) have studied the response of an accretion disk to tidal forces using a 2-D second order explicit hydrodynamic code. These computations show that tidal forces not only truncate the disk's outer radius and make it non-axisymmetric but also that they induce spiral shock waves in the disk (for further information on spiral shocks see e.g. Spruit 1991).

6. Accretion Disk Phenomena: Theory Versus Observations

Among the numerous phenomena that we associate now with the presence of an accretion disk, dwarf nova outbursts are certainly the most important and informative ones. Therefore, in this section, we shall mainly examine to what extent accretion disk theory in general and the disk instability model (DIM) in particular can account for the observed phenomena among CVs. However, before doing so, we shall summarize briefly the basic outburst properties of dwarf novae.

6.1. PHENOMENOLOGY OF DWARF NOVA OUTBURSTS

Dwarf novae (DN) are a major subclass of CVs. Their defining characteristic is that phases of low luminosity ("quiescence") alternate with phases of high luminosity ("outbursts") in a semi-regular way. Depending on characteristic details of the visual light curves, schematically shown in Figure 7, one distinguishes three main subclasses of dwarf novae, namely U Gem stars (UG), Z Cam stars (ZC) and SU UMa stars (SU). In the following we shall give a brief description of the associated phonomenology. We note in passing that apart from dwarf novae also other types of CVs (e.g. VY Scl stars, AM Her stars) show high (bright) and low (faint) states. However their light curves are sufficiently distinct from those of dwarf novae that no confusion should arise.

The main characteristics of the three dwarf nova subclasses can be summarized as follows (see also Figure 7):

1. *U Gem stars* (ordinary dwarf novae) show outbursts with an amplitude of $\Delta V \approx 2 - 5$ magnitudes which last for a few days. They repeat semi-regularly at intervals of a few weeks up to a few months. Usually the rise to outburst is steeper than the decline from it.

2. *Z Cam stars* are distinguished from U Gem stars by showing an additional phenomenon: the standstills. Occasionally on a decline from an outburst the object remains at an intermediate brightness, typically about 1 mag below maximum, for a more or less extended time, which can last from a few days to years, before it returns eventually to quiescence. Except for the standstills, Z Cam stars are like U Gem stars.

3. *SU UMa stars* are distinguished from U Gem stars by showing a second type of outbursts, the superoutbursts, in addition to normal ones. Superoutbursts occur less frequently than normal ones, typically every few months. During a superoutburst, which lasts typically for about two weeks and thus significantly longer than normal ones, the system is also brighter by about $\Delta V \approx 1$ mag. than during a normal outburst, hence the name superoutburst. The superoutburst light curve shows yet another phenomenon, one which has almost become the defining characteristic of this subclass: the superhumps. Superhumps are a periodic modulation of the system's brightness with an amplitude of a few tenths of a magnitude which appear only during a superoutburst. Most

remarkably, the period of this modulation, the superhump period P_{SH}, is in all cases longer by a few percent than the orbital period of the system. As the superoutburst progresses P_{SH} increases slightly whereas the amplitude of the modulation decays versus the end of the outburst (see e.g. Vogt 1980).

6.2. THE SITE OF THE OUTBURST

Before around 1970 it was not clear where in the system the brightening associated with a dwarf nova outburst occurs. Possible sites in discussion were the secondary star, the white dwarf or the accretion disk. However, with the introduction of high-speed photometry at the beginning of the 1970s, the ambiguity was quickly resolved. High-speed photometry of the eclipsing dwarf novae Z Cha by Warner (1974) and OY Car by Vogt (1983) have shown unambiguously that the site of the outburst is the accretion disk. Accordingly, an outburst is the result of a brightening of the disk which, in turn, is the result of an increased mass flow rate through the disk. Now, the next question to ask is which mechanism modulates the rate of mass flow through the disk? Basically there are two possibilities: In the first one the disk is unstable in the sense discussed in Sect. 5. In the second one, the mass flow from the secondary is not continuous but rather spasmodic. The first model is referred to as the disk instability model (DIM), the second one as the mass transfer burst model (MTBM). We shall discuss both of them, first, and in some detail, the DIM and second, at the end of this section, briefly the MTBM.

6.3. BASIC PREDICTIONS OF THE DISK INSTABILITY MODEL

Application of the disk instability discussed in Sect. 5 to dwarf nova outbursts yields immediately a number of very basic predictions that can be tested by observations. In line with the DIM, we shall assume that the rate of mass overflow from the secondary is not strongly varying on the time scale of an outburst. If this holds one can make the following basic predictions:
 1. Dwarf nova outbursts are restricted to CVs containing an accretion disk. In fact, as far as observations are available, there is no case known in contradiction to this.
 2. Given the system's orbital parameters (M_1, M_2, a), one can define two accretion rates $\dot{M}_A$ and $\dot{M}_B$ (see Figure 4) such that the disk instability occurs only if the rate of mass overflow $\dot{M}_{ov}$ falls in the interval $\dot{M}_B < \dot{M}_{ov} < \dot{M}_A$. If this is the case, a stationary disk with $\dot{M} = \dot{M}_{ov}$ would contain extended zones of partial H-ionisation (see Sect. 5).
 3. Systems containing a disk will show no outbursts if either $\dot{M}_{ov} > \dot{M}_A$ or $\dot{M}_{ov} < \dot{M}_B$.
    Although determination of $\dot{M}_{ov}$ from observations is a tedious task and the results are plagued with huge uncertainties, observational verification of predictions 2) and 3) has been attempted (e.g. Smak 1983) and yielded satisfactory results.

4. Dwarf nova outbursts can occur almost independently of the type of secondary star. Since almost all secondaries are unevolved (e.g. Ritter 1986), this means that the occurrence of outbursts does not depend on the spectral type, i.e. effective temperature, of the secondary and thus not on the orbital period. This is also consistent with observations.
5. The brightness of the hot spot does not change significantly neither at the beginning of nor during an outburst. Also this prediction is confirmed by observations.

   We note already here that concerning points 4) and 5) the MTBM comes to roughly opposite predictions.
6. Because of the redistribution of mass and angular momentum in the disk during an outburst and the supply of mass in between the outbursts with specific angular momentum that is lower than that at the disk's outer rim, the radius of the disk and thus the location of the hot spot undergo systematic and predictable changes during an outburst cycle. Such changes have been observed (Smak 1984a; O'Donoghue 1986; Zola 1989; Wood et al. 1989b) and found to be in agreement with model expectations.

6.4. DWARF NOVA OUTBURSTS IN THE LIGHT OF THE DISK INSTABILITY MODEL

The DIM with its basic predictions listed above provides the natural framework for understanding the significance of the various subclasses of non-magnetic CVs. In the following we shall describe briefly how these subclasses fit into this framework.

1. *Novalike systems and old novae* have presistently bright, nearly stationary disks and thus a relatively high mass overflow rate $\dot{M}_{ov} > \dot{M}_A$. Observational estimates of the mass transfer rates in these systems do indeed give high values of $\dot{M}_{ov}$ that are consistent with DIM predictions. However, because of the high uncertainties of the measured $\dot{M}_{ov}$ values, no strong conclusions can be drawn. However, if one accepts the basic predictions of the current model for the secular evolution of CVs (see e.g. Ritter 1986; King 1988), according to which mass transfer in systems with periods $P \lesssim 2^h$ is driven by gravitational radiation alone and that, therefore, typical values of $\dot{M}_{ov}$ are on the order of a few $10^{-11} M_\odot \text{yr}^{-1}$, the DIM predicts that none of these systems could have a persistently bright, nearly stationary disk. In fact, all non-magnetic CVs with 80 min $\lesssim P \lesssim 2^h$ are dwarf novae (e.g. Ritter 1990).
2. *Dwarf novae of the U Gem type:* The outbursts of these systems are understood as a consequence of the limit cycle instability discussed in Sect. 5. This means that in these systems $\dot{M}_B < \dot{M}_{ov} < \dot{M}_A$. Numerous quantitative models for specific, well-observed objects, have been computed in the recent past.
3. *Z Cam stars:* Their defining property, i.e. the occurence of standstills, is a bit harder to understand in the framework of the DIM. A plausible explanation has been put forward by Meyer and Meyer-Hofmeister (1983a), hereafter MMH. Accordingly Z Cam stars are systems in which $\dot{M}_{ov}$ is only a little smaller than

$\dot M_A$. Thus, a slight increase of $\dot M_{ov}$ would allow the disk to become stable. In the model of MMH that slight increase is provided by irradiation of the secondary by the disk after it has gone in outburst. Thus the system could be trapped in a state of permanent outburst, i.e. a standstill, as long as the irradiated secondary could sustain an increased $\dot M_{ov}$. Since this time scale is much longer, i.e. of the order 10^5yr, than the observed duration of standstills (typically weeks — few years), the return from standstill to quiescence must be triggered by another mechanism, one which, however, has not been identified with certainty. A star spot moving over the inner Lagrangean point and thus reducing $\dot M_{ov}$ despite irradiation might be a possibility.

4. *SU UMa stars:* Here two separate phenomena, i.e. the superoutbursts and the occurrence of superhumps during them have to be explained. For both phenomena, a satisfactory explanation has been given only very recently. According to Osaki (1989) the occurrence of superoutbursts can be understood as follows: During a normal outburst angular momentum is transported to the outer regions of the disk while matter is accreted on the white dwarf. Thus with every normal outburst the outer radius of the disk increases. This growth of the disk is ultimately limited by tidal interaction with the secondary which limits the disk radius to about 0.8–0.9 of the primary's Roche radius (Paczyński 1977; Papaloizou and Pringle 1977). Now, Whitehurst (1988) discovered that an accretion disk becomes tidally unstable, i.e. its outer parts are then subject to increased tidal dissipation, if the outer radius grows beyond the 3:1 resonance radius. While in systems with moderate mass ratios $M_1/M_2 \lesssim 3$ the 3:1 resonance is always outside the tidal truncation radius, it is the other way around in systems with a high mass ratio $M_1/M_2 \gtrsim 3$. The model of Osaki (1989) postulates that a superoutburst is triggered when the disk radius reaches the 3:1 resonance. Thus the occurrence of superoutbursts is restricted to systems with a high mass ratio, i.e. a small secondary mass and a short orbital period. In this model, the 3:1 resonance radius is reached in the course of a normal outburst. Subsequent increased tidal dissipation initiates a superoutburst during which most of the angular momentum stored in the disk since the last superoutburst is transferred back to the orbit and the disk is essentially emptied. After a superoutburst the disk's outer radius is again small and grows thereafter slowly with each subsequent normal outburst until the 3:1 resonance is reached again. One particularly nice property of this model is that it involves the same key ingredient, i.e. the 3:1 resonance, as Whitehurst's (1988) model for the superhump phenomenon: Numerical simulations performed by Whitehurst and by Hirose and Osaki (1989) show that a disk whose outer radius reaches the 3:1 resonance becomes eccentric and starts precessing in the frame corotating with the binary. As an example we show in Figure 6 results obtained by Hirose and Osaki (1990) where the retrograde precession of the disk is clearly seen. Tidal dissipation is strongest in the tidal bulge of the eccentric disk and thus results in a bright area, i.e. the superhump light source, which precesses around the

white dwarf with a synodic period P_{SH} that is longer by a few precent than the orbital one. In this way the model accounts in a natural way for the fact that the superhump phenomenon appears only in connection with superoutbursts. Both phenomena are related to the 3:1 resonance and thus restricted to systems with a high mass ratio, i.e. a small M_2 and a short orbital period, in full agreement with what is observed.

Apart from the dwarf novae discussed above, the DIM has also been of relevance to two other classes of binaries, namely the LMXBs and the symbiotic stars. In the following we summarize briefly the application of the DIM to these systems:

a) *Low-mass X-ray binaries:* Apart from the fact that in LMXBs the compact component is a neutron star (or a black hole) rather than a white dwarf as in CVs, LMXBs and CVs are very similar. One might therefore ask whether the equivalent of dwarf nova outbursts are observed in LMXBs. This is, however, not the case. The standard explanation for the lack of outbursts is that the strong X-ray/UV-irradiation of the disk by the central source raises the disk temperature everywhere to values above $\sim 10^4$K so that no zones of partial ionisation appear. In fact, computations of the vertical structure of irradiated disks show that the triple-valued $f - \Sigma$ relations disappear if the irradiation is sufficiently strong (see e.g. Meyer and Meyer-Hofmeister 1990; Mineshige *et al.* 1990).

b) *Symbiotic stars:* In these systems the mass donor is a giant. As a consequence the disk around the accretor is very extended. Therefore, such disks are likely to be rather cool, at least in their outer parts and thus particularly susceptible to disk instabilities. Thus, the equivalent to dwarf nova outbursts may occur in some systems. In fact, Duschl (1986a, b; 1988) has successfully modelled outbursts of symbiotic stars by applying the DIM, though it appears that the outbursts observed in some systems are better explained in terms of transient nuclear burning on an accreting white dwarf (Paczyński and Rudak 1980).

As we have shown above, the DIM is quite successful in accounting for numerous observed phenomena associated with dwarf novae. Nevertheles, we should also mention that there are also observations which are not easily accounted for by the DIM.

One case in point is the so-called UV-lag of dwarf nova outbursts, i.e. the fact that for several observed dwarf novae (e.g. VW Hyi) going into outburst, the UV radiation rises 0.5–1 day after the onset of the optical outburst (Verbunt 1987). Early versions of the DIM were unable to reproduce this behaviour. A number of more recent outburst models were computed aiming at exploring this phenomenon. The most recent investigation is by Duschl (1991), where also references to related work can be found.

The other case to be mentioned is the post-outburst light problem, i.e. the fact that some dwarf novae, after having reached quiescence brightness in the optical light, remain relatively bright in X-rays and UV-light for quite some time, indicating that some accretion is still going on, at rates that cannot be accounted for by the DIM. A possible solution has been proposed by Meyer (1990) in terms of a disk evaporation model.

Finally, we note that the fact that early DIMs could not adequately account for the above-mentioned problems has led some authors to reject the DIM of dwarf nova outbursts altogether and rather to favour the alternative MTBM. In our opinion such a conclusion is not well justified since the predictive power of MTBM is not nearly as high as that of the DIM, as we shall see in Sect. 6.5.

6.5. THE MASS TRANSFER BURST MODEL

In contrast to the DIM, the MTBM postulates that dwarf nova outbursts are a consequence of spasmodic mass transfer from the secondary which, in turn, is due to a dynamical instability rooted in the superadiabatic convection zone on the cool secondary. According to Bath (1969, 1972, 1975) and Papaloizou and Bath (1975) this instability is driven by the energy released when ionized hydrogen and helium recombine as the matter in the envelope expands and cools in response to mass loss. In addition to the papers just mentioned, the dynamical stability of mass loss from low-mass lobe-filling stars was also investigated in various degrees of approximation by Wood (1977), Gilliland (1985) and Edwards (1985, 1987, 1988). Unfortunately, these efforts yielded ambiguous results and it is, therefore, not yet clear whether or not the basic mechanism of the MTBM indeed works. Because of this, the MTBM could not be worked out in as much detail as the DIM. Nevertheless, if the basic mechanism of the MTBM works, a number of interesting predictions can be made:

1. Outbursts are possible irrespective of whether the system contains a disk or not. Accordingly some type of outbursts should also occur in AM Her systems which do not contain a disk. This, however, is in contradiction to observations.
2. At the onset of an outburst, the hot spot at the outer rim of the disk should brighten considerably (i.e. roughly in proportion to $\dot{M}$). Since this is in contradiction to observations, proponents of the MTBM (Bath *et al.* 1983) have sought a way out, however, without making the model thus more convincing.
3. The mechanism underlying the MTBM depends on the presence and the extent of a superadiabatic convection zone on the donor. In the absence of such a zone, i.e. if the secondary is sufficiently hot, mass transfer ought to be stable. Thus the occurence and the strength of mass transfer bursts is expected to depend on the donor's spectral type, i.e. on the orbital period of the system.
4. Since giants have much more extended superadiabatic convection zones than low-mass main sequence stars, cool donors of long-period semi-detached binaries, such as Algols, should be much more susceptible to the dynamical instability. Yet there is not a single compelling case of spasmodic mass transfer known among the Algols (see e.g. Hall 1988 for a more detailed discussion).

Apart from these predictions which are either unconfirmed or at variance with observations, there are two, more fundamental, reasons which make the MTBM unsatisfactory. The first one is the simple fact that within the framework of the

MTBM there is no scheme equivalent to the one provided by the DIM (discussed in Sect. 6.4) which allows us to understand, at least qualitatively, the great variety of CVs. In particular, the MTBM does not offer an explanation for the fact that some systems do not show outbursts although others, having virtually identical secondaries (same mass, spectral type), do.

The second and even more fundamental reason is that the DIM is not invalidated by the MTBM. So we must ask, if the MTBM accounts for dwarf nova outbursts, what does the disk instability account for? Or, to turn the argument around, since the DIM can account for virtually all phenomena associated with dwarf nova outbursts, there is no need for the MTBM.

7. Accretion Disk Spectra

7.1. REQUIREMENTS OF DISK SPECTRA

The fact that an accretion disk covers a wide range of photospheric temperatures makes the determination of a disk spectrum more complicated than the determination of a stellar spectrum which results from an atmosphere above the same effective temperature everywhere. For geometrically thin disks one might expect that one could find a reasonable approximation if one takes instead of the disk a series of concentric rings and constructs the disk spectrum from a superposition of ring spectra, each ring taken with its effective temperature (follows for stationary disks from the mass flow rate $\dot{M}$) and its gravitational acceleration (follows from the distance r, the height of the photosphere and the central mass).

But disk spectra differ from stellar spectra due to a number of important features:
1. the gravitational acceleration changes in a different way with the height in the atmosphere,
2. the gas in the disk has a rotational velocity,
3. the differential rotation in line forming regions introduces effects like microturbulence,
4. friction is a source of energy everywhere, also in the atmosphere,
5. the observed radiation depends on the viewing angle under which we see the disk,
6. a disk can be optically thin in a part of its radial extent or even as a whole,
7. irradiation from a central star or the secondary or a corona can be important.

7.2. COMPUTATION OF DISK SPECTRA

7.2.1. *Approximations with Black Body or Stellar-Type Flux Distributions*

In all approximations the disk is replaced by a series of concentric rings. Such rings can then be taken as radiating according to their effective temperature like a black body or a stellar atmosphere. Smak (1984c) used the black body approximation to determine the monochromatic fluxes and compared the theoretically predicted values of B-V versus visual magnitude during an outburst with observations for dwarf novae. An investigation of synthetic spectra of CVs based on black body radiation or alternatively on stellar atmosphere radiation was carried out by Wade (1984, therein also references to earlier work on model disk spectra). The stellar fluxes were adopted from Kurucz's (1979) stellar spectra.

Unfortunately, neither approximation provides a completely satisfactory representation of observed spectra of CVs, but the stellar atmospheres give a better fit of the observed power-law ultraviolet continua. The grid of Kurucz atmospheres has also been used for the modelling of dwarf nova outbursts by Pringle *et al.* (1986), one-temperature Kurucz spectra for a comparison with data for VW Hydri (Hassall 1984). The work by Wade was continued in an investigation of spectra of novalike variables and post-classical novae. For these luminous CVs relative and absolute energy distributions are known and it was tested (Wade 1988) whether black body or stellar atmosphere disk spectra describe the observed ultraviolet energy distribution accurately. The comparison shows that the stellar atmospheres give a better agreement in the colour-colour diagram, than the ones based on black body radiation. But they fail to fit the colours and the flux simultaneously (where the black body models are better).

In their investigation of the spectra of dwarf nova outbursts Cannizzo and Kenyon (1987) used for synthetic disk spectra absolute fluxes and broad band colour indices from a series of observed main sequence stars. These spectra, and also black body spectra, were used to investigate the UV delay (see Section 5.2).

Complementary to the mentioned approximations, radiation from cool optically thin disk regions might be included in the disk spectrum, mainly for dwarf novae during quiescence. For the treatment of optically thin disks see Williams (1980), Tylenda (1981) and for recent work on a non-LTE model Williams and Shipman (1988). A hot atmosphere above the disk (Schwarzenberg-Czerny 1981) yields essential contributions to the high energy part of the spectrum.

The approximations with the help of black body or stellar atmospheres spectra cannot account for the features (1) to (4) listed in Section 7.1. A dependence on the viewing angle, (5), can be included.

7.2.2. *Approximation of the Radiation Transfer in a Disk Atmosphere*

An ideal theoretical model for an accretion disk spectrum would include all the disk-specific features as mentioned in Section 7.1. Obviously this is a task even more complicated than solving the radiation transfer equations for a stellar atmosphere. It is therefore clear that simplifications are necessary to be able to find a solution to the problem.

Kříž and Hubeny (1986) determined the radiation field consistent with the disk structure. They used concentric rings, taken as plane-parallel slabs. The numerical calculations of the continuum spectra are performed for a stationary disk, applied to the observed fluxes from VW Hydri during quiescence. Friction is included according to a simplified description.

In a different approach la Dous (1989) determined continuous and line spectra of accretion disks in CVs for the optical and also the ultraviolet range. The computations are restricted to stationary disks. The radiation from the concentric rings is computed in a LTE approximation as for stellar atmospheres, but the gravitational acceleration appropriate for disks is included. No energy generation by friction is assumed for the atmosphere. Line broadening due to the rotational velocity is included.

Other work along this line are the investigations of Adam *et al.* (1988). Shaviv and Wehrse (1989) computed continuum spectra of accretion disks. They included an energy generation due to viscous forces. The results are compared with the observed energy distribution from RW Sex.

The theoretical modelling of Algol disks is discussed by Hubeny (1989).

7.3. THE PURPOSE OF SYNTHETIC ACCRETION DISK SPECTRA

In many cases of dwarf nova observations the spectra had been compared with theoretical spectra since this allows to estimate the mass flow rate $\dot{M}$ through the disk. It is only possible for a stationary accretion disk. According to our theoretical understanding this method for the determination of $\dot{M}$ should be applicable for nova-like variables, dwarf novae during the high luminosity phase in an outburst (stationary hot disk) and maybe to some dwarf novae in the quiescent state (for low mass accretion rates the cool disks are close to stationary disks).

Generally the comparison of observed and sufficiently good theoretical spectra enables us to learn about the physics of the accretion disk. This is specially interesting for disks which are not stationary, as disks in dwarf novae during the rise to an outburst and the later decline. The observed spectrum mirrors the physical state of the different regions in the disk, the changes between hot and cool structure, as predicted from the limit cycle instability (see Section 5.2). It is expected that the effect of energy generation by friction everywhere in the disk is responsible for main features of spectra from disks.

The improvement of our knowledge of disks in binaries will also help us to understand better the physics in disks on larger scales.

Acknowledgements

The authors wish to thank Y. Osaki for providing us with the original of Figure 6 and H.C. Arp for improving the language of the manuscript.

MAGNETIC CATACLYSMIC VARIABLES

A. R. KING and J. P. LASOTA

1. Introduction

Cataclysmic variables (hereafter CVs) are close binaries in which a white dwarf accretes from a low mass companion, usually close to the main sequence. If the white dwarf has a magnetic field of $\gtrsim 10^5$ G, the accretion flow will differ markedly from that expected in the non-magnetic case (see the chapter by Meyer-Hofmeister). In particular, matter will be channelled by the field on to restricted regions of the white dwarf surface, so that much of the accretion luminosity will be modulated at its spin period. Such systems are thus relatively easy to pick out observationally, and fall into two distinct groups. In one group, the *AM Herculis* or *polar* systems, the white dwarf spins synchronously (or very nearly synchronously) with the binary orbit, i.e. $P_{\rm spin} = P_{\rm orb}$. In the other group, the *intermediate polars* (IPs), the white dwarf spins more rapidly, so that $P_{\rm spin} < P_{\rm orb}$; typically $P_{\rm spin}/P_{\rm orb} \sim 0.1$. (They are sometimes called DQ Herculis systems after a possible prototype, which however has rather atypical properties: see Sect. 4 below.) As we shall see, there has been great progress in understanding these systems in the last decade, particularly the nature of the accretion process. A unique feature of the period distribution (Sect. 6) means that the AM Hers probably offer the best chance of understanding the evolution of all short-period mass-exchanging binaries (all CVs and low mass X-ray binaries, for example).

The nature of the relation between the two groups of magnetic CVs is still not understood, and is one of the major areas of current debate. The evidence which fuels the debate comes in two forms: (i) all AM Her systems are observed to have strong polarized cyclotron emission in the optical, while none of the IPs does, and (ii) AM Hers have $P_{\rm orb} \lesssim 3$ hr, while IPs have $P_{\rm orb} \gtrsim 3$ hr (cf Figure 1). The simplest interpretation of (i) is that the AM Hers have systematically stronger magnetic fields than the IPs. However, since CVs probably reduce $P_{\rm orb}$ from some value $\gtrsim 4 - 10$ hr to ~ 80 min during the course of their evolution (see Sect. 6), the simplest interpretation of (ii) is that IPs evolve into AM Hers (King, Frank and Ritter, 1985) once the binary separation becomes small enough for their spins to synchronize with the orbital rotation (see Sect. 2). Clearly this interpretation is incompatible with a systematic difference in magnetic fieldstrengths, as suggested by the first line of argument. Quite recently we suggested a way out of this difficulty (King and Lasota 1990b), by supplying a reason why non-synchronous systems

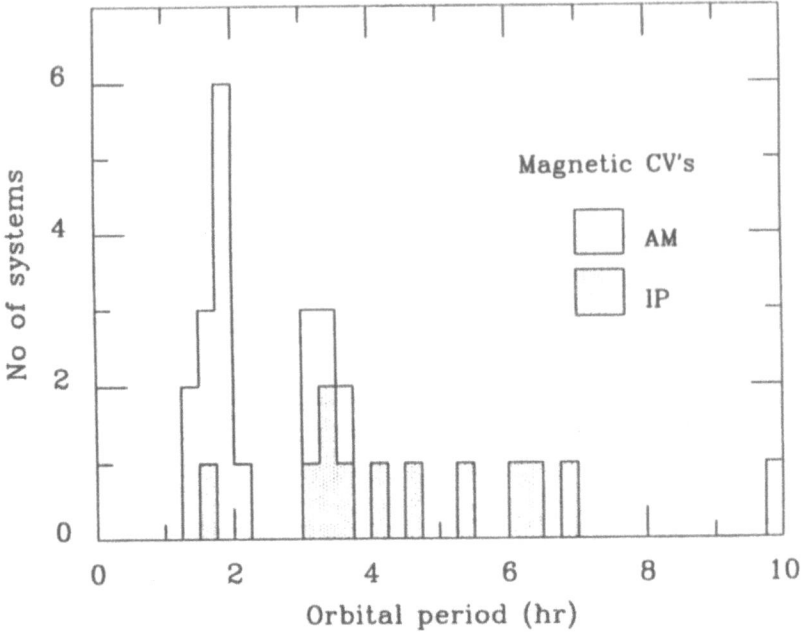

Fig. 1. The period distribution of magnetic CVs.

with fieldstrengths characteristic of AM Hers might be unobservable (see Sect. 7).

The plan of this chapter is as follows. Sections 2 and 3, 4 and 5 respectively discuss the accretion flows and binary geometry in the AM Her and IP systems. Sect. 6 describes the unique insight into binary evolution offered by the AM Hers. Sect. 7 and 8 discuss the relation between synchronous and non-synchronous systems, while Sect. 9 gives reasons to expect rapid progress in the near future. We make no attempt to present a comprehensive review, which would require a much longer chapter: we apologise in advance for the omission of some important topics.

2. The AM Herculis Systems

The AM Her systems have recently been thoroughly reviewed by Cropper (1990). Here we very briefly summarise their main features; the interested reader is referred to Cropper's article for details. The prime observational characteristic of phase-dependent optical polarization implies that the white dwarfs possess magnetic fields of the order $\sim 1 - 3 \times 10^7$ G. This is confirmed by the detection of cyclotron harmonics and Zeeman splittings in various systems. These fieldstrengths imply a radically different accretion geometry from that of most CVs. The magnetospheric radius is much too large for any accretion disc to form, and instead the gas stream from the secondary star must hit the fieldlines, with accretion occurring down fieldlines thereafter. The strength of the field is evidently also the reason that the white dwarf rotation is locked to that of the binary: we consider this question in further detail below.

Most AM Her systems are detected in soft ($\lesssim 0.1$ keV) X-rays, and are the strongest soft X-ray sources amongst CVs. Rather fewer systems are detected in hard ($\sim$ keV) X-rays. Both emissions are thought to arise where the accretion flow impacts the white dwarf surface. Because of fieldline-channelling, the (highly supersonic) inflow is essentially radial, and strong shocks must be involved in decelerating it and allowing it to settle. In the simplest model, (e.g. King and Lasota, 1979; Lamb and Masters, 1979) the gas stream is uniform and a single standing shock forms above the white dwarf surface. Most of the infall energy is then initially released as radiation characterised by the shock temperature

$$T_s = 3GM\mu m_H/8kR. \tag{1}$$

Here M and R are the white dwarf mass and radius respectively, so that (1) gives $T_s \sim 10^8$ K, implying hard X-ray emission. One half of this is directed down towards the white dwarf surface. While some is reflected, a substantial fraction is thermalised. The resultant emission is characterised by the effective temperature

$$T_{\text{eff}} = (L_{\text{acc}}/4\pi R^2 f\sigma)^{1/4}, \tag{2}$$

where

$$L_{\text{acc}} = GM\dot{M}/R \tag{3}$$

is the accretion luminosity and f the fraction of the white dwarf surface which is accreting. Because of the power 1/4, T_{eff} is always of order $\gtrsim 10^5$ K for realistic parameters, giving soft X-ray and EUV emission. The simple uniform stream picture predicts rather more hard X-rays than soft, whereas the opposite is observed, i.e. *the AM Hers have strong soft X-ray excesses* (see e.g. Watson, 1986, Osborne, 1988 and Beuermann, 1988 for reviews). By now it is widely accepted that the reason for this is that the accretion flow is highly non-uniform, much of it being in the form of dense blobs which penetrate the photosphere before being stopped (Kuipers and Pringle, 1982; Frank, King and Lasota, 1988). Under these circumstances most of the emission will be in soft X-rays, as observed. This picture allows an understanding of the abrupt changes sometimes observed in AM Her systems in the character of the soft X-ray light curves; flows with many blobs produce rather smooth sinusoidal modulations as the white dwarf rotates, while flows with rather few (~ 20) blobs give noisy light curves with sharp (sometimes square-wave) edges (Hameury and King, 1988). The fact that AM Her systems can have large phase intervals of zero soft X-ray emission strongly suggests that the blobs accrete over a fairly small area, i.e. that the accreting fraction f is quite small, typically $\sim 10^{-4}$.

The synchronous rotation of the white dwarf is by now well attested from observation of eclipsing systems; the spin and orbital periods differ by less than a few parts in a million. As the accreting matter carries angular momentum and tries to spin up the white dwarf, there must be a torque, evidently magnetic in origin, which resists it. The tight limits on the relative rotation strongly suggest this must be a non-dissipative torque, the most likely candidate being the interaction of the white dwarf's field with that of the secondary (Campbell, 1985). The latter is expected to have a surface field of order 10^3 G, typical for late-type stars (Marcy, 1984), so that the white dwarf and the secondary have magnetic moments $\mu_1 \sim 10^{34}$ G cm^3, $\mu_2 \sim 10^{33} - 10^{34}$ G cm^3. The dipole-dipole torque then transmits the accretion torque on the white dwarf back to the binary orbit. An interesting result emerges from an analysis of the dipole-dipole torque (King, Frank and Whitehurst, 1990): unless both dipoles lie in the orbital plane there is an unbalanced magnetic torque component about an axis in the plane. This would lead to precession of this plane in space with a period of order ~ 100 y, a prediction not supported by observations of eclipsing systems. Since accretion is observed to take place on regions well away from the white dwarf equator, this suggests that higher field multipoles may control the accretion flow near the white dwarf, with the fields becoming dipolar at larger distances from the stars and indeed lying in the orbital plane. Alternatively, the observed dipole positions may not represent torque balance, but result from librations about equilibrium. In this case, the dipoles would oscilate with periods ~ 30 y. Observations over the next few years should settle this question.

AM Her systems display irregular long-term variations in luminosity, in some cases spending months at a brightness level an order of magnitude less than normal. In other cases the systems spend most of their time in faint states with occasional excursions to brighter states. The reason for this behaviour is not understood, although it may be connected with irradiation of the secondary by soft X-rays and EUV from the white dwarf (King, 1989).

3. The Intermediate Polars

Rather less is known about the intermediate polars than the AM Her systems reviewed above. Even the belief that the white dwarfs were magnetic lacked direct observational support until Penning, Schmidt and Liebert (1986) discovered weak optical and infrared polarization in BG CMi, suggesting fieldstrengths at least of order $2 - 10 \times 10^6$ G (Chanmugam et al., 1990) for this system. If this system is typical, IPs evidently have magnetic moments $\mu_1 \gtrsim 10^{33}$ G cm^3, no more than a factor 10 smaller than AM Hers.

Figure 1 shows that here is little overlap in period with the AM Hers. Thus IPs in general have larger orbital separations than AM Hers; even with the same magnetic fields $B_1 \sim 1 - 3 \times 10^7$ G as the AM Hers, simple estimates suggest that the white dwarf could not be locked to the orbital rotation (see Figure 5).

When IPs were discovered, it was naturally assumed that they were the white dwarf analogues of the pulsing X-ray binaries, where a magnetic neutron star accretes from its companion via a disc (see Warner 1983). However, Hameury,

King and Lasota (1986) (hereafter HKL) pointed out that the disc assumption was much less firmly based for the IPs than for X-ray binaries, where the neutron star's magnetic moment is much smaller ($\sim 10^{30}$ G cm^3). The simplest criterion for disc formation is that the initial gas stream from the companion should be able to orbit the obstacle presented by the accreting star, rather than crashing directly into it (see e.g. Frank, King and Raine, 1991 for a review), i.e. the free stream's minimum approach distance $R_{\min}$ to the accreting object should exceed the object's effective size. In non-magnetic systems the latter is just the stellar radius R, but in the magnetic case it is some length R_{mag} given by the gas stream's interaction with the magnetosphere. This is obviously a difficult quantity to calculate, and it is not in any case clear what happens even if the criterion fails, i.e. $R_{\text{mag}} > R_{\min}$. Note that R_{mag} depends on the *instantaneous* accretion rate $\dot{M}$, not its average over evolutionary timescales. HKL suggested that for accretion rates typical of IPs, disc formation would be inhibited in systems with $P_{\text{orb}} \lesssim 5$ hr if $\mu \gtrsim 10^{33}$ G cm^3. In this case, the accreting matter gives all of its angular momentum to the white dwarf on interacting with its magnetosphere, whereas if a substantial disc is able to form, the matter loses most of its angular momentum (ultimately back to the companion star via viscosity and tides) before encountering the magnetosphere.

The direct observational evidence for the existence or non-existence of discs in IPs is not conclusive (see King, Mouchet and Lasota 1991; Hellier 1991). In particular it is perfectly possible that there are disclike objects in IPs, although most of the accretion flow does not go through them. We give an explicit example of such a "non-accretion disc" below.

The most important observational constraint bearing on this problem is the fact, first noted by Barrett *et al.* 1988 and later by Warner and Wickramasinghe 1991 (see also Wickramasinghe, Wu and Ferrario 1991), that there is a significant clustering of spin periods in IPs around the relation $P_{\text{spin}}/P_{\text{orb}} \sim 0.1$. Figure 2 shows a histogram of the observed values of $P_{\text{spin}}/P_{\text{orb}}$ for IPs. The clustering near a ratio ~ 0.1 is noticeable, although if all systems are included, not statistically significant.

This is however highly suggestive that $P_{\text{spin}}/P_{\text{orb}} \sim 0.1$ represents some kind of spin equilibrium for a subclass of the white dwarfs. Support for this idea comes from the observed values of $\dot{P}_{\text{spin}}$ which show both signs for systems in the cluster, and from the rather small measured values of $|\dot{P}_{\text{spin}}|$. Any such equilibrium must result from the exchange between spin and orbital angular momentum in the binary, and will differ according to whether a disc exists or not. We shall show that equilibrium at $P_{\text{spin}}/P_{\text{orb}} \sim 0.1$ arises naturally if most of the accretion, averaged over timescales $\sim 10^4$ y, is not through a disc.

In accreting from a disc, the white dwarf accretes specific angular momentum

$$j_{\text{acc}} = (GM_1 R_{\text{in}})^{1/2}, \tag{4}$$

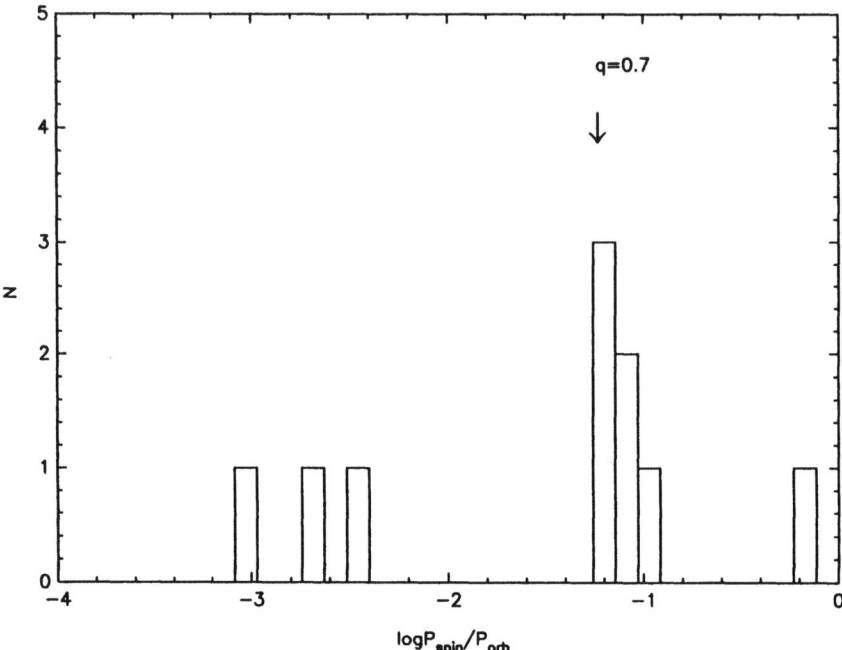

Fig. 2. Histogram of IPs versus $P_{\rm spin}/P_{\rm orb}$. The arrow shows the theoretical relation (15) for $q = 0.7$.

whereas it accretes

$$j_{\rm acc} = (GM_1 R_{\rm circ})^{1/2}. \tag{5}$$

from matter which has not passed through a disc. In both cases, the white dwarf will spin up until centrifugal forces allow it to expel the angular momentum at the rate it accretes; that is, until

$$j_{\rm acc} \simeq (GM_1 R_{\rm co})^{1/2}, \tag{6}$$

where

$$R_{\rm co} = (GM_1/\Omega_{\rm spin}^2)^{1/3} \tag{7}$$

is the corotation radius, with $\Omega_{\rm spin} = 2\pi/P_{\rm spin}$ the spin frequency. Comparing (4) and (5) with (6) we see that spin equilibrium is given by

$$R_{\rm co} \simeq R_{\rm in} \quad ({\rm disc}) \tag{8}$$

and

$$R_{\rm co} \simeq R_{\rm circ} \quad ({\rm no\ disc}) \tag{9}$$

in the two cases. Introducing a dimensionless factor into (9) and (8) would be equivalent to defining a "fastness parameter" (cf Ghosh and Lamb, 1979); since this is currently thought to have a value very close to unity (Wang, 1987), we omit it.

Remarkably, (9) gives a relation between P_{spin} and P_{orb} which is *independent* of the magnetic field (although it must of course be strong enough to prevent the formation of a disc for most of the time). We give an approximate analytic form of this relation by noting that for the mass ratios $q = M_2/M_1 \gtrsim 0.7$ typical of IPs (almost all of which are above the CV period gap) the specific angular momentum characterized by R_{circ} is approximately equal to that leaving the secondary at the inner Lagrange point L_1 (King, Frank and Whitehurst, 1990); only a small fraction is returned to the secondary by its tidal pull on the accretion stream. Thus

$$(GM_1 R_{\text{circ}})^{1/2} \simeq b^2 \Omega_{\text{orb}} \tag{10}$$

where $\Omega_{\text{orb}} = 2\pi/P_{\text{orb}}$ is the orbital frequency and b the distance of L_1 from the white dwarf. We replace the latter by the analytic approximation of Plavec and Kratochvil (1964):

$$b \simeq (0.5 - 0.227 \log q) a, \tag{11}$$

where a is the binary separation. Replacing a by use of Kepler's 3rd law

$$a^3 = GM_1(1+q)/\Omega_{\text{orb}}^2, \tag{12}$$

the combination of (10), (11) and (12) reduces (9) to

$$P_{\text{spin}}/P_{\text{orb}} \simeq (0.5 - 0.227 \log q)^6 (1+q)^2. \tag{13}$$

The relation (13) is shown by an arrow (for $q = 0.7$) on Figure 2. for various values of q. As can be seen, most observed IPs lie very close to equilibria given by (13) for reasonable values of q (~ 0.7). The relation (13) is effectively identical to that given by Warner and Wickramasinghe 1991. However their derivation is based on the incorrect belief that the spin rate of the white dwarf affects the possibility of disc formation, and in any case implies a different range of applicability of the result: our derivation requires μ large enough that $R_{\text{mag}} > R_{\text{circ}}$, whereas theirs requires μ small enough to ensure exactly the opposite.

We see that the observed relation (13) follows naturally if IPs do not accrete predominantly via discs. Episodes of disc accretion will not significantly disturb the relation, provided that they are $\lesssim 0.1$ of the spinup time $t_{\text{spin}} \sim 10^5$ y. Systems with $P_{\text{spin}}/P_{\text{orb}}$ smaller than given by (13) must accrete less specific angular momentum, and hence must accrete mainly from discs. This is very plausible in all three known cases: DQ Her and AE Aqr have very short spin periods (71 and 33 s respectively) strongly suggesting very small μ. In contrast GK Per has $P_{\text{spin}} = 351$ s, suggesting a fieldstrength similar to other IPs, but has a very long orbital period (48 h); there is then ample room for a disc to form ($R_{\text{circ}} \gg R_{\text{mag}}$). The present argument gives no information about what fraction of the accreted mass in systems with $P_{\text{spin}}/P_{\text{orb}}$ larger than given by (13) has passed through a disc or not.

The reasoning of the last Section shows that accretion mainly avoiding a disc naturally reproduces the observed spin equilibrium in many IPs. It does not by itself exclude disc flow as a way of achieving this. The deciding criterion for the formation of a disc is whether or not there is room for the accreting gas stream to circularize without interference from the magnetic field, i.e. whether $R_{\rm circ} > R_{\rm mag}$ or not. We note that this criterion is completely independent of the spin rate of the white dwarf (see 2 d) below). Once a disc forms, the usual treatments (Ghosh and Lamb, 1979; Wang 1987) show that the equilibrium $P_{\rm spin}$ is a function of μ and $\dot{M}$ only, and does not obey the relation (13), giving instead

$$P_{\rm spin} \propto \mu^{6/7} \dot{M}^{-3/7}. \tag{14}$$

It is clear that this type of dependence is not very promising for IPs, as there is no relation between $P_{\rm orb}$ and the *instantaneous* accretion rate $\dot{M}$. The *average* (over evolutionary timescales) of $\dot{M}$ actually increases with $P_{\rm orb}$, giving entirely the wrong connection between $P_{\rm spin}$ and $P_{\rm orb}$. Another problem for this type of model is that GK Per has very similar μ and $\dot{M}$ to the other IPs (see above) yet does not lie on the relation (10). Indeed it is hard to see how any theory of the accretion of spin angular momentum depending only on *local* disc properties can give this relation, as the following thought experiment shows. Imagine that some such local theory predicted that a given IP obeyed the observed relation (13). Now double its orbital period, leaving all other properties, in particular $\dot{M}$, unchanged. Since $P_{\rm spin}$ must stay fixed, the ratio $P_{\rm spin}/P_{\rm orb}$ is halved, and can no longer obey (13). Any permanent disc theory of this equilibrium must somehow introduce information about the binary orbit directly. If most of the accretion flow joins the disc at its outer edge, as in the usual picture, this information will inevitably be lost, as the accreting angular momentum is characterized by $R_{\rm mag}$ rather than $R_{\rm circ}$. Further, the equilibrium spin period will differ from that given by (13) by the factor $(R_{\rm in}/R_{\rm circ})^{3/2}$. It is difficult to see how a disc flow with $R_{\rm in} > R_{\rm circ}$ can ever be set up, and indeed the usual treatments predict $R_{\rm in} < R_{\rm circ}$ and $P_{\rm spin} \ll 0.1 P_{\rm orb}$.

One way of reconciling the required dependence on $R_{\rm circ}$ with the permanent presence of at least some kind of disc is to allow most of the accretion to skim over the upper and lower faces of the disc before interacting with the white dwarf magnetosphere. This kind of picture has been suggested before in several contexts, (Frank, King and Lasota 1987, Mason, Rosen and Hellier 1988, Hellier *et al.* 1989, Lubow 1989, Marsh and Horne 1990). In general it is not clear what fraction of the incoming stream is stripped off and accretes via the disc, although evidently this cannot become large without destroying the desired $R_{\rm circ}$ dependence. In the HKL picture a full accretion disc never develops, but there may be matter circulating outside the magnetosphere, which may carry off angular momentum from the white dwarf. Some such mechanism is in any case required, as the white dwarf is supposed to spin in equilibrium (another possibility would be a tenuous wind blown away from the magnetosphere).

Some of these situations can be described by suitable modifications of the inner boundary conditions for a Keplerian disc. In a steady state we let the mass flow rate through the disc be $\dot M_{\rm disc}$, so that

$$\dot M_{\rm disc} = 2\pi R \Sigma(-v_R). \tag{15}$$

Note that there is no prejudice about the magnitude or even the sign of $\dot M_{\rm disc}$ here, except that it is no bigger than the total accretion rate $\dot M$. In (15) Σ and v_R are as usual the surface density and radial drift velocity in the disc (for a review of standard disc theory, see Frank et al., 1991). We assume that the angular momentum flow from the secondary is all transmitted back from the white dwarf to the disc, and thence back to the secondary via the tides. Thus in the steady-state angular momentum balance equation

$$-\nu \Sigma \frac{d\Omega}{dR} = \Sigma(-v_R)\Omega + \frac{C}{2\pi R^3} \tag{16}$$

we replace the integration constant C by $+\dot M (GMR_{\rm circ})^{1/2}$ (note the sign!) and use (15) to eliminate v_R. The Kepler value $(GM/R^3)^{1/2}$ for the angular velocity Ω now gives

$$\nu \Sigma = \frac{\dot M_{\rm disc}}{3\pi} + \frac{\dot M}{3\pi}\left(\frac{R_{\rm circ}}{R}\right)^{1/2}. \tag{17}$$

In general the first term will be negligible compared with the second (realistically it is likely to be small and positive, although a negative $\dot M_{\rm disc}$ (an *excretion* disc) is a theoretical possibility, see e.g. Pringle 1991). Henceforth we drop this term, so that this is a *non-accretion* disc. From the standard relation

$$v_R = -\frac{3}{\Sigma R^{1/2}} \frac{d}{dR}[\nu \Sigma R^{1/2}]$$

we see that v_R is self-consistently zero: *there is no mass flow through the disc*, which acts as a fluid flywheel transmitting the accretion torque away from the white dwarf and back to the binary orbit. From (17) the surface density in this disc is very similar to that in a disc where mass is accreting in the usual way at the rate $\dot M$. The effective temperature is similar to the standard case, although its gradient is rather steeper ($\propto R^{-7/8}$). All of the radiated energy is supplied by the rotation of the white dwarf, not by accretion. The disc is optically thick, and its continuum is likely to differ only in details from the standard case.

Our derivation of the relation (12) uses only the *average* specific angular momentum fluxes, taken over timescales $\sim 0.1 t_{\rm spin} \sim 10^4$ y. Since $R_{\rm mag}$ decreases for higher $\dot M$, episodes of higher mass transfer may allow a disc to form. These change $P_{\rm spin}$ by amounts $\Delta P_{\rm spin} \sim (\Delta t/t_{\rm spin}) P_{\rm spin}$, where Δt is the length of the episode of disc accretion. Thus for $\Delta t \lesssim 0.1 t_{\rm spin} \sim 10^4$ y, there would be little perceptible deviation from the line (12). There is even an obvious selection effect favouring the discovery of systems undergoing brief disc episodes, because they are brighter. The surprisingly different properties of TX Col and TV Col, given that $P_{\rm spin}$ and $P_{\rm orb}$ are virtually identical for the two systems, may well be attributable to this kind of effect. TX Col shows very strong X-ray pulsing at the beat period between $P_{\rm spin}$ and $P_{\rm orb}$ (Hellier 1991), which is very suggestive of accretion without a disc. TV Col shows no such beat, and indeed the presence of extra periodicities in the optical (Hellier, private communication) may indicate the presence of a disc and accompanying bright spot.

4. X-ray Emission from IPs

Since all IPs are hard X-ray sources, we may be reasonably confident that matter accretes quasi-radially down fieldlines on to the white dwarf surface. The spin-modulated hard X-ray light curves may be crudely described as quasi-sinusoidal, in the sense that they do not have phase intervals of zero or constant X-ray flux (for a recent compilation, see Norton and Watson, 1989a). King and Shaviv (1984) noted this, and also the fact that at photon energies $\gtrsim 4$ keV the most likely cause of the modulation is occultation by the white dwarf body (as in AM Her systems). For a collection of systems large enough to rule out special viewing geometries, quasi-sinusoidal modulation only results if the emitting region has at least one dimension comparable with the white dwarf radius R. It is straightforward to show that this dimension is likely to be the "horizontal" extent of the accretion region on the white dwarf surface, which cannot be very dissimilar in any direction without giving rise to peculiar light curve shapes which are not observed. Thus one expects the fraction f of the surface occupied by the accretion region to be considerably larger than in AM Her systems, i.e

$$f \gtrsim 0.25. \tag{18}$$

King and Shaviv (1984) also pointed out that photoelectric absorption in cool parts of the accretion flow would increase the amplitude of the spin modulation at low energies. This general picture was confirmed by the study of Norton and Watson (1989a), who also found that the absorption modulation was generally in phase with the occultation. This follows from the formulae of King and Shaviv (1984) if $f \gtrsim 0.09$. An alternative picture was proposed by Rosen, Mason and Córdova (1988) for EX Hya in particular. They suggested that the modulation was caused by photoelectric absorption in quite distant parts of the accretion stream, where it rises out of the orbital plane from the inner edge of a disc. Recent GINGA observations of FO Aqr (Norton et al. 1989) however show spin modulation at energies 10–20 keV. This cannot be absorption, and could only be electron scattering if the column density N_H considerably exceeded the value $\sim 10^{23}$ cm^{-2} measured from the low-energy absorption.

Few of the IPs show any direct evidence of the ultrasoft X-ray component associated with $T_{\rm eff}$ (2) which is so dominant in AM Hers. This cannot simply be a result of photoelectric absorption, as the absorber itself must reradiate the absorbed flux at its own effective temperature. The most likely reason (King and Lasota, 1990a) is that $T_{\rm eff}$ is too low to allow detection of this component. For plausible spectra this requires

$$T_{\rm eff} \leq T_{\rm crit} \sim 10^5 \text{ K}. \tag{19}$$

From (2) it is clear that this amounts to a condition on f. King and Lasota (1990) combine this condition with the relation

$$N_{\rm H} = 0.15 \times 10^{23} \, \dot{M}_{17} \, f^{-1} (M_1 R_9)^{-1/2} \text{cm}^{-2} \tag{20}$$

(King and Shaviv 1984), where $\dot{M}_{17}, M_1, R_9$ are $\dot{M}, M, R$ in units of 10^{17} g s^{-1}, $M_\odot$ and 10^9 cm respectively, giving the absorption column density through the cool material above the accretion shocks. As can be seen, (19) favours larger f, whereupon (20) becomes harder to satisfy (for the observed values $N_{\rm H} \sim 10^{23}$ cm^{-2}) unless $\dot{M}$ is increased. The resulting limits are

$$f \gtrsim 0.1, \quad \dot{M} \gtrsim 10^{17} \text{ g s}^{-1}, \tag{21}$$

while imposing the condition $f \lesssim 1$ implies an upper limit on the white dwarf mass, $M \lesssim 0.6 - 1 \, M_\odot$. The limit on f is in line with expectations described above from the spin-modulated X-ray light curves, while that on $\dot{M}$ suggests that most IPs accrete at rates rather above the evolutionary mean value for their periods (see below). From (20) the accretion luminosity $L_{\rm acc}$ must exceed $\sim 10^{34}$ erg s^{-1}, whereas the observed hard X-ray luminosity L_X is always less than $10^{32} - 10^{33}$ erg s^{-1} (Norton and Watson, 1989b). These results can only be reconciled if most of the accretion luminosity is hidden at photon energies $\lesssim 0.1$ keV, i.e. *the IPs have very strong (ultra)soft excesses*. This suggests that the accretion process in IPs is basically similar to that in AM Hers, where a soft excess is well established. It is worth noting that an excess is required in IPs quite independently of the above considerations, unless we are prepared to assume that they all accrete well *below* their evolutionary means, since $L_X \ll$ the expected value of $L_{\rm acc}$.

5. The Evolution of AM Her Systems

The period distribution of the AM Her systems has one extraordinary feature: out of 15 known systems in the range 81 min$\leq P_{\rm orb} \leq 228$ min, no less than 6 lie in the narrow range 113.6–114.8 min. This highly significant feature, the "period spike", lies close to the lower edge of the well known CV period gap (e.g. Ritter, 1987) which is apparent in Figure 1. In current theories of the secular evolution of CVs (see e.g. King, 1988 for a recent review), angular momentum losses shrink the binary and cause mass transfer, reducing the binary period in the process. The mass loss proceeds on a timescale which becomes comparable to the secondary star's thermal time. This star's radius then does not shrink rapidly in response to the mass loss, and the secondary becomes progressively oversized for its mass compared to the main sequence since the star is out of thermal equilibrium. At a period ~ 3 hr, angular momentum losses are apparently severely reduced; this may be connected with the fact that the secondary becomes fully convective at around this period. The secondary is now able to shrink faster than the Roche lobe, and mass transfer ceases (Rappaport, Verbunt and Joss, 1983; Spruit and Ritter, 1983). Mass transfer eventually resumes because the star cannot get smaller than its main-sequence radius, and the separation and Roche lobe radius are continually reduced by gravitational radiation and any other angular momentum losses. For a plausible degree of thermal disequilibrium at $P_{\rm orb} \sim 3$ hr mass transfer restarts near $P_{\rm orb} \sim 2$ hr, as observed. Hameury et al. (1988) (hereafter HKLR) proposed to fit the AM Her period spike into this scheme by identifying it as the point where mass transfer restarts for a large class of these systems. They showed that the relative discovery probability

$$p(P_{\rm orb}) \propto \dot{M}^{3/2} / |\dot{P}_{\rm orb}| \qquad (22)$$

is considerably enhanced at this period. This occurs because of a combination of two effects working in the same direction. Since the secondary is fully convective and has re-established thermal equilibrium in passing through the period gap, its initial response to the resumption of mass loss is to expand adiabatically. Once it is sufficiently far out of thermal equilibrium it will begin instead to shrink in response to the mass loss. Because of Roche geometry, this means that $P_{\rm orb}$ will first *increase* slightly before decreasing. At the same time, the adiabatic expansion means that the mass transfer will occur at about twice the rate appropriate whenced the stellar radius shrinks instead. This follows from the evolution equation (e.g. King, 1988)

$$\dot{M}/M_2 = (-\dot{J}/J)[5/6 + \beta/2 - M_2/M_1]^{-1}, \qquad (23)$$

where $M_2, J, \dot{J}$ are the secondary mass, orbital angular momentum and its loss rate respectively, and β is the index in the secondary's mass-radius relation $R_2 \propto M_2^\beta$. The mass ratio M_2/M_1 is small, so the denominator on the rhs of (23) is effectively given by $5/6 + \beta/2$. For adiabatic expansion, $\beta \simeq -1/3$, while for a star shrinking on mass-loss β has approximately the main-sequence value ~ 1, giving denominators $\sim 2/3$ and $\sim 4/3$ in the two cases. Thus the system spends rather longer near its initial period and is somewhat brighter than at shorter periods (Figure 3), resulting in an enhanced discovery probability there (Figure 4).

This argument shows that CVs are more likely to be discovered as they resume mass transfer than at smaller $P_{\rm orb}$. But to give the observed period spike, we must require that this resumption period is the same for a substantial fraction of AM Her systems (to within the width of the spike, i.e. $\Delta P_{\rm orb} \lesssim 1$ min.). This requires them all to have secondaries which become fully convective at the same period, and to show the same degree of thermal disequilibrium at this period, so that they all restart

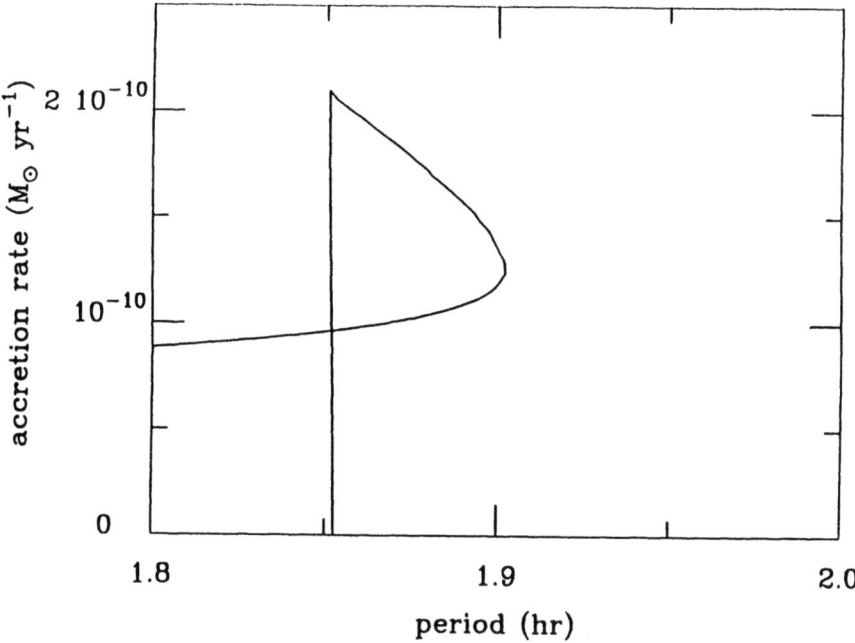

Fig. 3. The evolution of AM Her systems in the period spike.

mass transfer near ~ 114 min. Since the latter is measured by comparing the star's thermal timescale with the mass-transfer timescale $t_{\dot{M}} = M_2/\dot{M}$, (23) shows that M_1 must be the same for all the systems in the spike. (The only exceptions would be those rare systems which resumed mass transfer at longer periods and are now at 114 min purely by chance.) Since from (23) $\dot{J}/J$ must also be identical, we see that the angular momentum loss mechanism above the gap must be "universal" in the sense that its long-term average value can only depend on M_1, M_2 and $P_{\rm orb}$.

After HKLR made their proposal, a new AM Her system (EXO 033319-2554.2, now called UZ For) was discovered by Giommi et al. (1987) and Beuermann and Thomas (1987), with the longer period of 127 min. Unless this was taken as one of the rare CVs "born" in the period gap, it was clear that UZ For could only fit into HKLR's picture if it had passed through a rather narrower gap than the systems in the period spike. This requires its secondary to have had a smaller degree of thermal disequilibrium when it became fully convective, which in turn requires a longer mass transfer timescale $t_{\dot{M}} = M_2/\dot{M}$ above the period gap. For any reasonable form of angular momentum loss rate $\dot{J}$, (23) shows that the main sensitivity of $t_{\dot{M}}$ is to the factor $[5/6 + \beta/2 - M_2/M_1]$. Since M_2 and β are very similar for all systems just above the gap, we see that *a narrower period gap corresponds to a higher white dwarf mass*. The connection can be made quantitatively precise, and led Hameury, King and Lasota (1988) to predict a substantially larger M_1 for UZ For than in the spike, the most likely values being $\sim 0.6\, M_\odot$ for the spike and $\gtrsim 1.2$

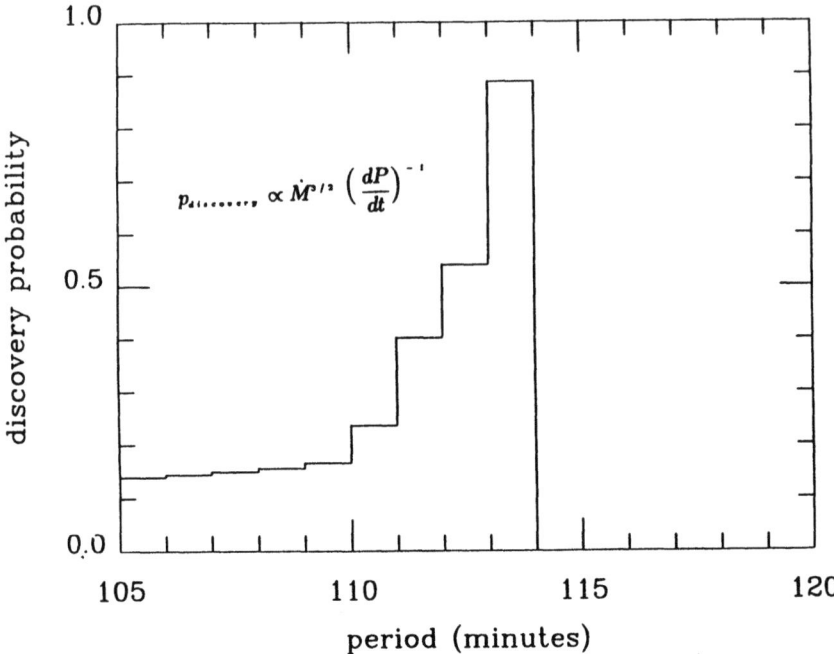

Fig. 4. Theoretical discovery probability of AM Her systems at the period spike.

$M_\odot$ for UZ For. The latter prediction was confirmed by Beuermann, Thomas and Schwope (1988) and Ferrario *et al.* (1989). A recent survey (King, Hameury and Lasota, 1990) of the available mass estimates concludes that they are compatible with the theoretical predictions.

It is clear that the mass constraints described in the last paragraph place very tight restrictions on physical processes and evolution in AM Her systems. For example, nova explosions are frequently thought to carry off rather more mass from the white dwarf than is accreted between outbursts, thus gradually reducing the value of M_1. If this occurred in most AM Her systems however, the value of M_1 at any epoch would depend on how long the system had been exchanging mass, and hence on the initial period, or equivalently the initial mass of the secondary. To get a large number of systems to pass through $P_{orb} = 114$ min with the same M_1 would require some kind of conspiracy between the initial white dwarf and secondary masses, which seems very unlikely. The only plausible way of arranging things is to assume that most AM Her systems have almost the same initial M_1, ($\simeq 0.6$–$0.7\ M_\odot$), and that this remains unchanged (Hameury, King, Lasota and Livio, 1989). There must also be a smaller group with M_1 allowed to have larger values. The latter would contain UZ For and V1500 Cygni, the only known nova amongst AM Her systems. Nova theory provides tentative support for the idea that M_1 might stay fixed for the lower-mass group, since the magnetic field prevents shear mixing of the accreted matter with the white dwarf envelope and so inhibits mass loss. The lack of this constraint may explain why no period spike has yet been identified for non-magnetic CVs, and why the evolution of the AM Her systems is particularly easy to follow. The low- and high-mass groups amongst the AM Hers may represent the products of case C and case BB evolution respectively.

As can be seen, the restrictions placed on evolutionary theory by the period spike are already severe. Sect. 9 shows how future observations (especially by ROSAT) will give still tighter constraints, allowing stringent tests of the whole current picture of the secular evolution of compact binaries. The remarkable regularity revealed by the existence of the AM Her period spike means that these systems will play a central role in understanding the evolution of close binaries.

6. The Relation Between AM Hers and IPs.

As has been discussed in Sect. 5 the accretion process in IPs is basically similar to that in polars: blobby accretion probably occurs in them too.

Let us consider the accretion of an individual blob of density ρ in the accretion stream. On leaving L_1 this generally follows a ballistic trajectory, since its ram pressure $\sim \rho v^2$ exceeds the local magnetic pressure $\sim \mu^2/8\pi r^6$; here v is the blob velocity and r is the distance from the white dwarf. Accordingly v is close to the free-fall value $(2GM_1/r)^{1/2}$, and ρ changes little as v is highly supersonic. As the blob falls towards the white dwarf, the magnetic pressure increases rapidly, and eventually forces the blob to follow fieldlines at a threading radius r_{thr} where ram and magnetic pressures are comparable:

$$r_{\text{thr}} \simeq (\mu^2/16\pi GM_1 \rho)^{1/5}. \tag{24}$$

Clearly, if the accretion is taking place in a non-synchronous system, e.g. an IP, the blob will be flung out by centrifugal repulsion if $r_{\text{thr}} > R_{\text{co}}$. Using (9) implies a lower limit to the density which can thread to the fieldlines and be accreted rather than expelled in this way:

$$\rho > \rho_{\text{thr}} = \mu^2 \Omega_{\text{spin}}^{10/3}/16\pi(GM_1)^{8/3}. \tag{25}$$

As is well known (Kuipers and Pringle, 1982; Frank *et al.*, 1988), blobs which follow fieldlines are stretched by tides and squeezed by the field, so that their density increases as $r^{-5/2}$. Thus (25) implies that the minimum density of a blob accreting on to the white dwarf surface is

$$\rho_{\min} = \rho_{\text{thr}}(r_{\text{thr}}/R_1)^{5/2}. \tag{26}$$

Using (24) and (25) gives

$$\rho_{\min} = \mu^2 \Omega_{\text{spin}}^{5/3}/16\pi(GM_1)^{11/6} R_1^{5/2}. \tag{27}$$

Numerically, for a $0.7 M_\odot$ white dwarf ($R_1 = 8.7 \times 10^8$ cm), (27) is

$$\rho_{\min} \simeq 4.6 \times 10^{-10} \mu_{33}^2 P_3^{-5/3} \quad \text{g cm}^{-3}, \tag{28}$$

where P_3 is P_{spin} in units of 10^3 s.

The limit (27), (28) is thus likely to characterize accretion at the white dwarf surface in an IP. It could be violated if the accretion flow were very homogeneous, or if there were some way in which ρ could decrease after the blob is threaded by the fieldlines. The first possibility is not supported by observation, as we have seen, and it does not seem likely that the second possibility could involve a substantial fraction of the accreting matter. Note that the limit does not apply in AM Her systems, where the corotation radius is outside the binary orbit.

The importance of the limit (27), (28) is that its μ^2 dependence shows that there is a tendency for stronger fields to inhibit hard X-ray production in IPs: the low-density material which would produce it is linked to fieldlines outside corotation and expelled. The fact that hard X-rays are observed in IPs implies that $\rho_{min} \lesssim 10^{-7}$ g cm^{-3} for them. From (28) this implies that $\mu \lesssim 10^{34}$ G cm^3. By contrast, AM Her systems have $\mu \gtrsim 2 - 3 \times 10^{34}$ G cm^3. We conclude that *intermediate polars must have weaker magnetic fields than AM Her systems*. This is in agreement with the simplest interpretation of their lack of optical polarization. The limit (28) now finally clears up the difficulty in understanding the secular evolution of magnetic CVs which was always presented by this type of explanation, namely the lack of observed non-synchronous progenitors for the AM Her systems at longer orbital periods. By the reasoning of Sect. 4 above, these systems must obey the relation (15) and have have spin periods similar to those of the known IPs. But then the limit (28) with $\mu \gtrsim 2 - 3 \times 10^{34}$ G cm^3 shows that they cannot be hard X-ray sources. In common with the known IPs they are probably not observable as soft X-ray sources either, possibly because the accretion region has a large area and hence a low effective temperature (see Sect. 5). Since most magnetic CVs are discovered via X-ray observations (including all of the systems with $P_{spin} \sim 0.1 P_{orb}$) we conclude that *the probability of discovering CVs with magnetic moments characteristic of AM Hers is extremely low if they are non-synchronous*. This conclusion applies also to AM Her systems which become non-synchronous because of upward fluctuations in the accretion rate which last longer than $\sim t_{spin}$ (Hameury, King and Lasota, 1989a, see Figure 5). The P_{spin} - dependence of (28) does allow the possible discovery of non-synchronous polars in hard X-rays provided P_{spin} is somewhat longer than in observed IPs ($\gtrsim 3000$ s). Since they must obey (15) this in turn requires longer orbital periods $\gtrsim 8$ h. Such systems must be intrinsically rare, as the secondary stars must be quite massive ($\gtrsim 0.9 M_\odot$). Mass transfer stability then requires massive white dwarfs ($> 1 M_\odot$): these are probably present in $\lesssim 10\%$ of AM Her systems (Hameury, King and Lasota, 1988, 1989b). It is quite likely that ROSAT will discover some systems of this type.

Another way of finding non-synchronous polars would be to abandon the restriction to X-ray emitting systems. Optical photometry readily reveals the presence of a spin modulation or its orbital sideband, but at present such studies are only carried out for known hard X-ray sources, thus ruling out strong-field systems if $P_{orb} \lesssim 5$ h. Any novalike systems which are not X-ray sources but show a pho-

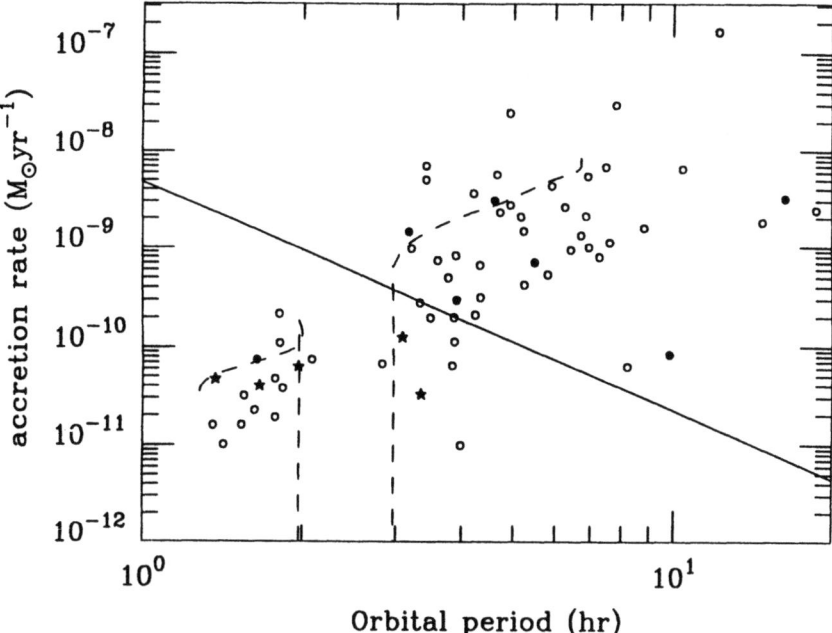

Fig. 5. Observed mass transfer rates of CVs, taken from Patterson (1984). AM Hers are represented as stars, IPs as filled circles and non-magnetic CVs as open circles. The diagonal lines represent the synchronization criterion (27) for $M_1 = 0.6\ M_\odot$, $\phi = 1$ and magnetic moments $\mu_{33} = 4,\ 10,\ 30$ (dashed, solid and dash-dotted respectively). The dashed curve is the evolutionary mean relation $\dot{M}\ (P_{\rm orb})$ predicted by the calculations of Hameury, King and Lasota (1988). Clearly fluctuations in $\dot{M}$ around this relation can carry systems across the diagonal lines: hence AM Her systems can become temporarily asynchronous in the range $3 \leq P_{\rm orb} \leq 4$ hr.

tometric period $\sim 10\%$ of $P_{\rm orb}$ are strong candidates, and could be confirmed by polarimetry. The absence of X-rays makes the presence of HeII 4686 unlikely, and for the system to be asynchronous at all requires $P_{\rm orb} > 3$ h. For these reasons the non-discovery of such systems in the sample observed polarimetrically by Cropper (1986) is not surprising. Indeed it is clearly quite difficult to find polarized systems from optical observations alone: the vast majority of AM Her systems were first found via X-rays.

We note finally that if a novalike shows a second period *not* $\sim 0.1 P_{\rm orb}$, especially if it is rather longer than this, one should not uncritically adopt an identification as an IP.

7. Evolution of Magnetic CVs

We have shown that the clustering of IPs about the relation $P_{\rm spin}/P_{\rm orb} \sim 0.1$ results naturally if the white dwarf accretes exactly the specific angular momentum of the secondary star. This implies the absence of a fully-developed accretion disc, which would destroy any obvious connection between P_{spin} and P_{orb}, and probably lead to equilibrium spin rates with systematically smaller values of P_{spin}/P_{orb}. We showed also that the magnetic moments of the IPs must be significantly less than those of the AM Her systems: the latter are effectively unobservable whenever the white dwarf rotates asynchronously, as their X-ray emission is quenched. It is possible that ROSAT may detect some long-period non-synchronous systems which will be found to show spin-modulated optical polarization and thus have AM Her fieldstrengths, and some high-field systems may lurk unrecognised amongst CVs which are not X-ray sources.

These results reconcile straightforward interpretations of the polarization data for magnetic CVs with a coherent picture of their secular evolution. We may summarize this (compare Lamb and Melia 1987) according to the strength of the magnetic moment: (i) 10^{34} G cm^3 $\lesssim \mu \lesssim 5 \times 10^{34}$ G cm^3. The upper limit here is probably set by cooperative magnetic braking via the white dwarf magnetic field (Hameury, King and Lasota, 1989b; see also Liebert and Stockman, 1985; King, 1985). Systems in this group evolve as suggested by Hameury, King, Lasota and Livio (1989), with the important modification that *they are unobservable whenever they are asynchronous*. When synchronous, these systems are observable as AM Hers. Criteria for synchronism are all effectively equivalent to the requirement $r_\mu \gtrsim a$ (Hameury *et al.*, 1987). All of the systems in this group are asynchronous ($r_\mu < a$) for $P_{orb} \gtrsim 4$ h, and thus unobservable. Between 3 and 4 h this is still largely true, but fluctuations of the mass transfer rate below the secular mean probably last long enough that at a given epoch a small fraction achieve synchronism ($r_\mu > a$) and become observable as AM Hers. All of these systems achieve synchronism during the passage through the period gap, as the accretion torques are very weak ($r_\mu >> a$). They emerge at the AM Her period spike (Hameury *et al.*, 1988, 1989) at $P_{orb} = 114$ min, and thereafter remain synchronous ($r_\mu \gtrsim a$), and observable as AM Her systems. Because relatively few are observable above the period gap, the period distribution of AM Hers is significantly different from that of non-magnetic CVs.

(ii) 10^{33} G cm^3 $\lesssim \mu \lesssim 10^{34}$ G cm^3.

The majority of observed IPs, lying on the $P_{spin}/P_{orb} \sim 0.1$ relation, are in this group. They do not have accretion discs for $P_{orb} \lesssim 5$ h. If $\mu \gtrsim 4 \times 10^{33}$ G cm^3, they would become synchronous in and below the period gap. While this may be true of BG CMi (Chanmugam *et al.*, 1990), it is unlikely that it holds for the majority. Synchronous systems with these fields would have effectively the same X-ray properties as the AM Hers, and hence be readily observable as resembling them in every respect except for a lack of optical polarization. The fact that no such systems are known at all already shows that the space density of systems with $4 \times 10^{33} \lesssim \mu \lesssim 10^{34}$ G cm^3 is considerably lower than that of the AM Hers ($\mu \gtrsim 10^{34}$ G cm^3). Thus the known IPs probably have $10^{33} \lesssim \mu \lesssim 4 \times 10^{33}$ G cm^3. This agrees with the simplest interpretation of the fact that their optical polarization is at least an order of magnitude lower than in the AM Hers, and would imply a bimodal distribution of magnetic moments amongst magnetic CVs. However, a better understanding of the discovery probabilities above and below the period gap, as well as the greatly enlarged IP sample which ROSAT will provide, are both needed before we can be confident of this.

(iii) $10^{31} \lesssim \mu \lesssim 10^{33}$ G cm^3.

These systems always have accretion discs and never synchronize. The known examples are DQ Her and AE Aqr.

We note that GK Per ($P_{orb} \simeq 2$ d) does not fit into the above scheme, as it contains a subgiant secondary and must be evolving to longer orbital periods. It has an accretion disc because the binary separation, and hence R_{circ}, is much larger (Watson, King and Osborne, 1985) than any likely magnetospheric radius.

If it is assumed to be in spin equilibrium its magnetic moment must be of order $3 \times 10^{32} - 10^{33}$ G cm^3.

8. Conclusions: The Future

We have seen that there has been very considerable progress in understanding both classes of magnetic CVs in recent years. Some fundamental problems remain. In particular the distribution of magnetic moments amongst the white dwarfs is still unclear; and it does not appear to be the same as for isolated white dwarfs.

The sucessful launch of the ROSAT/WFC X-ray – EUV satellite is likely to transform the subject dramatically: because of their strong soft X-ray emission, AM Her systems in particular will be turned up in large numbers. A simple scaling from the serendipitous discovery of 3 new AM Hers by EXOSAT in about 5 percent of the sky suggests that at least 60 such systems will be found. There are realistic hopes that these will be optically identified in a relatively short time. In this event magnetic CVs with known orbital periods will outnumber non-magnetic ones. In view of the great importance of the AM Her period distribution in particular (Sect. 4) this is a uniquely fortunate circumstance. Hameury, King and Lasota (1990) and King, Hameury and Lasota (1990) show how this sample will allow a clear test of the whole picture of CV evolution, especially the formation of the period gap, as well as deciding a number of other evolutionary issues. One need hardly add that we will be able to study a host of other important questions about these systems with vastly heightened insight.

SYMBIOTIC STARS

A. A. BOYARCHUK

1. Introduction

Among a wide variety of stellar spectra there is a group of spectra with remarkably peculiar features. In these spectra one can observe simultaneously strong emission lines of highly ionized elements and absorption features typical of M type cool giants (Figure 1). Very often these stars experienced outbursts and were therefore called nova-like stars. But now this group of objects is commonly called symbiotic stars. This biological term implies that in these stars there coexist objects of different nature. Symbiotic stars have been thoroughly investigated. Several reviews (Allen 1979; Boyarchuk 1984; Sahade 1976 and many others) and one monograph by Kenyon (1986) (a lot of important references to earlier publications can be found there), as well as two IAU Colloquia (No. 70, 1982 and No. 103, 1987) have been devoted to them. As a result of these activities our understanding of the nature of these objects has greatly improved.

Until the eighties there were discussions as to whether symbiotic stars were binary or single stars with specific combinations of photospheric and outer atmospheric layers. It now seems certain that symbiotic stars are interacting binaries, one of the components being an evolved cool giant, and the other a hot star of low luminosity or a hot dwarf with an accretion disk.

Although models of a hot corona around cool stars or a cool extensive envelope around a hot star can account for some observations, they are in contradiction with others and have been repudiated (Boyarchuk 1970, 1986). Below we consider observations of symbiotic stars from the point of view of a binary system model as well as the evolutionary status of the stars.

2. Cool Component

The first spectral observations of symbiotic stars in the visible region gave evidence for the existence of a cool component. The presence of the visible bands of TiO in their spectra allowed one to classify them as M2 or later spectral class. But it was quite difficult to determine the luminosity class of the cool components spectroscopically because of the low dispersion and superposition of the emission lines on the continuum. From the analysis of distances and proper motions of the stars it was discovered that the cool components might be of luminosity class III.

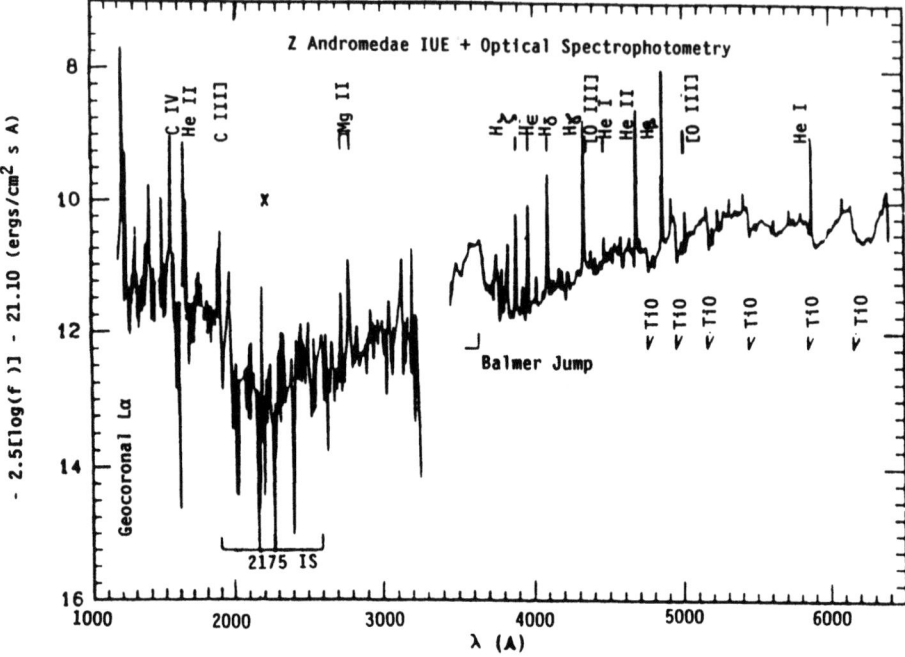

Fig. 1. Ultraviolet and optical spectra of the S-type symbiotic star Z And. The strongest emission and absorption lines are indicated (after Slovak and Code 1988).

IR observations of symbiotic stars provided more information on the properties of their cool components. Kenyon and Gallagher (1983) were the first who investigated IR-spectra for a relatively large number of stars for the purposes of spectral classification. They investigated the CO band at $\lambda 2.3 \mu$m. (According to Baldwin, Frogel and Persson (1973) the higher the stellar luminosity, the stronger the intensity of this band, though in this case noticeable scattering takes place). All of the 13 stars analyzed, except CI Cyg, may be classified as normal giants. CI Cyg is likely to have luminosity class II and may be similar to bright stars of the asymptotic giant branch (Kenyon and Fernandez-Castro 1987). The high luminosity of the cool component of CI Cyg is also confirmed by analysis of the eclipse light curve from which it follows that radius of the cool component is 300 $R_\odot$, which is several times as large as the size of a normal giant (Belyakina 1983).

The first photometric IR observations of symbiotic stars were obtained by Swings and Allen (1972). They discovered that for the majority of the stars the IR-colors did not differ from those of normal M-giants, though they observed some excess at long wavelengths for some stars. The authors assumed the excess to be due to the presence of dust in the systems. Glass and Webster (1973) confirmed these results.

Webster and Allen (1975) divided symbiotic stars into two (S and D) types. For the S-type stars the sources of infrared radiation are red giants, and in the

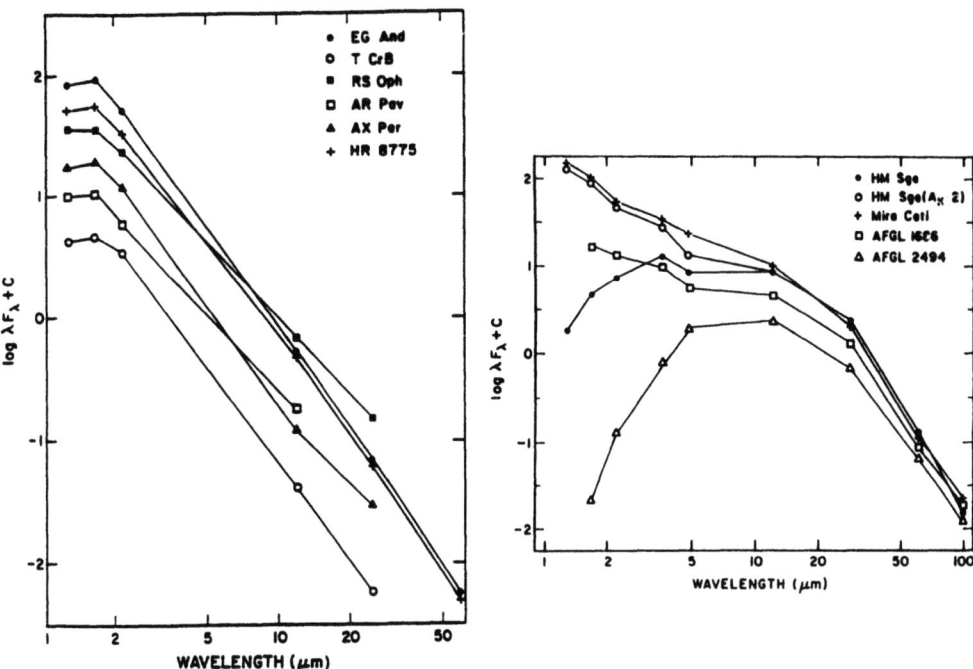

Fig. 2. Infrared energy distributions for the S-type symbiotic systems (left panel) and for the D-type symbiotic star HM Sge (right panel). Data for the M2 III star HR 8775, Mira Ceti and two obscured giants (AFGL 1686 and AFGL 2494) are plotted for comparison. Dereddened energy distribution of HM Sge ($A_K=2.0$) is also plotted (open circles; after Kenyon et al. 1986).

case of D-type stars the IR-excess (even in near-IR) is believed to be due to dust emission with T = 1000 K (see Figure 2). Later IR observations provided new data for symbiotic stars, in particular for D-type stars; at the same time more and more evidence was obtained in favor of the view that IR features of S-type stars do not differ from those of single red giants.

Feast, Robertson and Catchpole (1977) found that in the infrared region D-type stars undergo changes typical for Miras while S-type stars have constant brightness, though changes of brightness in the visual region can be large (Szkody 1977). Note that Taranova and Yudin (1981, 1982) found small variations in the infrared for S-type stars. Later on many observers investigated IR-radiation of D-stars and confirmed that the cool components of almost all the stars were Miras (Whitelock 1988). Their periods (from 280 to 580 days) are on the average somewhat longer than periods of the single Miras and are close to those of OH/IR sources. But the IR-colors observed are not similar to those of Miras. Allen (1983) made a suggestion based on IRAS observations and confirmed later by Kenyon et al. (1986) that this difference could be explained by extremely strong absorption (A_K up to 2^m, which corresponds to A_V up to 12^m) in dusty (silicate) envelopes. It is a very important fact that only cool component radiation is subject to such

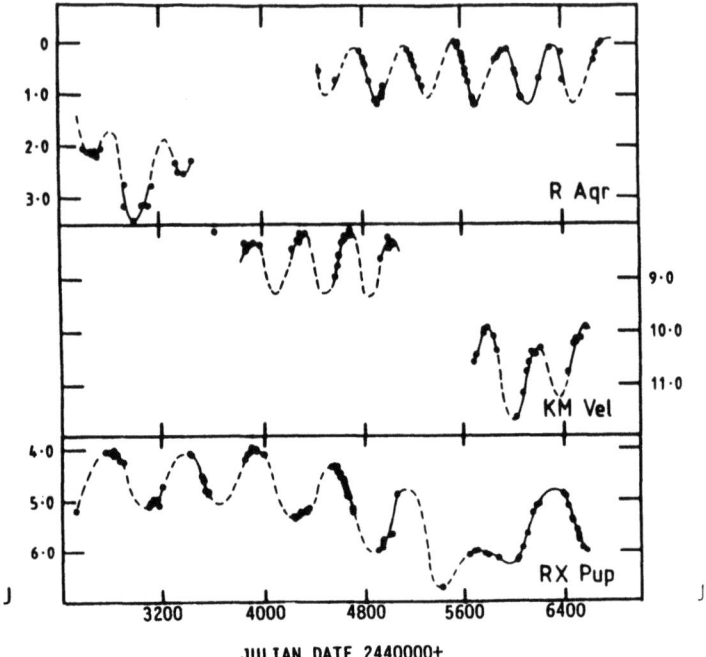

Fig. 3. J-light curves for some D-type symbiotics. In addition to the Mira pulsations, large changes in mean light level exist (after Whitelock 1987).

strong absorption. Neither the ratios of intensities of emission lines arising in the gaseous nebulae, nor the intensity of interstellar band λ 2170 Å indicate such strong absorption. It follows from these observations that $A_V < 2^m$. So it seems certain that the dust is mainly concentrated in the circumstellar envelope of the cool component. This can also account for the fact that in the visual spectral region of many D-type stars we observe only weak indications of a cool component: TiO bands, etc. (Kenyon et al. 1986). With such strong absorption of stellar radiation by the dust envelope the emission of the envelope itself would be self-absorbed. Since the absorption coefficient of dust depends on wave length, it will result in a non black-body character of flux distribution from the dust envelope. The color temperature of these dust envelopes in the near-IR is $\sim$ 1000 K and is much lower in the far-IR. Suggestions were made that we are dealing with two different envelopes. But after correction for self-absorption it is possible to describe the observed energy distribution in the IR-region as thermal emission of dust with temperature $T \approx 250 - 450 K$ (Kenyon et al. 1988).

Apart from the changes of brightness typical of Miras, Feast et al. (1983) discovered in approximately half of the analyzed stars general changes of brightness with amplitude 1-2^m in the form of waves of several years duration (Figure 3).

When the star is fainter it lies in the region of stronger absorption in the two-color diagram (Whitelock 1988). Thus, there is extra dust absorption occuring episodically. This phenomenon is not observed in single Miras. The nature of this additional dust absorption remains unclear. This dust cannot be concentrated near the hot component, for there have not been observed any changes in the λ 2170Å band intensity. There have been several suggestions that a dusty nebula forms along the orbit of the cool component which may cause eclipses from time to time; though the mechanism of its formation and stability remains unclear. Other explanations are based on the assumption that there is a sudden increase of mass loss by the cool component followed by enhanced dust formation. There seems to exist a connection between the increase of radio emission and the corresponding behavior of IR emission (Seaquist and Taylor 1987; Whitelock 1988).

Suggesting that cool components should have the same near IR-colors as single giants, Kenyon (1988a) separated the far IR-radiation of the dust for ten S-stars. From the calculated colors $(25-12)\mu$m he determined dust temperatures. Then assuming the dust to be heated by the cool component, he determined the size of the dust envelopes. Typical parameters of the dust in S-type stars are given in Table I.

Since the distance between the components is $\sim 1 - 4$ AU, the role of the hot component in the heating of dust should also be taken into account. So one must treat the above values with caution.

If the dust temperature is known, we can determine the mass M_d of the dust envelope using the measured flux $F(25)$ and, assuming the distance to the symbiotic star to be typically 1 kpc, we find an estimate $M_d \sim 10^{-8} M_\odot$. The gas-to-dust ratio is considered to be about 100 (Herman et al. 1986), so that M_{env} is $\sim 10^{-6} M_\odot$. For a wind velocity of 30 km/s a sphere of the radius 100 AU will be filled up in 10 years which implies a mass loss rate $10^{-7} M_\odot$/yr. Approximately the same value was obtained by Kenyon et al. (1988) for the two stars AG Peg and CH Cyg from [12/60] and [22/60] colors on the basis of calibrations of Thronson (1987) and Knapp and Wilcots (1987), respectively. The resulting mass loss by the cool components of S-type symbiotic stars is somewhat higher than that of single giants of the same spectral class.

Analysis of far IR-colors of D-type stars obtained by the IRAS satellite was performed by Kenyon et al. (1988). In this work self-absorption effects were taken into account. The main parameters of the dust in the D-type stars are given in Table I. Note again that the mass loss rate is somewhat higher than that for bright single long-period Miras.

3. Hot Component

Unlike the cool component, no direct indications of the hot component are ever observed in the spectra of symbiotic stars, making spectral classification impossible. This may be caused by many circumstances. To begin with, the strong spectral lines such as those of hydrogen, neutral and ionized helium and others are completely blended with emission lines, which originate in the gas component of the symbiotic

TABLE I
Main parameters related to dust in the symbiotic stars.

Parameter	S-type stars	D-type stars
T_d (K)	300	300
R_d (AU)	50–100	100–550
M_d ($M_\odot$)	10^{-8}	2×10^{-7}–10^{-5}
M_{env} ($M_\odot$)	10^{-6}	10^{-5}–10^{-3}
M($M_\odot$/yr)	10^{-7}	10^{-6}–10^{-4}
P_{orb} (yr)	1–5	> 1000
$a_1 + a_2$ (AU)	1–4	> 100

system. The only exception is AG Peg for which Boyarchuk (1966) observed broad emission lines due to the hot component which was classified as WN6. In all other cases the presence of the hot component is revealed by the energy distribution in the continuous spectrum and by emission lines formed in the gas component under the influence of radiation from the hot component.

Analyzing the continuous spectra in the region 3300-5000 Å, Boyarchuk (1967a) managed to separate the radiation of the hot source from the total radiation of the system for seven symbiotic stars. In Figure 4, a model of the hot components was determined from their energy distribution. They range from 0.7×10^5 to $1.4 \times 10^5 K$. Assuming the cool components to be normal giants, luminosities of hot components were determined from the flux ratio of the cool and hot components. Though the analysis seems to be rather uncertain, it is safe to say that the hot components of symbiotic stars are in the subdwarf region and not in the white dwarf region of the HR diagram.

OAO-2, ANS, IUE and Astron observations which extended the observation wavelength range down to 1200 Å, added much to our knowledge about hot components of symbiotic stars. By now more than 20 stars have been thoroughly investigated by many scientists, mainly with the IUE facilities. A most complete analysis of the hot component in the UV region was made by Kenyon and Webbink (1984). They combined the observed energy distribution of the cool giant with the theoretical energy distribution of the hot component and ionized gas in order to obtain a synthetic spectrum from 0.1 to 3.5 μm. They concluded that the temperature of the hot star is $\geq 50,000$ K and its luminosity $\geq 25 L_\odot$. These values are in good agreement with the data obtained by Boyarchuk (1970) from analysis of spectra in the visible range. If the hot radiation originates in an accretion disk, the accretion rate should be $> 10^{-6} M_\odot$/yr if the secondary component is a main sequence star, and $> 10^{-9} M_\odot$/yr, if it is a white dwarf.

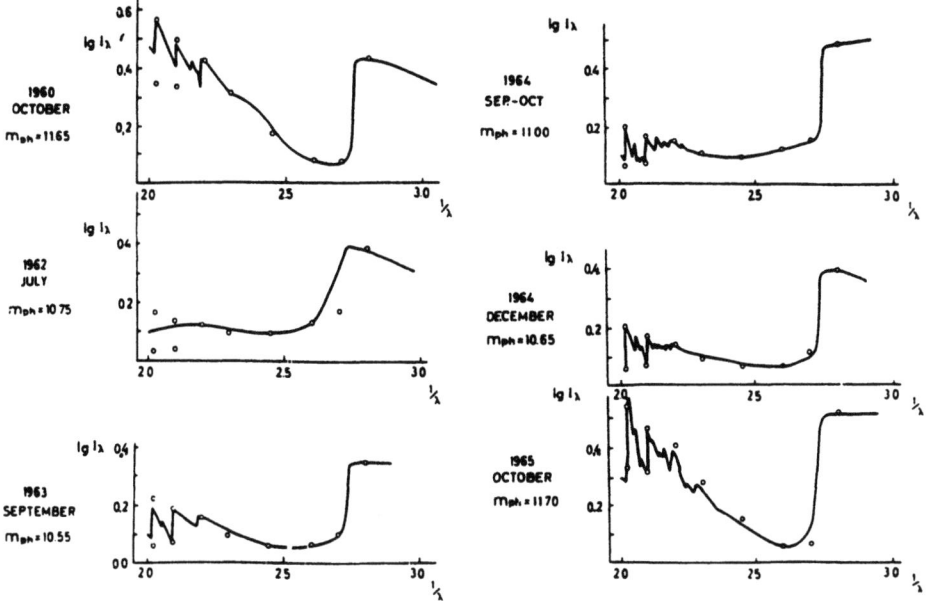

Fig. 4. Energy distribution in the optical spectrum of Z And. The solid lines represent the observed continuum; open circles - computed total radiation of the cool M2 III giant hot dwarf ($T \approx 10^5 K$) and ionized hydrogen ($T_e = 17000, n_e > 10^6$ cm^{-3}).

In order to be able to compare the results of theoretical estimates with observations, Kenyon and Webbink (1984) introduced color indices based on various combinations of fluxes at $\lambda\lambda 1300, 1700, 2200$ and 2600 Å. The comparison showed that in most cases the hot component is a subdwarf with temperature 40,000-160,000 K. In some symbiotic stars the hot component may be an accretion disk around a main sequence star with an accretion rate of $10^{-5} M_\odot$/yr. The observed colors do not fit the model of the accretion disk around a white dwarf for any symbiotic star. The same conclusion was drawn by Nussbaumer (1988) and Vogel (1988). They calculated the radiation field in the disk if the accretion is onto a white dwarf and noted that the temperature reaches values as high as $10^6 K$ in the boundary layer. In this case [Fe X] 6374 and [Fe XIV] 5303 should be observed. But these lines are absent in the spectra of symbiotic stars.

Boyarchuk (1985) used UV-flux data to determine luminosities of hot components. He suggested that in the J-color band the cool component (being a normal giant) completely dominates. At $\lambda 1300$ Å it is the hot component that dominates and the radiation of the ionized gas can be neglected. Interstellar absorption was estimated by the intensity of the $\lambda 2170$ Å band. The intensity ratio of the HeII 4686 and H_β lines were used for a qualitative estimate of the temperature of the hot component. This analysis led to the conclusion that the hot components of sym-

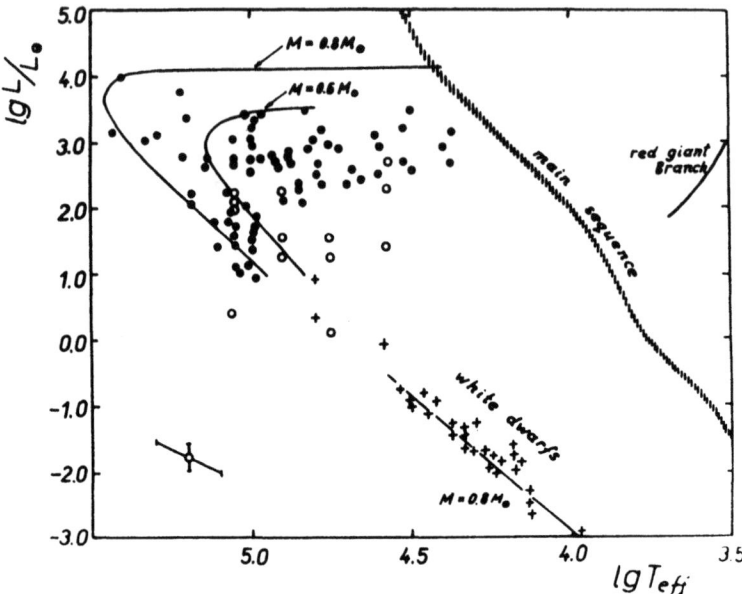

Fig. 5. The temperature-luminosity diagram (after Boyarchuk, 1985). The hot components of symbiotic stars are represented by open circles, central stars of planetary nebulae by dots, and white dwarfs by crosses. Solid lines are evolutionary tracks.

biotic stars are subdwarfs and lie in the region of planetary nebulae nuclei in the HR diagram (Figure 5).

In some papers the temperatures of the hot sources were determined by the Zanstra method (Iijima 1981; Blair *et al.* 1983; Kenyon 1983). The temperatures ranged about 160,000 K. It is difficult however, to estimate to what degree these temperatures characterize the field of the hot source radiation and its effective temperature. First of all, Balmer lines arise in an optically thick medium, and it is incorrect to use the case B approach here (Boyarchuk 1967b). The ratio of H_β line intensity to that of the continuum is three times smaller than in case B. As a result, the temperature which is obtained from the ratio I(HeII 4686/I(HI 4681) is too high. Also the region in which emission lines arise is extremely inhomogeneous and has a very complicated geometry. It is very unlikely that the optical thickness will be more than unity in all directions for the ionizing radiation of various ions. Hence from the analysis of emission line intensities by the Zanstra method one can only draw the qualitative conclusion that the temperature of the hot component of symbiotic stars is high.

Thus, one can regard the hot components of most symbiotic stars to be subdwarfs with temperatures ≈ 0.5 to 1.5×10^5 K. It is possible that accretion occurs, but this does not play an important role in energy production. For some symbiotic stars an accretion disk around a main sequence star could be the hot source, although it is difficult to explain how to provide the necessarily high rate of accretion.

4. Diffuse Gas

The third component of a symbiotic star is the nebula. Strong emission lines formed in these nebulae are the main features of symbiotic stars as are TiO absorption bands. Of course, the nebula does not consist of gas only. There is also a dust component which we mentioned above when we described the cool star. Below we shall focus our attention on the gas component.

Rather diverse sets of emission lines, from those of hydrogen and ionized metals to lines of highly ionized ions such as [NeV] and [FeVII], are observed. This is evidence of a strong inhomogeneity of physical conditions. That is why when we determine the ionized gas parameters, e.g., electron density n_e and temperature T_e, we must be aware of the fact that the derived values refer only to some parts of the emitting volume.

The character of the emission spectrum and the intensity ratios show that the main ionization mechanism is radiative and not collisional (Nussbaumer 1988). Consequently due to energy losses via L_α line excitation caused by electron collisions, electron temperatures cannot exceed 20,000 K (Hummer 1963). Since forbidden lines are observed, collisional cooling will be stronger, and electron temperatures lower. Boyarchuk (1968) assumed that T_e = 17,000 K would well represent the thermal conditions in the gaseous nebulae of symbiotic stars. Similar T_e values are given by Kenyon (1986).

Symbiotic star spectra yield rather large values of the ratio of the [OIII] line intensities: $I(4363)/(I(5007) + I(4959)) \approx 0.2 - 0.5$. In this range the value of the ratio does not depend on temperature only, as it does in the case of the planetary nebulae. On the contrary, it depends strongly on n_e and very weakly on T_e. Taking T_e = 17,000 K, from the observed values of [OIII] line ratios, we found the value of n_e for several symbiotic stars. For S-type stars it proved to be $\sim 10^7 cm^{-3}$ on the average. For the D-type star V1016 Cyg, $n_e = 2.5 \times 10^6 cm^{-3}$ (Boyarchuk 1968).

UV-observations made possible the use of semiforbidden lines to estimate n_e Kenyon (1986) gives n_e values for 16 symbiotic stars derived from IUE spectra. Despite significant scatter, one may regard most S-type systems to have $n_e \sim 10^9 cm^{-3}$, while for D-type stars $n_e \sim 10^6 cm^{-3}$, i.e., it follows from the observations that the latter objects have less dense nebulae. It should be emphasized also that for S-type systems n_e determined by optical and UV-lines differ approximately by two orders, while for D-systems they coincide. This implies that D- and S-type stars have different degrees of inhomogeneity.

In the case of symbiotic stars estimation of the mass of the emitting gas becomes rather difficult. It is not possible to use forbidden lines because of strong inhomogeneity. Helium lines and lines of heavier elements are of little use because of the uncertainties of ionization zone sizes. Great differences between the observed and theoretical (case B) Balmer decrements evidence a high self-absorption in these lines, at least in some of the regions of their formation. This makes hydrogen lines of little use for evaluation of the mass of the emitting gas.

The most reliable results are obtained by using observations of the continuous recombination radiation of hydrogen (Boyarchuk 1970). The total energy, E_λ, radiated from the nebula is:

$$E_\lambda = \epsilon(T) n_e n^+ V$$

where $\epsilon(T)$ is the emission coefficient of ionized hydrogen, and V is the volume of the nebula. Assuming the continuous spectrum is the sum of the radiation of the hot component, ionized component, and normal red giant, one can find all the ratios of fluxes of each source at any wave length. Knowing the absolute value of the red giant radiation flux, it is possible to determine the nebular radiation flux. Then the nebular mass, M_{neb} can be estimated from:

$$M_{neb} = \frac{E_\lambda}{\epsilon_\lambda(T)} \frac{\mu m_H}{n_e},$$

where m_H is the hydrogen atom mass and μ the molecular weight. The size R of the nebula is

$$R = \left\{ \frac{4}{3\pi} \frac{E_\lambda}{\epsilon_\lambda n_e^2} \right\}^{1/3}$$

Assuming $T_e = 17{,}000$ K and n_e determined from [O III] forbidden lines, Boyarchuk (1984) found that for six S-type stars the nebulae have masses which average about $10^{-4} M_\odot$ and sizes of the order of 100 AU. Using the same method for the D-type star V1016 Cyg, he determined the nebular mass to be $10^{-2} M_\odot$ and $R \sim 650$ AU. Of course, one should bear in mind that these parameters depend strongly on the adopted value of n_e. For S-type stars, n_e values determined by semi-forbidden lines are 100 times as high as n_e determined by [O III] lines. If we take the first value of n_e, the nebular mass will be $10^{-6} M_\odot$ and its size ~ 5 AU Thus the variations of n_e that take place in the nebula make estimates of the mass and size of the ionized region rather uncertain. It is necessary to know the ratio of the amounts of energy emitted from "dense" and "tenuous" parts of nebulae. If this ratio is 10:1 then $M_{neb} \sim 10^{-5} M_\odot$ for S-type stars.

Radio observations also provide information on the properties of the gaseous nebulae of symbiotic stars. The radio fluxes from symbiotic stars are low. Purton et al. (1973) measured the radio flux from V1016 Cyg at 10.63 GHz. Later, mainly as a result of the survey by Wright and Allen (1978), radio emission from other symbiotic stars was discovered. In 1981 upper limits of 10 mJy were found for the radio fluxes of 13 symbiotic stars (Boyarchuk 1984).

Later VLA observations decreased the upper limit to 0.5 mJy for 30 systems (Seaquist et al. 1984). By 1988 radio emission from 33 symbiotic stars had been detected (Seaquist, 1988). First it should be emphasized that on average the radio fluxes of D-type stars are ten times higher than those of S-type stars. Radio emission was detected in almost all D-type stars which were observed with the VLA. Radio flux variations proved to be larger than 30% on a time scale of several years in few of the stars.

Many symbiotic stars have been observed in the radio region at several frequencies and we can study the energy distribution which usually of the form

$$F_\nu \sim \nu^\alpha,$$

where α is spectral index. In the case of free-free transitions for optically thin gas $\alpha = -0.14$, and for optically thick gas (black body flux distribution) $\alpha = 2$. If density decreases with distance as r^{-2}, then $\alpha \approx 0.6$ (Wright and Barlow 1975).

The indices observed range from -0.25 to +1.25 and appear concentrated between 0.50 and 1.0 (Seaquist 1988). This means that most stars have a stable spherically symmetric outflow of matter with constant velocity, although in some stars the ionized gas could have variable optical thickness. The optical thickness appears to be larger for S-type stars because the observed fluxes are lower by several orders than those calculated from the H_β line intensity for optically thin gas. This supports the idea that S-type star nebulae are compact and most of the H_β emission comes from accretion disks, intersystem gas flows and chromospheres. For D-type stars the observed and predicted radiation fluxes are comparable which agrees well with the lower electron density and larger size determined from optical lines.

In only ten of the symbiotic stars observed with the VLA was it possible to measure angular sizes of nebulae (Taylor 1988). They are within 0.2 arcsec. These stars have all experienced nova-like outbursts in the past. Kenyon (1986) estimated mass loss rates of $10^{-7} - 10^{-6} M_\odot$/yr for four stars, for which α is close to 0.6 (the case of continuous outlow).

Although the excitation of this gas in the inhomogeneous envelopes of symbiotic stars has not yet been investigated in detail, many authors have considered some aspects of its dynamics and evolution.

Wright and Barlow (1975) supposed that the nebula is formed by a stellar wind from the cool component and that is is completely ionized by the radiation of the hot star. In this model the observed radio emission is related to the parameters of the stellar wind. The assumption of complete ionization of the nebula is apparently incorrect. Therefore, Taylor and Seaquist (1984) considered the two-dimensional problem of ionization of the nebula formed by a stellar wind from the cool star. The ionization depends on the UV-luminosity of the hot source, the separation of the components and the mass loss rate from the cool star. At low ionizing capability, X, the ionization zone is a comparatively small sphere surrounding the hot source. As X increases the ionization zone grows and at high X practically the whole nebula is ionized with the exception of a small cone surrounding the cool star.

Thermonuclear flashes on the surface of the hot component give rise to the

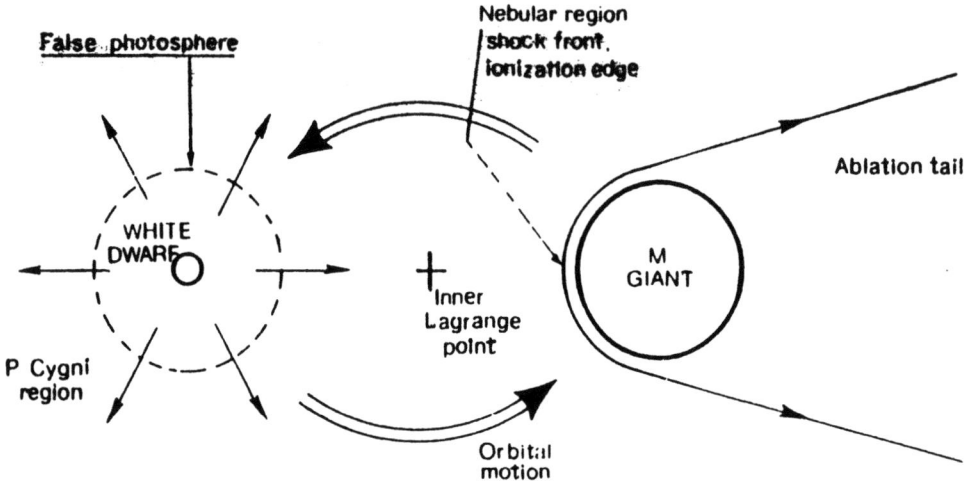

Fig. 6. Schematic diagram of the AG Peg system (after Penston and Allen 1985).

ejection of matter and the subsequent stellar wind. The two winds interact and the wind from the hot component will strongly distort the originally spherical outflow from the cool component. This phenomenon was studied by Penson and Allen (1985) for AG Peg and their model is illustrated in Figure 6. One can see that the cool gas forms a tail behind the cool component.

It should be noted that gas dynamical processes in symbiotic stars are usually studied in a qualitative manner and a detailed treatment is one of the main tasks in modelling symbiotic stars.

5. Orbital Motion

Orbital motion in stars can manifest itself in the form of periodic changes of the radial velocity and/or brightness if orbital plane inclination is favorable.

Merrill (1929) first discovered that the radial velocities of AG Peg change with a period of about 800 days. It was ascertained that the radial velocities measured for different elements, though having one period, exhibit different amplitudes and phases. This shows that different lines are formed in different regions of the binary system. Similar periodic changes of radial velocities have been discovered in other symbiotic stars.

In this connection there arises a very important question: how can one choose from such a large variety of radial velocity curves those which best represent the orbital motion of both components. Obviously, a radial velocity curve from the absorption lines of the cool star represent the orbital motion of this component. It is more difficult to determine radial velocities which characterize the motion of the hot component. As was mentioned above, absorption lines which arise in the atmospheres of hot components are not observed in the spectra of symbiotic stars. Permitted lines of highly ionized elements can presumably be such candidates, because they can occur in relatively dense parts of the envelope of a symbiotic binary that are located close to the hot component (Boyarchuk 1969). Naturally, the variations of the radial velocity of these lines should be in anti-phase to the changes in radial velocity determined from the absorption lines. In Figure 7, radial velocity curves for four symbiotic stars with the periods from 370 to 9740 days are shown. The amplitude ratio of the radial velocity curves is $\sim 1 : 3$.

If we assume the cool giant mass to be $2 - 3 M_\odot$, the hot component mass will be $0.6 - 1.0 M_\odot$, which corresponds to the masses of sub-dwarfs and white dwarfs.

The radial velocity amplitudes of the cool giants are not very large, and to determine them the radial velocities must be measured with very high accuracy. The best measurements of the radial velocities of cool components appear to have been made by Garcia and Kenyon (1988) who measured them for nine symbiotic stars with a reticon combined with an echelle spectrograph (Figure 8). Mean square errors of their measurements are $\pm$ 0.5 km/sec. The orbital periods within $\sim$ 200-800 days and the semi-amplitudes of the radial velocity curves of the cool components are from 5 to 9 km/sec which is in general agreement with the measurements made earlier.

Eclipses have been discovered in several symbiotic stars. They are most distinctly observed in CI Cyg (Belyakina 1983). In this system eclipses of the hot component by the cool one are shown in Figure 9. When the hot component is a very compact star, its contribution to optical radiation is small and we do not observe its eclipse. According to photometric observations (Belyakina 1979) activity started in 1973 and resulted in the formation of an extended atmosphere or a dense envelope around the hot component. As a result, the radiation of the hot component corresponded to that of an F5 star and completely dominated the optical spectrum. Thus, very deep eclipses were observed, especially in 1975. Out of eclipse the F-type spectrum is observed, and during the minimum the M-type spectrum is observed, i.e. spectral observations confirm that the hot component is eclipsed by the cool giant. It may be concluded from the duration of the eclipse that the size of the cool component is about 300 $R_\odot$, i.e., it is of luminosity class II. After 1985 CI Cyg activity stopped, and the hot component returned to its normal state. Its intensity in the optical region became negligible and we observe brightness variations of the cool component which are due to the reflection effect.

Leibowitz and Formiggini (1988) carried out a rather thorough study of the light curves of the three symbiotic stars AG Dra, AX Per and AG Peg in a quiescent state and showed that the brightness variations of these stars are mainly due to the reflection effect. In addition, a shallow eclipse was found for AG Dra (Figure 10).

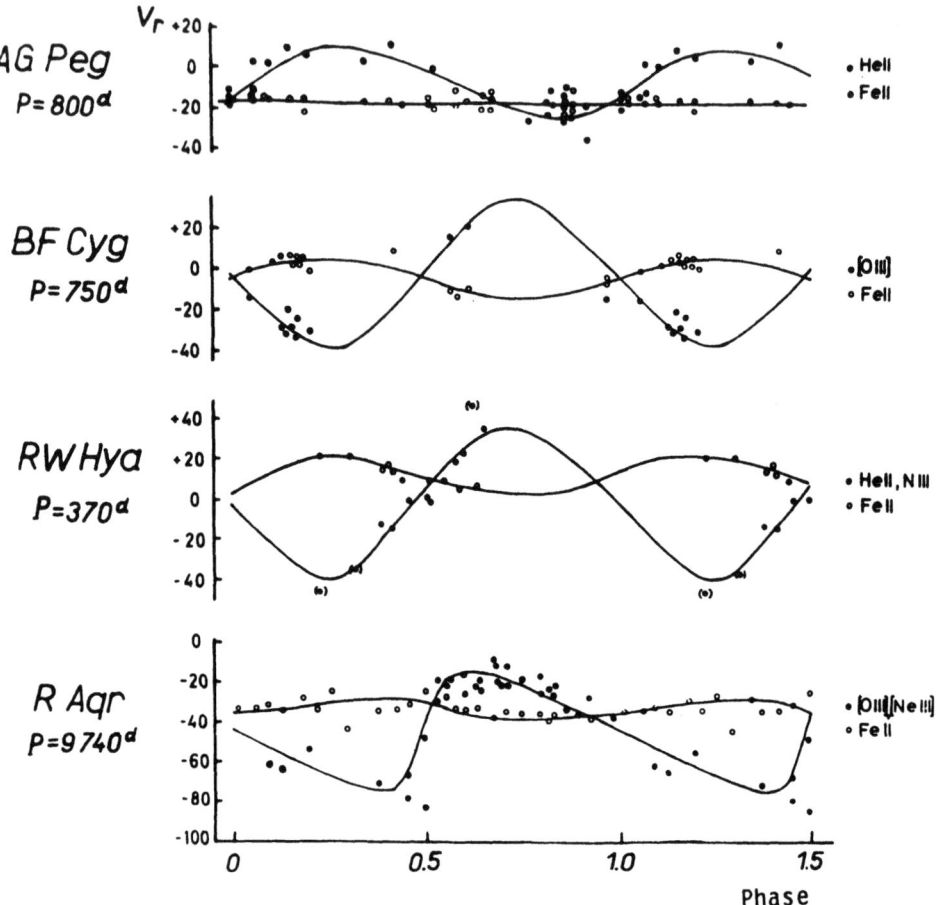

Fig. 7. Radial velocity curves of symbiotic stars.

Summing up the observations of symbiotic stars, one can postulate an average model for symbiotic stars.

Observationally there are at least two types of symbiotic stars. S-type stars consist of a cool giant with mass $M_c \sim 2 - 3 M_\odot$ and a hot subdwarf with $M_h \sim 0.6 - 1 M_\odot$. The orbital period is 1.5-2 years and the distance between the components is about 3 AU. There is a gas envelope and in the far IR-region dust emission is observed.

D-types consist of a Mira-like cool component and a hot component. Orbital motion is not detected in this type of symbiotic star and we can only assume $M_c \sim 1 M_\odot$ and $M_h \sim 0.8 M_\odot$. The period probably exceeds 1000 years and the distance between the components is ~ 100 AU. There is an extensive gas envelope, and the presence of dust is evident even in the near IR-region.

There arises the question of why short-period D-type and long-period S-type

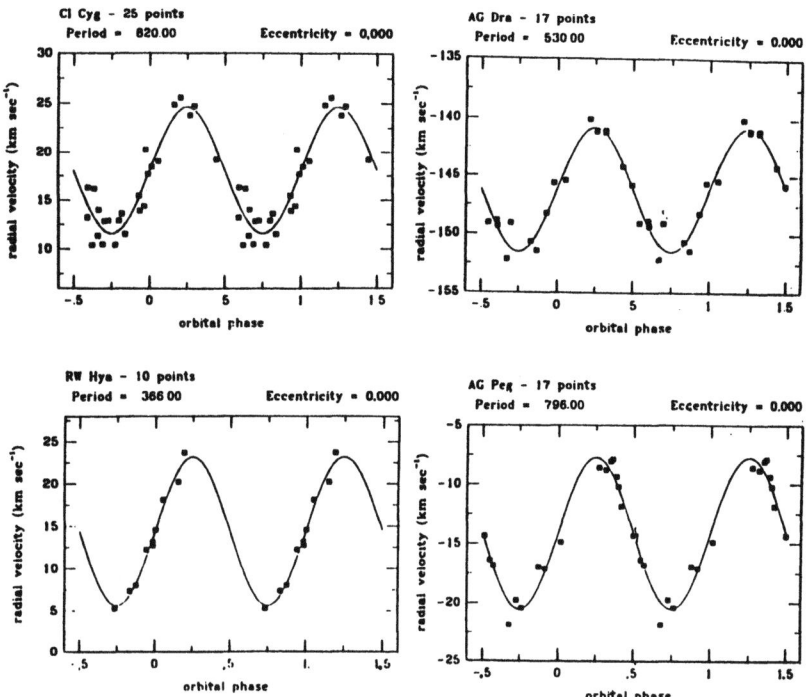

Fig. 8. Radial velocity curves of the cool components of symbiotic stars (after Garcia and Kenyon 1988).

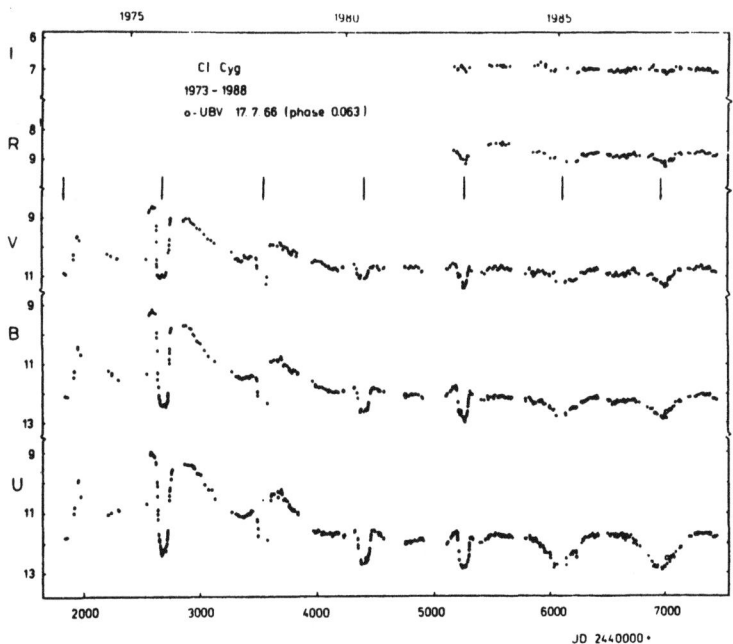

Fig. 9. Light curves of CI Cyg (after Belyakina 1990).

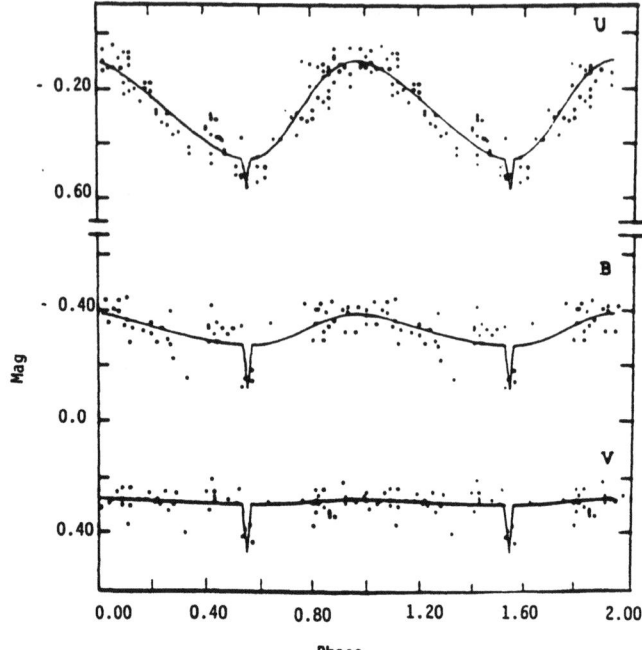

Fig. 10. Light curves of AG Dra, folded to its binary period of 554 days. Points are observations. Solid lines are the best fit solution of the reflection model (after Leibovitz and Formiggini 1988).

symbiotic stars are not found. In a close pair the cool component does not evolve to the Mira stage. Due to binarity it looses mass extensively in the red giant stage and passes into the subdwarf stage without becoming a Mira-like star at all. The size of Miras is 2-3 AU, which equals the distance between the components of symbiotic stars. Hence for relatively close pairs it would more likely evolve into an S-type system. On the other hand, in wide pairs the hot component is so distant from the cool star that it excites a comparatively weak stellar wind from the cool star with emission lines of sufficient intensity for symbiotic phenomena to be observed.

6. Active Phase of Symbiotic Stars

The model of symbiotic stars described above gives a satisfactory explanation of their quiescent state. Hot subdwarfs or white dwarfs excite the gaseous envelope. Since the characteristics of both types of stellar sources change on an evolutionary time scale, it is impossible to explain in the frame of this model the drastic changes of both spectrum and brightness that take place in practically all symbiotic stars. Hence, there must exist an additional energy source of a variable character with a typical time scale up to several tens of days.

Accretion of matter ejected by the cool component onto the hot one was proposed to be such a source. The mass loss rate is variable due to various instabilities of the cool giant. Hence the accretion rate onto the hot component should be variable too. As a result variability occurs. The idea that accretion energy is the main energy source of symbiotic stars was thoroughly developed by Bath and Pringle (1982) and Duschl (1988). In this case a white dwarf, subdwarf, or main sequence star surrounded by an accretion disk may be the companion of a cool giant. The critical factor is the value of the accretion rate. It is minimal for a white dwarf and maximal for a main sequence star. But in any case, the accretion rate necessary to provide the observed characteristics of symbiotic stars is much higher than could result from outflow from the cool component if we use estimates of mass loss rates for single stars. According to estimates, symbiotic star components do not fill their critical Roche lobes (Boyarchuk 1986). Therefore, outflow through the Lagrange point is likely to be negligible. Thus the assumption that accretion energy plays the main role in symbiotic stars seems to be problematical.

Tutukov and Yungelson (1976) proposed that accretion onto the degenerate dwarf is the source of symbiotic activity. The main energy source is not the accretion itself, but nuclear reactions in the hydrogen-rich surface layer consisting of matter accreted by the dwarf. Paczinski and Rudak (1980) showed that, subject to the accretion rate, various types of layer sources may be realized. Later these ideas were developed in many papers. Current ideas about this question are summarized by Kenyon (1988b).

In degenerate CNO- or He-dwarfs nuclear burning ceases and they cool down radiating energy. But if in a binary system hydrogen-rich matter is accreted onto their surface, a surface layer will be formed where hydrogen burning can occur. Depending on the rate of the accretion, different situations may arise.

1. The accretion rate $\dot{M}_{acc}$ is lower than the critical value $\dot{M}_{1c}$ Iben (1982) has determined this value as

$$\dot{M}_{1c} = 1.32 \times 10^{-7} M^{3.57} M_\odot/\text{yr},$$

where M is the mass of accreting star in solar units. For subdwarfs $M \simeq 0.8 M_\odot$, so that $\dot{M}_{1c} \simeq 6 \times 10^{-8} M_\odot/\text{yr}$. In this case initially the temperature at the bottom of the hydrogen envelope will not be high enough to ignite hydrogen. But due to the growth of the envelope the temperature increases and when it reaches a critical value, hydrogen burning begins and the outburst occurs. Then a quiet phase of hydrogen accumulation recurs.

2. The accretion rate is higher than the critical value $\dot{M}_{1c}$. In this case at any time there is enough matter on the surface of the star for hydrogen burning. A continuous stable burning without any outbursts takes place and we observe a very hot star.

3. The accretion rate $\dot{M}_{acc}$ is higher than the critical value $\dot{M}_{2c}$

$$\dot{M}_{2c} = 6 \times 10^{-7} (M - 0.52)^{3.57} X^{-1} M_\odot/yr,$$

where X is the mass fraction of hydrogen in the accreted envelope (Kenyon, 1988). For $M = 0.8 M_\odot$ and $X = 0.8$ the critical rate is $\dot{M}_{2c} = 1.7 \times 10^{-7} M_\odot/\text{yr}$. Now there will be too much accreted matter and it will not have time to burn but will be accumulated in the form of an extended atmosphere in equilibrium with the layer source. The spectrum of such a system will be similar to that of F-type or later supergiants.

Symbiotic stars are not a homogeneous group and the behaviour of many of them can be explained within this scheme.

As already mentioned, symbiotic D-type stars are wide pairs and accretion is not strong $\dot{M}_{acc} \ll \dot{M}_{1c}$. Therefore, for long periods hydrogen is accumulated and then an outburst takes place. Examples of such stars are HM Sge and V1016 Cyg. Their outbursts are reminiscent of very slow novae.

S-type stars are closer pairs and the accretion rate is higher. If $\dot{M}_{acc} < \dot{M}_{1c}$ accumulation of hydrogen with the subsequent outbursts takes place as for D-type stars. AG Peg is an example. If the accretion rate is high $\dot{M}_{acc} > \dot{M}_{1c}$, stable burning occurs and the symbiotic star is observed in a quiet phase. If the accretion is onto a white dwarf, the additional radiation of the layer source makes the observed flux of radiation similar to that of a subdwarf. With an increase of the accretion rate, the radiation of the hot component will increase and light oscillations will become detectable. Since the ratio $\dot{M}_{2c}/\dot{M}_{1c}$ is about 2-3, light variations at steady burning can reach one magnitude. With a considerable growth of $\dot{M}_{acc}$ it may happen that $\dot{M}_{acc} \gg \dot{M}_{1c}$ and the hot component will mimic an F5-type star which with a radius of several tens times that of the Sun. The brightness of the symbiotic star will grow by several orders with the main contribution to the emission being the hot component. Active phases of Z And in 1961 and CI Cyg in 1975 are examples of such phenomena.

7. Conclusion

It was shown in this review that symbiotic stars are a peculiar class of interacting binaries which consist of cool giants and hot stars of low luminosity. It is interesting to consider their evolutionary stage. First of all it is necessary to determine their age. The calculations of the evolutionary tracks of single stars are of dubious value due to uncertainty of the influence of component interaction on their evolution.

Boyarchuk (1975) assumed that the luminosities of the cool components are close to those of normal giants of the same spectal type and that the cool components contribute half of the total radiation from the symbiotic star in the V-band and then determined the distances to symbiotic stars taking into account the interstellar absorption. He considered the distribution of symbiotic stars in the galactic plane and in the z-coordinate. The stars are concentrated in the galactic plane and show no concentration toward the galactic centre. One can conclude that the symbiotic stars belong to the old disc population. One can also draw this conclusion from an analysis of radial velocities. Kenyon (1986) came to similar conclusions. The space distribution of symbiotic stars resembles very much that of planetary nebulae.

About 20 symbiotic stars are located within 2 kpc of the Sun. If symbiotic stars are distributed homogeneously in the galactic plane then the total number of symbiotic stars will be equal to ~ 1000. This estimate is rough, but the symbiotic stars are relatively rare objects.

It was shown that symbiotic stars and nuclei of planetary nebulae are related objects in many respects. If one considers nuclei of planetary nebulae as an evolutionary stage of single stars of a certain mass, then the symbiotic stars can be considered as the similar evolutionary stage for binaries in which initially more massive components have evolved into subdwarfs and initially less massive components are still red giants. A degenerate dwarf with a steady burning shell can imitate a subdwarf.

THEORY AND OBSERVATIONS OF CLASSICAL NOVAE IN OUTBURST

S. STARRFIELD

I have become convinced that the whole nova phenomenon must be studied; the variations of total light and continuum, of radial velocity, and the intensities and profiles of absorption and emission lines must be seen as connected parts of one physical phenomenon, rather than as isolated data that can be understood separately.

– C. Payne-Gaposchkin (1957)

1. Introduction

Novae in outburst have been the object of intense study for more than 100 years but it is only in the last two decades that we have begun to make serious progress in understanding the cause of both their outbursts and the extraordinary phenomena that they display during their outbursts. The multiwavelength observations of these fascinating objects exhibit such a diversity of behavior, both in outburst and in quiescence, that virtually all of the techniques and analyses of modern astronomy must be used to understand the observations. Our advances in understanding the classical novae outburst have come about because of observations with digital detectors, satellite observations at wavelengths unobservable from ground based observatories, and theoretical studies with large computers. For example, as a direct result of studies done with both the IUE and EXOSAT satellites we have:

1. been able to identify two major classes of outburst, one that occurs on an oxygen, neon, magnesium white dwarf and one that occurs on a carbon, oxygen white dwarf;
2. obtained abundances for a large number of recent novae;
3. identified and studied at least two classes of recurrent nova, one that has a giant for the mass-losing companion and one that has a compact but evolved companion;
4. found evidence at late times for a high temperature ($T > 3 \times 10^5$ K) source within the system;
5. found that the outburst lasts longer in the ultraviolet than in the optical; and
6. found that the outbursts of the fastest classical and recurrent novae exceed the Eddington luminosity, for a 1.0 $M_\odot$ white dwarf, at maximum light.

Studies of the infrared emission of novae done over the past two decades have shown that novae form grains during their outbursts and it has been possible to actually observe the grains form and follow the changes in their emission characteristics over time. These same studies have shown that all types of grains, found in the ISM (carbon, silicates, SiC, and PAH), are produced in the ejecta of novae. In addition, Nova QV Vul 1987 formed all four types of grain during the course of its outburst which seriously constrains the theories of grain formation. Infrared observations of novae have also been used to obtain black body expansion parallaxes twice during the outburst providing improved distance estimates to galactic novae.

As a result of theoretical hydrodynamic studies done over the last two decades, the nova outburst is now thought to be the result of a thermonuclear runaway (TNR) which occurs in the accreted hydrogen rich envelope of a white dwarf in a close binary system. The hydrodynamic simulations of the growth of the accreted layer on the white dwarf have been very successful in reproducing the gross features of the nova outburst: the amount of mass ejected, the kinetic energies of the ejecta, and the optical light curves. These studies have also shown that it is necessary to include both a nuclear reaction network and the physics of the infalling material in the calculations. More important, these calculations *predicted*: (1) that enhanced CNO nuclei would be found in the ejecta of fast novae, (2) that the isotopic ratios of the CNO nuclei would be far from solar, (3) that there should be a post maximum phase of constant luminosity lasting for years, (4) that fast classical novae would have larger CNO enhancements than typical slow classical novae, and (5) that the observed features of the outburst would be strong functions of the mass of the underlying white dwarf.

The most important summaries of the nova phenomena can be found in Payne-Gaposchkin (1957) and McLaughlin (1960). References to previous studies can be found in these publications and will not be repeated here. Reviews can also be found in articles by Gallagher and Starrfield (1978), Starrfield and Snijders (1987; reprinted in 1989); Starrfield (1986, 1987, 1988, and 1990), Gehrz (1988), and Shara (1989). More recently, a book has appeared which was edited by Bode and Evans (1989) and which discusses the behavior of classical novae both in outburst and quiescence. The two most recent IAU Colloquia devoted entirely to the nova phenomena were in Paris in 1976 (Friedjung 1977) and Madrid in 1989 (Cassatella and Viotti 1991). Bode (1987) edited a book on the 1985 RS Oph outburst and there were many papers on the nova outburst in the 1986 Bamberg Workshop on Cataclysmic Variable Stars (Drechsel, Kondo, and Rahe 1987).

2. Observations of the Outburst

2.1. PREMAXIMUM

A nova outburst is classified, according to the rate of decline from maximum, as either 'fast' or 'slow'. The initial eruption of a fast nova is very rapid, with the major part of the rise to visual maximum taking place in a day or less. During the rising branch of the light curve, a classical carbon, oxygen (CO) nova exhibits spectral features corresponding to an optically thick, expanding shell (Stryker *et al.* 1988; Starrfield 1990). The first spectra obtained in the optical are usually dominated by broad absorption lines and emission lines are either weak or absent. Spectral types are B to A, although some novae have been observed to have a later spectral class; for example, RR Pic was listed as having a class of F and Nova QV Vul 1987 was probably G or K (Gehrz and Starrfield 1992, in preparation). In addition, the optical spectrum of V1500 Cyg 1975, one day before maximum, was that of a B2Ia star with unusual absorption line strengths for the C, N, and O elements (Boyarchuk *et al.* 1977; Ferland, Lambert, and Woodman 1986a,b; Lance, McCall, and Uomoto 1988). The unusual strengths, as compared to normal stars, of the C, N, and O lines seen in Nova V1500 Cygni 1975, are rather common (McLaughlin 1960). Spectra obtained in the ultraviolet show a cool continuum but usually the FeII "forest" is superimposed on this continuum (Wehrse *et al.* 1990, 1991).

As the expansion is very rapid and the bolometric luminosity is either nearly constant or still rising, the effective temperature of a CO classical nova declines until it reaches a value of $\sim 4 \times 10^3$ K to 7×10^3 K at visual maximum (Gallagher and Ney 1976; Gallagher and Starrfield, 1978). However, this is not the case for the fastest recurrent (RN) or oxygen, neon, magnesium (ONeMg) novae which reach maximum with their effective temperatures exceeding 10^4 K (Shore, Sonneborn, and Starrfield 1990; Shore *et al.*, 1991). The most probable explanation for this difference in behavior is that considerably less material is ejected by these novae during their outbursts as compared to the other classes of novae (Starrfield, Sparks, and Truran, 1985; Starrfield, Sparks, and Shaviv, 1988).

If a nova is caught early enough in the outburst, an IUE spectrogram will show a continuum rising to the red broken both by emission and absorption lines (see, for example, Starrfield *et al.* 1988b, Stryker *et al.* 1988; Starrfield 1990). An important advance in our understanding of the spectra of novae at maximum, occurred when it was realized that the first UV spectra of SN 1987A strongly resembled early UV spectra of novae; except that the lines were much broader in the supernova. As a result, it was apparent that modern techniques in spherical, expanding, stellar atmospheres, developed to treat SN II (Hauschildt, Shaviv, and Wehrse 1989), could be applied to novae (Wehrse *et al.* 1990, 1991).

The first results of these studies are now being applied to the IUE spectra of novae at maximum. The initial results have shown that the actual effective temperature cannot be determined from the observed continuum but must be obtained from continuum fitting to a very broad region of the spectrum. This is because the expanding atmosphere of a nova has a very large extension and the observed continuum is formed by overlapping lines of FeI, II, and III. The initial analyses have found that a nova with an effective temperature of 25,000 K mimics the continuum of a more normal star with an effective temperature below 10,000 K. The intent of these analyses is to perform spectral syntheses of the observed spectra and determine elemental abundances. These data can then be compared, at a later time,

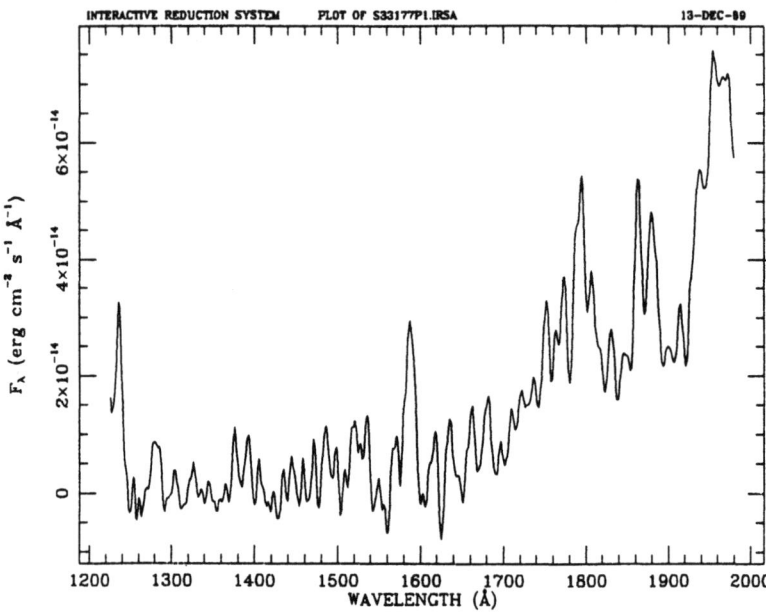

Fig. 1. Low dispersion IUE SWP spectrum obtained of LMC 1988 #1 on day 1988/64. This was a 30 minute exposure. Note the strength of the feature at 1590 Å.

with abundances determined from nebular emission line analyses, and, thereby, both methods can be checked (Wehrse *et al.*, 1990, 1991).

In order to illustrate the early spectral development of a classical CO nova, in the first few figures, I show the evolution over the first few weeks of the outburst of LMC 1988 #1. Figure 1 shows one of the first spectra that was obtained for this nova. It was a 30 minute SWP exposure obtained on March 30, 1988. Figure 2 shows the SWP spectrum of the nova obtained on April 8, 1988 (a 45 minute exposure). This nova continued to brighten in the ultraviolet for sometime after it had reached maximum light in the optical (Austin *et al.* 1990). This can be seen by noting the flux levels in the two spectra. An ultraviolet light curve can be found in Austin et al. (1990) and in Figure 10. This nova did not reach maximum ultraviolet light until about April 25, 1988 while it reached maximum light in the optical around the end of March 1988. This is characteristic of the light curves for all novae: optical maximum occurs before the maximum in the ultraviolet.

Figure 2 also shows the typical emission lines produced by a gas expanding and beginning to go optically thin. [NIII] 1750 Å and [OIII] 1660 Å, 1666 Å are present as is [CIII] 1909 Å. These lines are not generally seen in the ultraviolet spectrum this early in the evolution of a nova. The ultraviolet spectrum continued to evolve rapidly and Figure 3 (April 15, 1988: a 60 min. exposure) is very different from the spectrum shown in Figure 2. Note the strength of [NIII] 1750 Å as compared to all of the other lines that were present in the spectrum. [NIV] 1486 Å is now strong

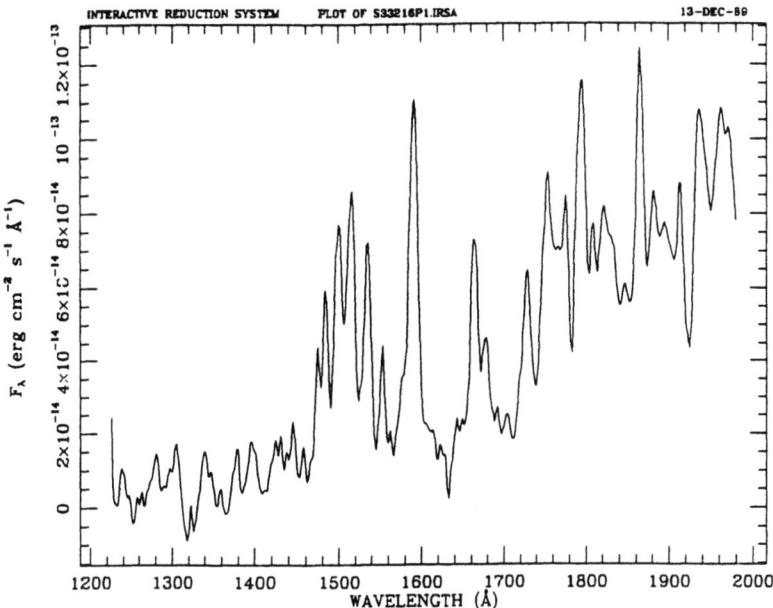

Fig. 2. A low dispersion IUE SWP spectrum obtained of LMC 1988 #1 on day 1988/97. This was a 45 minute exposure. NIII] 1750 Å is now present.

and CIV 1550 Å is present and exhibits a P-Cygni profile indicating mass outflow.

The last spectrum that I show for this nova was obtained on May 16, 1988 (Figure 4) and one can see by the marked change in the lines and their strengths that, in the month since Figure 3 was obtained, the densities in the nova ejecta have dropped considerably. Although a strong blue continuum is still present, emission lines from an optically thin gas are obviously present and strong. [NIII] 1750 Å has dropped in strength, relative to OI 1304 Å. HeII 1640 Å is now clearly present and CIV 1550 Å shows a strong P-Cygni profile indicating that mass is still being lost. For comparison, in Figure 5, I show a 2 minute SWP spectrum of OS And 1986 obtained on 14 December 1986. Note the strong resemblance of this spectrum to Figure 2. The strong feature at ∼1590 Å is unidentified.

In contrast to the spectral development of a moderate speed CO nova like LMC 1988 #1, fast ONeMg novae show a very hot continuum at maximum plus emission lines characteristic of a low density gas. In fact, maximum light in the ultraviolet occurs very soon after maximum optical light. This implies that fast ONeMg nova outbursts occur on very massive white dwarfs with small envelope masses. This same behavior is also seen in RN which eject much less mass than a classical CO nova.

The most recent ONeMg nova studied in outburst occurred in the LMC in 1990 and exhibited outburst behavior very similar to that of the galactic ONeMg nova V693 CrA 1981 (Williams *et al.* 1985). In the next few paragraphs, I discuss the

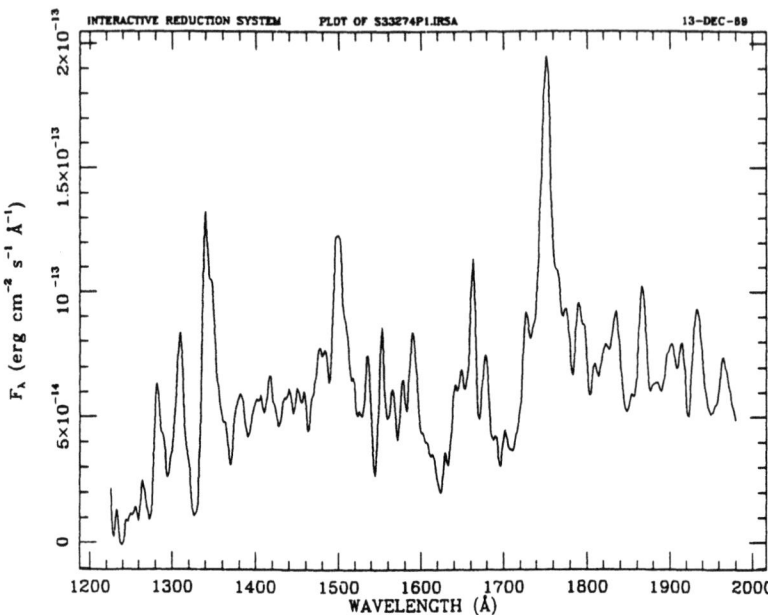

Fig. 3. A low dispersion IUE SWP spectrum of LMC 1988 #1 on day 1988/104. It was a 60 minute exposure. The continuum has begun to flatten but NIII] 1750 Å is still very strong.

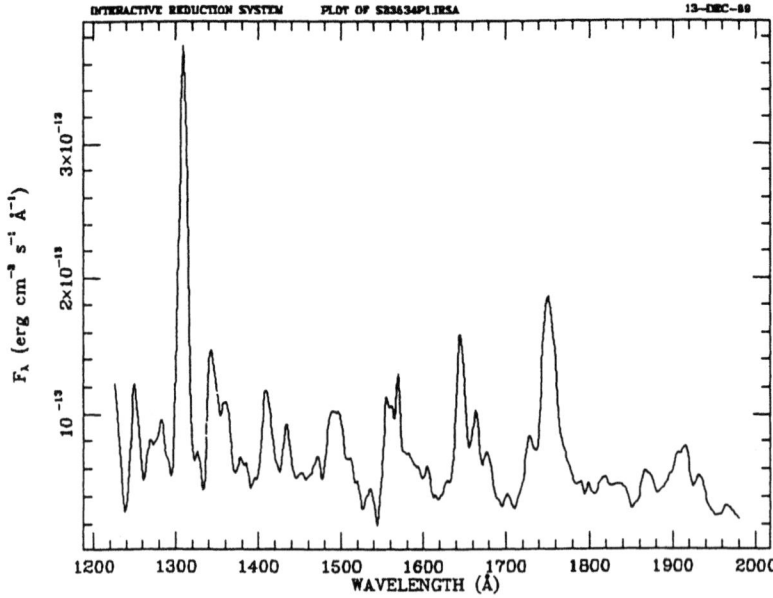

Fig. 4. A low dispersion IUE SWP spectrum of LMC 1988 #1 on day 1988/136. The nova has begun to fade and this was a 100 minute exposure. Note that OI has become the strongest line in the spectrum and that a number of low density lines have begun to appear.

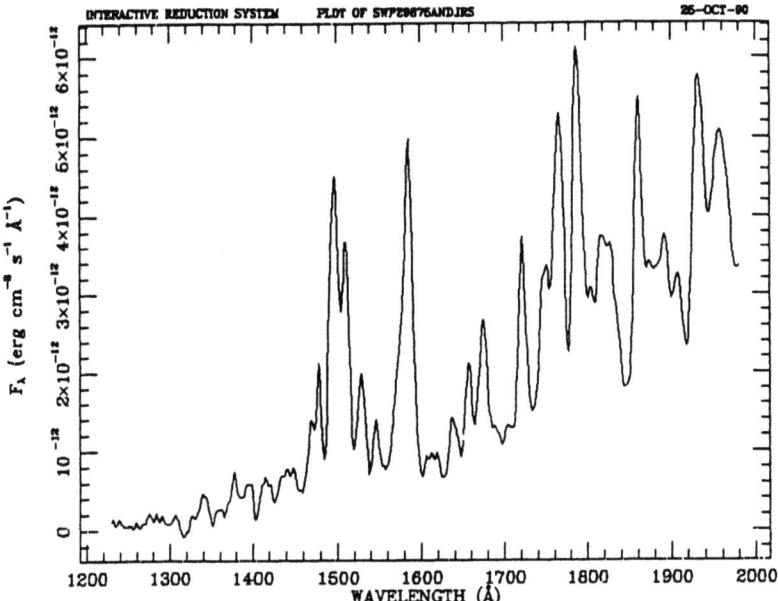

Fig. 5. This is a low dispersion IUE SWP spectrum of Nova OS And 1986 obtained on day 1986/348 which was shortly after discovery. It is a 2 minute exposure. It strongly resembles the spectrum shown in Figure 2.

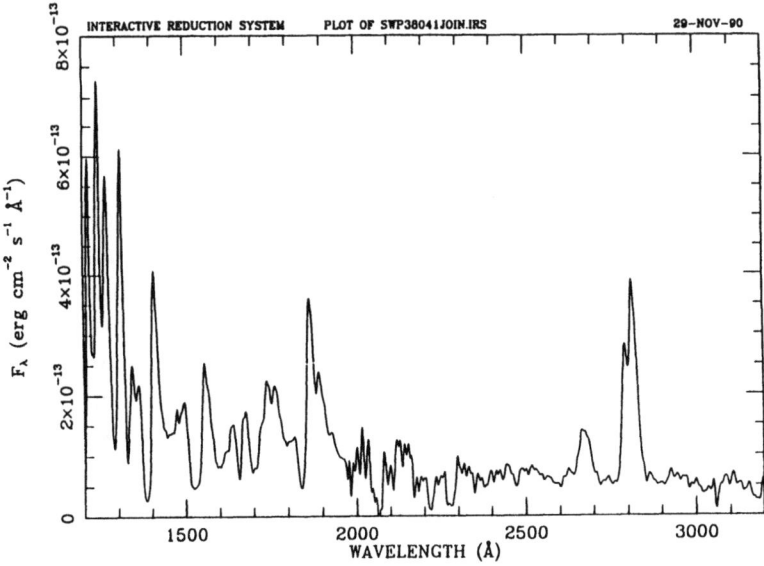

Fig. 6. This was the first IUE low dispersion SWP+LWP spectrum obtained of LMC 1990 #1 and was obtained on day 1990/18. The SWP spectrum was a 30 minute exposure.

observations of the outburst of this nova which is called LMC 1990 #1 because there was a second nova found in outburst in the LMC in 1990. This nova was discovered on 16.47 January 1990 at a visual magnitude of 11.5 and IUE observations began on 17.95 January, within hours of the announcement of the discovery. The initial IUE spectrum (Figure 6) showed a hot continuum and strong emission lines from NV, Si IV, C IV, and AlIII. As can be seen in Figure 6, they all exhibited P-Cygni profiles with flat-bottomed absorption troughs extending to more than -6000 km s^{-1}. There is no resemblance, whatsoever, to early spectra of slow CO novae. However, there is a strong resemblance to spectra of RN such as U Sco suggesting that fast ONeMg novae do not eject much material and the expanding shell has gone nearly optically thin at maximum light in the optical.

The optical spectra obtained by Dopita and Rawlings provided additional evidence that LMC 1990 #1 was an extragalactic analog of V693 CrA (IAUC 4964; note that this same circular also contained the announcement of the discovery of LMC 1990 #2 by Liller) and, thereby, an ONeMg nova. They reported that [NeIII] 3868 Å appeared on 22 January, and [Ne V] 3426 Å on 29 January. By 13 February 1990 [NeV] was the strongest emission feature in the optical spectrum. The ultraviolet forbidden neon lines seen at late stages in other ONeMg novae ([Ne IV] 1602 Å and 2420 Å) were not detected in the IUE spectra but this is not surprising since they did not appear in either V693 CrA or QU Vul until well into the nebular stage when the densities in the ejecta had dropped by a large factor from the values determined at maximum light. In Figures 7 and 8, I show spectra of LMC 1990 #1 obtained on 25 January and 4 February 1990. Note the marked change in the spectrum over this last time interval. In fact, the IUE spectra showed that the expanding shell became optically thin in the Lyman continuum about 30 January or a little over two weeks into the outburst. Analysis of these spectra indicated that about $10^{-5} M_\odot$ to $10^{-6} M_\odot$ were ejected during the outburst of LMC 1990 #1. This is in reasonable agreement with the results for V693 CrA (Williams *et al.* 1985).

Finally, the ultraviolet light curve suggests that the luminosity in the wavelength region from 1200 Å to 3300 Å reached a value of 3×10^{38} erg s^{-1} at maximum brightness. This value assumes a distance to the LMC of 55 kpc [E(B-V)=0.15, non-30 Dor extinction] and shows that the luminosity in this nova exceeded the Eddington luminosity for a $1.0 M_\odot$ white dwarf. Finally the nitrogen abundance appears to be larger than normal for the LMC. Spectra of a galactic fast ONeMg nova can be found in Williams *et al.* 1985. In order to demonstrate the difference between a fast ONeMg nova and a galactic *slow* ONeMg nova, in Figure 9 I show an early spectrum of QU Vul 1984 #2. It was a 15 minute exposure obtained on 11 January 1985 which was about two weeks after discovery. This nova is clearly still in the optically thick phase and more closely resembles the slower CO novae than the fast ONeMg novae. It clearly ejected more material than the fast ONeMg novae.

Infrared observations, when obtained early in the outburst, show that it is radiating like a black body expanding with time. Because it is optically thick in the

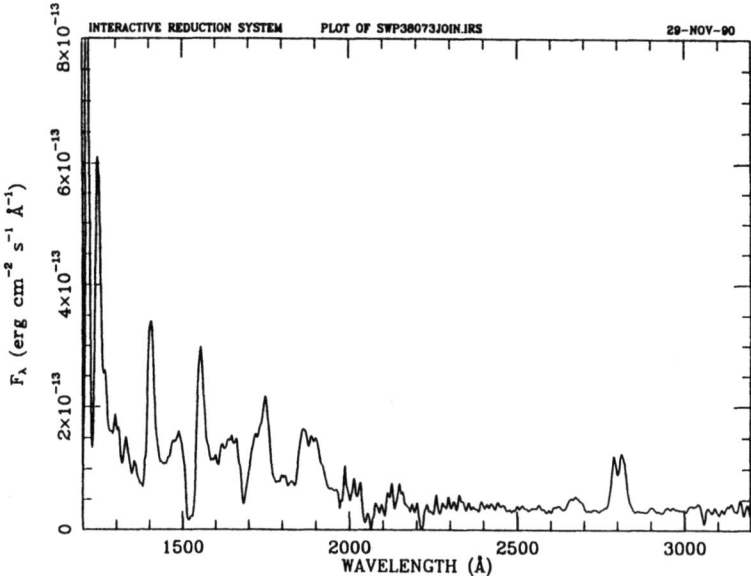

Fig. 7. This low dispersion IUE SWP+LWP spectrum was obtained on day 1990/25. The SWP spectrum was a 48 minute exposure. Note the prominent P-Cygni feature at CIV 1550 Å.

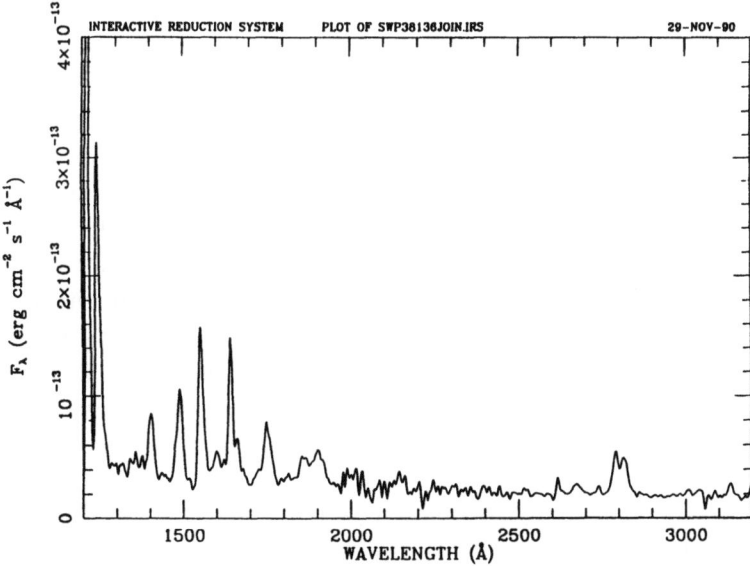

Fig. 8. This low dispersion IUE SWP+LWP spectrum was obtained on day 1990/36 and the SWP spectrum was an 85 minute exposure. The expanding material became optically thin some time between this spectrum and that shown in Figure 7.

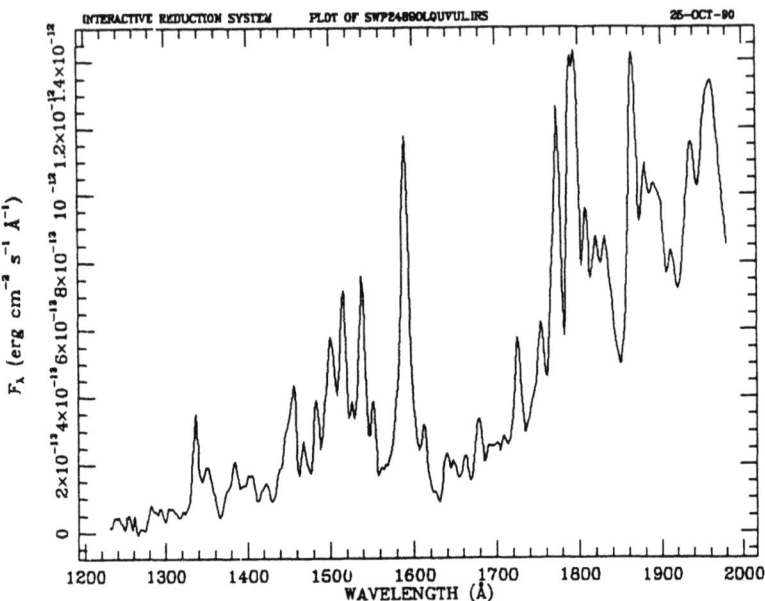

Fig. 9. This IUE SWP low dispersion spectrum was taken of nova QU Vul 1984 #2 on day 1985/11. It was a 15 minute exposure. This was a slow ONeMg nova and early spectra more closely resemble CO novae such as OS And 1986 and LMC 1988 #1 than fast ONeMg novae such as LMC 1990 #1 or V693 CrA 1981.

infrared, Ney and Hatfield (1978) refer to this as the pseudo-photospheric expansion phase. An excellent review of the infrared emission of novae can be found in Gehrz (1988). Gallagher and Ney (1976) used the existence of this phase to determine the distance to V1500 Cyg by measuring the rate of the angular expansion of the photosphere and combining that rate with the doppler velocity of the lines observed at that time. Given both that the measured velocities correspond to the material that is producing the black body emission and that this material is ejected with spherical symmetry, one can determine a very accurate distance. This expansion parallax method has now been applied to all novae that were observed early enough in the outburst to detect the expanding photosphere (Gehrz 1988).

At some time after the expanding material begins to go optically thin at infrared wavelengths, the nova enters the free-free expansion phase (Gallagher and Starrfield 1978). Gehrz *et al.* (1974), Ennis *et al.* (1977), Gehrz *et al.* (1980a,b), and Gehrz (1988) show that the density of the expanding shell can be determined from the wavelength where the optically thin free-free emission turns over into the Rayleigh-Jeans tail. This result is also well known from radio studies (Hjellming 1990).

2.2. MAXIMUM

The maximum in optical light occurs when the opacity in the outer layers of the expanding (and cooling) gas declines to a value where they begin to go optically thin and the pseudo-photosphere moves inward in mass as the material continues its expansion. Because the deeper layers, which are now becoming visible are hotter, the bolometric correction increases and the optical brightness declines. Although originally thought to occur because of the recombination of hydrogen, which would produce the required decrease in the opacity, it is now known that the presence of numerous lines of FeI, FeII, and similar ions in the ultraviolet are as least as important as hydrogen in determining the location of the photosphere in the expanding gas (Wehrse *et al.* 1990).

In addition, because the opacity is lower longward of the Balmer discontinuity at 3647 Å, radiation preferentially escapes there. This causes the expanding layers to cool rapidly and as the temperatures decline below $\sim 1.2 \times 10^4$ K important changes occur in the emergent spectrum. The most striking change is the appearance of strong absorption from FeII which acts to increase the opacity and shift the peak of the emitted radiation to about 2800 Å. Therefore, the ultraviolet spectra of novae obtained at or near maximum show, in general, a cool continuum with strong evidence for FeII features. In fact, it is not clear that any of the sharp "emission" features seen in the IUE spectra are really emission lines; they may only be regions of transparency between overlapping absorption lines. For example, the prominent feature at ~ 1590 Å in Figures 1, 2, 5, and 9 is ubiquitous in early spectra of optically thick nova shells but cannot be identified with any abundant element. In addition, at virtually the same time, nearly all of the Balmer lines develop P-Cygni profiles indicating that a large amount of material has been ejected during the outburst.

At maximum the Principal Spectrum (see McLaughlin 1960) appears in the optical. The features that are seen evolve continuously in velocity as the receding photosphere moves inward in mass and many of the absorption lines seen at maximum can eventually be identified with the emission lines in the nebular shell (McLaughlin 1960; Gallagher and Starrfield 1978). As the nova begins its decline, the continuum declines more rapidly than the emission lines and the P-Cygni profiles slowly disappear. After a short time, the emission lines dominate the spectrum. They have rather complex profiles and suggest that material has been ejected in clouds, blobs, or rings (Gallagher and Starrfield 1978). As was predicted in the mid-1970's (Gallagher and Starrfield 1976; Bath and Shaviv 1978); the fading continuum has now been shown, by ultraviolet observations done with the IUE satellite, to be caused by the hardening of the emergent intensity as the pseudo-photosphere moves inward to deeper and hotter layers (Austin *et al.* 1990).

For a typical classical nova, where the mass of the ejecta may range from $10^{-6} M_\odot$ to as large a value as $10^{-5} M_\odot$, the duration of the photospheric phase depends upon the speed class of the nova. In fast, optically luminous, novae the ejected shell may become optically thin in a few days; thereafter, the optical is dominated by bremsstrahlung and hydrogen bound-free emission (Gallagher and Ney 1976; Ennis *et al.* 1977; Gehrz 1988; Martin 1989). In slower, lower optical brightness novae, continuous mass loss can maintain a low temperature photosphere for several months to years (Bath 1978; Ney and Hatfield 1978). For example, LMC 1988 #1 took nearly two months for the expanding envelope to become optically thin while LMC 1990 #1 became optically thin within a few weeks of maximum light (Sonneborn, Shore, and Starrfield 1990). For smaller values of ejected mass, $10^{-7} M_\odot$, which is typical of RN and the fastest of the ONeMg novae, the optically thick phase is severely curtailed and may last only a few days if it occurs at all.

2.3. POST-MAXIMUM DECLINE

The post-maximum decline in the optical is the result of the relaxation of the hydrostatic remnant to a stable luminosity and the gradual redistribution of the luminosity into the UV and EUV as the shrinking photosphere moves inward in mass. The stable luminosity is set by the core mass of the underlying white dwarf and can be estimated from the core-mass luminosity relationship (Paczynski 1971). For the most massive white dwarfs the luminosity predicted from this relationship will nearly equal that of the Eddington luminosity (Starrfield *et al.* 1990a):

$$L_{\text{Edd}} = \frac{4\pi c G M_{WD}}{\kappa_{es}}$$

for pure electron scattering opacity. When expressed in magnitudes, L_{Edd} has a value of $M_{\text{Bol}} \cong -7$ for a typical nova near maximum. When this limit is applied to novae in outburst, it implies that nova systems contain massive white dwarfs.

It is now clear that the bolometric luminosities of the fastest and brightest novae initially exceed the Eddington luminosity (see for example, V1500 Cyg: Wu and Kester 1976; LMC 1988 #2: Austin *et al.* 1990; LMC 1990 #1: Sonneborn, Shore, and Starrfield 1990; LMC 1990 #2: Shore *et al.* 1991), they drop to the core-mass luminosity soon after maximum. In Figure 10, taken from Austin *et al.* 1990, I show the ultraviolet light curves of LMC 1988 #1 and #2 obtained from integrating under SWP+LWP IUE spectra. In order to obtain the luminosity, they assumed a distance to the LMC of 55 kpc. The depression in the light curve of LMC 1988 #1 was caused by the formation of dust. In contrast, novae with maximum absolute visual magnitudes fainter than -7 show little change in M_{Bol} after maximum; the initial decline in the visual is caused by flux redistribution into the UV (Bath and Shaviv 1978). At this time mass is being ejected from the white dwarf by a radiation pressure driven wind (Starrfield *et al.* 1990b) which will continue until most of the remaining accreted material (plus some core material from the white dwarf) on the white dwarf is lost.

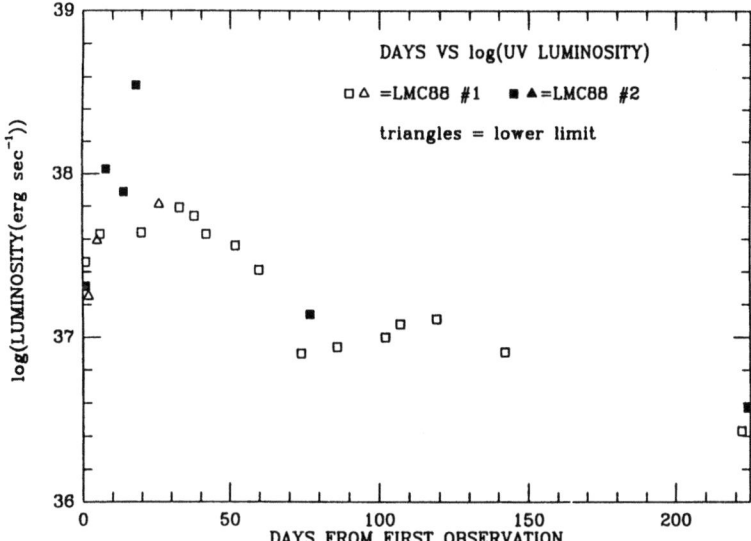

Fig. 10. This figure was taken from Austin *et al.* (1990) and shows the ultraviolet light curves of LMC 1988 #1 and #2. Each point is obtained by integrating under a combined SWP + LWP spectrum (minus Lα) and was put on an absolute scale by assuming a distance of 55 kpc to the LMC. LMC 1988 #2 clearly exceeds LEd at maximum. The depression in the decline of the light curve of LMC 1988 #1 marks the formation of dust.

During the declining branch of the light curve, novae undergo important changes in physical conditions. Because the hardening of the radiation field from the central source produces an ionization front moving out through the expanding ejecta, because the density of matter is high compared to most nebulae, and because the process is time dependent; the spectrum can become exceedingly complex, especially for novae with significant post-maximum stellar winds (Gallagher and Starrfield 1978). In addition, optical high-dispersion spectra (1 Å to 3 Å or less) show that the material is ejected in a very inhomogeneous fashion (Wagner *et al.* 1992; in preparation). Figures 11 and 12 are taken from this paper and show, respectively, the Hα profile (resolution 3 Å) for OS And 1986 and QV Vul 1987. These spectra were obtained in November 1990 with the Perkins 1.8-m telescope at Lowell Observatory using the Ohio State CCD spectrograph. It is interesting that the profiles for these two novae seem to show the same general shape.

New absorption line systems now appear (in the optical) at high velocities compared to the principal absorption line system. These systems are called the diffuse enhanced and consist of broad features that may separate (as time passes) into many sharp components. At still later times in many novae, very highly ionized features appear in absorption at even larger velocities than the diffuse enhanced systems. These systems are collectively called the Orion spectrum (based on their similarity to features seen in OB stars) and are optical analogues of the 'sharp, narrow, absorption' components seen in the ultraviolet spectra of the most luminous O and B stars which also exhibit strong stellar winds. However, the Orion features always seem to remain broad and do not separate into sharp components.

During this stage the ionization increases to levels of 50–60 eV and the electron densities decrease to values of 10^8 to 10^{10} cm^{-3}. We can now use the techniques developed for the analysis of planetary nebulae and quasars to determine the variation of electron density and temperature with time and, also, the elemental abundances (Ferland and Shields 1978; Ferland, Lambert, and Woodman 1986a,b; Lance, McCall, and Uomoto 1988). Since novae are time dependent, this phase represents an interesting exploration of nebular physics and allows us to use the time variation as an additional constraint in determining the abundances.

2.4. INFRARED EMISSION AND GRAIN FORMATION

Some two to three months into the outburst, most novae begin to develop a second phase of infrared emission. The assumption commonly made is that this excess is caused by the formation and growth of grains in the expanding shell which then reradiate the ultraviolet energy from the hot, luminous white dwarf (see Gehrz 1988 and Gehrz, Truran, and Williams 1991). When first detected, the infrared excess attributed to grains exhibits a continuum black body temperature slightly exceeding 10^3 K. If this material has just formed, then this temperature is considerably below $\sim 2 \times 10^3$ K at which the grains are normally expected to form. As time passes, the temperature of the grains slowly decreases to $\sim$800 K. One explains the variation in temperature by the formation of small particles which are inefficient radiators. They then slowly grow to a size that exceeds that of normal interstellar particles (0.01μm–0.03μm).

Until recently, the composition of the grains was thought to be amorphous carbon since the infrared excess failed to show any features (Gehrz 1988). However, infrared studies of V1370 Aql 1982, already known from IUE studies to be unusual because it ejected material with very strange abundances (Snijders *et al.* 1987), found an infrared excess with a broad continuum peaked at about 8μm. Superimposed on this very broad feature was narrow emission at 10μm that was attributed to SiC (Gehrz 1988). Two years later Nova QU Vul 1984 #2 was observed to form silicate grains since its infrared emission showed both the 10μm and 20μm features characteristic of SiO_2. It showed no continuum excess and its infrared emission never became optically thick (Gehrz *et al.* 1986). Since that time infrared features attributed to PAH's have been detected in Nova QV Vul 1987 (Gehrz 1990). This nova was extremely cool at maximum light and formed an extremely thick dust shell about six weeks into the outburst. Over the course of the next few months, it proceeded to exhibit infrared emission features characteristic of all of the four types of dust detected in the ISM and previously seen in other novae. However, current theories of grain formation predict that at most one type of grain can be formed in material of a given composition and the type of grain depends on the abundance ratio of carbon to oxygen in the ejecta since equilibrium condensation calculations indicate that carbon grains condense in a carbon rich environment and silicates condense if the material is oxygen rich. The nova observations are perplexing.

Another interesting result of the infrared studies of novae is that some novae show optically thick infrared emission from grains while others exhibit only optically thin emission. In both cases they develop the infrared emission that is characteristic of grain formation but, in the optically thin case, the amount of energy radiated by the grains never approaches the energy emitted in the optical at maximum. In contrast, in the optically thick case, the infrared emission in the novae increases until it is radiating the same amount of energy as was observed in the optical early in the outburst. In addition, an optically thick nova exhibits a very steep decline in the optical light curve, at the time of grain formation, with a slow recovery as the grains absorb less energy. This feature is called the transition phase in the optical. Optically thin emitting novae do not show a pronounced transition and sometimes the optical light curve shows no indication that dust has formed. A possible explanation for this phenomenon is that the region where the dust is forming has been ejected asymmetrically and does not completely block the ultraviolet light from the central source (Gehrz 1988). Both Nova V842 Cen 1986 and QV Vul 1987 formed optically thick dust shells with deep transition regions in their optical light curves. However, in contrast to other novae observed in the infrared, QV Vul emitted more energy from grains in the infrared than was seen in the optical at maximum (Gehrz 1989; private communication). A very thick expanding envelope, which reached radii of 10^{13} cm before becoming optically thin, would explain this observation.

Finally, Ney and Hatfield (1978), observing NQ Vul 1976, used the gradual growth in size of the dust forming region to apply the black body expansion parallax method to this nova. Since that time, it has been applied to all novae that showed an infrared excess characteristic of grain formation. Although one is still faced with the problem of determining the appropriate doppler velocity to use in the calculation, it is a very accurate technique for determining nova distances. It will be interesting to check the values determined from this method with the nebular expansion studies done in the optical (Cohen and Rosenthal 1983; Cohen 1985) when the expanding shells of these novae are finally resolved.

2.5. FINAL DECLINE

As the nova begins its final decline, the ionization level in the ejecta increases to moderately high levels and the electron density decreases to values approaching those found in planetary nebulae. This means that the analysis techniques used to obtain abundances in planetary nebulae can also be used for ovae ejecta at late times (Williams *et al.* 1981, 1985; Williams 1990; Stickland *et al.* 1981; Snijders *et al.* 1987; Saizar *et al.* 1991; Austin *et al.* 1991, in preparation; Andrea *et al.*

1990). It is not known when or how a nova ends its outburst, but on a time scale of (at most) a few years, mass loss decreases and the nova returns to its quiescent luminosity (Starrfield 1979; MacDonald, Fujimoto, and Truran 1985; Starrfield 1989; Starrfield et al. 1990a,b). It is at this time that coronal line emission is seen in the infrared spectra of novae.

It has been known for decades that the optical spectra of novae show coronal line emission but such lines were not detected in the infrared until the outburst of V1500 Cygni. Since that time other novae have been found to exhibit coronal line emission with the most notable case being Nova QU Vul 1984 #2 (Greenhouse et al. 1988). Greenhouse et al. (1990) have also found coronal line emission from [CaVIII], [SiVI], and [SiVII] in Nova V1819 Cyg 1986 and Nova V827 Her 1987. Benjamin and Dinerstein (1990) report that they found coronal line emission in V2214 Oph 1988 and they confirmed the emission already reported by Greenhouse et al. for V1819 Cyg and V827 Her. Gehrz (1988) speculates that coronal line emission is probably ubiquitous in novae but is generally hidden by the grain emission. The lines have an excitation temperature of $\sim 10^6$ K and their presence constrains the modeling of the physical conditions in the gas at the time that they are seen. Although Benjamin and Dinerstein (1990) discuss the possibility that the source of the coronal line emission is collisional ionization in a shock, the most likely explanation is photoionization by the hot central source. Numerical simulations of the turn-off of a nova show that the underlying white dwarf can approach temperatures of 10^6 K which seems a likely source of the photoionization (see Figure 18). In their successful attempts to model the outburst of the RN LMC 1990 #2, Shore et al. (1991) found that it was necessary to assume that the underlying white dwarf had a temperature exceeding 2×10^5 K a few days after maximum light. Source temperatures below 10^5 K were prohibited by the observations (see Section 4.2).

The requirements for a hot source are also strongly supported by the X-ray observations of novae in outburst. Novae were first detected in outburst at X-ray wavelengths by EXOSAT which discovered emission from Nova GQ Mus 1983 (Ögelman, Beuermann, and Krautter 1984). This result was followed shortly thereafter by detections of Nova PW Vul 1984 #1, QU Vul 1984 #2, and the recurrent nova RS Oph 1985 (Ögelman, Krautter, and Beuermann 1987; Mason et al. 1986). Clearly, an object radiating at a luminosity $\sim L_{\rm Edd}$ with a radius of $\sim 10^9$ cm, will be emitting copious amounts of soft X-rays. Therefore, the observations of the three classical novae are best understood if we assume that the X-ray flux is coming from the white dwarf which, according to the X-ray observations, has a temperature of $\sim 3 \times 10^5$ K and a luminosity which corresponds to a $\sim 1.0 M_\odot$ white dwarf radiating at the Eddington limit (Ögelman, Krautter, and Beuermann 1987). Recent optical spectra of GQ Mus suggest that the central source has continued to increase in temperature but the luminosity is uncertain (Krautter and Williams 1989).

The X-rays detected during the outburst of RS Oph were thought to come from the interaction of a blast wave with the wind from the red giant secondary (Mason et al. 1986; Bode and Kahn 1985). RS Oph is a RN which has outbursts about every 18 years. The system contains a white dwarf and a red giant that is losing mass both into the Roche lobe of the compact star and, in addition, into the region surrounding the binary. Unlike the classical nova systems, in which there appears to be little gas surrounding the system, the gas ejected from the compact object in RS Oph must penetrate the circumbinary material which comes from the red giant wind. As the ejected gas collides with the circumbinary material, a shock forms and will produce X-rays if the velocity of the expanding material is large (Bode and Kahn 1985; O'Brien, Kahn, and Bode 1986). In fact, the material ejected during the RS Oph outburst was found to be expanding with velocities exceeding 10^3 km s^{-1} (Shore, Sonneborn, and Starrfield 1992; in preparation) which is sufficient to produce X-rays from shock emission. Ultraviolet emission from this interaction can be seen in the high dispersion IUE observations obtained during the 1985 outburst of RS Oph (Shore, Sonneborn, and Starrfield 1990a,b; 1992 in preparation, see Figures 19, 20, and 21 which are obtained from that paper). Optical spectra also indicated expansion velocities for the ejected gas of $\sim 3 \times 10^3$ km s^{-1} (Bruch 1986) and exhibited coronal line emission from [FeX] 6374 Å and [FeXIV] 5303 Å which indicated temperatures of nearly 1 keV. At late times the observed X-ray flux fell more rapidly than predicted by Bode and Kahn (1985) and Mason et al. suggested that they had observed the breakout of the shock from the red giant wind. This suggestion is strongly supported by the changes in the CIV 1550 Å line strength with time (Shore, Sonneborn, and Starrfield 1992, in preparation).

The final EXOSAT X-ray data point for RS Oph was obtained more than 200 days after the optical outburst and showed that, at that time, the source still had a temperature $\sim 3 \times 10^5$ K. This is in excellent agreement with the temperatures measured for the three classical novae and is strong evidence for the existence of a white dwarf in this system. However, the X-ray luminosity of RS Oph was less than expected for a massive white dwarf radiating at the Eddington limit and suggests that the white dwarf had finally ejected all of the accreted layers and was now cooling back to quiescence. This determination of a cooling time scale for the white dwarf provides a strong constraint for calculations of TNRs on massive white dwarfs.

Finally, Ögelman, Krautter, and Beuermann (1987) may have measured the turn off time scale for Nova GQ Mus 1983 since their last observation, 900 days after maximum light, suggests that its X-ray flux had decreased by more than a factor of two from maximum. If correct, this is an important measurement because, previously, we could only guess how long the nova would take to turn off.

3. Explosions in Cataclysmic Variable Systems

In the previous section, I have described the observations of the outburst and implicitly assumed the model to be presented in this section. In the first part of this section I will discuss the structure of the nova binary system and then proceed to describe the current ideas about the cause of the nova outburst.

3.1. A Nova as a Cataclysmic Variable Binary

As has been known since the early 1960's, novae are members of the class of cataclysmic variable stars (Kraft 1964). These systems contain a cool secondary and a white dwarf primary. The secondary fills its Roche lobe and material is lost through the L_1 point into the lobe surrounding the white dwarf. This material forms an accretion disk and ultimately falls onto the white dwarf. The rate at which the secondary is losing mass is unknown but recent work by Shara (1989) and his collaborators suggests that this rate is time variable (see also Shaviv and Starrfield 1987b). For a number of reasons, discussed in the review by Shara (1989), he proposed that mass lost by the secondary could decline to very low levels between outbursts on the white dwarf. He refers to this time as the period of hibernation. Observational support for this hypothesis has been provided by Vogt (1990). In addition, Shara and his collaborators have been searching historical records for candidates for very old novae with some success. For example, he has recovered CK Vul and WY Sge, two novae that exploded more than 300 years ago and which are now much fainter than more recent old novae (Shara 1989). Shara has also been involved in an unsuccessful search for other historical old novae which implies that they are now very faint as compared to novae with more recent outbursts and, therefore, must have evolved to very low mass transfer rates. However, there are novae that exploded more than 150 years ago which are still bright (V841 Oph and Q Cyg for example) and it is not clear for how long after an outburst that the binary system exhibits a high mass transfer rate onto the white dwarf. At some time after the outburst the old nova must begin its slide into obscurity. Hibernation cannot last forever and after some time has elapsed a high rate of mass transfer begins again and the nova evolves to another outburst (Ford 1978).

Given that material is lost from the secondary and enters the accretion disk, the viscous process by which this material moves through the accretion disk and actually reaches the white dwarf surface is also unknown (Pringle 1981). In addition, it is not clear at what rate mass is being transferred from the accretion disk onto the white dwarf or whether it arrives vertically or nearly horizontally. For example, the theoretical studies done assuming purely spherical accretion show that if the rate of accretion onto the white dwarf is too high, then a nova outburst will probably not occur since thermonuclear burning starts when the envelope is not very degenerate (Prialnik et al. 1972). However, there are direct predictions of the energy that should be emitted when the gas falls onto the surface of the white dwarf and neither ultraviolet nor X-ray observations confirm these predictions (Cordova and Howarth 1987).

The masses of the white dwarfs are also not very well known but theoretical studies now require that these systems contain very massive white dwarfs in order to produce either a fast classical nova or RN outburst (Starrfield 1986). The cause of this requirement is that the more massive a white dwarf, the more degenerate the bottom of the accreted hydrogen rich envelope at the time of runaway. This will produce higher peak temperatures and a more explosive event as is observed for fast classical novae. Slow novae may have lower masses and DQ Her, for example, is thought to have a mass of $\sim 0.6 M_\odot$ (Young and Schneider 1980).

Once the material has arrived in the accretion disk, it must still be accreted onto the surface of the white dwarf. Although, it is still moving in Keplerian or near-Keplerian orbits close to the surface of the white dwarf, the actual disk-star interaction is not understood at all for the normal classical nova. In the case of V1500 Cyg, now known to be an AM Her variable (Stockman, Schmidt, and Lamb 1988), hydrogen rich material is predicted to arrive at the poles on trajectories that are nearly normal to the surface. This is because the magnetic field strength in an AM Her variable is high enough to prevent an accretion disk from forming and material flows directly from the secondary to the white dwarf. On the other hand, if the disk extends to the surface of the white dwarf, then spherical inflow cannot take place and there must be some kind of boundary layer where the infalling gas actually interacts with the outer layers of the white dwarf.

Nevertheless, most of the simulations of accretion onto a white dwarf have assumed spherical inflow because that is the simplest process to treat in a one-dimensional computer code. In contrast, Sparks and Kutter (1987) and Kutter and Sparks (1987) have tried to simulate the disk-star boundary by including shear instabilities in their calculations. They simulated the accretion of material with angular momentum onto the white dwarf but assumed no mechanism for removing the angular momentum of the accreted material. While this resulted in mixing of accreted material with core material, the added centrifugal pressure support from the material moving at large tangential speeds on the white dwarf produced only very weak outbursts. The material was not very degenerate at the time of the TNR.

Finally, if one assumes that the accretion disk has an inner boundary that lies above the surface of the white dwarf and that all material arrives vertically onto the white dwarf, then the spherical accretion studies are useful. However, there is a further complication pointed out by Shaviv and Starrfield (1988) which is that the infalling material contains gravitational potential energy. The infalling material must release an energy of:

$$L = \frac{GM}{R} \dot{M}$$

Because the binding energy for material in an orbit just above the surface of the white dwarf is $0.5 GM/R$, half of this energy must be released in the accretion disk and half must be released at the boundary between the disk and the star (King 1989). The calculations of spherical accretion onto white dwarfs show that the boundary layer consists of an accretion shock at the surface of the white dwarf and the virial theorem states that half of the remaining energy is radiated and half is transported into the interior of the white dwarf (Shaviv and Starrfield 1987a). In addition, the internal energy of the infalling material is GM/R when it arrives on the surface. These two effects have only recently been included in the spherical inflow calculations and are important (Shaviv and Starrfield 1987b).

3.2. THE CAUSE OF THE OUTBURST

The theoretical studies indicate that the accreted layer grows in thickness until it achieves a temperature at its base that is high enough for thermonuclear burning of hydrogen to begin. The further evolution of nuclear burning on the white dwarf then depends upon the mass and luminosity of the white dwarf, the rate of mass accretion, and the chemical composition of the reacting layer (Truran 1982; Starrfield 1989). Since there is observational evidence that the infalling material is mixed with the core (Starrfield 1988; Truran 1990), the chemical composition must also be a function of the above parameters. The simulations show that, if the material is sufficiently degenerate, a thermonuclear runaway (TNR) occurs, and the temperatures in the accreted envelope grow to values exceeding 10^8 K. The studies of mass accretion onto a white dwarf, under a variety of conditions, imply that sufficiently degenerate means that the pressure at the base of the accreted layer must reach values of about 10^{19} dynes cm^{-2} to 10^{20} dynes cm^{-2} prior to the TNR (MacDonald 1983).

During the early phases of nuclear burning, most of the energy comes from the proton-proton chain ($T_b < 10^7$ K; where T_b is the temperature at the base of the accreted layer) but once the temperature exceeds about 10^7 K, more energy is obtained from the CNO reactions. Note, however, that this is a composition dependant statement. Since the ^{12}C(p,γ)^{13}N reaction rate is dependent on the abundance of ^{12}C in the envelope, the temperature at which this reaction begins to produce a significant amount of energy depends on the amount of ^{12}C present in the envelope. During the early stages of the evolution to a TNR, the lifetimes of the CNO nuclei against proton captures are very much longer than the decay times for the β^+-unstable nuclei: ^{13}N (863 s), ^{14}O (102 s), ^{15}O (176 s), and ^{17}F (92 s), which insures that these nuclei can decay and their daughters can capture another proton in order to keep the CNO reactions cycling. As the temperatures increase in the shell source, the lifetimes of the CNO nuclei against proton captures continuously decrease until, at temperatures >10^8 K, they become shorter than the β^+-decay lifetimes; for these conditions, the β^+-unstable nuclei will become abundant as the rate of nuclear energy generation is constrained. All of the computer simulations further indicate that, during the evolution to peak temperature, a convective region forms just above the shell source and grows until it includes the entire accreted envelope. It follows that, at the peak of the outburst, the most abundant of the CNO nuclei in the envelope will be the β^+-unstable isotopes.

It also follows that the time to peak temperature is a function both of the initial luminosity of the white dwarf and the initial abundances of the CNO elements. This is because the nuclear burning time scale decreases for increased energy generation. A luminous white dwarf is hotter at the composition interface then a cool white dwarf which increases the rate of energy generation and enhancing the numbers of CNO nuclei in the envelope also increases the rate of energy generation. For a given accretion rate the envelope mass at runaway is a function of the time scale to runaway, so if the accretion time is short, then the envelope mass is small (Starrfield 1989).

The rapid rise to temperatures above 10^8 K has several effects on the subsequent evolution. First, since the energy production in the CNO cycle arises from proton captures followed by β^+-decays, the rate at which energy is produced at maximum temperature depends on both the half-lives of the β^+-unstable nuclei and the abundances of the CNO nuclei initially present in the envelope. Second, since the convective turn-over time scale is about 10^2 sec, near the peak of the TNR, a significant fraction of the β^+-unstable nuclei can reach the surface before decaying (Starrfield 1989).

Once peak temperature is reached and the envelope begins to expand, the rate of energy generation in the surface regions declines only as the abundances of the β^+-unstable nuclei decline, since their decay is neither temperature nor density dependent (Truran 1982; Starrfield 1989). Numerical simulations performed with enhanced abundances of the CNO nuclei show that these decays will release more than 10^{46} erg into the envelope after it has begun to expand and the envelope will reach radii of more than 10^{10} cm before all of the ^{13}N has disappeared (Starrfield, Truran, and Sparks 1978). Therefore, the decays of the β^+-unstable nuclei provide a delayed source of energy which is responsible both for assisting in the ejection of the shell and for powering the super-Eddington luminosity phase of the outburst.

Both optical and ultraviolet observations of novae ejecta indicate that there is mixing of a significant amount of core material into the accreted layer, such that the chemical composition determined for the ejected material will reflect a combination of core and accreted material. This implies that, for those novae which show the most extreme levels of abundance enrichments of CNO and ONeMg nuclei in the ejecta, the white dwarf must be losing mass as a result of the outburst. These same abundance studies further show that the core material must come from either a CO or ONeMg white dwarf. It now seems possible that some of the ONeMg novae may be responsible for synthesizing the ^{26}Al that is thought to be responsible for heating of the small bodies in the early solar system (Nofar, Shaviv, and Starrfield 1991; Weiss and Truran 1991).

3.3. THE THEORETICAL PHASES OF THE OUTBURST

Using the theoretical development presented in the previous paragraphs, it is possible to separate the outburst into four phases each of which marks an important change in the observable characteristics of the nova. These four phases are

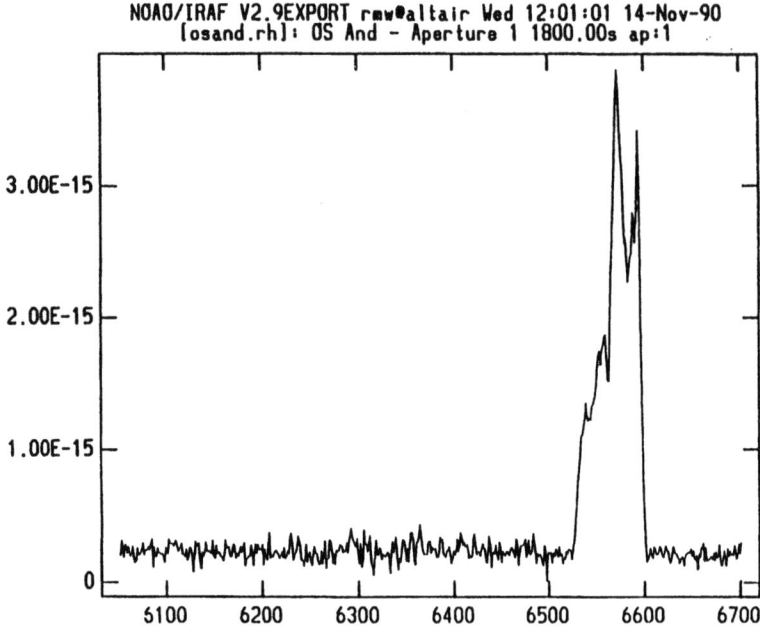

Fig. 11. This is a high dispersion (3 Å) optical spectrum taken at the Perkins 1.8-m telescope at Lowell Observatory. It shows the region around Hα for OS And 1986 and was taken in November 1990. Note the very castellated structure of this emission line.

1. the rise to bolometric maximum which occurs on a convective turn over time scale,
2. the rise to optical maximum which occurs on a   scale for the outer layers of the expanding envelope to reach a radius of $\sim 10^{12}$ cm,
3. the constant luminosity phase which lasts for a significant part of the remaining evolution of the nova, and
4. the turn-off phase in which nuclear burning ends and the nova returns to quiescence. In the following subsections I discuss each phase in turn.

3.3.1. The Rise to Bolometric Maximum

Although Robinson (1975) found that some novae do show evidence for the growing TNR at their surfaces, in most novae there is very little indication that anything unusual is occurring in the interior until the β^+- unstable nuclei reach the surface. At this time their decays cause the luminosity to rise to the Eddington luminosity, or even beyond (Austin *et al.* 1990), and the layers begin to expand. Since this rise is occurring on the convective turn-over time scale, 10^2 sec to 10^3 sec, it must be very fast and, because the radius of the white dwarf is still small, the effective temperature of the white dwarf can exceed 10^5 K for a very short time (Starrfield *et al.*, 1990a,b). It has been predicted that there will be a brief but intense period of EUV or soft X-ray emission at this time (Starrfield *et al.* 1990a). In Figure 13,

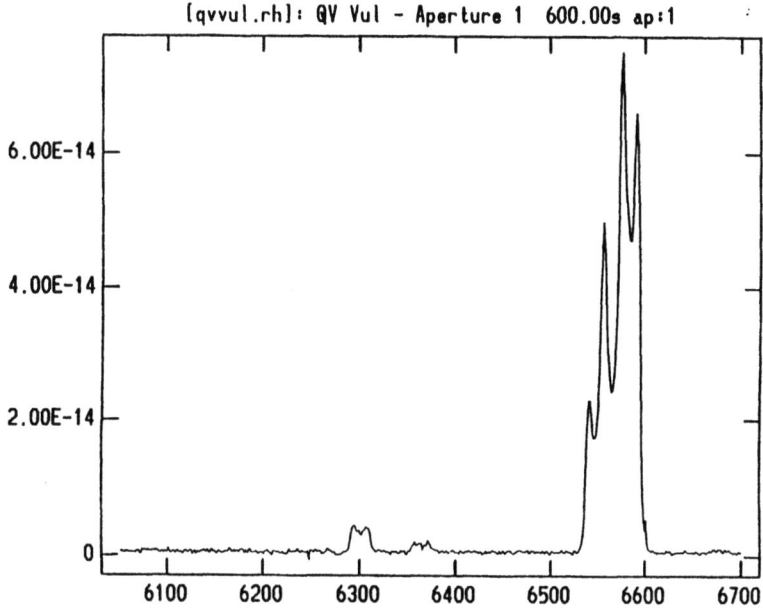

Fig. 12. This is a spectrum obtained for Nova QV Vul 1987 with the same equipment as was used for the spectrum shown in Figure 11. It also shows Hα with a 3 Å resolution. It is very interesting that it shows four major features with similar line ratios to the Hα profile of OS And.

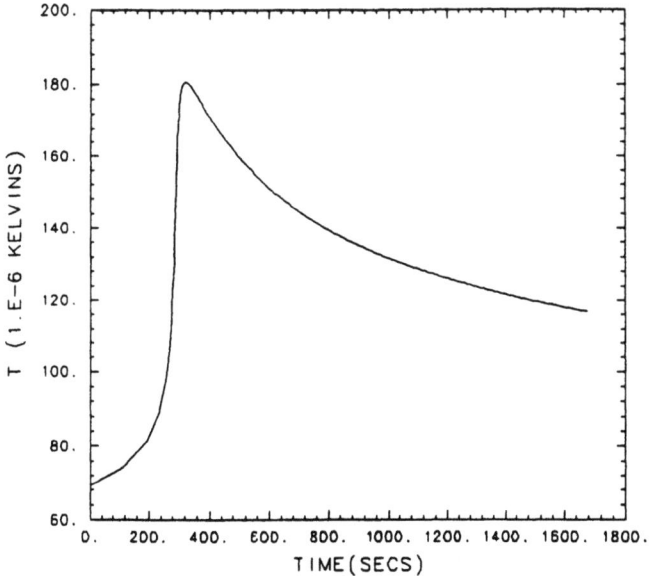

Fig. 13. The variation with time of the temperature in the shell source for a simulation with a white dwarf mass of 1.25 $M_\odot$. Note the steep rise and slower decline.

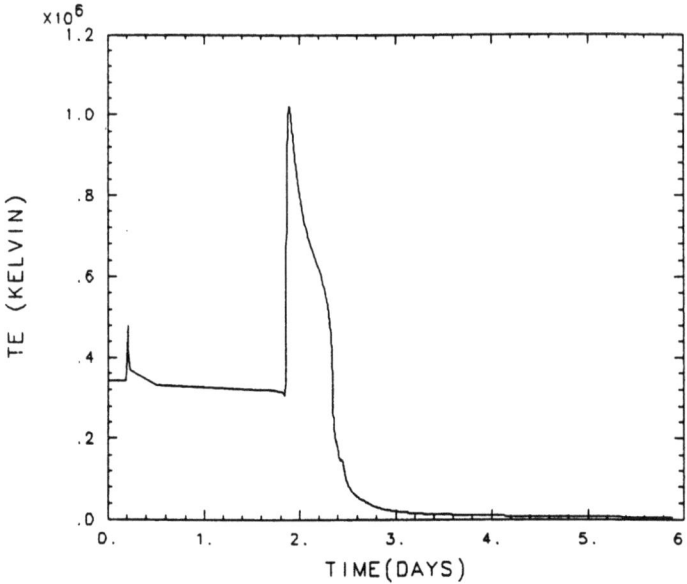

Fig. 14. The variation of T_e with time near the time of peak M_{Bol} for a simulation of a TNR on a 1.35 $M_\odot$ white dwarf. The rapid decrease after maximum is caused by the expansion of the outer layers.

I show the variation of temperature in the shell source (deepest hydrogen rich zone) with time for a TNR on a 1.25 $M_\odot$ white dwarf. The entire accreted shell is convective during the time displayed in this figure. Figure 14 shows the effective temperature as a function of time for a TNR on a 1.35 $M_\odot$ white dwarf. The details of the evolution are presented in Starrfield, Sparks, and Shaviv (1988). It can be seen that peak effective temperature exceeds 10^6 K. A compilation of results for a variety of white dwarf masses is given in Figure 18 where it can also be seen that the peak value of the effective temperature at this time is a measure of the mass of the white dwarf. The large amount of energy deposited in the outer layers on a short time scale, plus the fact that the luminosities can reach or exceed the Eddington luminosity, causes the outer layers to begin expanding and the effective temperature rapidly declines.

3.3.2. Rise to Maximum in the Optical

Once the layers begin expanding and the effective temperature starts to drop, the bolometric correction decreases until the time when the radius has reached about 10^{12} cm. If we assume that the luminosity has remained virtually constant or even increased slightly, then the temperature in the outer layers must have declined to a value around 10^4 K and hydrogen begins to recombine. This produces a drop in the opacity and the pseudo photosphere moves inward in mass. We have yet

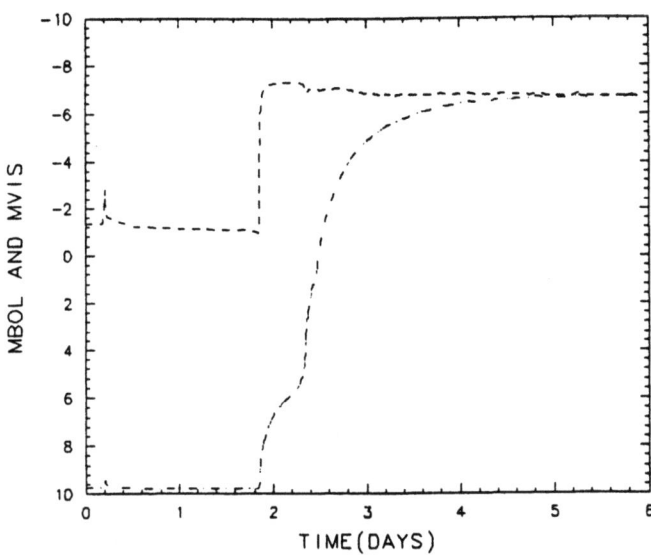

Fig. 15. This figure shows the early light curve for a simulation of a TNR on a 1.35 $M_\odot$ white dwarf. The upper curve is as a function of time and the lower curve is M_{vis} as a function of the same time. Note that M_{Bol} rises very rapidly as the β^+ unstable nuclei reach the surface and then stays virtually constant.

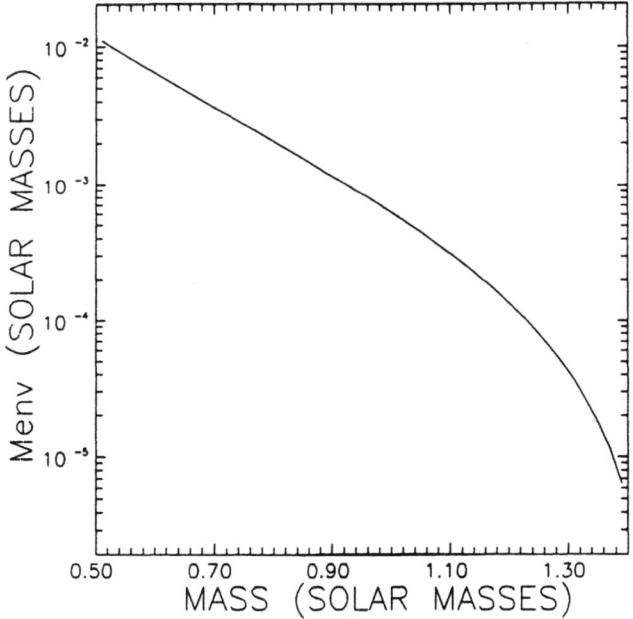

Fig. 16. This figure shows the variation in the envelope mass necessary to ignite a TNR on a white dwarf as a function of the mass of the white dwarf. It assumes a constant pressure at the base of the accreted envelope of 10^{19} dynes cm^{-2} (MacDonald 1983).

to determine what the principal source of opacity is at this time and it could be either hydrogen or low ionization metals such as iron. Nevertheless, whatever is providing the opacity at this time, maximum brightness in the optical is an opacity effect. In the ultraviolet, where the opacity is higher, we find that maximum light occurs after maximum light in the optical (Austin et al. 1990). In Figure 15, I show the light curve for one evolutionary sequence with the upper curve being the bolometric light curve and the lower curve the visual light curve.

The time scale for this phase of the outburst depends on the expansion velocity of the outer layers. The velocity must depend on the ratio of the total energy released into the shell, around the time of maximum energy generation, to the binding energy of the shell. The time scale must also depend on the mass of the envelope which is dependant on the mass of the white dwarf and the rate of accretion (Starrfield 1986). Since the mass of the envelope is much smaller for massive white dwarfs (this is shown in Figure 16), the expanding envelope should reach higher velocities for the same input energy. Therefore, it must be the case, if the abundances are enhanced, then more energy is produced at maximum temperature in the shell source and the accreted layers are ejected with higher velocities then if the abundances are not enhanced. I identify fast novae as those with large enhancements of core material in the envelope and slow novae as those with smaller or no enhancement of the envelope (for white dwarfs of mass greater than 1.0 $M_\odot$).

The effective temperature reached at optical maximum must also be a strong function of the mass of the expanding envelope. For low mass white dwarfs, with massive envelopes, we can expect that the expanding layers will reach radii of 10^{12} cm to 10^{13} cm before going optically thin. For example, in Nova QV Vul 1987 the spectral energy distribution fit a black-body curve of $\sim$4000 K (Gehrz 1989; private communication). In contrast, for RN which probably occur on very massive white dwarfs, the envelope masses are very small and the luminosities are either close to, or exceed, the Eddington luminosity so that the material is ejected at very high speeds. In this case, the effective temperature at maximum can exceed 10^4 K (Wehrse et al. 1990a,b; Shore et al. 1991). This can be seen in the ultraviolet spectra of RN (Shore, Sonneborn, and Starrfield 1990; Shore et al. 1991). Up to now all RN with compact secondaries, observed in the ultraviolet, show a very hot continuum at maximum light indicative of a small amount of mass being ejected.

3.3.3. *Constant Bolometric Luminosity*

This phase of the outburst was one of the first predictions of the TNR theory of the outburst. It arises because not all of the accreted material is ejected during the burst or explosive phase of the outburst and anywhere from 10% to 90% will remain on the white dwarf in hydrostatic equilibrium. For some years there have been problems with the life time of this phase. For 1.0 $M_\odot$ white dwarfs, the nuclear burning time scale for the envelope can be as long as 400 years (Truran 1982) which, obviously, disagrees with the observations that imply that most novae

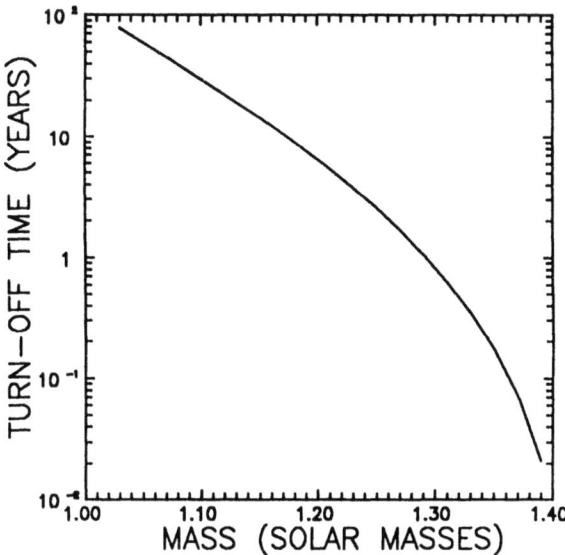

Fig. 17. This figure shows the time necessary to eject the entire accreted shell via a radiation pressure driven wind. It is a steep function of white dwarf mass because the equilibrium core mass luminosity of a white dwarf approaches L_{Edd} as the mass approaches the Chandrasekhar Limit. In addition, the envelope mass that must be ejected is a decreasing function of white dwarf mass as can be seen in Figure 16.

return to quiescence within 10 years.

However, the theoretical studies of this phase of evolution show that the equilibrium radius of the hydrostatic remnant is about 10^{11} cm, which is larger than the Roche Lobe radii of some cataclysmic variable stars. Therefore, slightly after the peak of the outburst, the binary will be revolving within the extended radius of the white dwarf. This cannot be a stable situation and Livio et al. (1990) have examined the consequences of the dynamical friction that arises from the motion of the binary within the extended envelope of the white dwarf. Dynamical friction seems capable of ejecting that part of the envelope that extends past the Roche Lobe of the secondary on short timescales (MacDonald 1980; Livio et al. 1990).

There is another process that must be acting at this time: radiation pressure driven mass loss. Not only is it driving mass off the remnant, it must also be acting on a short time scale. The studies of this phenomenon in novae imply very short time scales for ejecting the remaining material (Starrfield et al. 1990a,b). For white dwarf masses in excess of 1.2 $M_\odot$, the constant bolometric luminosity phase should last no longer than about 10 years. In Figure 17, I show the turnoff time as a function of white dwarf mass. The calculations upon which this figure is based (Starrfield et al. 1990a,b) assume both mass loss via the Castor, Abbott, and Klein (1975) theory and also that the envelope mass versus white dwarf mass relationship shown in

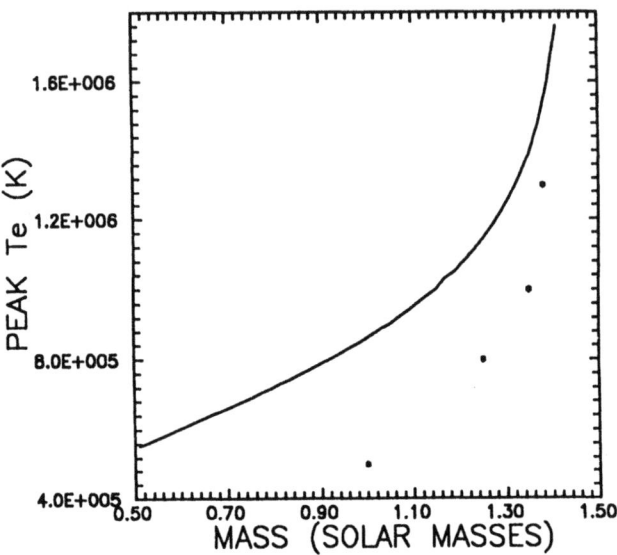

Fig. 18. This figure shows the peak effective temperature as a function of white dwarf mass. The curve is obtained by assuming the core mass luminosity (Paczynski 1971) and the equilibrium white dwarf radius as a function of mass. The points are from actual simulation of TNRs on white dwarfs of masses 1.0 $M_\odot$, 1.25 $M_\odot$, 1.35 $M_\odot$, and 1.38 $M_\odot$ and show that some expansion has occurred at the time of the first phase of Peak T_e.

Figure 16 is valid. The turnoff times shown in Figure 17 are probably upper limits for low mass white dwarfs since this calculation neglects the dynamical friction of the binary. On the other hand, the theory also breaks down for the most massive white dwarfs where the luminosity is very close to the Eddington luminosity. The recent ultraviolet observations of RN, which are predicted to occur on very massive white dwarfs (Starrfield, Sparks, and Truran 1985; Starrfield, Sparks, and Shaviv 1988), imply that the mass is ejected by a wind and the outburst is over within weeks after maximum light has occurred (Shore et al. 1991) as one would expect for white dwarfs with masses near the Chandrasekhar limit.

3.3.4. *The Return to Quiescence*

The final phase of the outburst marks the ejection (or conversion to helium) of all of the hydrogen in the accreted envelope and the cessation of nuclear burning on the white dwarf. It is not clear if this phase also marks the resumption of mass transfer by the secondary, if mass transfer has already resumed, or if it has been going on during the entire outburst. It is possible to make some predictions about the characteristics of the system during this phase since we can assume that the white dwarf is emitting at a constant (or near constant) luminosity and the radius is declining. Since it will eventually reach the equilibrium radius of the white dwarf, which is less than 10^9 cm, the most massive white dwarfs can reach temperatures exceeding 10^6 K during the last stages of the outburst (Starrfield et al. 1990a,b). This result can be seen in Figure 18 which shows the variation in effective temperature as a function of mass for white dwarfs that are radiating at a luminosity determined by the core mass-luminosity relationship (Paczynski 1971). In addition, their radius is obtained from the equilibrium radius for a white dwarf of a given mass. The four points plotted on this curve are taken from actual evolutionary sequences and show that some radius expansion has taken place.

Observational support for this prediction comes from the EXOSAT studies of GQ Mus, PW Vul, QU Vul, and RS Oph (Ögelman, Krautter, and Beuermann 1987; Mason et al. 1986). All four novae were observed to contain a hot source emitting at a temperature of $\sim 3 \times 10^5$ K. This value is somewhat low for a 1.25 $M_\odot$ white dwarf but calibration problems may have affected their temperature determinations. On the other hand, optical observations of GQ Mus in 1988 and 1989 (Krautter and Williams 1989) show that [FeX] 6374 Å is stronger than Hα which cannot occur unless there is a hot (T $> 5 \times 10^5$ K) photoionizing source inside the system. IUE observations of GQ Mus obtained in September, 1989, showed a very blue continuum with a number of features present in the spectrum (Krautter and Starrfield 1992; in preparation).

4. Recurrent Novae

The last few years have seen major advances in our understanding of the ultraviolet behavior of RN. In addition, in 1990 we observed a RN in the LMC (1990 #2: Shore, Sonneborn, and Starrfield 1990; Shore et al. 1991). LMC 1990 #1 was an ONeMg nova similar to Nova V693 CrA 1981 (Williams et al. 1985; Sonneborn, Shore, and Starrfield 1990).

RN are members of the nova subclass of cataclysmic variable stars. However, unlike classical novae which are observed to suffer only one major outburst on historical time scales, RN suffer less extreme (than classical novae) outbursts which repeat on observable timescales. Therefore, a nova will be classified as recurrent if it exhibits a second outburst and there are RN, such as T Pyx, U Sco, and RS Oph, that have experienced more than two outbursts in the last 100 years. I also note that the interoutburst time is not regular and can range from 9 years (U Sco) to longer than 50 years (T CrB). It is also possible that there are novae that can reoccur on time scales of 100 to 200 years and we have yet to observe the second outburst.

In order to understand the rapid recurrence time scale of these systems, it has been predicted: 1) that a RN system must contain an accreting white dwarf which is very near the Chandrasekhar mass limit so that the envelope mass required to reach runaway will be small; 2) that the white dwarf in RN binary systems must be more luminous than the white dwarfs in normal classical novae systems so that the evolution time to runaway will be short; and 3) that the secondary must be evolved so that the mass transfer rate onto the white dwarf will be higher for RN than for classical novae (Starrfield, Sparks, and Truran 1985; Starrfield, Sparks, and Shaviv 1988). These three conditions are necessary to produce TNRs that can reach maximum in less than 3 years.

There are interesting observational differences between RN and classical novae. For example, all well studied RN have evolved secondaries. This is very different from classical novae for which the secondary (mass loser) is commonly assumed to be on or close to the main sequence. In addition, the magnitude range during the outburst is much smaller for recurrent novae than for classical novae: about 8 to 10 mag. rather than 14 mag. or greater. Concomitant with this smaller range in brightness, a recurrent nova ejects less mass ($\sim 10^{-7}\ M_\odot$) during the outburst than does a classical nova ($\sim 10^{-5}\ M_\odot$). This is probably the cause of the smaller magnitude range in RN. They eject much less material and become optically thin before the outer layers have reached radii of 10^{12} cm. Therefore, most of the radiative energy will be emitted in the ultraviolet. This was certainly the case for LMC 1990 #2 and U Sco (Williams *et al.* 1981; Shore *et al.* 1991).

To complicate the situation even further, it turns out that there are at least two classes of RN. Those like U Sco and V 394 CrA, in which the secondary is small and the orbital periods are a few days or less (Schaefer 1990), and those like V745 Sco, T CrB, and V3890 Sgr in which the secondary is a giant (Starrfield and Wagner 1990). The differences in the size of the secondary will cause the observed characteristics of the outburst to differ even if the evolution of the white dwarfs are similar. Finally, there is RS Oph in which the secondary is not only a giant but it is losing mass in a dense wind so that the outburst takes place within the wind of the giant.

4.1. THE 1985 OUTBURST OF RS OPH

I will concentrate on the 1985 outburst of RS Oph because not only are there numerous high dispersion IUE data that have recently been analyzed and which are providing a great deal of information about the progress of the outburst (Shore, Sonneborn, and Starrfield 1992, in preparation), but RS Oph was also detected in outburst by EXOSAT. One of the most active of the RN, outbursts have been observed in 1898, 1933, 1958, 1967, and 1985. The 1985 outburst of RS Oph was exceptionally well covered in the ultraviolet with both low and high dispersion observations (Snijders 1986; see Bode 1986 for a complete discussion of the outburst behavior of this nova and data at a variety of wavelengths). Since the white dwarf is actually orbiting within the wind from the red giant companion, we must try to understand how the blast wave from the white dwarf will propagate through and interact with the red giant wind. Fortunately, we can compare the data for RS Oph with data for other RN, which have giants as secondaries but whose giants do not exhibit a thick wind, and then assume that any differences

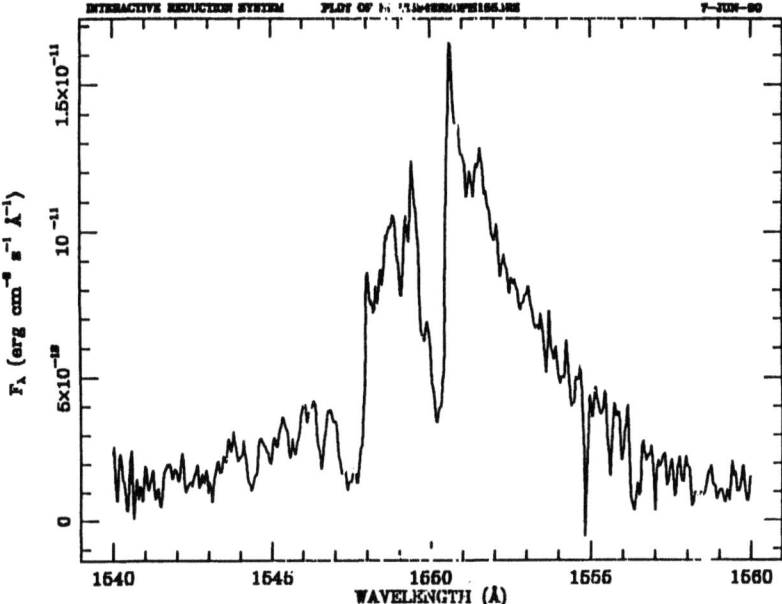

Fig. 19. This is a high dispersion IUE SWP spectrum of Nova RS Oph 1985 obtained on day 1985/45.5 and shows the CIV 1550 Å profile. The broad feature is from the expanding blast wave while the absorption comes from the ionized red giant wind of the companion.

in behavior are caused by the difference in the environment. Since a detailed discussion of the ultraviolet behavior of this recurrent nova will appear elsewhere (Shore, Sonneborn, and Starrfield 1992, in preparation), here I only present and discuss three high dispersion images that show some of the variation in the CIV 1550 Å profiles with time. Figure 19 shows the high dispersion profile of CIV obtained on 1985 day 45.5. There are two important features to be noticed in this spectrum which shows broad emission with superimposed absorption. First, the broad emission extends (FWHM) to velocities of about 1200 km s^{-1}, while FWZI is about 1900 km s^{-1}. This emission must be coming from the pseudo photosphere produced by the shock moving through the red giant atmosphere. The absorption components are much narrower and are probably coming from the photoionized wind from the red giant.

The presence of these absorption components, early in the outburst, has a very important implication. The material in a red giant wind should be too cool to have sufficient C^{+++} to produce the observed absorption unless it has been ionized to this stage by the outburst. It seems likely that the EUV phase at maximum M$_{Bol}$ was the original cause of this ionization and precursor UV photons from the blast wave are maintaining the ionization (Shore, Sonneborn, and Starrfield 1992; in preparation). There are more than enough ionizing photons emitted during the rise to bolometric maximum to ionize the entire shell if the outburst occurs as a result of a TNR.

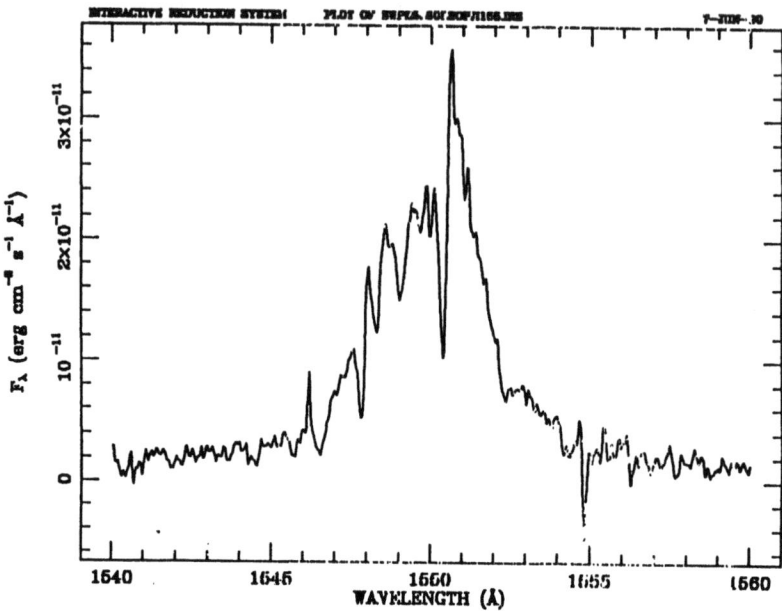

Fig. 20. This high dispersion IUE SWP spectrum of RS Oph 1985 was obtained on day 1985/52.5 and also shows the CIV profile. The expanding shock has continued to move through the envelope, which has caused the narrowing of the profile, and the absorption from the red giant wind has also narrowed because it is coming from farther out in the wind.

One of the differences between this nova and other RN is that the velocity of the blast wave is much smaller for RS Oph than was found for U Sco, V745 Sco, or V3890 Sgr. If we assume that the outburst has the same cause (a TNR), that the masses and luminosities of the white dwarfs are approximately equal, and the compositions are also about the same, then the lower velocities must be a direct result of the interaction of the blast wave with the red giant wind. We also expect the widths of the emission lines to narrow as the blast wave moves through the wind into lower velocity material. This can be seen in Figure 20 which shows the region around CIV on day 1985/52.5. The line has become narrower and the absorption is less pronounced since the blast wave has penetrated more of the shell.

The final high dispersion spectrum of RS Oph (Figure 21) was obtained on day 1985/82.3 and shows the emission from the red giant shell after the blast wave has reached the edge of the shell. We see the two components of the CIV doublet very clearly. I note that when the doublet was first seen, early in the outburst, both components were optically thick. As the outburst evolved, first the 1550 Å component became optically thin, then the 1548 Å component became optically thin.

X-ray observation of this outburst, obtained with the EXOSAT satellite showed a

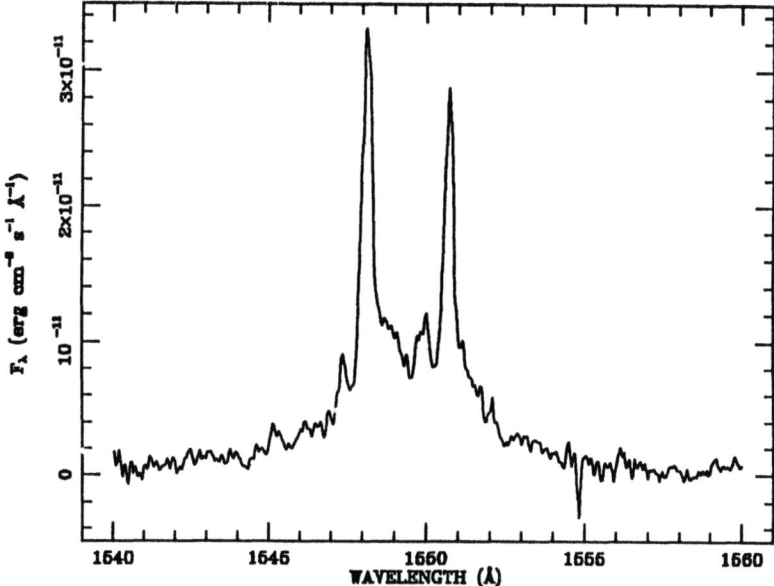

Fig. 21. This high dispersion IUE SWP image of RS Oph 1985 was obtained on day 1985/82.3 and shows the same region as in the previous two figures. This was taken about the same time as the EXOSAT X-ray data indicated that the blast wave had broken through the wind. In fact, the most prominent features are caused by the emission from the red giant wind.

strong decrease in emission beginning at about the same time as when the spectrum shown in Figure 11 was obtained. This behavior was also explained as break out of the shock at the edge of the red giant wind (Mason *et al* 1986; see Section 2.5).

4.2. THE OUTBURST OF THE RN LMC 1990 #2

Up to the time of the writing of this chapter there have been four novae in the LMC that have been studied in some detail with a variety of methods. In this subsection I will discuss the outburst of LMC 1990 #2 which was found to be a RN. Its outburst was very similar to the galactic RN: U Sco 1979, 1987 (Williams *et al.* 1981) and V394 CrA 1987. In contrast to RS Oph, T CrB, and other RN with giant secondaries, this class of objects have secondaries that are evolved but must be small in size since they have relatively short orbital periods (Schaefer 1990). The evidence for their secondaries being evolved is that U Sco, V394 CrA, and LMC 1990 #2 are transferring material that is hydrogen poor so that it must have undergone hydrogen nuclear fusion at some time in the past. In addition, LMC 1990 #2 has provided much important data about the RN class in general (Shore, Sonneborn, and Starrfield 1990; Shore *et al.* 1991). For the very first time, we know the distance to a RN and, thereby, can determine the energetics of the outburst to a high degree of accuracy. For example, as can be seen from the ultraviolet light curve presented in Shore et al. (1991), this nova radiated at super-Eddington luminosities at the peak of the outburst. Although predicted by the theoretical calculations (c.f., Starrfield, Sparks, and Shaviv 1988), this is the first observational confirmation of this prediction.

The optical outburst of Nova LMC 1990 # 2 was discovered on February 14.1 (UT) by Liller (1990) as the result of photographic nova patrol observations. In his announcement, Liller suggested a correspondence in position between this nova and one reported in the LMC in 1968 (Sievers 1970) which reached a maximum visual magnitude of about 10.7 mag. Shore et al. (1991) remeasured the original plates at Bamberg and found a coincidence in position to an acceptable degree of accuracy. They are extremely confident that it is recurrent. They also examined other plates of the area and were able to set limits on the brightness of the secondary that place it as fainter than the secondaries of galactic RN with *giant* secondaries.

Their principal result was that the luminosity in the ultraviolet exceeded 7×10^4 $L_\odot$ during the first few days of the outburst. An ultraviolet light curve can be found in Shore et al. (1991). This value is greater than the Eddington luminosity of a 1.0 $M_\odot$ white dwarf, assuming a solar mixture of the elements and electron scattering as the opacity. The ultraviolet luminosity rapidly declined after the first day, which appears to have been at maximum brightness in the ultraviolet. However, the optical outburst was discovered two days prior to the first IUE observation and those observations (IAU Circular 4964) suggested that this nova had already reached optical maximum when first discovered. Since it was super-Eddington when first observed in the UV and a nova cannot remain that bright for more than a day or two, Shore et al. suggested that it was at maximum when first observed with the IUE satellite. Therefore, this RN shows the same behavior seen in other novae studied both in the optical and ultraviolet: the ultraviolet light maximum follows optical maximum (Austin et al. 1990). By 23 February, approximately one week after ultraviolet maximum, the integrated luminosity had fallen to approximately 6.3×10^3 $L_\odot$

Analysis of the optical and IUE spectra of this RN provided some very interesting results. The mass of the ejecta was $\sim 10^{-7}$ $M_\odot$ which is in good agreement with the ejected mass in U Sco (Williams et al. 1981). Models of the continuum and emission line fluxes were able to place good limits on the effective temperature of the underlying white dwarf. This is because the NV 1240 Å line is very sensitive to the flux emitted in the Lyman continuum and, therefore, is a very good thermometer. Shore et al. report that in those models where they decreased the Teff of the (underlying) ionizing source from 2×10^5 K to 10^5 K, the predicted NV line strength decreased by more than an order of magnitude. Therefore, for temperatures below $\sim 2 \times 10^5$ K, no reasonable match could be obtained for the observed nitrogen line strengths for any consistent set of abundances. Finally, they also found that helium and nitrogen were enhanced in abundance over solar material by significant factors.

5. Nucleosynthesis in Novae

As has been continuously emphasized in various studies of the nova outburst, a proper treatment of nuclear energy generation is essential to understanding the cause of the classical nova and RN outburst. The super-Eddington character of the bolometric luminosity at maximum of fast novae is dependent on the violence of the thermonuclear runaway, which in turn depends on the CNO nuclear reactions and the composition of the accreted nova envelope. In addition, novae may be important contributors to the galactic abundances of the rarer isotopes ^{13}C, ^{15}N, and ^{17}O, as well as ^{7}Li. I also note that the same novae whose ejecta are enhanced in intermediate mass nuclei (V693 CrA, V1370 Aql, QU Vul, LMC 1990 #1) exhibit velocities of ejection exceeding 8×10^3 km s^{-1} so that this material is well mixed in the ISM.

In order to determine the important nucleosynthesis regimes as a function of temperature, density, and time, Nofar, Shaviv, and Starrfield (1990) and Weiss and Truran (1990) have independently calculated the nucleosynthesis expected to occur in nova explosions. The results of both their studies can be summarized as follows: (1) They confirm earlier findings of Hillebrandt and Thielemann (1982) and Weischer et al. (1986) that extremely low levels of ^{26}Al and ^{22}Na are expected to be formed in nova envelopes with a solar initial heavy element composition. This result implies that slow CO novae and RN are not expected to contribute significantly to the abundance of ^{26}Al in the galaxy although it still might be possible for slow CO or RN to contribute to some of the abundance anomalies detected in meteorites. (2) Enhancing only the CNO nuclei does not guarantee significantly increased ^{22}Na or ^{26}Al although CO novae may be responsible for production of some rare light nuclei. (3) Greatly increased ^{22}Na and ^{26}Al production does result from envelopes with substantial initial enhancements of elements in the range from neon to aluminum. For example, for the choice of an initial composition consisting of matter enhanced to a level of $Z = 0.25$ in the products of stellar carbon burning (Arnett and Truran 1969), their calculations predict that the abundances of ^{22}Na and ^{26}Al can be one to two orders of magnitude larger than for equivalent models where the initial composition is solar. I note that large enhancements of nuclei from nitrogen to sulfur have been observed in novae ejecta (Sonneborn, Shore, and Starrfield 1990; Starrfield 1988; Truran 1990). (4) Novae with ejecta rich in material from an ONeMg white dwarf may represent an important source of ^{26}Al in our Galaxy. Order of magnitude estimates indicate that their integrated contribution is within a factor of 3 to 10 of what is observed implying that more detailed calculations are necessary. (5) The abundances of ^{22}Na predicted for the ejecta of novae involving ONeMg white dwarfs are sufficiently high that we may expect relatively nearby ONeMg novae to produce detectible flux levels of ^{22}Na decay γ-rays. (6) The calculations also indicate that there should be a strong anti-correlation between ^{22}Na and ^{26}Al overproduction in nova outbursts. (7) The degree of enhancement of ^{22}Na and ^{26}Al is a sensitive function of the temperature

TABLE I
Partial List of Recent Classical and Recurrent Novae.

Nova	Year	Dust	Class
V1668 Cyg	1978	Yes	CO
U Sco	1978	No	REC
V693 CrA	1981	No	ONeMg
V1370 Aql	1982	Yes[4]	ONeMg
GQ Mus	1983	No	CO
PW Vul	1984	Yes	CO
QU Vul	1984	Yes[4]	ONeMg
RS Oph	1985	No	REC[1]
OS And	1986	No	CO
V842 Cen	1986	Yes	CO
V394 CrA	1987	No	REC[1]
QV Vul	1987	Yes[3]	CO[4]
LMC 88 #1	1988	Yes	CO
LMC 88 #2	1988	?	CO
V745 Sco	1989	No	REC[1]
Sco 1989	1989	?	?[2]
LMC 90 #1	1990	?	ONeMg
LMC 90 #2	1990	?	REC[1]
V3890 Sgr	1990	No	REC[1]

[1] Recurrent nova
[2] Unknown
[3] No late time spectra
[4] Silicate dust

history (assuming the same initial concentration of nuclei) and, therefore, detection of ^{22}Na would provide useful constraints on the evolution of the thermonuclear runaway.

6. Conclusions

The studies of novae done over the past two decades have far advanced our understanding of the cause and evolution of the nova outburst. Nova outbursts are not rare phenomena as can be seen from Table I which shows all of the novae that have been studied in the past few years. The importance of satellite studies to novae can be understood if we note that the photospheres of novae are small and very luminous so that they emit most strongly in the ultraviolet and soft X-ray. As a direct result of IUE studies, we have discovered that there are two compositional classes of novae: those that occur on an ONeMg white dwarf and those that occur on a CO white dwarf. Theoretical studies, using the existence of ONeMg nova explosions, indicate that these novae are the source of ^{26}Al in the solar system. In studies of recurrent novae, we have found evidence for the passage of a blast wave through the envelope of a red giant and can show that the luminosities of both fast classical novae and recurrent novae are super-Eddington at maximum brightness.

Given the successes in the studies of novae, what are the reasons for continued observations of novae. Although we have identified broad classes of novae, detailed studies of novae have shown that every well studied outburst is unique and the expanding, optically thin shell in each object samples a different regime of electron density and electron temperature parameter space. This is to be expected since we must realize that the explosions take place on white dwarfs with a range in masses, luminosities, accretion rates, and compositions. Continual studies of novae are improving and expanding our knowledge of the plasma diagnostics that can be applied to other objects (symbiotic variable stars and AGN's are two important examples) which exhibit evidence for expanding gas. Finally, we remind the reader that studies of novae provide time-dependant information about the evolution of stellar winds, grain growth, emission lines, and stellar atmospheres that can be compared with objects whose features do not change rapidly with time.

It is clear that obtaining high quality data on novae at maximum, or before maximum, can provide very important information about the outburst. Unfortunately, such data have been obtained for only a few novae and recent studies of nova spectra in the IUE archives have shown that it is very easy to overexpose the continuum of a nova and render the exposure useless. Observations of novae require very broad wavelength coverage and early spectra show that the nova evolves very rapidly near maximum. It is necessary to obtain numerous spectrophotometric data over the first few days of the outburst. It is also important to follow the evolution of the spectra as the nova evolves from optically thick to optically thin. Such an observing campaign requires repeated observations and very broad wavelength coverage over a long period of time

In the past two years, we have obtained important data for novae in the Large Magellanic Cloud. Their brightnesses at maximum are well suited for multiwavelength studies with a variety of ground based and satellite telescopes. Obtaining the distance to the LMC, by as many different methods as possible, is one of the fundamental steps in determining the size scale of the Universe. One of the fundamental indicators is the brightnesses of novae a few days past maximum. We have found that novae exhibit a variety of brightnesses at maximum and detailed study of each outburst is required to make them useful as distance indicators.

The studies of these 4 novae in the LMC show that both classical and recurrent novae can become super-Eddington at maximum but that it should still be possible to obtain a very good determination of the rate of decline-absolute magnitude relationship. However, up to the present there have been only 4 novae that have been studied with the IUE satellite and each of them has suffered a unique outburst (a slow CO nova, a fast CO nova, an ONeMg nova, and a recurrent nova).

The explosion of a recurrent nova in the LMC has provided us with an unparalleled opportunity to understand the outbursts of these enigmatic objects. As a direct result of knowing the distance to the LMC, from other techniques, we can firmly state that the peak luminosity of this nova exceeded the Eddington limit for a 1.0 $M_\odot$ white dwarf with solar abundances. Its luminosity becomes even more super-Eddington if we note that the ejected abundances were not solar, but helium (He/H$\sim$1) and nitrogen (N $\sim$ 30$\times$ solar), at least, were enhanced over an equivalent solar mixture by large factors. In addition, the actual opacity in the shell probably far exceeded the electron scattering opacity which is usually used to calculate the Eddington luminosity. Unfortunately, because of the differences in the outbursts of the 4 novae in the LMC, we must obtain more data on such novae in order to not only obtain more information on these distance indicators but, in addition, to obtain abundances for objects that are sampling the interiors of white dwarfs in an external galaxy.

Acknowledgements

I am very pleased to acknowledge valuable discussions with S. Austin, A. Cassatella, G. Ferland, R. Gehrz, P. Hauschildt, S. Kenyon, J. Krautter, I. Nofar, P. Saizar, G. Shaviv, S. Shore, E. Sion, G. Sonneborn, W. Sparks, L. Stryker, J. Truran, R. Wade, R. M. Wagner, R. Wehrse, and R. Williams. I am grateful for the hospitality of the Los Alamos National Laboratory and a generous allotment of computer time. Some of the material presented in this chapter was obtained through the facilities of the Boulder RDAF which is supported by NASA grant NAS5-28731 to the University of Colorado. I would also like to acknowledge continuing support from the National Science Foundation through grant AST88-18215 to Arizona State University, from NASA through grant NAG5-481 to Arizona State University, and from the DOE.

INTERACTING BINARIES AND TYPE I SUPERNOVAE

K. NOMOTO

1. Type I Supernovae and Subtypes

Supernovae have been spectroscopically classified into Type I and Type II according to the absence and presence of hydrogen in their optical spectra. Type I supernovae (SNe I) are further subclassified into Ia, Ib, and Ic. As shown in Figure 1 (Branch *et al.* 1991), the early–time ($t \approx 1$ month past maximum) photospheric spectra of SNe I define such subtypes. SNe Ia are characterized by the presence of a deep absorption trough near 6150Å produced by blueshifted SiII λ6355. SNe Ib and Ic, by contrast, do not show this line. Moderately strong HeI lines, especially HeI λ5876, distinguish SNe Ib from SNe Ic at early times, i.e., SNe Ib exhibit absorption lines of [HeI], whereas these lines are absent in SNe Ic (Harkness and Wheeler 1990 for a review). Type II supernovae (SNe II) are subclassified into SNe II-P (*plateau*), SNe II-L (*linear*), and SNe II-BL (*bright linear*) according to the light curve shape (Branch *et al.* 1991).

The late-time ($t \approx 5 - 10$ months) optical spectra of SNe provide additional constraints on the classification scheme (Figure 2; Branch *et al.* 1991). SNe Ia show strong blends of hundreds of Fe emission lines, mixed with some Co lines as well. SNe Ib and Ic, on the other hand, are dominated by relatively unblended emission lines of intermediate–mass elements such as [OI], [CaII], and CaII. SNe II are dominated by the strong Hα emission line.

The supernova classification has become more complicated since the discovery of SN 1987K whose spectral classification changed from Type II to Type Ib/Ic as it aged, thereby being called as SN IIb (Filippenko 1988). A hydrogen feature has also been identified in the early time spectrum of SNe Ic 1987M (Jeffery *et al.* 1991) and 1991A (Filippenko 1992). Interestingly the early time spectra of SN 1987K are very similar to SNe Ic (1983V and 87M) (Figure 3: Filippenko *et al.* 1990; Wheeler and Harkness 1990).

The lack of strong hydrogen lines implies that the progenitors of SNe I have lost most of their hydrogen-rich envelope at the time of explosion. Two cases are possible: (1) mass loss over to the companion star during the evolution of a close binary system, and (2) stellar wind type mass loss from a single star. For these cases, we have basically three candidates for the progenitors of SNe I, i.e., (1) white dwarfs, (2) helium stars in binaries, and (3) single Wolf-Rayet stars. Because

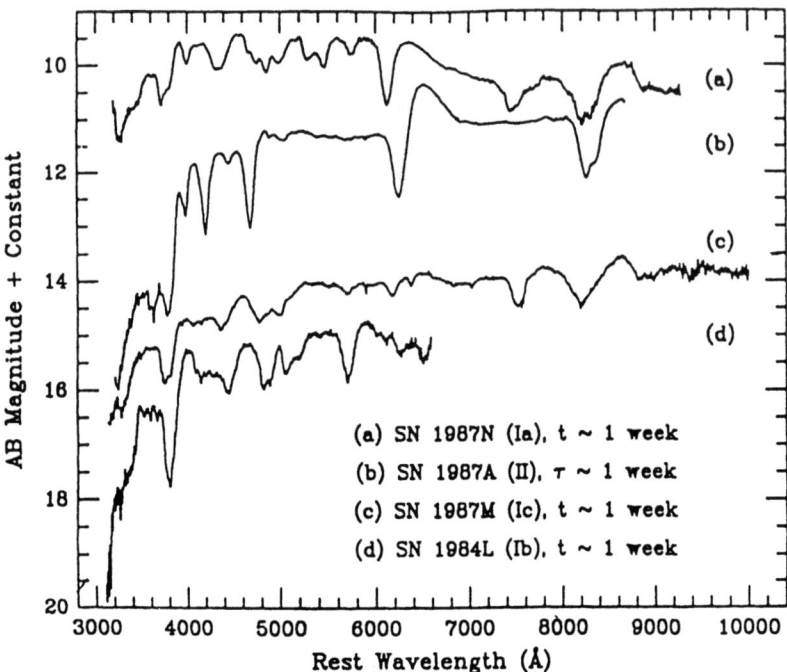

Fig. 1. Spectra of supernovae, showing the early-time distinctions between the four major types and subtypes (Branch et al. 1991). The variable t is taken to indicate time after observed visual maximum, whereas τ represents time after core collapse. AB magnitude $= -2.5 \log f_\nu - 48.6$, where the units of f_ν are ergs s^{-1} cm^{-2} Hz^{-1}.

of the complicated mass loss processes, it has not been easy to identify the exact evolutionary origin of SNe I.

2. Type Ia Supernovae

For SNe Ia, accreting white dwarfs have been considered to be promising candidates of their progenitors. The explosion mechanism originally suggested by Hoyle and Fowler (1960), i.e., the thermonuclear explosion of electron-degenerate cores, basically has been confirmed by extensive numerical modeling and comparison with observations (Nomoto 1986a; Woosley and Weaver 1986b).

2.1. WHITE DWARF PROGENITORS

Isolated white dwarfs are simply cooling stars that eventually end up as invisible frigid stars. The white dwarf in a close binary system evolves differently because the companion star expands and transfers matter over to the white dwarf at a certain stage of its evolution. The mass accretion can *rejuvenate* the cold white dwarf (e.g., Nomoto and Sugimoto 1977), which could lead to SNe Ia events in some cases.

The mass accretion onto the white dwarf releases gravitational energy at the white dwarf surface. Most of the released energy is radiated away from the shocked

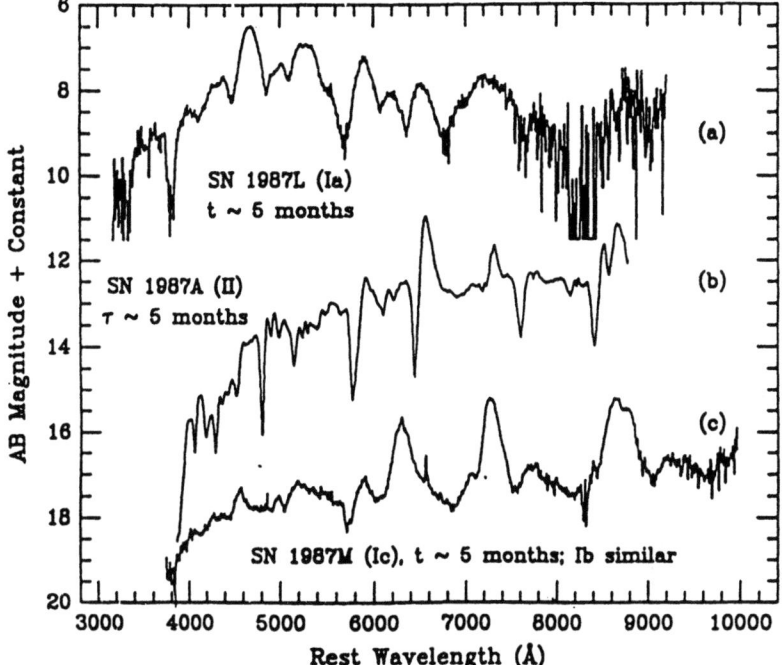

Fig. 2. Spectra of supernovae, showing late–time distinctions between different types and subtypes. Notation is the same as in Figure 1. The SN Ia 1987N (Figure 1) was spectroscopically similar to SN 1987L, shown above. At even later phases, SN 1987A was dominated by Hα. The SN Ic 1987M as well as SNe Ib was dominated by strong emission lines of [OI], [CaII], and the CaII near–infrared triplet, with only a weak continuum.

region as UV and does not contribute much to heating the white dwarf interior. The continuing accretion compresses the previously accreted matter and releases gravitational energy in the interior. A part of this energy is transported to the surface and radiated away from the surface (radiative cooling) but the rest goes into thermal energy of the interior matter (compressional heating). Thus the interior temperature of the white dwarf is determined by the competition between compressional heating and radiative cooling; the white dwarf is hotter if the mass accretion rate $\dot{M}$ is larger, and vice versa (e.g., Nomoto 1982a).

The scenario that possibly brings a close binary system to a SN Ia explosion is as follows (although the exact evolutionary origin has not been understood): Initially the close binary system consists of two intermediate mass stars ($M \lesssim 8$ $M_\odot$). As a result of Roche lobe overflow, the primary star of this system becomes a white dwarf composed of carbon and oxygen (C+O). When the secondary star evolves, it begins to transfer hydrogen-rich matter over to the white dwarf.

When a certain amount of hydrogen is accumulated on the white dwarf surface, hydrogen shell burning is ignited (Figure 4; Nariai and Nomoto 1979; Nomoto 1982a). Its outcome depends on $\dot{M}$: For slow accretion ($\dot{M} \lesssim 1 \times 10^{-8} M_\odot$ yr^{-1}), hydrogen shell burning is unstable and tends to *flash*, which leads to the ejection

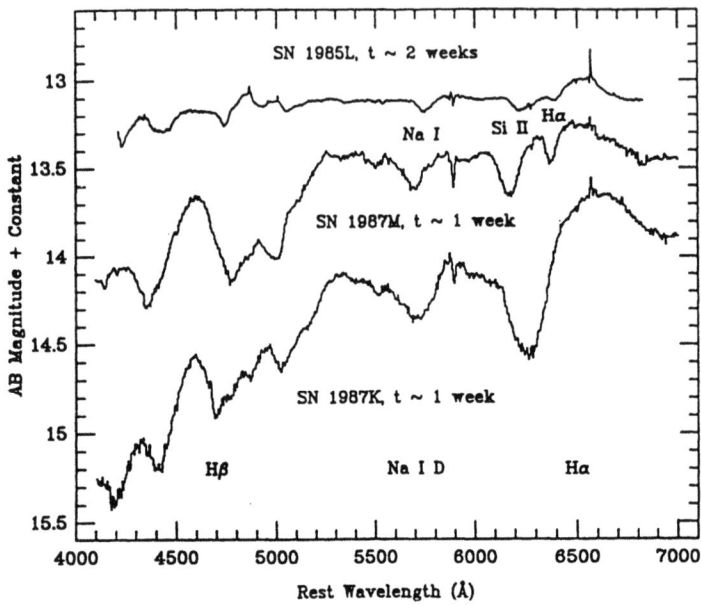

Fig. 3. Comparison of the observed spectra of SN Ic 1987M, SN IIb 1987K, and SN II 1985L (Filippenko *et al.* 1990; Filippenko 1992).

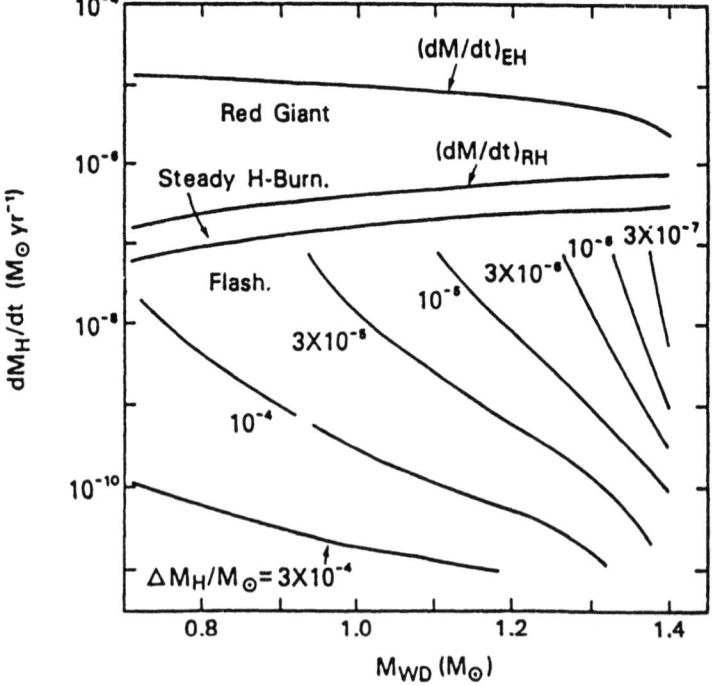

Fig. 4. Types and strength of hydrogen shell burning as a function of accretion rate and the white dwarf mass (Nomoto 1982a).

of most of the accreted matter from the white dwarf; the strongest flash grows into a *nova* explosion (e.g, Nariai *et al*. 1980 and references therein). For these cases, the white dwarf does not become a *supernova* since its mass hardly grows. In other words, it seems rather unlikely that novae are the precursors of supernovae.

For intermediate accretion rates ($3\times 10^{-6} M_\odot$ yr$^{-1} \gtrsim \dot{M} \gtrsim 1\times 10^{-8} M_\odot$ yr^{-1}), on the other hand, hydrogen flashes and the subsequent helium flashes are of moderate strength, thereby increasing the C+O white dwarf mass toward the Chandrasekhar mass. When the white dwarf mass becomes 1.4 $M_\odot$ and the central density reaches $\sim 3 \times 10^9$ gcm^{-3}, explosive carbon burning starts in the white dwarf's center. The precursor systems might be observed as symbiotic stars (Nomoto 1982a,b).

If the accretion rate is higher than $\sim 3 \times 10^{-6} M_\odot$ yr^{-1}, the accreted matter is too hot to be swallowed by the white dwarf (Nomoto *et al.* 1979b). The matter forms a common envelope, which is eventually lost from the system. As a result of mass and angular momentum losses from the system, some binaries form a pair of C+O white dwarfs. Further evolution of such a double white dwarf system is driven by gravitational wave radiation and leads to a Roche lobe overflow of the smaller mass C+O white dwarf (Iben and Tutukov 1984; Webbink 1984). The fate of these merging white dwarfs is not clear yet but would be either a Type Ia supernova explosion or a collapse to form a single neutron star.

The evolution and the final fate of accreting white dwarfs depend further on the several other parameters in close binary systems, namely, the mass, age, composition of the white dwarf and the nature of the companion star; this will be described in Section 3.

2.2. CARBON DEFLAGRATION MODEL

When carbon is ignited at the white dwarf's center, carbon burning is so explosive as to incinerate the material into iron-peak elements; the central temperature reaches $\sim 10^{10}$ K. The resulting shock wave is not strong enough to ignite carbon in the adjacent layer; in other words, *a detonation* wave that propagates at supersonic speed does not form. Instead, the interface between the burned and unburned layers becomes convectively unstable. As a result of mixing with the hot material, fresh carbon is ignited. In this way, a carbon burning front propagates outward on the time scale for convective heat transport (Nomoto *et al.* 1976, 1984; Woosley and Weaver 1986a,b). This kind of explosive burning front that propagates at a subsonic speed is called *a convective deflagration* wave. In the standard model, the propagation speed of the convective deflagration wave is on the average about one-fifth of the sound speed. It takes about one second for the front to reach the surface region, which is significantly slower than the supersonic detonation wave. Hence the white dwarf expands during the propagation of the deflagration wave (Figure 5).

Behind the deflagration wave, the material undergoes explosive nuclear burning of silicon, oxygen, neon, and carbon depending on the peak temperatures. In the

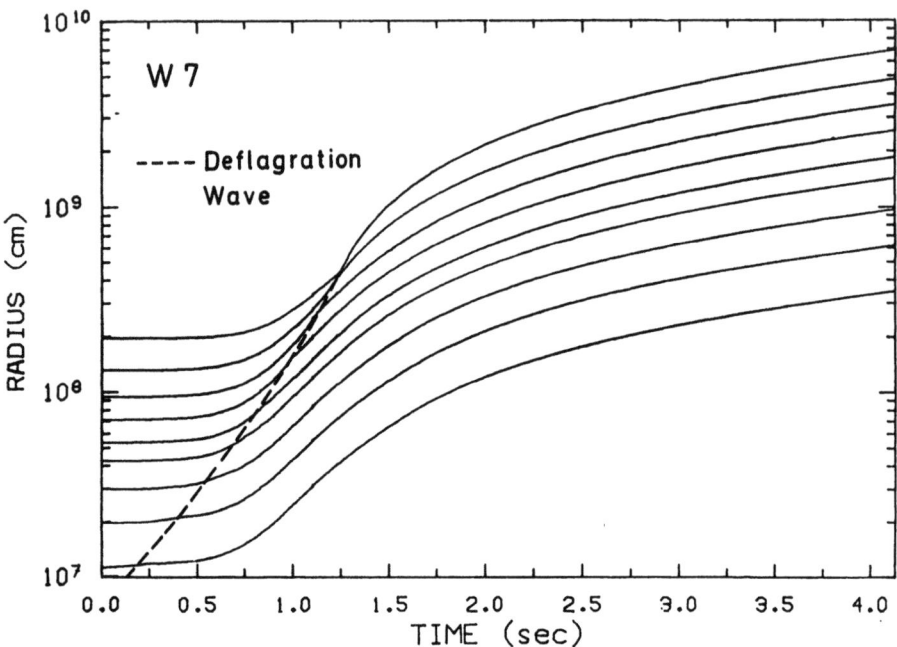

Fig. 5. Propagation of the carbon deflagration wave (dashed line) and the associated expansion of the white dwarf (model W7) (Nomoto et al. 1984). The solid lines corresponds to $M_r/M_\odot$ = 0.007, 0.03, 0.10, 0.25, 0.41, 0.70, 1.00, 1.28, 1.378, respectively.

inner layer, nuclear reactions are rapid enough to incinerate the material into iron-peak elements, mostly ^{56}Ni. When the deflagration wave arrives at the outer layers, the density it encounters has already decreased due to the expansion of the white dwarf. At such low densities, the peak temperature is too low to complete silicon burning and thus only Ca, Ar, S, and Si are produced from oxygen burning. In the intermediate layers, explosive burning of carbon and neon synthesizes S, Si, and Mg. In the outermost layers, the deflagration wave dies and C+O remain unburned. The composition structure after freeze-out is shown in Figure 6 (Thielemann et al. 1986).

In the standard carbon deflagration model W7 (Nomoto et al. 1984), the amount of ^{56}Ni produced is $M_{Ni} = 0.6\ M_\odot$, and the explosion energy is E = (Nuclear energy release − Binding energy of the white dwarf) = 1.3×10^{51} erg. The nuclear energy release is large enough to disrupt the white dwarf completely and no compact star is left behind.

2.3. LATE DETONATION MODELS

The preceding standard model W7 accounts well for the observed light curve and spectra at maximum brightness and late times of SN Ia as will be shown in sections 2.4 and 2.5. The pre-maximum spectra of recent SNe Ia, however, have revealed a

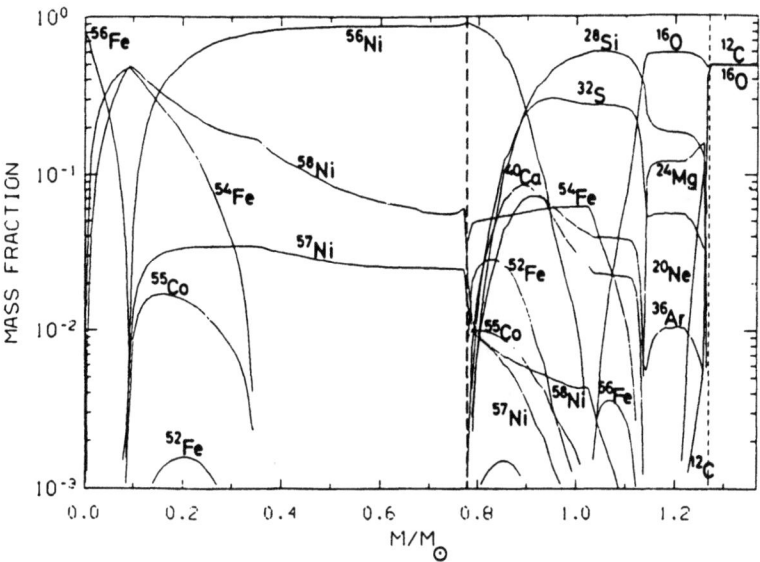

Fig. 6. Composition of a carbon deflagration model W7 for Type Ia supernovae as a function of interior mass. The white dwarf undergoes incineration into iron-peak elements (mostly ^{56}Ni) at $M_r < 0.7 M_\odot$, and partial explosive burning in the intermediate region at $0.7 M_\odot < M_r < 1.3 M_\odot$; in the outermost layer, carbon and oxygen remain unburned (Thielemann et al. 1986).

significant variation of the composition of the outermost layers, which is difficult to interpret with W7. Two such examples are as follows:

(i) The outermost layers of SN 1990N are composed of Si, Ca, Fe, and Co with the expansion velocities as high as $v_{\text{exp}} \sim 20,000$ km s^{-1} (Leibundgut et al. 1991).

(ii) The composition structure of SN 1991T is quite unique (Filippenko et al. 1992; Ruiz-Lapuente et al. 1992) consisting of (I) the outermost layer composed of Ni and Fe, (II) the intermediate layer rich in Si/Ca, and (III) the central layer dominated by Fe. In other words, the Si/S/Ca-rich layer (II) is sandwiched by the two Fe layers (I and III). Such a composition *inversion* with Fe above Si is very different from that inferred from typical SNe Ia. On the other hand, the composition of the inner layers (II and III) is very similar to those of typical SNe Ia including SN 1990N (Filippenko et al. 1992), thereby being accounted for by the carbon deflagration model W7.

For the composition of the outermost layer of SN 1990N, possible models includes mixing of Si, Ca, and Fe to the outer layer of W7 (Branch et al. 1985), the delayed detonation models (Khokhlov 199 ; Woosley & Weaver 1992), and the carbon detonation models for low mass white dwarfs (Shigeyama et al. 1992). They may, however, be difficult to cause a composition inversion with Fe/Ni in the outermost layers as in SN 1991T.

A model that can account for SN 1990N and 1991T in a unified manner is the formation of a detonation in the outermost layers, induced by a fast propagation of

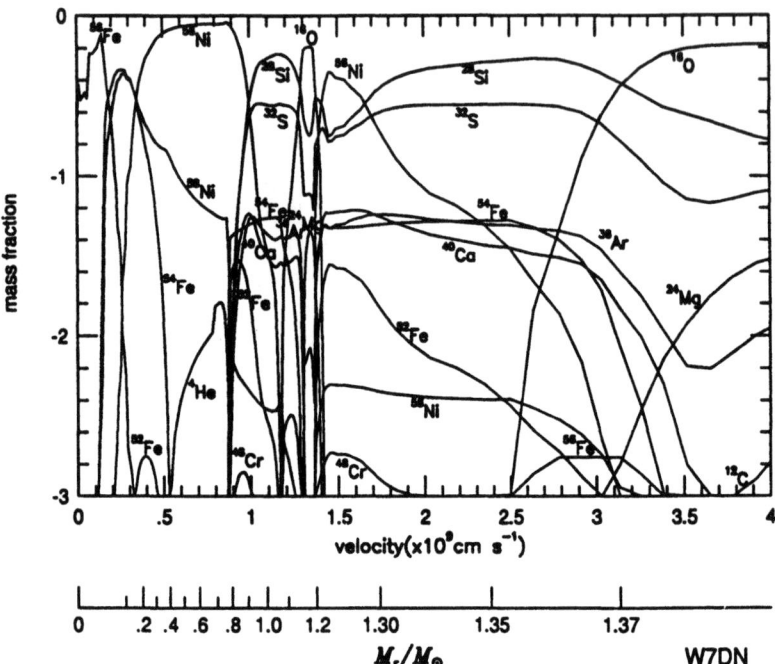

Fig. 7. Composition of the late detonation model W7DN as a function of the expansion velocity and M_r. The transition from deflagrtion to detonation takes place at $M_r = 1.20\ M_\odot$ where the density ahead of the burning front is $\rho_0 = 1.7 \times 10^7$ g cm^{-3}.

a deflagration wave. The transition from a deflagration to a detonation can occur when the propagation velocity of the deflagration wave v_{def} is accelerated to the Chapman-Jouguet velocity v_{CJ} (e.g., Williams 1985). The transition is more likely to occur at lower densities, because the ratio v_{CJ}/v_s (v_s denotes the sound velocity) is smaller due to a larger density jump across the burning front (Khokhlov 1991a), while v_{def}/v_s is larger. Since the convective deflagration front may well be quite turbulent and the acceleration of v_{def} may take place in an indeterministic manner, the condition of the transition from deflagration to detonation has not been well understood and it could occur even for $v_{def} < v_{CJ}$.

Yamaoka et al. (1992) presented several hydrodynamical models where the carbon deflagration (W7), producing a central Fe/Co/Ni core and an intermediate Si/S/Ca layer, is later transformed into a detonation in the outermost layers; this is called a *late detonation*. Nucleosynthesis in the detonated matter depends sensitively on the density at the transition from deflagration to detonation; for higher density, more Fe relative to Si is produced. The composition structures of two late detonation models W7DN (Figure 7; see the caption) and W7DT (Figure 8) can account for the basic features of SN 1990N and SN 1991T, respectively. Figure 9 shows the model W8DT where the deflagration W8 (Nomoto et al. 1984) being ~ 1.2 times faster than W7 is transformed into detonation, which may also reproduce the basic features of SN 1991T. In these models, the pre-maximum Fe and

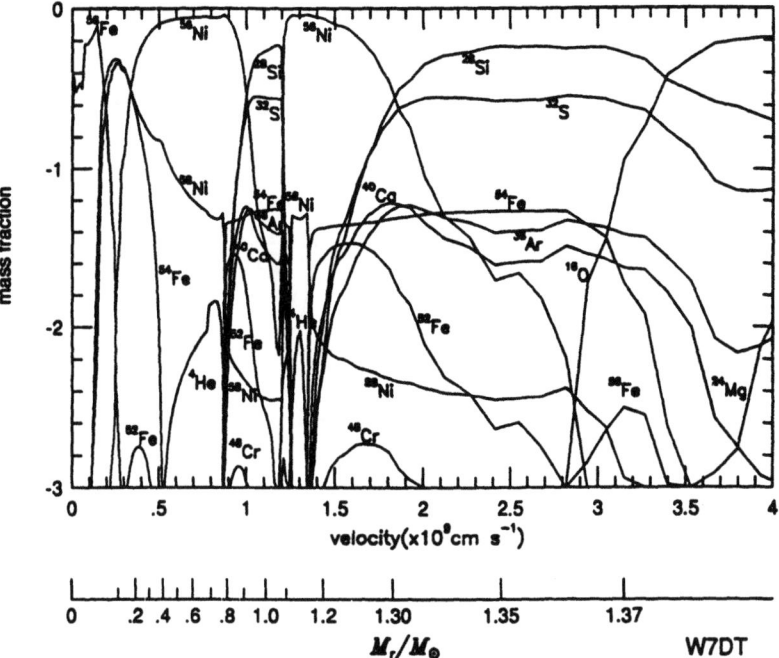

Fig. 8. Same as Figure 7 but for W7DT with the transition at $M_r = 1.13\ M_\odot$ with $\rho_0 \sim 3.5 \times 10^7\ \text{g cm}^{-3}$.

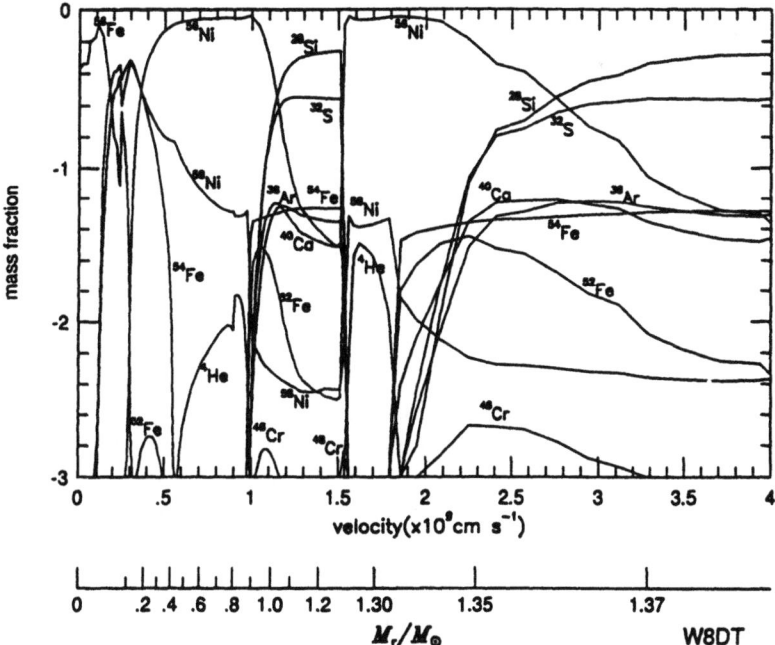

Fig. 9. Same as Figure 7 but for W8DT with the transition at $M_r = 1.25\ M_\odot$ with $\rho_0 \sim 3.5 \times 10^7\ \text{g cm}^{-3}$.

Fig. 10. *a: left* The expanding supernova matter is excited by the decay of ^{56}Co into ^{56}Fe. *b: right* Collapse of an O+Ne+Mg core is induced by electron capture (H. Nomoto 1989).

Si/Ca are the products of the late detonation while the features near maximum light and later phase are due to the deflagration.

With these models, the following scenario for the variation among SNe Ia is suggested: There exist some variations in the ignition conditions, such as the central density and temperature of the white dwarf, due to variations of the white dwarf age and the mass accretion rate. This would result in a difference in v_{def}, which in turn leads to a somewhat larger difference in the density at the transition from deflagration to detonation. This results in a similar composition in the inner regions of (III) and (II), but a large variation of the composition in the outermost layer (I) such as in SN 1990N and SN 1991T. Cases like W7DN would be more common among SNe Ia, since the transition to detonation may occur more easily at lower densities because of smaller v_{CJ}/v_s and larger v_{def}/v_s. Some variations in the expansion velocities of Si may be due to the variation of the transition density.

2.4. LIGHT CURVE

The explosion energy goes into the kinetic energy of expansion, and without a late time energy source the exploding white dwarf could not be bright. However, during the expansion phase, ^{56}Ni decays into ^{56}Co with a half-life of 6.6 days and ^{56}Co decays into ^{56}Fe with a half-life of 77 days. These radioactive decays produce γ-rays and positrons whose energies power the light curve as follows (Figure 10).

Gamma-rays originating from radioactive decays are degraded into X-rays by multiple Compton scatterings. The photoelectric absorption of X-rays and the collisional ionization due to energetic electrons eventually heat up the expanding materials and produce the optical light as clearly observed in SN 1987A (e.g., Kumagai *et al.* 1989; Shigeyama *et al.* 1988; Shigeyama and Nomoto 1990).

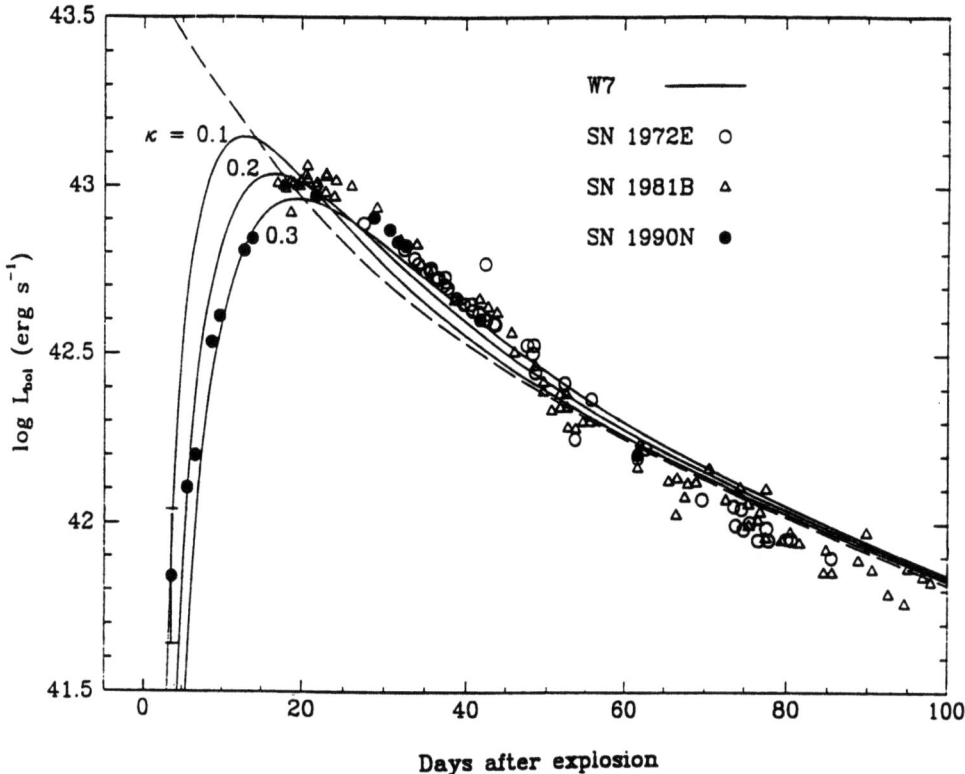

Fig. 11. The calculated bolometric light curves of the carbon deflagration model W7 for the optical opacity of $\kappa = 0.1$, 0.2, and 0.3 cm^2 g^{-1} (solid lines) are compared with the observed visual light curves of SN 1972E, SN 1981B, and SN 1990N.

The light curve powered by the radioactive decays reaches its peak at about 15–20 days after the explosion and declines because of the increasing transparency of the ejecta to γ-rays as well as the decreasing number of radioactive elements. The calculated curve is in good agreement with the observed bolometric light curves of SN 1972E, SN 1981B and SN 1990N, if the optical opacity in the outer layers is as large as $\kappa = 0.3$ cm^2 g^{-1} due to the steep velocity gradient (Figure 11; Nomoto et al. 1992).

2.5. SPECTRA

Because SNe Ia do not have a thick hydrogen-rich envelope, elements newly synthesized during the explosion can be observed in the spectra; this enables us to diagnose the internal hydrodynamics and nucleosynthesis in SNe Ia.

Synthetic spectra are calculated based on the abundance distribution and expansion velocities of the standard model and are found to be in excellent agreement with the observed optical spectra of SN 1981B (Branch et al. 1985) and 1989B as seen in Figure 12 (Harkness 1991). The material velocity at the photosphere

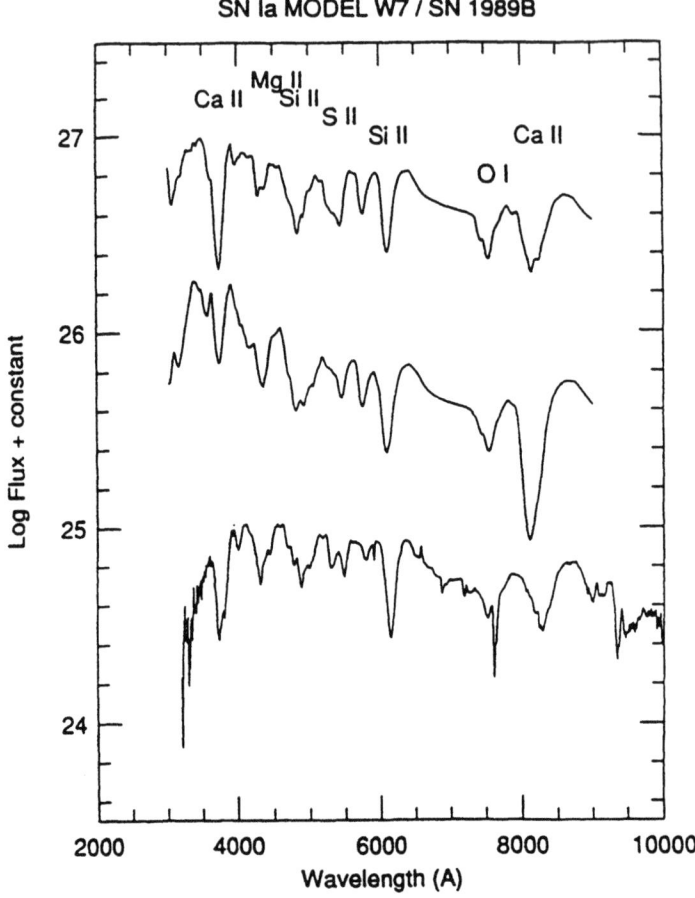

Fig. 12. The maximum light spectrum of SN Ia 1989B is compared with a synthetic spectrum for the carbon deflagration model W7 with no mixing (top) and mixing for $v > 11000$ km s^{-1} (middle) (Harkness 1991).

near maximum light is $\sim 10{,}000$ km s^{-1} and the spectral features are identified as P-Cygni profiles of Fe, Ca, S, Si, Mg, O.

At late times, the spectra are dominated by the emission lines of Fe and Co. The outer layers are transparent and the inner Ni-Co-Fe core is exposed. Synthetic spectra of emission lines of [FeII] and [CoI] agree quite well with the spectra observed at such phase (Axelrod 1980; Woosley and Weaver 1986b). The agreement implies that both explosion energy and nucleosynthesis in the carbon deflagration model are consistent with the observations of SNe Ia.

3. Accretion–Induced Collapse (AIC) of White Dwarfs

3.1. SOLID C+O WHITE DWARFS

It is possible that the accreting C+O white dwarfs could collapse rather than explode, depending on the conditions of the white dwarfs. As described in Section 2.1, compression of the white dwarf by the accreted matter first heats up a surface layer and, later, heat diffuses inward (Nomoto et al. 1984). If the initial mass of the white dwarf, M_{CO}, is smaller than $1.2 M_\odot$, the entropy in the center increases substantially due to the heat inflow and thus carbon ignites at relatively low central density ($\rho_c \sim 3 \times 10^9$ g cm^{-3}). On the other hand, if the white dwarf is initially more massive than $1.2 M_\odot$ and cold at the onset of accretion, the central region is compressed only adiabatically and thus is cold when carbon is ignited in the center. In the latter case, the ignition density is as high as 10^{10} g cm^{-3} (e.g., Isern et al. 1983) and the white dwarf may well have a solid core. For such a case, it is necessary to determine the critical condition for which a carbon deflagration induces collapse rather than explosion.

In the AIC models, collapse of the white dwarf is induced by electron capture that effectively reduces the Chandrasekhar mass. However, since the white dwarf contains nuclear fuel, whether the white dwarf undergoes collapse or explosion depends on which is faster behind the deflagration wave, nuclear energy release or electron capture. The energy generation rate is determined mainly by the propagation velocity of the deflagration wave, v_{def}, while the electron capture rate depends on the density. If v_{def} is lower than a certain critical speed, electron capture induces collapse. If on the other hand v_{def} is sufficiently high, complete disruption results. Nomoto and Kondo (1991) and Canal et al. (1990) have shown that the critical velocity that divides collapse and explosion is $v_{\mathrm{crit}} \sim 0.04\, v_s$ for $\rho_c \sim 10^{10}$ g cm^{-3}.

For solid white dwarfs, a *conductive* deflagration wave propagates unless the solid layer is melted earlier. Since the realistic value of conductive deflagration speed is $v_{\mathrm{def}} \sim 0.01 v_s$ (Woosley and Weaver 1986a), the collapse is the most likely outcome for the solid white dwarf (Nomoto 1986b; Canal et al. 1990; Nomoto and Kondo 1991).

3.2. O+NE+MG WHITE DWARFS

The O+Ne+Mg white dwarfs are fomed with the initial masses being as large as 1.1 - 1.37 $M_\odot$ from stars with initial masses of 8 - 12 $M_\odot$ in close binaries (Nomoto et al. 1979a; Nomoto 1984). After mass accretion from the companion star, the mass of the white dwarf increases toward the Chandrasekhar mass for a certain range of accretion rate (Figure 14). When ρ_c exceeds 4×10^9 g cm^{-3}, the O+Ne+Mg white dwarf undergoes electron captures ^{24}Mg$(e^-, \nu)\,^{24}$Na$(e^-, \nu)^{20}$Ne and ^{20}Ne$(e^-, \nu)^{20}$F$(e^-, \nu)\,^{20}$O (Figure 10). Electron capture not only reduces the effective Chandrasekhar mass but also releases heat due to γ-ray emission which eventually ignites oxygen deflagration at a central density of $\rho_{\mathrm{ig}} \sim 1 - 2.5 \times 10^{10}$

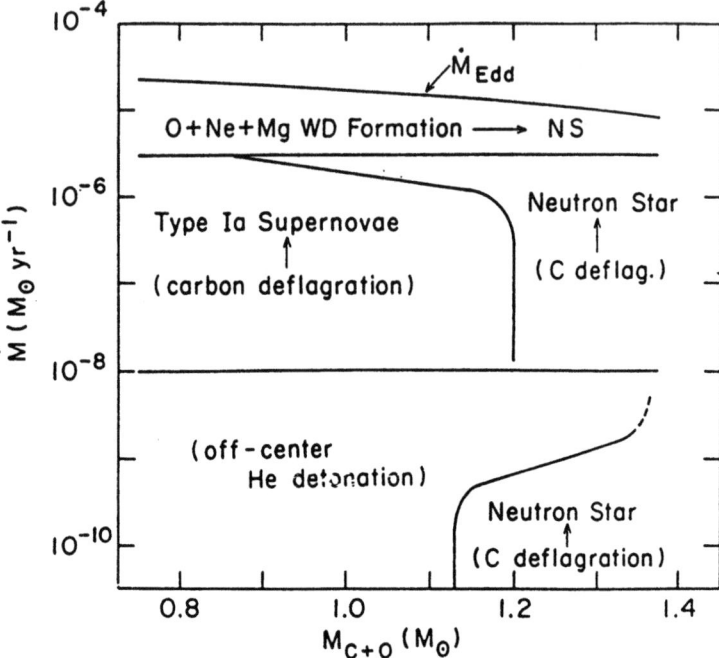

Fig. 13. The final fate of accreting C+O white dwarfs expected for their initial mass and accretion rate $\dot{M}$.

g cm^{-3} (Miyaji et al. 1980; Miyaji and Nomoto 1987; Nomoto 1987). Nomoto and Kondo (1991) have shown that collapse is more likely outcome than explosion from oxygen deflagration.

3.3. CONDITIONS FOR AIC AND SN IA

To make estimates for the occurrence frequency of AIC and SN Ia, we draw boundaries for AIC in a diagram of mass accretion rate $(\dot{M})$ versus mass of the white dwarf at the onset of accretion (M_{CO} and M_{ONeMg}) in Figures 13 and 14 (Nomoto 1986b; Nomoto and Kondo 1991).

In these figures, the fate of accreting white dwarfs depends mainly on the mass accretion rate as discussed in Section 2.1. We note that the boundaries must be regarded as optimistic ones for the growth of white dwarfs since wind-type mass loss associated with shell flashes of hydrogen and helium is not fully taken into account (e.g., Kato and Hachisu 1989).

1) For $\dot{M} \gtrsim \dot{M}_{det}$, both hydrogen and helium flashes are weak and recur many times to increase the white dwarf mass.

2) For $\dot{M}_{det} \gtrsim \dot{M} \gtrsim 10^{-9} M_\odot$ yr^{-1}, off-center helium detonation prevents the white dwarf mass from growing (Nomoto 1982b; Woosley et al. 1986). Here we adopt $\dot{M}_{det} \sim 1 \times 10^{-8} M_\odot$ yr^{-1}, since the ^{14}N$(e^-, \nu)^{14}$C$(\alpha, \gamma)^{18}$O (NCO) reaction ignites weak helium flashes (Hashimoto et al. 1986) if the mass fraction of CNO

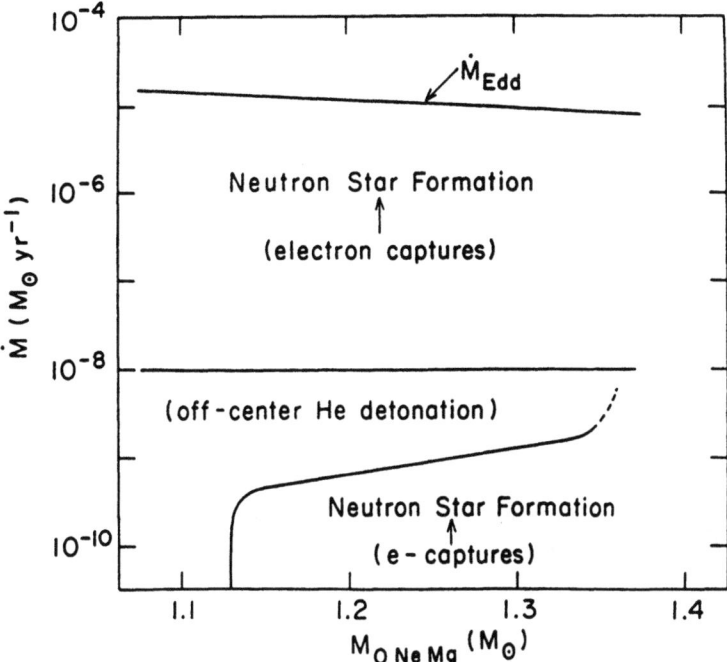

Fig. 14. Same as Figure 13 but for O+Ne+Mg white dwarfs. Collapse is triggered by electron capture on ^{24}Mg and ^{20}Ne.

elements in the accreting material exceeds 0.005. For smaller CNO abundances, the NCO reaction is not effective and thus $\dot{M}_{det} \sim 4 \times 10^{-8} M_\odot$ yr^{-1} (Nomoto 1982a).

3) For $\dot{M} \lesssim 10^{-9} M_\odot$ yr^{-1}, accretion of hydrogen-rich material gives rise to nova-like explosion which would not allow the white dwarf mass to grow. If the accreting material is helium, on the other hand, the material is too cold to ignite helium burning, thereby increasing the white dwarf mass. An exception is the case with $M_{CO} \lesssim 1.1 M_\odot$ where pycnonuclear helium burning is ignited (Nomoto 1982a,b).

For $\dot{M} \gtrsim 2 \times 10^{-6} M_\odot$ yr^{-1} in Figure 13, we adopt the following scenario. First merging of double C+O white dwarfs forms a thick disk around more massive component (Benz et al. 1990). Subsequent heat generation at the boundary layer ignites off-center carbon burning (Mochkovitch and Livio 1990), which burn the entire C+O white dwarf into O+Ne+Mg quietly (Saio and Nomoto 1985). Eventually the O+Ne+Mg white dwarf collapses. This is an optimistic scenario for AIC since such merging white dwarfs are also the possible candidates of SN Ia progenitors (Iben and Tutukov 1984; Webbink 1984).

4. Type Ib/Ic Supernovae

Wolf-Rayet stars with a wide range of masses have been proposed for the progenitors of SNe Ib and Ic, since most of SNe Ib/Ic are associated with star-forming regions (Wheeler and Harkness 1990 for a review). Recently Shigeyama *et al.* (1990), Hachisu *et al.* (1991), Nomoto *et al.* (1990), and Yamaoka and Nomoto (1991) have calculated the progenitor's evolution, nucleosynthesis, Rayleigh-Taylor instabilities, and optical light curves of exploding helium stars. They have suggested that the helium stars of 3–5 $M_\odot$ (which form from stars with initial masses $M_i \sim 12-18 M_\odot$ in binary systems) are the most likely progenitors of typical SNe Ib/Ic and that SNe Ic progenitors may be slightly less massive than those of SNe Ib. Such low mass helium star models can account for the observations that: (1) the light curves of SNe Ic decline faster than SNe Ib, and (2) the early time spectra of SNe Ic show the presence of hydrogen (Jeffery *et al.* 1991), while hydrogen is absent in SNe Ib. It remains an open question how the presence of hydrogen causes the difference between SNe Ib and Ic in their early time spectra.

4.1. Evolution of Interacting Binaries

The difference in the spectral feature between SNe Ic and Ib may be due to the presence of a thin envelope of hydrogen in SNe Ic immediately prior to the explosion. By evolving massive stars in close binary systems, we examine whether hydrogen can be left on the helium stars after mass exchange and wind-type mass loss. Followings are some preliminary results for two cases, where the initial masses of the primary stars are $M_i = 13 M_\odot$ and $18 M_\odot$, and their Roche lobe radii are $50 R_\odot$ (Yamaoka and Nomoto 1991).

After hydrogen exhaustion, the star undergoes Roche lobe overflow forming a helium star of $\sim 3.4\ M_\odot$ ($13 M_\odot$) and $\sim 5 M_\odot$ ($18 M_\odot$). Figure 15 shows the composition structure at the onset of mass transfer during core helium burning (upper) and after the Roche lobe overflow (lower). It is seen that significant amount of hydrogen remains in a relatively thick layer below the surface (0.5 $M_\odot$ for $13 M_\odot$ and 1 $M_\odot$ for $18 M_\odot$).

Whether such a hydrogen layer will be further lost from the helium-rich star depends on the Roche lobe radius and wind mass loss rate. If the helium star is detached from the Roche lobe during helium burning, the star loses its masses in a wind. If the mass loss rate depends on the mass such as $\dot{M} \propto M^{2.5}$ (Langer 1989), it may lead to a difference of the surface abundance between $13 M_\odot$ and $18 M_\odot$. In the case for $13 M_\odot$ hydrogen still may remain in the layer down to $\sim 0.2 M_\odot$ below the surface, whereas for $18 M_\odot$ all hydrogen may be lost in a wind (Yamaoka and Nomoto 1991). If a common envelope forms or the Roche lobe radius has become small enough for the mass and angular momentum to be lost from the system, then all hydrogen will be lost from the star.

Although more parameter study is needed, the present results suggest that more

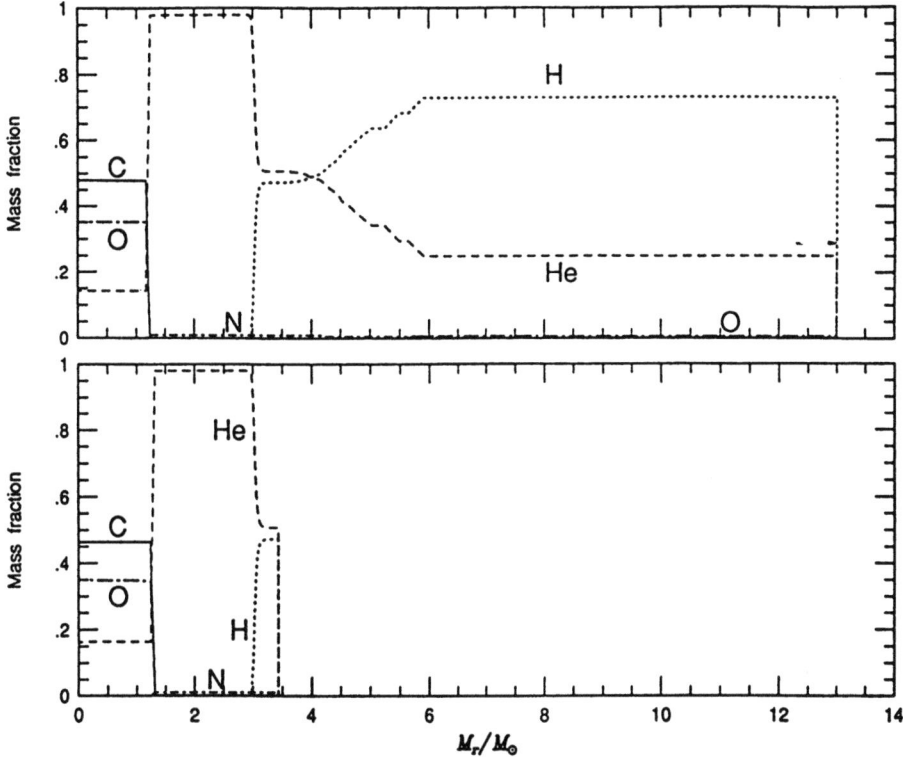

Fig. 15. Change in the composition structure during the binary evolution for the star with $M_i = 13 M_\odot$ (Yamaoka and Nomoto 1991).

hydrogen may remain on the stellar surface if the helium star had initially smaller main-sequence mass. Also more compact binary systems may lose more hydrogen through common envelope evolution.

4.2. NUCLEOSYNTHESIS

Assuming that the helium star progenitors of SNe Ib/Ic are formed in close binary systems as described above, Shigeyama et al. (1990) performed hydrodynamical calculations of the explosion of helium stars with masses $M_\alpha = 3.3, 4, 6,$ and 8 $M_\odot$. These are presumed to form from the main-sequence stars of masses $M_i \sim 13, 15, 20,$ and $25 M_\odot$, respectively. These stars eventually undergo iron core collapse as in SNe II. A shock wave is then formed at the mass cut that divides the neutron star and the ejecta.

Behind the shock wave that propagates outward, materials are processed into nuclear statistical equilibrium (NSE) composition, mostly ^{56}Ni, if the maximum temperature exceeds 5×10^9 K (e.g., Hashimoto et al. 1989). As derived from the approximate relation $E = 4\pi r^3/3 \, aT^4$ with E being the final kinetic energy of explosion (e.g., Thielemann et al. 1990), such a high temperature is realized in a

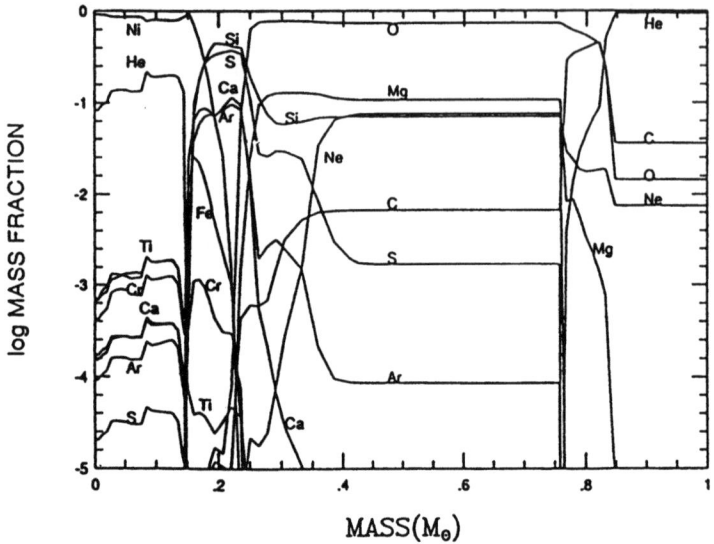

Fig. 16. Explosive nucleosynthesis in the 4 $M_\odot$ helium star (Shigeyama et al. 1990). Composition of the innermost 1 $M_\odot$ of the ejecta is shown. (The outermost 1.72 $M_\odot$ helium layer and the 1.28 $M_\odot$ neutron star are not included in the figure.) About 0.15 $M_\odot$ ^{56}Ni and 0.43 $M_\odot$ oxygen are produced.

sphere of radius $\sim 3700 \, (E/10^{51}\mathrm{erg})^{1/3}$ km. With $E = 1 \times 10^{51}$ erg, this region contains a mass M_{NSE} ($\sim$ 1.44–1.46 $M_\odot$ for M_α = 3.3 and 4 $M_\odot$). Then the mass of ^{56}Ni plus neutron-rich iron peak elements is given by $M_{\mathrm{NSE}} - M_{\mathrm{NS}}$.

The adopted presupernova models (Nomoto and Hashimoto 1988) have the iron core masses as small as 1.18 $M_\odot$ and 1.28 $M_\odot$ for M_α = 3.3 $M_\odot$ and 4 $M_\odot$, respectively, being significantly smaller than 1.4 $M_\odot$ in the 6 $M_\odot$ star due to the larger effect of Coulomb interactions during the progenitor's evolution. If M_{NS} is approximately equal to the iron core mass, the *upper limit* to the possible ^{56}Ni masses are obtained as 0.26 and 0.15 $M_\odot$ for M_α = 3.3 and 4 $M_\odot$, respectively.

Figure 16 shows the abundance distribution after explosive burning for M_α = 4 $M_\odot$ (Shigeyama et al. 1990). The masses of oxygen produced in the outer layers are 0.21 and 0.43 $M_\odot$ for M_α = 3.3 and 4 $M_\odot$, respectively. These masses could be consistent with those inferred from the late time spectra of SNe Ib/Ic in view of the strong dependence of the oxygen mass on the temperature of the ejecta (e.g., Uomoto 1986).

4.3. Light Curves

Figure 17 shows the observed bolometric light curves of SNe Ia 1972E and 1981B (Graham 1987), SN Ib 1983N (Panagia 1987), and the approximate bolometric light curve of SN Ic 1987M constructed from flux-calibrated spectra (Filippenko et al. 1990; Nomoto et al. 1990). In each case, the observed light curve has been

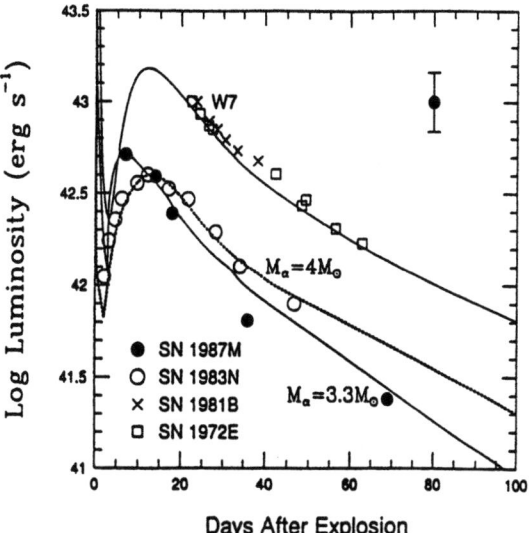

Fig. 17. Approximate bolometric light curve of SN Ic 1987M, and the bolometric light curves of SNe Ia 1972E and 1981B and of SN Ib 1983N. The predicted curves of the 3.3 $M_\odot$ model for SN 1987M, the 4 $M_\odot$ model for SN Ib, and the W7 model for SN Ia are indicated by solid and dotted lines. The error bar illustrates the 2σ photometric uncertainty in the SN 1987M points.

shifted along the abscissa to match the corresponding theoretical curve. The peak bolometric luminosities assume $H_0 = 60 \text{ km s}^{-1} \text{ Mpc}^{-1}$. The previous Wolf-Rayet star models have some difficulties (1) in reproducing the light curves of typical SNe Ib which decline as fast as SNe Ia (Panagia 1987), and (2) in producing enough ^{56}Ni to attain the maximum luminosities of SNe Ib in relatively low mass helium star models (Ensman and Woosley 1988). In particular, Figure 17 demonstrates an important feature of SN Ic 1987M, i.e., its brightness fell somewhat more rapidly than that of SNe Ia and SN Ib 1983N (also SN Ic 1983I reported by Tsvetkov 1985). Maximum brightness of 1987M is not significantly different from those of SNe Ib if we take the extinction estimated by Jeffery et al. (1991).

Figure 17 also shows the calculated bolometric light curves of the exploding helium star models with $M_\alpha = 3.3\ M_\odot$ for SN Ic and 4 $M_\odot$ for SN Ib as well as the white dwarf model W7 for SNe Ia (Nomoto et al. 1984). The amount of ^{56}Ni is 0.58 $M_\odot$ (W7), 0.15 $M_\odot$ (SN Ic), and 0.15 $M_\odot$ (SN Ib). The helium star models assume uniformly mixed distribution of elements from the center through the layer at 0.2 $M_\odot$ beneath the surface for both cases. Such a mixing may be due to the Rayleigh-Taylor instability during the explosion (Figure 18; Hachisu et al. 1990, 1991).

The calculated bolometric light curves of helium stars are powered by the radioactive decays of ^{56}Ni and ^{56}Co (Figure 10a). Maximum brightness is higher if the ^{56}Ni mass is larger and the date of maximum earlier. After the peak, the optical light curve declines at a rate that depends on how fast γ-rays from the radioactive

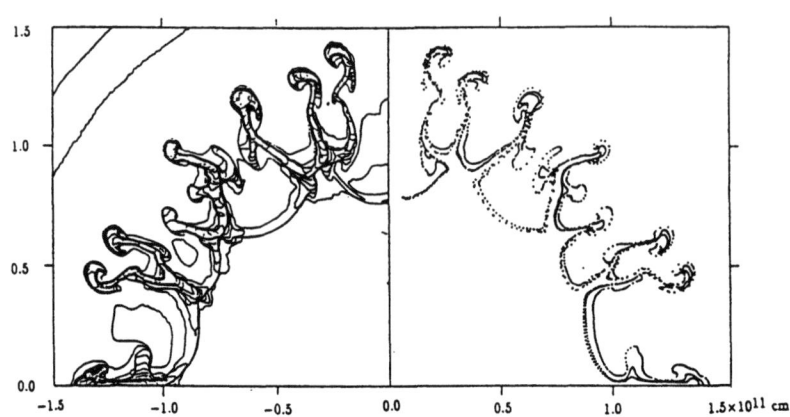

Fig. 18. Rayleigh-Taylor instabilities in the exploding helium star of $M_\alpha = 3.3\ M_\odot$ ($t = 180$ s). Shown are the density contour map (*left*) and the marker particles at the composition interfaces, He/C+O, O/Si, and Ni/Si from the outer layers (*right*) (Hachisu et al. 1991).

decays escape from the star without being thermalized, thereby declining faster if the ejected mass is smaller and if ^{56}Ni is mixed closer to the surface.

The resulted bolometric light curves of $M_\alpha = 3.3 M_\odot$ and $4 M_\odot$ are in good agreement with SN Ic 1987M and SN Ib 1983N, respectively. Compared with the $4 M_\odot$ model, the light curve of the 3.3 $M_\odot$ model declines faster due to the smaller ejected mass, just as observed in SN Ic 1987M.

5. Evolutionary Origin of Binary Pulsars and SN Ib/Ic

The low mass helium stars considered for the progenitors of SNe Ib/Ic would mostly occur in close binaries, because the 12–18 $M_\odot$ stars would not lose their entire hydrogen-rich envelope by wind mass loss. Meurs and van den Heuvel (1989) predicted that more than 70 percent of massive star explosions would occur in close binaries. This estimate predicts that the occurrence frequencies of SNe Ib/Ic are higher than SN II, which might be consistent with the increasing number of SNe Ic recently discovered (Muller et al. 1992).

The binary scenario suggests that SNe Ib/Ic might be closely related to the formation of binary pulsars and X-ray binaries. If the binary system is not disrupted by the supernova mass ejection, a neutron star is left to orbit around a various types of a companion star (a main-sequence star or a helium star). Many of them would become Be X-ray binaries.

Recently the masses of the component stars in the binary pulsar system 1532+12 has been determined, which strongly suggests a neutron star companion (Wolszczan 1991). Their masses, eccentricity, semi-major axis, and orbital period are summarized in Table I together with those of the first binary pulsar 1913+16 (Taylor 1992).

A possible evolutionary scenario for such systems are as follows (e.g., van den

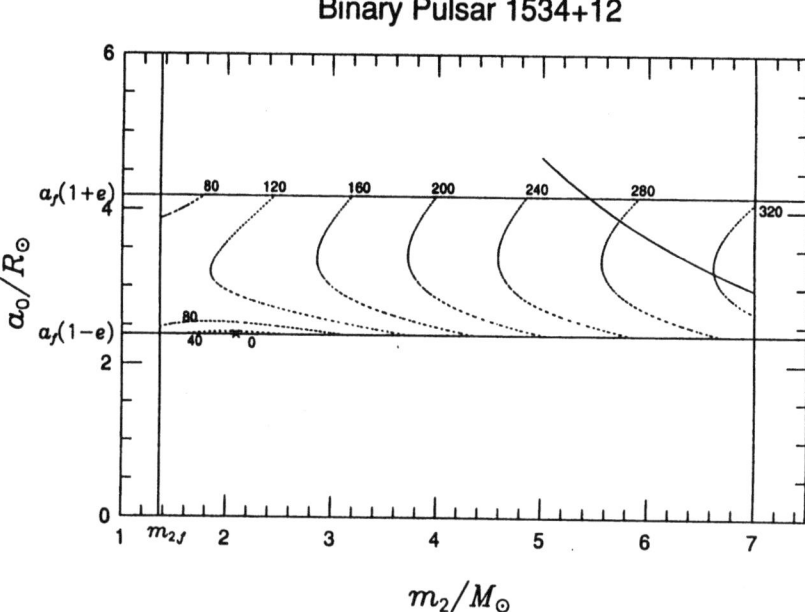

Fig. 19. Kick velocity imparted to the neutron star at the explosion of the helium star 2 of mass M_α as a function of M_α and the initial orbital radius a_0 before the explosion (Yamaoka *et al.* 1992). Here $a_f(1-e) < a_0 < a_f(1+e)$. For the solid line, the radius of the helium star is equal to its Roche lobe radius, so that only the upper-right part of the parameter space is allowed. For PSR 1534+12, $v_{\rm kick} \sim 160\text{--}260$ km s^{-1} is necessary to avoid the disruption of the binary system.

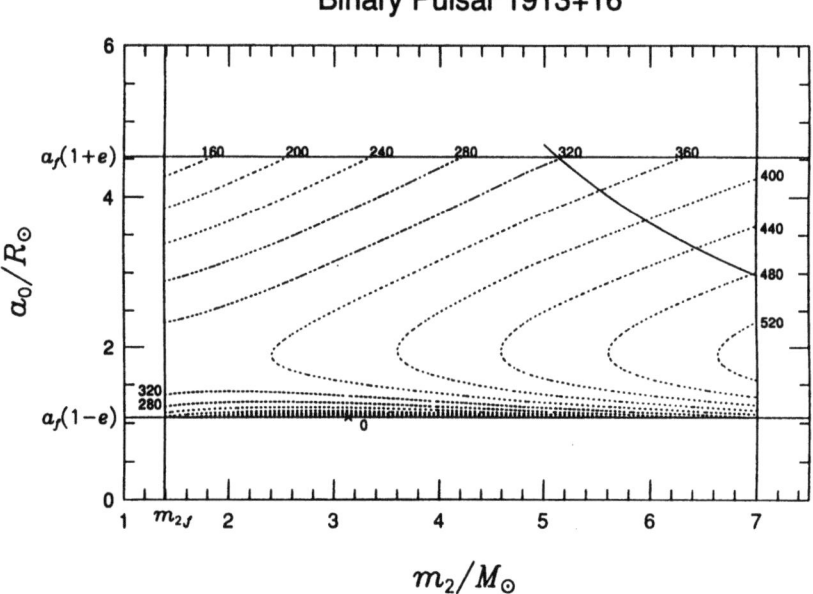

Fig. 20. Same as Figure 19 but for PSR 1913 + 16, where $v_{\rm kick} \sim 300\text{--}460$ km s^{-1} is necessary (Yamaoka *et al.* 1992).

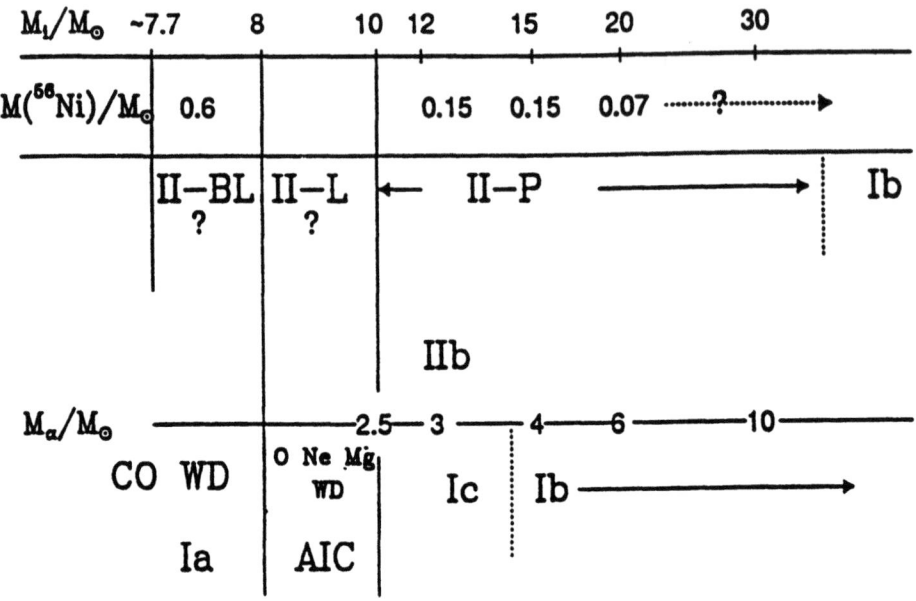

Fig. 21. Hypothetical connection between supernova types and their progenitors for single stars (upper) and close binary stars (lower). M_i and M_α are the initial mass and the helium star mass, respectively. AIC stands for accretion-induced collapse of white dwarfs.

Heuvel 1991):
1. Two main-sequence stars 1 and 2,
2. Roche lobe overflow of star 1, which becomes a helium star 1,
3. the first supernova explosion of the helium star 1 to form a neutron star 1,
4. Roche lobe overflow of star 2, which leads to a spiral-in of the neutron star 1 into star 2 and thus to a considerable shrink of the system due to the losses of angular momentum and mass from the system; the system now consists of the recycled neutron star 1 and a helium star 2,
5. the second supernova explosion of the helium star 2; this forms a two neutron star system in an eccentric orbit.

Given the observed orbital parameters in Table I and the assumption of a circular orbit for the pre-explosion helium star 2–neutron star 1 system, the mass of the helium star 2 M_α and the possible kick velocity v_{kick} at the explosion can be calculated (Nomoto et al. 1991; Wijers 1991). If the explosion is spherical (i.e., $v_{kick} = 0$), $M_\alpha \sim 2.1\ M_\odot$ (Wolszczan 1991). This is smaller than the minimum mass of the helium star that can form a neutron star; $\sim 2.5\ M_\odot$ (Nomoto 1984) – $\sim 2.2\ M_\odot$ (Habets 1986) depending on the treatment of overshooting and semi-convection. This suggests that either the explosion is not spherical or the exploding ⋯ had lost even its helium layer before the explosion.

If we introduce a finite v_{kick} for the explosion of the helium star 2, the kick elocity and its direction can be calculated as functions of assumed M_α and the

TABLE I
Binary Pulsars

PSR	1913+16	1532+12
M_p ($M_\odot$)	1.4421±0.0012	1.32±0.03
M_c ($M_\odot$)	1.3875±0.0012	1.36±0.03
e	0.617127	0.274
a_f ($R_\odot$)	2.8	3.28
P_b (hours)	7.752	10.1
$P_b / \dot{P_b}$	3×10^8	$\sim 1 \times 10^9$

initial orbital radius a_0 before the explosion where $a_f(1 - e) < a_0 < a_f(1 + e)$ (Figure 19; Yamaoka et al. 1992). It should be noted that smaller helium stars have larger radii at the collapse which is $\sim 4~R_\odot$ for $M_\alpha \sim 3.3 M_\odot$ (Nomoto and Hashimoto 1988). For the solid line in Figures 19 and 20, the radius of helium star is equal to its Roche lobe radius. As far as the star 2 is a helium star, therefore, M_α should be larger than 5 $M_\odot$ to underfill the Roche lobe. If this is the case, a kick velocity of $v_{kick} \sim$ 160–260 km s^{-1} is necessary to avoid the disruption of the binary system. The same relation is obtained for 1913 + 16 in Figure 20, where $v_{kick} \sim$ 300–460 km s^{-1} is necessary (Nomoto et al. 1991; see also Burrows and Woosley 1986).

Two possible extreme cases are: (1) the star 2 is a helium star of more massive than $\sim 5~M_\odot$, for which the explosion produces a large kick velocity, and (2) the star 2, being initially a helium star of smaller than 5 $M_\odot$, loses its helium envelope to become an almost bare C+O star. The masses of C+O stars are: 6.0, 3.8, 2.1, and 1.8 $M_\odot$ for M_α = 8, 6, 4, 3.3 $M_\odot$, respectively (Nomoto and Hashimoto 1988). For $M_\alpha \lesssim 4 M_\odot$, therefore, the explosion of star 2 could be spherical with v_{kick} = 0.

If the main difference between SNe Ib and Ic are the absence and presence of helium, the first explosion could be SN Ib because of possible presence of hydrogen, while the second explosion might be SN Ic if star 2 loses even its helium envelope.

6. Concluding Remarks

Figure 21 summarizes the initial masses M_i of the progenitors for the various types of supernovae currently proposed by several groups (e.g., Branch et al. 1991). The upper and lower rows respectively show the cases of single stars and helium stars of masses M_α (or white dwarfs) in close binary stars. The produced masses of ^{56}Ni are inferred from light curves based on the radioactive decay model.

Single stars more massive than $\sim 8\ M_\odot$ would retain their hydrogen-rich envelope, thereby ending up their lives as SNe II. Among them, SNe II-BL and SNe II-L are tentatively assumed to be the explosions of AGB stars having degenerate C+O cores (carbon deflagration) and O+Ne+Mg cores (electron capture induced collapse), respectively (Swartz *et al.* 1991). Interacting binaries are the likely progenitors of various types of SNe I as has been discussed in the preceding chapters. It must be emphasized that the supernova types vs. progenitor's mass relation presented in Figure 21 is still highly hypothetical and will be tested by future observations of light curve tails and good spectra (Branch *et al.* 1991).

Acknowledgements

I would like to thank T. Shigeyama, S. Kumagai, H. Yamaoka, T. Tsujimoto, M. Hashimoto, F.-K. Thielemann, A.V. Filippenko, D. Branch, D. Jeffery, and Y. Kondo for collaborative work on the subjects discussed in this paper. This work has been supported in part by the grant-in-Aid for Scientific Research (01540216, 02234202, 02302024, 03218202, 04234203, 04640265) of the Ministry of Education, Science, and Culture in Japan, and by the Japan-U.S. Cooperative Science Program (EPAR-071/88-15999) operated by the JSPS and the NSF.

BINARY SYSTEMS WITH COMPACT COMPONENTS

V. TRIMBLE

1. Introduction: Formation, Detectability, and Fate of Binaries with Compact Components

Binaries with compact components (neutron stars and black holes) are a natural outcome of standard close binary evolution, and the supply can be augmented by capture and accretion-induced collapse of white dwarfs under suitable circumstances (Sect. 1). Our basic understanding of the formation, evolution, and deaths of such systems provided a good match to the properties of the classes of objects normally attributed to them. These include the X-ray binaries (of high and low mass, and a few with black hole primaries) and the "recycled" binary and millisecond pulsars (Sect. 2 & 3). Statistical issues of which processes dominate the production of low-mass and pulsar systems are still under debate. The thermal radiation from the X-ray systems is much better understood than is the non-thermal radiation of the pulsars (Sect. 4), though the former present the additional complication of gas transfer between the stars (Sect. 5). Some relatively unusual types of systems (Sect. 6) may or may not be part of the basic evolutionary picture.

Normal single star evolution produces neutron stars (and perhaps black holes) from the 1–3% of stars initially exceeding 6–8 $M_\odot$. In a close binary system, mass transfer will more than reverse the mass ratio before the primary completes its evolution; thus a spherically symmetric supernova explosion will not unbind the system (Trimble and Rees 1971). A certain number of binaries with compact components should, therefore, result from standard processes.

The supply can be augmented, first, by collapse of a white dwarf in an existing system when accreted material drives it above the Chandrasekhar limit (Van den Heuvel 1981) and, second, by capture of an existing neutron star in a dense environment (Fabian, Pringle and Rees 1975). The signature of the first process should be a young-looking neutron star (meaning one with dipole $B \geq 10^{11}$G) in an old system, and the signature of the second an old neutron star in a young-looking system (meaning an uncircularized orbit or, statistically, one that will evolve into something else on a short time scale). Neutron star rotation period is not a good age indicator, because at least some are born as slow rotators (Emmering and Chevalier 1989) and they can be spun both up and down by accretion and magnetic torques in close binaries.

Not all conceivable binaries with compact components will be detectable as such. A non-accreting neutron star or black hole in orbit around a massive secondary will be indistinguishable from a main sequence star of the same mass. (Notice the convention that the initially more massive star continues to be labeled the primary, no matter what happens to it later.) Accretion onto the compact component from either a wind or Roche lobe overflow by the secondary will yield detectable X-rays for accretion rates in the range 10^{-12}–10^{-7} $M_\odot$ yr^{-1}. Lower rates will be very faint, and higher ones, in excess of that producing the Eddington luminosity, will become optically thick and radiate primarily visible and ultraviolet radiation.

Pulsar radio emission falls below the plasma frequency of even the weak stellar wind of a main sequence companion. Thus all known binary pulsars orbit degenerate dwarfs or other neutron stars. Since "true" pulsars draw their energy from rotation and magnetic fields and have luminosities that scale as $B^2 P^{-6}$, most will again fade from sight in less than the age of the universe, leaving their systems again inconspicuous. There seems to be nothing that would call attention to a binary black hole unless it walked into the room with you.

Classes of objects normally attributed to binaries with compact components include the X-ray binaries with X-ray luminosity exceeding 1–10 $L_\odot$. binary pulsars, and millisecond (single) pulsars believed to be their descendants. Some additional objects and phenomena probably or possibly associated with these main classes are addressed in the final section. The X-ray binaries are further divided into high and low mass systems, on the basis of their less compact secondaries ($\geq$ 8–10 $M_\odot$ vs. $\leq$ 1.0–1.5 $M_\odot$). Most or all of the massive ones (which include the handful of black hole candidates) come from direct core collapse in pre-existing systems, while many or most of the low mass ones appear to arise from induced collapse and/or capture. The binary pulsars left after accretion stops consist of two neutron stars in the massive case and of neutron star plus degenerate dwarf in the low mass one. No binary pulsars with companion masses suggesting black holes are known, and we do not yet necessarily statistically expect any among the dozen or so (epoch 1990.0) published systems.

Final states of binaries with compact components can include the combinations *WD+NS, NS+NS, NS+BH,* and *BH+BH*, all undetectable (once pulsar emission dies away) unless gravitational radiation carries off enough angular momentum to cause spiralling in (more accretion X-rays) and merger (burst of gravitational radiation and, perhaps, a spurt of neutron-rich material). Alternatively, a second supernova explosion (or sufficiently energetic shedding of planetary nebula) may unbind the system, liberating high velocity, single compact objects, including, presumably, detectable old and young pulsars. Since a second supernova will throw off more than half the total system mass in many cases (unless common envelop evolution has already greatly eroded the secondary), this unbinding should be common. Curiously, the predicated correlation of pulsar age and velocity is opposite to the one seen (Radhakrishnan and Shukre 1985).

A scenario of this general sort underlies most work done during the past decade or more on X-ray binaries and related systems. The basic scheme meets the ob-

servational test of producing all the kinds of objects we see and
should see but do not. A second, statistical, test of producing the right numbers of different kinds of systems in the right places leads to a few reservations noted in later sections.

The literature of this subject is extensive; a search of the main archival journals for 1988–89 located more than 330 relevant papers, relatively few of which can be cited here. Among the important reviews and conference proceedings are Lamb and Petterson (1983), Lewin and Van den Heuvel (1984), Fabian (1985), Truemper et al. (1986), Helfand and Huang (1987), Giovannelli and Mannocchi (1987), Mason et al. (1988), White and Filipov (1988), White (1989), and ESLAB (1990). Other discussions by the present author include Trimble (1991, 1992). These each cover a fairly wide range of topics. Articles and proceedings that address more specific objects and phenomena are noted in individual sections below.

A curious gap in the current literature is the lack of a reasonably complete catalogue of X-ray binaries and related objects, to update that of Bradt and McClintock (1983). Nagase (1989) tabulates 30 massive *XRBs* and Van Oijen (1989) a slightly smaller number with optical identifications; Parmar and White (1988) list 25 low mass systems (*LMXRBs*) with well-established orbits; and Rappaport et al. (1989) present properties of 14 binary and millisecond pulsars (but at least five more have appeared in *IAU* Circulars between their cutoff and mine).

2. Types of Systems

2.1. A Few Words about Names

The nomenclature of individual binaries with compact components and of the various categories is at least as much of a mess as that of any other subdiscipline in astronomy. The brightest few X-ray sources, found in early rocket flights, were called Cyg X-1, X-2, X-3 etc. (the LMC and SMC being regarded as constellations for this purpose); and a few objects are known by the names of their optical identifications, like HZ Her, γ Cas, and X Per.

The commonest sort of designation consists of four digits indicating Right Ascension in hours and minutes, followed by a + or − sign and two digits for the declination in degrees. Rounding can be either uniformly downward or to the nearest whole number (so that a single source can appear as 1915−05 and 1916−05 in the same paper). Two sources that would otherwise have the same designation can be distinguished with tenths of a degree (2030+375 or 2030+37.5) or as A,B,..., e.g. 0021−72A and 0021−72B, two millisecond pulsars in the globular cluster 47 Tuc.

A dozen or so sources are better known by their galactic (longitude and latitude) coordinates in degrees, as GX 1+4, GX 304−1, etc. These are readily distinguished by the number of digits before the + or - sign, which is always 1, 2, or 3; while the RA-Dec designations always have four digits before the sign, even if one or more is a zero.

Some letters in front of the numbers provide further information about the object. X is an X-ray source, B a burster, P a pulsator, and T a transient. XPT 0332+53 is a pulsating transient X-ray source, and so forth. Other letters indicate the satellite in whose catalog the source appears: A=Ariel, U=Uhuru, H=HEAO-1, G=Ginga, E=Einstein, S=SAS, M=OAO. The only possible source of confusions here (so far) seems to be G for galactic coordinates and G for Ginga, but the number of digits distinguishes them. Further refinement is possible – 4U is the fourth Uhuru catalog, for instance.

Mercifully, the binary and millisecond pulsars have all been discovered late enough and been faint enough to get called PSR 1913+16 etc. from the beginning, though the ambiguity in direction of roundoff remains.

A few classes of objects are well enough defined to tempt authors to abbreviations. Commonest is LMXRB (or LMXB) for low mass X-ray binaries. The logical opposite, HMXRB (high mass X-ray binary) occurs occasionally (White 1989), but MXRB (for massive X-ray binary) is commoner, and MXB not unknown (but unfortunate, owing to the possible confusion with a hypothetical burster in the OAO catalog). There is no real ambiguity because massive systems essentially never burst, but only the true afficionado knows enough of the six-digit numbers by heart to sort out which is meant in every possible case.

The present article attempts no rationalization of any of the systems. Where a specific reference is cited, the name is that used there. Otherwise it is merely one of the commoner possibilities.

2.2. THE MASSIVE X-RAY BINARIES

These X-ray sources are optically identified with O and B stars of $\geq$ 8–10 $M_\odot$, and constitute a young (disk) population. They are moderately high velocity objects, given their youth (Van Oijen 1989), reflecting some combination of recoil from asymmetric supernova explosions and expulsion from clusters (Leonard and Duncan 1989). Binary nature is revealed sometimes by eclipses and always by Doppler shifts of lines in the visible spectrum of the companion and/or of regular pulsations of the X-rays. When both are seen, we get an estimate of the mass of the compact component (usually in the range expected for a neutron star, e.g. Kruzina and Cherepashchuk 1986 on Vela X-1).

X-ray properties nearly universal in the class include hard spectra, rather stable pulsation at a period attributed to the rotation of the neutron star, and an absence of bursting, though some are transients. The range of rotation periods (for about 20 objects, Tawara et al. 1989) is 0.06 to 835 sec and slightly overlaps the range of orbit periods of lower mass objects. The lower X-ray luminosities ($\leq 10^{37}$ erg/sec) are associated with erratic flaring activity, in which L_x can change by a factor of 100 in tens of minutes.

The optical identifications can be subdivided into OB supergiants of luminosity classes I–II (and orbit periods 1.4–41 days) and emission line stars of classes III–V (and orbit periods 16–188 days). Further subdivisions and correlations are possible and may have evolutionary significance (Corbet 1987).

The inventory of the supergiant objects is thought to be reasonably complete in the senses

1. that searches for relatively faint X-ray emission at the positions of known O stars yielded no new candidates (Chlebowski *et al.* 1989), and
2. that there are few unidentified galactic plane sources sharing the properties of the class.

New systems with Be star identifications are being found continuously (Koyama *et al.* 1989) and many must remain to be discovered more than 2–3 kpc from us.

All of these systems apparently evolved from initially massive, rather close binaries (Van den Heuvel and Habets 1985) in which the neutron star was formed by direct core collapse of the primary at the end of its hydrostatic evolution. Both mass and angular momentum are typically lost from the progenitor system (Vanbeveren 1989), presumably for the most part during a phase when the stars share a common envelope, whose properties and effects still require considerable work (Bodenheimer and Taam 1986). Because mass transfer erodes the primary more rapidly than would a single star wind, the initial mass required to yield a neutron star rises from 6–8 $M_\odot$ (for single stars) to about 10 $M_\odot$ (Van den Heuvel 1986). The upper limit to the masses of stars that can leave neutron star remnants in interacting binaries is at least 40 $M_\odot$, from the present mass of the secondary in 4U 1223-62. The corresponding limit for single stars is unknown.

Though only about 100 massive X-ray binaries have been found in the Milky Way, their lifetimes are short, near 5×10^4 yr (Van den Heuvel and Habets 1985), so that the birthrate must be about one per millenium for those with supergiant secondaries, and perhaps as much as one per century for those with *Be* star secondaries (the optically identified ones being nearly all within a couple of kpc, Van Oijen 1989). Thus the birthrate of binary neutron stars may be nearly as large as the 1/60 yr rate for single (pulsar) ones (Narayan 1987).

The short lifetimes reflect the rather special conditions in which the two stars must find themselves. The neutron stars still have the strong magnetic fields of their youth (near 10^{12} G for ages $\leq 10^{6-7}$ yr), strong enough to channel accreting gas down onto the polar caps; hence the regular X-ray pulsation. (Please reserve the word pulsar for objects powered by rotational energy rather than by accretion!) But, more constraining, the secondary must be somewhat evolved (to have a wind strong enough to provide gas to be accreted), but must not yet have filled its Roche lobe (at which point mass transfer on the thermal time scale will choke off the X-rays).

In addition to the rotation and orbit periods, rapid erratic flaring (presumably associated with lumpiness in the accreting gas), and secular evolution over 10^{4-5} yr, massive X-ray binaries display several other interesting temporal variations. First, many (especially those with Be secondaries) are recurrent transients. The recurrence period (10–200^d) is that of the orbit, which is typically quite eccentric, and maximum luminosity is the result of increased accretion rate at periastron (Priedhorsky and Holt 1987). Variable mass outflow from the Be stars themselves can also produce recurrent transients (Whitlock *et al.* 1989; Taam *et al.* 1988, Janot-Pacheco *et al.* 1989). Second, a rather heterogeneous collection of sources (LMC X-4, Her X-1, SS 433, Cyg X-1) show milder cyclic variations with 0.1–2 yr periods that can perhaps be attributed to precession of an accretion disk or of the neutron star itself.

Third (and the subject of the most voluminous literature) nearly every rotation period that has been measured several times has changed with a characteristic time scale of 10^2 (Dotani *et al.* 1989) to 10^5 yr. Some objects spin up systematically (most of them showing evidence of accretion disks, Henrichs 1984); some spin down; and a good many alternate between the two, spinning up when they are bright and accreting vigorously, and down in between (Whilte 1989; Makashima *et al.* 1988). Spin-up from an aligned accretion disk and spin-down by pulsar-type emission are fairly straightforward. In addition, spin down can be driven by a wind with a velocity gradient (Taam and Fryxell 1989), by a wind-fed disk that switches between being aligned and counter-aligned (Taam and Fryxell 1988), and even by an aligned accretion disk (owing to magnetic torques, the propellor mechanism of Ghosh and Lamb 1978). The details are exceedingly complex in all cases, and the reason for preponderance of very slow rotation (the median period is close to 100 sec) is not transparently obvious (Henrichs 1984).

2.3. THE BLACK HOLE CANDIDATES

Establishing the presence of a black hole in a binary system requires

(a) high luminosity (or highly variable) X-rays (so you know there is something compact there),

(b) orbit parameters that require the compact object's mass to exceed the maximum for a stable neutron star (a poorly-known number, but 3 $M_\odot$ is safe for this purpose),

(c) absence of periodic variability attributable to rotation of a solid surface, and

(d) exclusion of alternatives.

The *ur*-candidate, Cyg X-1 (= HDE 226868) met tests (a), (b), and (c) from the start and survived assorted semi-serious challenges to (d) that invoked triple system (Fabian *et al.* 1974) or subdwarf primaries (Trimble *et al.* 1973). All recent analyses concur that the X-ray component falls somewhere between 5 and 10 $M_\odot$ (Fabian *et al.* 1989; Dolan and Tapia 1989; Sokolov 1988), based on optical velocities, light curves, and polarization, which helps to determine the angle of inclination of the system.

Other X-ray properties of Cyg X-1 that have been taken as possible black hole signatures include chaotic variability (with no low-dimensional attractor, Lochner *et al.* 1989) and a variable, complex spectrum (Basani *et al.* 1989) especially the soft component (White 1989). These signatures have not, so far, led to reliable pure X-ray identifications of additional systems with black holes (White 1989), but have rather rung false alarms, like Cir X-1 (which later displayed X-ray bursts of the nuclear burning type) and 0332 + 53 (which later showed a 4.4 sec rotation period, Ilovaisky 1988).

The other two persuasive cases, LMC X-3 (Buchkarev 1989; Kuiper *et al.* 1988) and the soft transient A 0620-00 (Johnson *et al.* 1989) also rely heavily on optical data. The same is true for LMC X-1 (White 1989) for which a plausible case can also be made. All have massive optical components except A 0620-00, which seems to be a K5 star now near 0.7 $M_\odot$, but which must have passed through a massive XRB state (De Kool *et al.* 1987). At any given time, a few other objects are likely to be under consideration, like the soft transient GX 2000+25 (Tsunemi *et al.* 1989).

It is worth noting that the black hole, at least in Cyg X-1, must have formed directly from core collapse. To drive a neutron star past stability and "grow" it up to the present mass with accretion at the Eddington rate would take longer than the past lifetime of the secondary star.

2.4. LOW MASS X-RAY BINARIES

The low mass X-ray binaries (LMXRB) are optically identified with late type stars (some main sequence, some evolved to red giants or white dwarfs) of $M \leq 1.0$–1.5 $M_\odot$. One transition object, HZ Her(= Her X-1, with a slightly evolved A type secondary near 2 $M_\odot$ is generally included in the class. The systems show a galactic distribution associated with the central bulge, globular clusters, and the thick disk. Their kinematic age is between that of the halo and that of the thick disk (Cowley *et al.* 1988). Their binary nature was initially somewhat difficult to establish, since very few eclipse, and only three show rotationally modulated variation that can be used as a clock. But about 25 objects (Parmar and White 1988) have now been shown to have orbit periods between 11 minutes and 10 days. The evidence includes some optical Doppler shifts; a number of optical light curves (which show variations attributable to distortions of the shape or surface temperature of the secondary, Van Paradijs *et al.* 1989); half a dozen X-ray light curves with partial eclipses or other periodic variability of X-rays scattered to us by coronal gas above the accretion disks; and a number of dips, meaning partial occultation of the X-rays by wind or accretion disk material.

The paucity of total X-ray eclipses is believed to be an orientation effect – in systems nearly-enough edge on to eclipse, an optically thick accretion disk keeps us from seeing a bright X-ray source (Milgrom 1978). The masses of the accreting components are generally not well determined in these sources. The distinction between white dwarf and neutron star systems is, therefore, made on the basis of ratio of X-ray to optical luminosity (higher for neutron stars) and on the basis of short term variability (neutron stars burst; cataclysmic variables flicker). A few systems hover between the categories. The 47 Tuc X-ray source (Aurière *et al.* 1989) has the luminosity of a LMXRB but 120 sec pulsation like a cataclysmic variable (Cordova and Mason 1984). LSI+61°303, on the other hand, has a low L_x/L_{opt} but other properties like a neutron star system (Mendelson and Mazeh 1989).

X-ray phenomena common in the class include soft spectra, bursts, and quasi-periodic oscillations (Sect. 6.2). Apart from Her X-1 (1.24 sec), GX 4 + 1 (122 sec), and X 1626 – 67 (Levine *et al.* 1988; 7.7 sec), none shows persistent evidence for rotational modulation. This presumably results from decay of the neutron star magnetic fields over $\sim 10^9$ yr and has made it hard to establish what the typical rotation periods are. A few sources in outburst may reveal short periods (most recently Aql X-1 at 0.132 sec, Kelley *et al.* 1989) that are arguably rotations; and the "beat frequency" interpretation for QPOs provides an indirect argument for the rapid rotation expected after many years of accretion from a disk aligned with the binary orbit.

The better studied LMXRBs are variable at optical, ultraviolet, and radio wavelengths as well (Hasiger *et al.* 1990, Vrtilek *et al.* 1990, Van Paradijs *et al.* 1990, and Hjellming *et al.* 1990 on Cyg X-2). The correlations are complex and not yet fully understood. Radio emission from LMXRBs is probably fairly common (Lehto *et al.* 1989), often displaying coherent alignment on a variety of scales (Strom *et al.* 1989 on Cyg X-3; Velusamy and Subrahmanya 1989 on Sco X-1). This seems to come from jets oriented by accretion disks (Achterberg 1989; Molnar *et al.* 1988; Hjellming and Johnson 1988). Also arguably to be associated with the disks are 0.5–2 yr brightness cycles in half a dozen objects (including Her X-1, the source in NGC 6624, and the rapid burster 1730 – 335). These act like the outbursts of dwarf novae and may indeed come from disk instabilities (Priedhorsky and Holt 1987).

Such disks are inevitable. The low mass secondaries, whether main sequence, subgiant, or degenerate dwarf, will not have winds sufficiently powerful to produce detectable accretion X-rays, except in a few cases where the wind may be boiled off by the X-rays and so be self-sustaining (Chevalier *et al.* 1989 on Cen X-4). Thus the donors must be filling their Roche lobes, and gas will arrive at the neutron star with far too much angular momentum to flow smoothly downward. It flows into a disk in which mass is continuously transported inward and angular momentum outward.

The existence of the disks is not in doubt. They dominate the continuous and line spectra of most LMXRBs (only the longer-period systems with evolved secondaries, like Cyg X-2, Her X-1, and GX 1 + 4 show stellar spectral features). Somewhat less certain is the process that keeps the mass transfer going and feeds the disk (Taam 1986; Van den Heuvel 1986). Because the donor is the less massive star, its Roche lobe tends to expand as material is transferred. Contact can be maintained

(a) by radiation-driven expansion, which is rather rare, but may apply to Her X-1 during quiescent (Khruzina and Cherepashchuk 1989) and to Cyg X-1 (Tavani et al. 1989),

(b) nuclear evolution of the donor, which is relevant for those systems with periods longer than 0.5 days and secondaries massive enough to evolve in a Hubble time, and

(c) loss of angular momentum to the system by magnetized winds from the secondary (plausible for main sequence stars) or gravitational radiation (the only alternative left for systems with degenerate companions).

Her X-1 presents an additional complication of a 35 day brightness modulation that is neither the rotation period (1.24 sec) nor the orbit period (1.7 days). Precession of the accretion disk, periodically blocking our view, was the first suggestion, and still has supporters (Bisnovatyi-Kogan et al. 1989), but Truemper (1986) has presented persuasive data from EXOSAT, showing that the pulse/interpulse ratio of the 1.24 sec rotational modulation changes through the 35 day cycle. It is hard to see how the properties of the disk out at 10^5 stellar radii could do this, leading him to favor precession of the neutron star as the underlying mechanism. Even the precessional nature of the cycle has been doubted (Sheffer 1988).

For the LMXRBs collectively, the chief remaining uncertainty is in how they form. The process is entitled to be rather a rare one, since the 100 or so systems (believed to be a fairly complete galactic inventory down at least to 10^{34} erg/sec) have lifetimes of anything up to 10^9 yr (Van den Heuvel and Habets 1985). Evolution from initially fairly wide, but interacting, binaries via a common envelope phase is certainly possible (Joss and Rappaport 1979); but the neutron stars should then be about as old as the present secondaries, and their magnetic fields might be expected to have decayed by e^{100} or so from their initial values.

Somehow this has not happened. Rather, fields are in the range 10^{8-10} G for the binary and millisecond pulsars (thought to descend from LMXRBs, Sect. 2.6) and for some of the LMXRBs themselves (Taam and Van den Heuvel 1986 on Sco X-1). Two explanations are possible. Either the neutron stars are much younger than the secondaries, having formed recently by collapse of white dwarfs driven above the Chandrasekhar mass (Van den Heuvel 1986), or magnetic field decay slows down considerably after a few million years, so that the field effectively bottoms out at intermediate values. This is possible for several different physical pictures of the production and decay of neutron star fields (Jones 1988; Usov 1988; Blandford et al. 1983; Srinivasan et al. 1990).

2.5. The X-Ray Binaries in Globular Clusters

The globular clusters include about 10^{-4} of the mass of the Milky Way, but host nearly 10% of its bright X-ray sources, far more than their fair share (a peculiarity displayed to even greater extent in M31). The sources are not conspicuously different from LMXRB found in the galactic disk and bulge, and the dozen brightest are now generally regarded as close neutron star binaries, though the clusters are otherwise deficient (or at any rate not overstocked with) close binaries (Margon and Cannon 1989). Widespread belief that the less luminous cluster X-ray sources ought to be accreting white dwarf binaries seemingly founders on the total lack of optical evidence for cataclysmic variables at the relevant positions (Shara *et al.* 1988; Margon and Boltz 1987).

Evidence for the LMXRB nature of the bright cluster sources includes

1. X-ray bursts seen in most of them and well modelled as episodic nuclear burning in accreted gas,
2. positions within their clusters that, statistically, imply masses near 2 $M_\odot$ (Lewin and Joss 1983), and
3. optical or X-ray evidence for plausible binary periods in several, including 11.4 minutes for MXB 1826–30 in NGC 6624 (Sansom *et al.* 1989), 32 minutes for 0021–72A in 47 Tuc (Wijers 1989), and 8.46 hours for 4U 2127+11 in M15.

Analogy with other LMXRB and some optical evidence indicates that the very brightest sources ($\geq 10^{38}$ erg/sec) have evolved, low-mass giants as donors of the accreting gas, while most of the others are main sequence stars (Cudworth 1988). The very short orbit period of the NGC 6624 system means that the donor must be degenerate. Evolutionary scenarios yield such systems at two phases (Fedorova and Ergma 1989).

The excess of X-ray binaries in the clusters compared to the field immediately suggests an extra mechanism for forming them. Clark (1975) proposed binary tidal capture, facilitated by the high star density and low velocity dispersion in many cluster cores. Even where we see a giant secondary now, careful calculation shows (Bailyn 1988) that capture must have occurred during the main sequence phase. Working backwards from the numbers of sources seen leads to the conclusion that the average globular cluster should harbor about 100 (single) neutron stars. This implies either a rather steep initial mass function or loss of many neutron stars because they formed with high velocities (Van den Heuvel 1983; Verbunt and Meylan 1983). The number of single neutron stars required rises proportionally as you scale your estimated XRB lifetime down from 10^9 yr.

If most globular cluster X-ray binaries are capture products, they should be concentrated in the clusters with the highest capture probabilities – those nearest to, or already past, core collapse. This is, on the whole, true (Lewin and Joss 1983, who tabulate the bright cluster sources and properties of their hosts), though not exclusively so. Further exploration of this topic would lead us too far afield into the exceedingly complex territory of dynamics of star clusters (Goodman and Hut 1985). Grindlay (1988) has proposed an alternative, that most of the globular cluster X-ray binaries formed by accretion-induced collapse of white dwarfs, capture events then leading to their presence in hierarchical triple systems.

The high concentration of LMXRBs in globular clusters and the difficulties found in accounting for the field population have led to proposals that the field sources might originally have formed in clusters and have been either kicked out (the optical gamma velocity of the M15 source is -150 km/sec, though this may not represent the centroid of the binary, Naylor and Charles 1989; Bailyn *et al.* 1989) or liberated when the cluster was torn apart by tidal forces. The current existence of optically-obscured globular clusters around sources in the galactic bulge has been largely ruled out through infrared searches (Van Paradijs and Isaacman 1989).

2.6. BINARY AND MILLISECOND PULSARS

"The" binary pulsar (1913+16) and "the" millisecond pulsar (1937+21), both with small rates of period increase, implying weak magnetic fields, remained unique for some time after their discoveries (Hulse and Taylor 1975; Backer *et al.* 1982). Development of techniques for de-dispersing short periods in distant objects led to rapidly growing inventories. The last published tabulation (Rappaport *et al.* 1989) included eleven binary pulsars (three in globular clusters) and three single, short period, weak field ones (two in globular clusters). Reported since are the binary 1820−11 (Lyne and McKenna 1989), and five globular cluster pulsars: 1639+36 in M13 (P=0.01s, IAUCirc 4819), 1310+18 in M53 (P=0.033s, IAUCirc 4853), 2127+11 B and C in M15 (P=0.056 and 0.030s, IAUCirc 4762 and 4772) and a 0.29 sec object in NGC 6440 (IAUCirc 4905). 1310+18 is apparently in a long-period orbit, and 2127+11 C in an 8 hour, high eccentricity orbit; the others are single or in very long period orbits. The orbit periods were discussed in a poster paper at the 175th American Astronomical Society meeting and have not yet been published (Anderson *et al.* 1989). Most of the known systems are rather faint (even as pulsars go), and about 10% of a volume-limited sample would be short period, weak field (that is, millisecond) pulsars (Taylor 1987).

The preponderance of globular cluster sources among the recent discoveries partly reflects the search strategies adopted, but the real excess is at least comparable with that of low mass X-ray binaries. Some of the field binary pulsars must, however, have closer affiliations with the massive XRBs, since the companions are also neutron stars (Taylor and Weisberg 1989 on 1913+16; Taylor and Dewey 1988 on 2303+46, which also has the strongest magnetic field of the binary pulsars, nearly 10^{12} G). Most of the magnetic fields (determined from slow-down rates) are between 10^8 and 10^{10} G.

The majority view is that the combination of rapid spin and weak field means that an old neutron star, whose field had decayed and whose initial kinetic energy had been largely radiated away, has been spun back up by accretion in an X-ray binary (Alpar *et al.* 1982; Van den Heuvel 1988) to produce a recycled or born-again pulsar. The 1.6 msec rotation periods of 1937+21 and 1957+20 are, in fact, very close to the shortest that can be so achieved (Wasserman and Cordes 1988). To complete the scenario, some of the rejuvenated neutron stars must be liberated to account for the single millisecond objects. Within globular clusters, this can be accomplished through a close encounter with another star (Wolszczan 1989; Rappaport *et al.* 1989). But for most of the field objects, the companion must have mass-transferred, evaporated, or merged itself out of existence. Some of the LMXRBs are well on their way to the first, and 1957+20 (Sect. 6.4) is apparently working hard on the second form of suicide.

There remain one difficulty and one opportunity. The difficulty is that, after correction for the completeness of searches and the probable luminosity function of the recycled pulsars, they seem to be far too numerous to descend only from X-ray binaries like those known. The problem exists for field objects, especially short period ones (Kulkarni and Narayan 1988; Cote *et al.* 1989), but is even more acute in the globular clusters (Baily and Grindlay 1990; Kulkarni, Narayan and Romani 1990), unless our usual understanding of the relative lifetimes (Van den Heuvel 1988a) is wrong by at least an order of magnitude, and the X-ray binary phase persists for only 10^7 yr (Verbunt *et al.* 1989; Tavani 1989).

A possible resolution of this statistical discrepancy between millisecond-binary pulsars and their putative progenitors comes from accretion-induced collapse of white dwarfs in either primordial or capture binaries (Van den Heuvel and Taam 1984; Van den Heuvel 1988a; Bailyn and Grindlay 1990; Kulkarni, Narayan and Romani 1990). The distribution of neutron star magnetic fields initially resulting from this process depends on the fields of the white dwarfs upon which it is inflicted (Chanmugam and Brecher 1989). Most of these are, at any rate, less than 1 MG, so most of the pulsar fields should be less than 10^{12} G. Experienced investigators differ on whether a particular millisecond pulsar, like that in M28, is more likely to have arisen from accretion induced collapse (Verbunt *et al.* 1987) or capture (Romani *et al.* 1987). Several issues remain to be sorted out. One is the need for the donor star to be stripped to a degenerate core quite soon after collapse occurs (or we will get an X-ray binary anyhow and be no better off than before). Another is the range of white dwarf masses and accretion rates that lead to induced collapse. It is clearly narrow (Van den Heuvel 1986), possibly so narrow as to be useless for this problem, because the white dwarf typically gets spun up to its break-up speed before much mass is accreted (Narayan and Popham 1989).

The opportunity arises because the millisecond and binary pulsars provide very stable clocks, some of them in gravitational potential wells so deep that general relativity is not a negligible correction factor. The short orbit period and high eccentricity of 1913+16 make it especially useful. Important results (Rawley *et al.* 1989; Taylor and Weisberg 1989; Backer and Hellings 1986) include

(a) good measured masses of 1.44 and 1.39 $M_\odot$ for the two neutron stars,

(b) confirmation of the quadrupole formula for intensity of gravitational radiation from the measured decay of the orbit,

(c) an upper limit to the rate of change of the gravitational constant, $\dot{G}/G \leq 1.2 \pm 1.3 \times 10^{-11}\ yr^{-1}$, and

(d) an upper limit to the energy density of gravitational radiation in the universe of less than 4% of the closure density, for waves with frequencies between 10^{-12} and 10^{-9} Hz.

3. Evolutionary Phases and Processes

The basic scenario described in Sect. 1 appears to account for most of the properties of binaries with compact components. This section gives brief attention to several side issues and dissenting viewpoints.

The distribution of orbit periods must reflect both initial conditions of the systems and changes that occur as a result of mass transfer and loss. The time scales expected for change are long, $\geq 10^{6-7}$ yr, even for massive systems (Kelley 1986), and rather few detections have been reported. The period increases seen in Cyg X-3 ($\tau = 4.5 \times 10^5$ yr, Molnar 1988) and the LMXRB 1922–371 ($\tau = 1.8 \times 10^6$ yr, Hellier and Mason 1989) are acutally the opposite sign from the expected period decreases for main sequence companions. Tutukov (1988) has attributed the near-gap between 1 and 4 hours in the distribution of LMXRB periods to the turn-off of magnetized stellar winds (hence turn-off of mass transfer until nuclear evolution or gravitational radiation takes over). A similar explanation is often advocated for the gap in the period distribution of cataclysmic variables (elsewhere in this volume).

Two interesting dissenting views are

(a) that neutron star magnetic fields decay only as a direct result of accretion (Shibazaki et al. 1989), hence accounting for the strong ones in gamma ray burst sources (Sect. 6.3) and

(b) that there is a systematic progression among X-ray binaries from faint, flaring ones like A0620 – 00 to bright stable ones like Sco X-1 to bursters (Amnuel et al. 1989).

Supplementary to the basic scenario are

(a) calculations of mergers of pairs of neutron stars, the events producing r-process material and gamma ray bursts (Eichler et al. 1989) and

(b) calculations of the effects of asymmetric supernova explosions on correlations of system ages with eccentricities (Habets 1987) and on the expected correlations of pulsar velocities with their magnetic fields (Bailes 1989).

This latter reconciles prediction of the correlation (high velocity = strong field) with that seen.

Triple systems (neutron star + donor in close orbit, third star further out) have been invoked to account for irregular periods, multiple periods, and mismatches between X-ray and optical periods, especially by Grindlay (1986 on GX 17+2, 1989 on 4U 1915+05). The clearest case is perhaps 4U 2129+47, whose optical eclipse disappears during times of low X-ray luminosity, leaving the steady light of an FV-star (Thorstensen et al. 1988; Kaluzny 1988). The F star is too large to be the secondary in the X-ray system with its 5.24 hour period, and chance superposition is fairly unlikely. A 30 day optical period has been reported for the third star (Garcia et al. 1989) but presents the problem that the ratio of large to small separation would not have been large enough for stability at the time the neutron star was a red giant.

4. Radiation Processes: X-Ray Bursters

While the primary energy sources in X-ray binaries and recycled pulsars are accretion and rotation respectively, the radiation we actually detect has necessarily undergone a good deal of reprocessing since its liberation as gas kinetic energy or Hz-kHz radio radiation.

4.1. Recycled Pulsars

Binary and millisecond pulsars radiate by the pulsar radiation mechanism, whatever that is. Taylor and Stinebring (1986) provide a relatively optimistic description of our present state of incomprehension, and Rankin and Gil (1989) a more pessimistic one. No further elaboration will be attempted here. Optical counterparts of a couple of the binary systems are apparently white dwarfs whose surface ($\approx$ black body) temperatures decline with age in the usual way (Kulkarni 1986; Callanan et al. 1989).

4.2. X-Ray Binaries

Virtually all well-studied objects display continua more complex than a single power law or black body curve, many have emission or absorption features, and nearly all the spectra vary on most of the time scales (rotation, orbit, flicker, flare, burst, transient...) on which total luminosity varies. It is often, though not always, possible to describe the variability in terms of a couple of components that act more or less independently (for instance, radiation from the accretion disk plus radiation from the boundary layer between it and the neutron star). The situation is further complication in massive XRBs by the possibility of non-thermal processes in the strong neutron star magnetic field and, in systems with disks, by the possibility of disk coronae. While the situation is messy, there do not seem to be any fundamental problems, which is perhaps why no comprehensive review of the subject appears to have been published. The following paragraphs address components thought to be important in largish numbers of systems.

The cyclotron energy, $h\nu_B$, is $12(B/10^{12}G)$ keV and so is in the X-ray range for the fields of young neutron stars. Features between 20 and 60 keV in the spectrum of Her X-1 were the first to be understood in terms of cyclotron resonances (Kirk and Truemper 1984). GX 1+4 and 4U 0115+63 show similar emission and absorption structure indicating fields of 2–5×10^{12} G. More generally, a large portion of the continuous spectra of massive XRBs can be modeled as cyclotron emission from a range of fields around the star and subsequent scattering by hot electrons (Alexander 1989).

The K line of iron at 6.4–7.1 keV (depending on how many other electrons the atom has) is the strongest discrete emission feature seen in X-ray binaries, occurring both in high mass systems like LMC X-4 and in many low-mass ones (Hirano et al. 1987; Dennerl 1989). The emission is usually attributed to a hot corona around an accretion disk, if only because most other possibilities have been ruled out (Kallman 1989). The coronae are supported by X-ray heating and are optically thin. A recent calculation of the exptected radiation yielded lines about 0.5 keV wide (from a blend of ionization states) but still narrower than those seen (Kallman and White 1989).

Emission lines from other elements should arise in the same gas, and K lines of ionized nitrogen and oxygen are seen in grating spectra of the few sources bright enough to have been studied this way (Vrtilek et al. 1986). Because the lines are closely spaced and fainter than the iron one, larger X-ray collecting areas and higher spectral resolution than has so far been available will be needed before they can be widely used as plasma diagnostics.

Continua require at least two components to fit observed spectra. These can be described as photosphere plus disk (Makishima et al. 1989), black body (i.e. boundary layer) plus inverse Compton scattering (White 1988), black body plus power law (Balucinski and Hasinger), Comptonized black body from boundary layer plus multi-temperature black body from disk (Mitsuda and Tanaka), and so forth. These are not physically distinct situations, but different attempts to describe the same horribly complex one of radiative transfer in a gas distribution that is neither spherical nor slab-like. The same difficulty arises in modelling the ultraviolet and optical emission from disks of cataclysmic variables (elsewhere in this volume). Both problems are rather far from being solved at the moment.

4.3. X-Ray Bursters

X-ray bursters are a (large) subset of the low mass X-ray binaries that brighten up to near the Eddington luminosity in seconds or less and fade back in ≤ 1 sec to minutes. They repeat, some more or less regularly, over hours to days. Bursting occurs preferentially in the fainter LMXRBs and in ones of relatively high metallicity (Motch *et al.* 1989). Peak spectrum is roughly a black body which cools as the burst fades. Averaged over time, the energy emitted in bursts is only about 1% of the steady luminosity. In the sources with optical identifications, the bursts can be followed in visible light. Lewin and Joss (1984) provide an excellent summary of the work up to that time and of the evidence that the sources are indeed LMXRBs.

The "best buy" model for the phenomenon is closely related to the nova mechanism. Hydrogen gas accreted by a neutron star, like that accreted by a white dwarf, eventually undergoes nuclear processing. On the white dwarf, hydrogen burns steadily at some accretion rates, but accumulates and explodes episodically at others. On a neutron star, the hydrogen burning shell is always stable, but the helium-burning one will flash for a wide range of conditions (Lewin and Joss 1984). These thermonuclear explosions heat the surface of the star more or less uniformly, so as to produce roughly black body emission which will cool and fade as the fuel is used up, in agreement with the observations. The burst itself can increase the mass transfer rate, so that accretion luminosity may also contribute (Walker and Meszaros 1989). There may be a further analogy with nova explosions in the importance of mixing of the accreted fuel with the stellar surface (Fujimoto *et al.* 1988).

Accepting this model, we find one central puzzle and one central disappointment. The puzzle is that the bursts sometimes happen closer together than the time over which accretion could possibly provide enough fuel for the second burst. X1608-52, for instance, had two comparably strong bursts only 10 minutes apart (Nakamura *et al.* 1989). Attempts to understand this have proceeded along the lines of burning hydrogen in some but not all bursts or of hoarding unburned helium between bursts (Gottwald *et al.* 1987; Van Paradijs, Pennix and Lewin 1987; Gottwald *et al.* 1989).

The disappointment is that one might quite reasonably expect to measure the radii and masses of the neutron stars from their black body temperatures and luminosities plus the presumed redshift (gravitational plus transverse Doppler) of a common line near 4.1 keV. And it does not work. First, the line comes at so nearly the same energy in many sources that, if it is a redshifted iron K line, the neutron stars are implausibly similar to each other (Magnier *et al.* 1989; Nakamura *et al.* 1988). Second, and even more serious, the continua are not really black bodies, and the conversion from color temperature to effective temperature is extremely uncertain (Kaminker *et al.* 1988; Damen 1989; Pennix 1989). A subsidiary disappointment is that, if we do not know the effective temperatures accurately, we cannot use the requirement of peak luminosity $\leq$ Eddington luminosity to get a distance scale and an independent measure of our distance from the center of the Galaxy. Ignoring all the problems leads to the interestingly small value of 6 kpc for R_0 (Haberl *et al.* 1987).

A single object, the rapid burster MXB 1730-335, bursts at closer intervals of second to minutes (though not continuously) and does not cool as it fades. The phenomenon is fairly confidently attributed to instabilities in accretion rate (Lewin and Joss 1984; Hanami 1988).

5. Accretion and Accretion Disks

Accretion disks first received serious astrophysical attention in discussions of the formation of the solar system in the early 20th century. The most unambiguous evidence for them comes from the cataclysmic variables, where their sizes are measured by the eclipses they cause and their velocity structures are probed by the profiles of the lines they emit (Wade 1985). They are also invoked to explain behavior of active galactic nuclei and some X-ray binaries. The first models for the three applications are roughly contemporaneous (Lynden-Bell 1969 for QSOs, Smak 1971 for Cataclysmic Variables; Prendergast and Burbidge 1968 for XRBs). The notation used to describe energy and momentum transport in accretion disks was introduced by Shakura and Sunyaev (1973) who collected many of the uncertainties into a single viscosity parameter. Pringle (1981) and Petterson (1984) have reviewed disk structure and stability; and early important papers are reprinted with commentary by Trèves et al. (1989).

Disks must exist, at least in low mass X-ray binaries (because gas coming from Roche lobe overflow will have too much angular momentum to fall directly onto the neutron star) and in massive systems with black holes (because spherically accreting gas carries its kinetic energy down with it rather than radiating, and we do see the systems). Observational, as opposed to logical, evidence for the disks comes from spin-ups of neutron star rotation periods, variable optical emission explicable as coming from gas illuminated by the X-rays, and periodic dips in X-ray intensity (whose depth varies with wavelength and from cycle to cycle), lasting as much as 1/3 of the orbit period and so implying the presence of significant amounts of material well away from the accreting star.

To model such a disk, one writes down differential equations (roughly equivalent to the equations of stellar evolution) to describe the conservation of mass, energy, and momentum (in three dimensions), the equation of state, the cooling (radiation) process, and the energy release process (normally parametrized as a viscosity that lets gas flow inward, liberating gravitational potential energy). And of course one has to do something about the boundary conditions – the outer edge of the disk where the incoming gas stream from L_1 hits it, and the inner edge where it encounters a magnetosphere, a neutron star surface, or a last stable orbit. If your disk is massive enough to be self-gravitating or if a black body at the local disk temperature is not an adequate approximation to the radiation flux, further complications ensue. Not surprisingly, all existing solutions make a number of approximations!

The simplest possible case, steady mass flux through the disk and black body emission perpendicular to it, leads to a unique expression for disk temperature vs. radius,

$$T = \left(\frac{3\,G\,M\,\dot{M}}{8\,\pi\,R^3\,\sigma}\,(1 - R_*/R)^{1/2}\right)^{1/4}. \tag{1}$$

Convolving this in turn with a black body spectrum gives an expression for emitted flux proportional to $\nu^{1/3}$ at low frequency (Lynden-Bell 1969). Something of the sort is seen for many cataclysmic variables.

These models are called α disks, because Shakura and Sunyaev (1973) suggested hiding our total ignorance of what causes the outward angular momentum transport in a parameter, α defined by saying, first, that the torque of an outer annulus acting on an inner one, $G(R)$, is given by

$$G(R) = 2\,\pi\,R^2\,\varepsilon\,\nu\,R\,\frac{d\Omega}{dR}, \tag{2}$$

and second that ν, the kinematic viscosity (which must have units R^2/t to make the previous equation come out right) can be written as

$$\nu = \alpha\,c_s\,H \tag{3}$$

where c_s is the local speed of sound, H is the thickness of the disk, ε is the local areal mass density, and $d\Omega/dr$ is the gradient of the angular velocity. If the viscosity were actually due to turbulence, α could not exceed unity, since the eddies are unlikely to be supersonic or bigger than the disk thickness. Magnetic stresses ought to yield a value of α somewhat smaller than one (Shakura and Sunyaev 1973; Petterson 1984).

In addition to displaying a $\nu^{1/3}$ spectrum (ν is habitually used both for frequency and for the viscosity parameter in discussing these objects), α disks are geometrically thin, optically thick, large in radius, but small in total mass. They are also unstable to just about any sort of perturbation you care to impose (Papaloizou and Lin 1989; Blaes and Hawley 1988).

Pringle (1981) notes that the disks should not really be described as unstable, but rather as inconsistent. That is, if you assume a stationary disk with a given prescription for viscosity, cooling, etc. but find that thermal and viscous perturbations grow in it, then you have chosen the wrong viscosity (or something) and should try again. It is not known whether there are circumstances under which no viscosity law leads to stationarity.

Real disks do, in fact, change with time in several ways, and they are not (or at least not always) very thin. For instance, dips and other structure in X-ray light curves imply that some disks are both thick and structured (Parmar and White 1988) and that this extended part of the disk can come and go in hours (Smale et al. 1988). Transient X-ray emission in LMXRBs can be explained by disk instabilities (switching between optically thin and thick) analogous to those often blamed for dwarf nova outbursts (Huang and Wheeler 1989; Mineshiga and Wood 1989; Mineshiga and Wheeler 1989). The high and low states of Cyg X-1 may have a similar explanation (Hoshi and Inoue 1988).

A disk need not necessarily form in systems where a neutron star accretes material from the wind of a massive companion. The neutron stars in such systems are young and typically have strong magnetic fields. Thus the accretion process is dominated by chanelling along the field lines (Lyubarskii and Sunyaev 1989; Kraus 1989; Parmar et al. 1989). The structure is sufficiently simple that the X-ray variability can be used to probe wind structure (White 1989 on X1700-37).

6. Puzzling Objects and Phenomena

6.1. GEMINGA

Compact galactic sources of gamma rays of 0.1–1 GeV include (Bignami 1987) the Crab and Vela (single) pulsars, Cyg X-3 (an LMXRB) and few other unambiguous identifications, partly because the error boxes are typically 1° or more across and partly because some of the sources may be genuinely diffuse (that is, regions of high interstellar gas density or high cosmic ray flux). Geminga, which is both the Gemini gamma ray source and Milanese dialect for "it does not exist", has the smallest of these error boxes. Within it, there is only one interesting X-ray source, 1E 0630+178.

The X- and γ-ray objects share (a rather tentative) periodicity near 60 sec, which may be lengthening on a time scale of only 1000 yr. Absence of X-ray absorption indicates a distance of no more than 100–200 pc. At this distance, even the fainter LMXRBs would have apparent magnitudes of 10–15; but the brightest peculiar star in the X-ray error box is a 25th magnitude object of unusual, blue color (Bignami 1989; Halperin and Tytler 1989). The ratio of X-ray to optical luminosity is, therefore, 1000 or more, an unprecedentedly high value in the absence of obscuration. Since the X-ray source (at least) is very compact, some sort of neutron star is distinctly implied. Both single and binary models have been suggested (Bignami 1987; Halperin and Tytler 1989), and it is not clear whether Geminga actually belongs in this chapter.

6.2. QUASI-PERIODIC OSCILLATIONS (QPOS)

Many astronomical objects (from the earth's atmosphere and solar chromosphere outward) display variability with Q values ≥ 1 to 100 that could plausibly be described as quasi-periodic oscillations. Particularly complex ones occur in some cataclysmic variables (elsewhere in this volume). Those in question here (elegantly reviewed by their co-discoverer, Van der Klis 1989) occur, mostly, in low mass X-ray binaries, some of which also display bursts, orbital modulation, and other variability. The phenomenon thus belongs unambiguously in this chapter.

Typical properties include frequencies of 5–60 Hz, peak widths near half the centroid frequency, amplitudes of 1–10%, persistence for $\geq 10^5$ cycles, recurrence, and remarkably complex (but repeatable, describable, and arguable explainable) correlations of their properties with source intensity and spectral hardness (Hasinger and Van der Klis 1989; Schulz et al. 1989).

Several of the best-studied sources, including Cyg X-2 and Sco X-1 (Hasinger et al. 1989) present (at least) two discrete modes, associated with different spectral states (labeled normal, horizontal, and flaring branches, from the trajectory of the varying sources through an X-ray color-color diagram). Others, including 1820−30 in NGC 6647 (Dotani et al. 1989) have different correlations, which may or may not amount to discrete modes. Extra white or red noise comes and goes with some of the modes and must also be explained by a complete model.

In the most thoroughly developed model, the quasi-period seen is a beat between the periods of the inner edge of the accretion disk and the neutron's stars rotating magnetosphere. It shows up because clumps of gas find it easiest to enter the magnetosphere at its poles (inclined to the rotation poles; Alpar and Shaham 1985). Many properties of QPOs on the horizontal branch are thereby explained (Van der Klis 1989; Fortner et al. 1989), though there may be some difficulty with the phase lags between QPOs at different X-ray energies (Mitsuda and Dotani 1989).

If the beat model is the right answer, then the periods seen to confirm the expectation that the old neutron stars in LMXRBs have been spun up by accretion to millisecond rotation periods. Some of them must also have accreted an appreciable fraction of a solar mass over their lifetimes and should have masses considerably larger than the 1.4 $M_\odot$ associated with massive X-ray binaries and binary pulsars. Thus the eventual disentanglement of QPO properties should allow us to investigate neutron stars and their equations of state over a wide range of masses and radii.

The beat model does not account well for normal branch QPOs or for the behavior of sources with other kinds of correlations. As instability in the accretion process itself has been widely advocated, and might be associated with the marginally stable orbit at $r = 6 M$ (Paczyński 1989), with accretion disk coronae (Stollman and Kuperman 1988), or with Rayleigh-Taylor instabilities of gas already in the magnetosphere (Hanami 1988).

Phenomena that are probably related to QPOs appear in other sorts of objects, including the black hole candidate LMC X-1 (Ebisawa et al. 1989), the 1983 August 1 gamma ray burster (Kuznetsov et al. 1988), the rapid burster, and several massive X-ray binaries (Van der Klis 1989). The case of GX 2030+375 (Angelini et al. 1989) is particularly instructive, because the beat frequency, 0.2 Hz, and the known NS rotation period, 42 sec, permit calculation of the size of the magnetosphere (assuming a beat model). The implied surface dipole field strength of $10^{12-13} G$ is perfectly plausible, providing support for the basic model.

6.3. SS 433

This object has been honored by good reviews (Margon 1984; Zwitter 1989) and, earlier, by its very own conference (Beer 1981). Its binary nature is not in question, since narrow hydrogen and helium emission lines display a stable period of 13.1 days and a velocity amplitude of 195 km/s (Margon and Anderson 1989). But the really interesting velocity variation belongs to wider red and blue shifted Balmer emission lines, which have a period of 164 days and amplitudes of ±40,000 km/sec!

The high velocity emission was quickly (and stably) modelled as a pair of oppositely directed jets, moving at $0.26c$ in directions that precess (in 164^d) around an axis tilted at 79° to the plane of the sky. Total energy output, including the kinetic energy of the jets, is about 10^{39} erg/sec, like that of the very brightest massive X-ray binaries. Both the X-ray and the radio emission seem to be closely correlated with motion of material in the jets (Band 1989; Kawai et al. 1989).

The jets must have been dumping their kinetic energy into the surrounding interstellar medium for at least thousands of years, and cannot have left it unscarred. There is, in fact, a surrounding extended synchronotron radio source called W50. It is generally catalogued as a supernova remnant (conceivably left from the event that formed the compact component) but is probably largely the product of continuous energy input from the jets (Grindlay et al. 1983).

A number of points remain to be understood. First is the nature of the compact component. A black hole of 10 $M_\odot$ (Zwitter and Calvani 1989) or even 60 $M_\odot$ (Antokhina and Cherepashchuk; 1988) would seem to be implied by the orbital velocity amplitude and system orientation. But it is not certain just where the emitting gas is in the system or whether the extended star fills its Roche lobe (a constraint on orbit size), as a result of which, a mass in the neutron star range cannot really be excluded (Brinkman et al. 1989; Kawai et al. 1989). Resolution of the ambiguity requires picking out lines that move with the center of mass of one star or the other, for instance from the accretion disk around the compact primary (Filippenko et al. 1988).

A second puzzle is the jet acceleration mechanism (Zwitter 1989; Liang 1988), though the jet velocity (for which $v/c = (1216 - 912)/1216$) strongly suggests that radiation pressure acting through locking of Lyman α to the Lyman limit must be important. Third, we do not know the nature of the instability that drives the jet precession (Margon and Anderson 1989; Zwitter 1989).

6.4. PSR 1957+20

This recently-discovered binary pulsar (Fruchter et al. 1988), with orbit period 9.17 hr and neutron star rotation period 1.6 msec, is in some ways the logical outcome of processes occurring in low mass X-ray binaries like Cyg X-3 and 1820-30 (Klúzman et al. 1988; Molnar 1988; Phinney et al. 1988; Van den Heuvel and Van Paradijs 1988). The corresponding optical object shows a well-defined light curve, dominated by pulsar heating of the secondary (Kulkarni et al. 1988; Van Paradijs et al. 1988; Fruchter et al. 1988a; Djorgovskii and Evans 1988; Callanan et al. 1989a), which, in combination with the orbit parameters coming from the Doppler-shifted pulsar period, implies a degenerate (probably helium) secondary of 0.02 $M_\odot$ and 0.15 $R_\odot$.

The catches are several. First, the radio pulsar is occulted for 50 minutes out of each orbit, and the occultor is both significantly larger (factor >2) than the companion star and essentially gray. A cloud of particulate matter would act this way (Michel 1989), but there is no particular reason to expect one out there, while the expected phenomena – free-free absorption in a wind or reflection at the edge of a dense wind from the companion (Rasio et al. 1989) – don't really match the data very well.

The second is the nature of the process that drives the wind and erodes the companion (assuming that the system is on its way to becoming a single millisecond pulsar). Evaporation of the secondary by some form of energy from the pulsar must be involved (Cheng 1989; Ruderman et al. 1989a,b). Unfortunately, it is not at all clear that the pulsar actually provides enough energy to do this in the particular case of 1957+20 (Eichler and Levinson 1988; Ruderman et al. 1989a).

The third difficulty arises if an ablation wind is producing the extended eclipse – the system is perilously close to becoming a single millisecond pulsar, having a future life expectancy of 10^8 yr (Van Paradijs 1988; Cheng 1989) or even only 10^7 yr (Ruderman et al. 1989b), though earlier phases must have lasted much longer (Czerny and King 1988). The surprise is only a statistical one, and should perhaps not be regarded as worrisome as long as only one such object is known.

6.5. TeV and PeV Gamma Rays

The set of phenomena described in this section clearly belongs to the class of binaries with compact components, but it is possible to doubt that the phenomena actually occur as reported.

An individual neutral particle of $\geq 10^{12}$ eV (1 TeV), hitting the earth's upper atmosphere, produces a cascade of relativistic secondaries and, hence, a flash of Čerenkov light. Above $\geq 10^{15}$ eV (1 PeV), the secondaries reach the ground as extensive air showers. Detectors for one or both of these on every continent including Antarctica have recorded events associated with X-ray binaries and related objects (Lamb and Weekes 1989; Shapiro and Wefel 1989). Putative sources include Cyg X-3 (for which the evidence is the strongest), Her X-1, Vela X-1 (not to be confused with the Vela pulsar, which has, however, also been reported as a very high energy gamma ray source), Cen X-3, 4U 0115+53, and 1822-37 (Ciampa et al. 1989).

Neither detection technique has high angular resolution; thus the evidence for the existence of the sources over and above the known background (Cassiday et al. 1989b) includes correlation of arrival times with known rotation or orbital periods of the X-ray binaries (Cassiday et al. 1989a and Tonwar et al. 1988 on Cyg X-3; Raubenheimer et al. 1989 and Carramiñana et al. 1989 on Vela X-1; Ciampa et al. 1989 on 1822–37; Dingus et al. 1988a and Lamb et al. 1988 on Her X-1).

Though the positive reports are quite numerous, reservations at some level are appropriate because

(a) most of the detections are at about the 3σ level,

(b) the sources are not continuously seen, even by a single group using the same techniques over time (Lawrence *et al.* 1989; Cassiday *et al.* 1989b), and

(c) the periods are not always quite the same as the X-ray ones (Lamb *et al.* 1988).

The period differences and the intermittancy can, however, be modelled (Cheng and Ruderman 1989), and the real difficulty lies elsewhere.

For both TeV and PeV incident particles, there is a signature that should distinguish photon-initiated from hadron-initiated events – narrowness of the light cone for Čerenkov flashes and paucity of muons in extensive air showers. Now, while the TeV gamma rays from the Crab pulsar join onto its spectrum at lower energies and exhibit the photon signature (Weekes *et al.* 1989), the PeV events from the X-ray binaries are muon rich (Dingus *et al.* 1988a,b), as if hadron-initiated (though no known hadron could get here intact from there without getting hopelessly lost in the galactic magnetic field). Nor can we simply claim that new physics causes very high energy photons to produce excess muons. This easy out is closed off because the events attributed to the X-ray binaries are also anomalous in data from Los Alamos at TeV energies (Haines *et al.* 1990), where photons from the Crab pulsar are known to behave like photons. Either the TeV particles from the two classes of objects are physically different (and only one is anomalous), or some aspect of the data are not reliable. The most extreme view is that none of the TeV and PeV detections should be regarded as definitely established.

A number of theorists has devised ways of accelerating particles (usually protons) to 10^{16} eV near magnetic neutron stars, so that TeV and PeV secondaries can be produced (Katz and Smith 1988; Eichler and Ko 1988; Kiraly and Meszaros 1988; Berezhinsky 1988; Schlickeiser 1989). The problem is what sort of secondaries they can produce that will act like the high energy particles required here.

Trevor C. Weekes deals with the question of TeV radiation from binary stars in the next chapter.

6.6. CTB 109=E2259+586 (The Fahlman-Gregory Object)

Demonstrable associations between single (pulsar) neutron stars and supernova remnants remain rare (Weiler and Sramek 1988). For binary neutron stars, they are even rarer. Cir X-1 and LSI +61°303 are surrounded by small synchrotron nebulae (Frail *et al.* 1987). But X2259+586 remains the only accretion-powered X-ray source inside what looks like a normal radio and X-ray emitting supernova remnant, CTB 109. Fahlman and Gregory (1981) reported X-ray pulsations with a 3.5 sec period. It now seems more likely that the neutron star rotates every 6.98 sec (Koyama *et al.* 1989; Carlini and Treves 1989).

The rotation period is lengthening, but only on a time scale of 3×10^5 yr (Koyama et al. 1989a), meaning that the X-ray emission must be accretion powered. The stability of the observed rotation period also indicates that the companion must be less massive than $0.13\ M_\odot$ and, in the absence of a convincing optical identification (Davies et al. 1989), we have no direct evidence for the other star or for the orbit period. The extended radio structure includes jet-like features connecting the compact X-ray source with the extended emission, suggesting that CTB 109 (like W50) may be more the product of the central binary than of the supernova that originally made the neutron star (Gregory and Fahlman 1983).

6.7. Gamma Ray Bursters

The existence of these is not in doubt. When suitable detectors are in orbit, they record a couple a day. But the association with neutron stars is not absolutely certain, and that with binaries entirely hypothetical. The bursts are not a uniform class of events, and the range of properties includes (Lamb 1984; Hurley 1987; Epstein and Hurley 1989)

(a) duration of 0.05 to 100 sec,

(b) continuous spectra that look thermal with $kT = 30-300$ keV,

(c) spectral features including emission at about 400 keV and absorption at 20–60 keV in many bursts, and

(d) temporal structure with anything from a single peak to very complex subpulses and a few quasi-periodicities.

The distribution in the sky is indistinguishable from isotropy (Hartmann and Epstein 1989; Hartmann and Blumenthal 1989) and numbers of bursts vs. flux density (when selection effects are allowed for) is indistinguishable from that of a homogeneous distribution (Kudryavtsev and Svertilov 1988; Paczyński and Long 1988; Jenning 1988; Schmidt et al. 1988). The sources must, therefore, be either in the solar neighborhood or cosmologically distant. Most puzzling of all is the absence of concurrent or steady emission at other wavelengths, down to 10^{-13} erg cm^{-2} sec^{-1} in soft X-rays (Boer et al. 1988) and 22^m in the optical, imposing severe limits on any possible companion and on accretion onto the (presumptively neutron star) source. Only the atypical soft repeater of 1979 March 5 occurred in an interesting direction in the sky – towards N49, a supernova remnant in the Large Magellanic Cloud.

The strongest evidence for a magnetized neutron star source comes from the 20–60 keV absorption features seen in about one-fifth of the bursters observed in adequate detail to detect them (Melia 1989). Interpretation of these as cyclotron transitions between Landau levels of electrons in a $10^{12-13} G$ field is now advocated by many theorists (Murakami et al. 1988; Van Paradijs 1989; Hammeury and Lasota

1989; Brainerd 1989; Harding and Preece 1989; Meszaros *et al.* 1989; Alesander and Meszaros 1989; Wang *et al.* 1989). This interpretation implies quite young neutron stars, and it is difficult to see how these can be numerous enough in the solar neighborhood to provide all the recorded bursts.

The gamma rays must come from an optically thin region above the photosphere (Mitsuda *et al.* 1989; Hammeury and Lasota 1989; Ho and Epstein 1989; Dermer 1989), and the most likely emission process is inverse Compton scattering. Energy must be resupplied to the emitting zone during the bursts, and just how accretion, nuclear, or internal energy of the neutron star can be converted to the right form remains something of a mystery.

The only evidence for a companion is the need for a source of material to accrete or burn and the need for a place to reprocess gamma rays to make optical flashes that have been advertized as coming from the error boxes of a few of the sources (Greiner *et al.* 1987). None were at the same times as the gamma ray bursts, and some at least were really reflections from satellites (Schaefer *et al.* 1989). The optical flash phenomenon need have no connection with gamma ray bursts. Alternative models to magnetized neutron stars have ranged as far afield as superconducting superstrings (Paczyński 1988) and gravitational lensing at cosmological distances (Paczyński 1987a). The case for gamma ray bursters as neutron stars in binary systems is, at best, not proven.

6.8. EXTRAGALACTIC OBJECTS

No new kinds of things have yet appeared among the extra-galactic binaries with components, but there may be some statistical differences. M31 apparently has a larger fraction of its X-ray luminosity coming from low mass systems than does the Milky Way (Makishima *et al.* 1989a), and its globular clusters are more likely to have sources brighter than 10^{37} erg/sec (Long and Van Speybroeck 1983). The LMC recurrent transient 0538-66 rises above 10^{39} erg/sec, brighter than anything in the Milky Way (Samle and Chevaler 1989; Stevens 1989), and its shortest known rotation period among neutron stars in massive X-ray binaries (0.069 sec) has only recently been bested by a galactic source at 0.06 sec (Caraveo *et al.* 1989).

That the known LMXRB in the LMC (LMC X-2, CAL 83, CAL 87, Callanan *et al.* 1989) are all rather bright is presumably a selection effect. But that the LMC contains 1/3 to 1/2 of the persuasive black hole candidates (LMC X-3 and X-1) is more surprising, though of low statistical significance. The SMC inventory includes one pulsating source and a couple of transients, all with probable optical identifications (Bradt and McClintock 1983). The only other galaxy with an X-ray binary candidate is M33, where Peres *et al.* (1989) have recently found that a suggested optical counterpart varies with a 1.78 day period.

Note added in proof

The gamma ray bursts recorded by the GRO satellite are definitely not uniformly distributed in space. Rather, we are seeing the edge of the distribution while still being in the middle of it. This combination has prompted renewed interest in sources bright enough to be seen at cosmological distances (B. Paczynski 1991, *Acta Astron.* 41, 257).

TeV RADIATION FROM BINARY STARS

T. C. WEEKES

1. Introduction

One of the major surprises of X-ray astronomy was the discovery of galactic binaries as highly luminous sources emitting as much as 10^{38} erg s^{-1} via thermal emission caused by accretion onto the compact companion. That similar amounts of energy might be emitted by non-thermal processes in the form of particles with energy as great as 10^{15} eV (1 PeV) is even more surprising and is one of the most striking results to come from the new discipline of Very High Energy gamma-ray astronomy.

Based on observations at longer wavelengths there was little to suggest that close binaries might be sources of cosmic radiation. In the past decade there has been a sufficient number of reported detections of gamma rays of energy 1 TeV and greater that the phenomenon cannot now be ignored. These observations suggest that the physics of compact objects in binaries is not nearly as well understood as previously assumed, that transient non-thermal acceleration processes are prevalent in many close binaries, and that binaries are major sources of cosmic rays. Furthermore the peculiarities in the observed signals may have important implications not only for the astrophysics of multiple star systems but also for high energy particle physics. If only a fraction of the implied results are verified this will be a rich field for further study. The problem is that there is such a wide diversity of results (often in conflict with one another) that it is hard to know what to believe. This is particularly so since many of the results have marginal statistical significance. It becomes a question of personal taste as to whether one regards the whole phenomenon as still tentative or whether one regards it as one of the most important new areas in high energy astrophysics. It certainly must be included in any contemporary discussion of binary stars; the question is whether a decade from now the phenomenon is a major part of binary star research or has become just a historical footnote.

2. Techniques

For most of the X-ray and gamma-ray spectrum the earth's atmosphere is completely opaque, which means that cosmic observations are only possible from space vehicles. However at energies above 100 GeV a new window opens with the development in the earth's atmosphere of an electromagnetic cascade whose secondary components penetrate to ground-level where they can be detected by a variety of methods. The phenomenon is illustrated in Figure 1 where the two principal detection methods are highlighted. At energies above 100 TeV the particles in the electromagnetic cascade (the so-called air shower) can be detected by standard particle detectors e.g. Geiger counters, scintillation detectors, water Cherenkov detectors. At lower energies there are not enough particles reaching detector level, even at high mountain altitudes, to make detection possible; however the Cherenkov light given off in the atmosphere by the passage of the relativistic shower of particles is detectable under clear dark night-sky conditions. The light detectors used are simple and inexpensive. Angular resolutions of less than a degree can be achieved. The spread of the secondary component of the showers gives collection areas in excess of 10,000 m^2. Although the techniques are limited by the large background of isotropic charged cosmic rays (mostly protons), the sensitivity is such that fluxes that are only a few percent of the background are detectable as point source anisotropies. In the second-generation systems now coming into use at both TeV and PeV energies the differences in the photon and proton initiated air showers are exploited to reject the background showers and enhance the flux sensitivity to gamma-ray sources. For a more complete review of detection techniques see [1,2].

3. Observations

The atmospheric Cherenkov and air shower techniques were used for gamma-ray astronomy as long ago as 1960 but the first report of the detection of a binary did not come until 1973 [3]. This discovery did not attract much attention until a decade later when it was confirmed by other groups and seen at energies reaching as high as 10 PeV. Since that time there have been a number of reports of binary source detections; Table I lists the first reports of detected TeV or higher emission and the confirmations, if any. Since many of the reported detections are unconfirmed we shall only concentrate on three sources whose existence can be treated with some confidence. For a more complete review see [2,4]. The principal parameters associated with these sources are listed in Table II [5].

3.1. Cygnus X-3

In as many as 15 published papers this mysterious object has been reported as a source of gamma rays with energies of 100 MeV to 0.5 EeV. It has been the subject of a number of reviews, one of them critical of its very existence at energies above 1 MeV [6]. On the basis of its very great variability at radio wavelengths (which certainly points to the presence of non-thermal processes), this source was selected as a prime target for gamma-ray emission. Hence its identification as a very high energy source should come as no surprise.

Cygnus X-3 is the only x-ray binary that has been reported as a source at

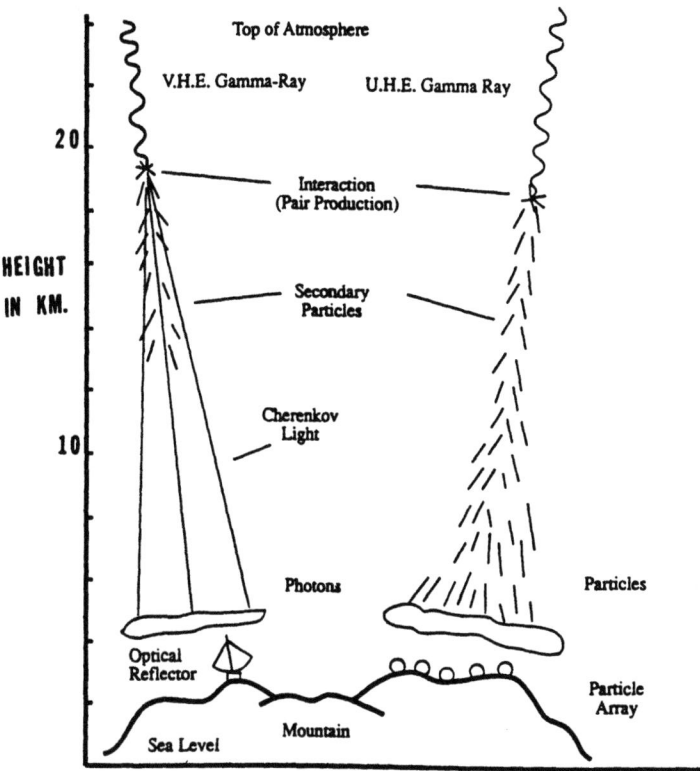

Fig. 1. Schematic development of showers in the 1 TeV and 1 PeV energy ranges and the techniques used to detect them.

100 MeV energies. It is also the only source that is reported at energies in excess of 0.5 EeV [7]. The spectral index deduced from the measured fluxes (not all of which are averaged over many years) is very flat (integral spectrum -1.0), with only a steepening at the highest energies (Figure 2).

Although estimates of absolute fluxes and energy thresholds are notoriously difficult at these energies, there is strong evidence of variability from experiments at all energies (including evidence from single experiments where the uncertainty in absolute calibration is unimportant). The time-scale for variability appears to be months to years with evidence for a decline in intensity since the initial discovery. There is some evidence that the episodes of gamma-ray emission are associated with radio outbursts.

The flat power-law spectrum implies that Cygnus X-3 is a powerful source of high energy particles [5]. It is easy to calculate the power required to continuously replenish the supply of cosmic rays of 10 to 100 PeV in the Galaxy ($\sim 3 \times 10^{38}$ erg s^{-1}). The gamma-ray flux from Cygnus X-3 suggests that the source radiates 4×10^{36} erg s^{-1} in 1 to 10 PeV gamma rays; assuming that these are secondary to hadrons of ten times greater energy and produced with 10% efficiency,

TABLE I
Very High Energy Gamma-ray Binaries.

Source	Classification	Discovery	Group	Energy	Comment
Close Binaries with Neutron Stars					
Cygnus X-3	Binary?	1973	Crimea	GeV,TeV, PeV,EeV	Confirmed
Hercules X-1	Low Mass	1983	Durham	TeV,PeV	Confirmed
4U0115+63	Massive	1985	Durham	TeV	Unconfirmed
Vela X-1	Massive	1984	Adelaide	TeV,PeV	Confirmed
Centaurus X-3	Massive	1988	Durham	TeV	Unconfirmed
LMC X-4	Massive	1985	Adelaide	PeV	Unconfirmed
Cataclysmic Variables					
AE Aqr	C.V.	1989	Potchefstroom	TeV	Confirmed
AO Psc	C.V.	1989	Potchefstroom	TeV	Unconfirmed

TABLE II
Parameters of the most important Tev sources.

| Source | Time scale | | | Luminosity (erg s^{-1}) | |
	Spin	Orbit	Precession	X-ray	VHE gamma-ray
Hercules X-1	1.24s	1.7d	35d	10^{35} to 10^{37}	10^{35} to 10^{37}
Cygnus X-3	12.59ms?	4.8h	19.2d?	10^{37} to 10^{38}	10^{35} to 10^{37}
Vela X-1	283s	8.97d		10^{34} to 10^{37}	10^{34} to 10^{35}

then one can conclude that only one source like Cygnus X-3 is needed to supply the Galaxy with UHE cosmic rays. The report by the University of Durham group [8] of the detection (with high statistical significance) of a periodicity of 12.5908 ms in data taken on Cygnus X-3 is indicative of the presence of a fast pulsar in the system. This would suggest that the rotation of the neutron star is the source of the high energy particle acceleration, a result of major astrophysical significance. The effect has not received an independent confirmation and three groups have reported null results in their attempt to verify the effect at similar energies; also the periodicity has not been seen at any other wavelength. To complicate the issue the Crimean group have reported the detection of a periodicity of 9.2 ms in archival data with high statistical probability [9]. Obviously further confirmation is required before the presence of a pulsar in the system is taken as established.

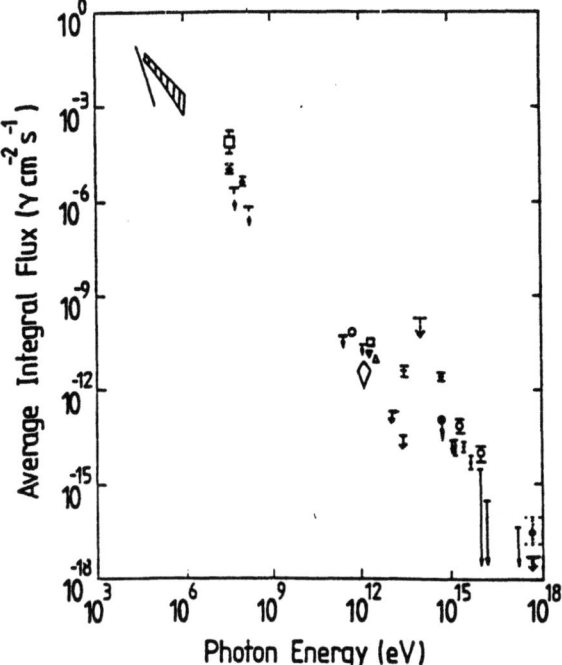

Fig. 2. The measured gamma-ray energy spectrum of Cygnus X-3 from 1 keV to 1 EeV [2].

3.2. HERCULES X-1

Six independent groups have reported the detection of a flux of energy in excess of 0.1 TeV from this classical x-ray binary [10]. In general the putative gamma-ray signal has the periodicity of the neutron star rotation.

The methodology of source detection is as follows. The source is tracked for a period of hours (for the ACT this consists of the telescope physically tracking the source; in the all-sky viewing air shower array, the source is tracked through the database). If an increase in counting rate is noted from the source direction, the data is tagged for future periodicity analysis. In general these increases are not of themselves statistically significant. In the periodic search, usually using the Rayleigh analysis, a range of periods is searched around the known period of the pulsar. If strong power is found near this value, the observation is thought to have detected a high energy flux from the Hercules X-1 system.

The most interesting Hercules X-1 observations occurred in 1986. These imply both unusual astrophysics and unusual physics. The observations were reported by three groups [11,12,13]. The observations have two striking features: (i) the period detected is shifted to the blue from the canonical neutron star rotation period by 0.16%; (ii) the primaries do not appear to be photonic in nature.

The shift in period is clear from the composite periodograms shown in Figure 3

[14]. It should be emphasized that these are not overlapping observations but are spread over a time interval of three months.

The shift in period can be accommodated in some models of the source; the gamma-ray production takes place under certain conditions at a location in the rotation disk which is not corotating with the neutron star. It is difficult to understand how these conditions could persist for more than 100 days and then disappear.

3.3. VELA X-1

This is unlike the other binaries in that the massive system appears to be powered by a stellar wind. It is subject to rapid spin-ups and spin-downs. There have been six reports of the detection of gamma-ray emission but these differ from one another in several respects [4]. Three of the reports are at TeV energies and in these cases emission is seen at or close to the neutron star rotation period. The emission is apparently steady although there is also evidence for episodes of enhanced emission. At PeV energies where there are also three reported detections the situation is less satisfactory; only the orbital period is seen but the phase of the emission is different in all three reports even though two of them are contemporaneous.

4. Interpretation

The production of particles of energy in the PeV range requires progenitors of energy 10 PeV or greater. It is unlikely that electrons can survive to these energies because of radiative losses so that the particles must be protons and hence the sources must be relevant to the whole problem of the origin of the cosmic radiation. There is no intimation from the study of binaries at lower energies that these objects are cosmic ray producers.

However following the reports of TeV and PeV gamma ray observations from close binaries there has been no shortage of models to explain the phenomenon; none of these have received universal acclaim. They have been reviewed by Hillas [15] and Harding [16].

The energy source for the pulsed emission is assumed to be either the rotation of the neutron star or the accretion unto it. The acceleration mechanisms are listed by Harding [16] as dynamo, shock, reconnection and plasma turbulence. To account for the wide diversity in the observed emission it is probable that more than one emission mechanism is involved.

5. Conclusions

It is apparent that very high gamma-ray astronomy, at the very end of the electromagnetic spectrum, has opened a new perspective on the physics of close binaries. Because of the weak fluxes, the relatively primitive nature of the detection techniques used to date, and the apparently sporadic nature of the emission it is not

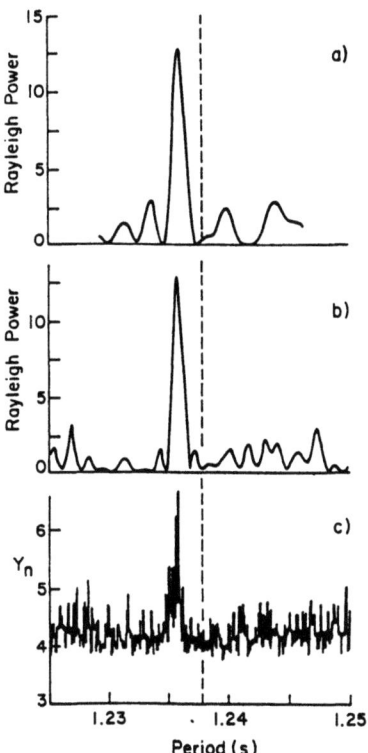

Fig. 3. Power as a function of period for three sets of observations of Hercules X-1 in 1986: (a) Haleakala Observatory, 13 May, 1986 [11]; (b) Whipple Observatory, 11 June, 1986 [12]; (c) Los Alamos experiment, 23 July, 1986 [13]. The X-ray period is shown by the dashed line.

surprising that most observations are severely limited in their statistical significance. There is not even certainty that in all the observations discussed above that the observed quanta are, in fact, gamma rays. Where there is information on the nature of the primaries in these experiments, it would appear that the showers detected are more hadronic in character. Given the distance to the sources and the observed periodicity, there is no known particle that could produce the observed effects.

While there is still uncertainty as to the identity of the particle phenomenon responsible for these detections of the close binaries, there is solid evidence that, at least at TeV energies, the gamma-ray shower develops in the atmosphere as expected. There is now one source (the Crab Nebula [17]) which can used as a standard candle of VHE gamma rays to test new detection techniques which claim to distinguish gamma rays from the background cosmic rays.

This review of VHE gamma-ray observations of binary star systems was supported by the U.S. Dept. of Energy.

ζ AURIGAE ATMOSPHERIC ECLIPSING BINARIES

I. A. AHMAD

1. Introduction

ζ Aurigae stars are defined as eclipsing binary systems consisting of a cool supergiant and a hot dwarf companion. Such stars may exhibit interactive phenomena. These interactions are of a low intensity compared to more dramatic systems such as, for example, cataclysmic binaries. Nonetheless, the ζ Aurigae systems are of primary importance in the field of interacting binaries.

The dwarf companion serves a a probe of the atmosphere of the supergiant in a fashion not possible in more intensely active systems because the activity in latter systems complicates both the spectrum and the models to such a degree that analysis is quite difficult. Despite the relative simplicity of their interactions, ζ Aurigae systems involve processes common to the more complicated systems. Thus, an interaction is taking place, available for analysis with the aid of the probe provided by the atmospheric eclipse of the secondary, but at a level of intensity low enough that the interaction may be treated as 'perturbations' on a single star. Differences from a single star include radiative effects of the secondary on the chromosphere of the primary (Hjellming and Newall 1983, Eaton 1988, and Ahmad 1987, 1990) and the dynamic and gravitational interactions between the stars.

In the classical ζ Aur systems, a supergiant of type G or later is paired with a B dwarf secondary (Wright 1970). In recent decades, systems in which the primary is of type M have been treated as a separate class called VV Cep systems after the prototype (Bidelman 1954; Cowley 1969). In this paper that distinction is respected and we restrict the discussion to systems with an evolved primary of type G or K. We shall also restrict our discussion to systems in which the secondary is a B star, although the systems might be defined to include cases with A star secondaries (e.g., Griffin *et al.* 1990).

The following terms are useful in discussing the interaction phenomena in ζ Aur type systems:

Primary eclipse: the passage of the dwarf behind the supergiant.

Atmospheric eclipse: Attenuation of light from the dwarf by the lower chromosphere of the supergiant.

Obscuration phase: Attenuation of light from the dwarf by the shock cone and/or accretion column.

Shock cone: The wake of shocked material in the supergiant wind as it collides with the dwarf and/or its wind.

Accretion column: Relatively cool material within the shock cone falling onto the dwarf.

Boundary layer: A disk, ring, shell or accreting material more or less axisymmetrically surrounding the dwarf.

2. Historical Review

The early optical work on these systems was summarized by Wilson (1960) and by Wright (1970). The unmistakable atmospheric eclipse in these systems was the mother lode of provocative data, posing several fascinating problems to researchers.

Rotational velocities of the chromospheric gas unmistakably demonstrated that the chromosphere was not in simple axisymmetric rotation. Disparities between the ionization states of spectral lines due to calcium and iron were problematical and it seemed doubtful that chromospheric excitation was simply due to irradiation from the hot companion.

The basic conception of the nature of ζ Aur systems was established early. As summarized by Sahade and Wood (1978), there is a complex chromospheric structure including the suggestion of dense condensations as small as 10^3 km in the upper atmosphere. Asymmetries between ingress and egress persisted from one eclipse to the next but the gradients changed.

Saito (1970) made an early attempt at detailed modeling the effect of UV radiation from the dwarf on the supergiant chromosphere. He noted that gas on that side of the ζ Aur K star chromosphere facing the B component would be ionized creating an H I, H II region boundary that would rotate through the chromosphere at the rate of the system's revolution. Using Kiyokawa's (1967) analysis of the ζ Aur light curve, he demonstrated that a shock front would form on one side of the H II region as the region revolved through the atmosphere. He concluded that this effect, combined with slow rotation and random gas cloud motions account for the egress velocities.

Within the last ten years data from the International Ultraviolet Explorer (IUE) has made possible a much more detailed and sophisticated analysis of the nature of interaction in these particular systems (Dupree and Reimers 1987; Hack and Stickland 1987). Stencel *et al.* (1979) presented evidence that the B star in 32 Cygni resides inside the upper atmosphere of the supergiant.

A watershed development was Chapman's (1981) analysis of the prototype ζ Aurigae. The MgII and C IV resonance lines provided the basis for deriving a model of the K star wind far from the surface. Chapman was able to demonstrate the existence of an accretion column within an accretion shock cone around the dwarf companion (cf. Jackson 1975). The wind speed was estimated at $100 \, \text{km} s^{-1}$ and the density at the B star at 3×10^6 cm^{-1} for an implied mass loss rate of $2 \times 10^{-8} \, M_\odot \, \text{yr}^{-1}$.

In recent years the list of ζ Aur systems has grown. To ζ Aur, 31 Cygni and 32 Cygni we may now add 22 Vulpeculae (Ake, Parsons and Kondo 1985). 22 Vul differs from the three classical cases in that its luminosity class falls between a supergiant and a bright giant. Analysis of the chromospheres of these systems has now been enhanced by the development of techniques for disentangling the spectra by computer subtraction (Griffin and Griffin 1986).

Griffin (1988) identified HR 6902 as a ζ Aur system in the title of his paper. The primary of HR 6902, however, is distinctly of luminosity class II. We shall see how these differences in luminosity class seem to relate to the observables which are due to interaction.

3. Chromosphere and Wind

Interaction in the systems is made possible by the extension of the primary chromosphere to the orbit of the secondary and/or the presence of a wind from the primary.

Schröder (1985, 1986) derived a chromospheric density distribution for 32 Cyg, 31 Cyg, and ζ Aur from ultraviolet observations. Optical data is used to analyze lower heights in the chromosphere (Schröder, Griffin and Griffin 1990).

Che, Hempe and Reimers (1983) developed a nonspherical three-dimensional line transfer code for determining mass loss by line synthesis of the Fe II, Si II, and S II resonance lines. They applied the technique to ζ Aur, 32 Cyg and 31 Cyg. Reimers and Che-Bohnenstengle (1986) subsequently applied the method to 22 Vul. Taking into account observed wind parameters and the empirical chromospheric density model, Schröder (1985) inferred a velocity acceleration model which implied a thicker acceleration zone than previously accepted. Ahmad and Stencel (1988) reached a similar conclusion using two different methods.

Kuin and Ahmad (1989) have presented a *prima facie* case for Alfvén waves as the wind driving mechanism. They found that the damping scale length of the waves changes from very small values near the star to larger values, of the order of the local radius, further out into the wind. This implied a necessary constraint for the existence of cool massive winds and suggested that the "dividing line" in the H–R diagram marks the locus at which this constraint is violated. The requirement is effectively a lower limit on stellar wind densities for which radiative losses can balance heat dissipated into the wind.

4. Accretion and Shock

Chapman's (1981) discovery of an interaction between the supergiant wind and the B star was confirmed by Ahmad, Chapman and Kondo's (1983; ACK) study of MgII, C IV, Si IV, and N V lines in both ζ Aur and 32 Cyg. The Jackson-type model suggested by ACK was subsequently refined by Ahmad (1989) in order to

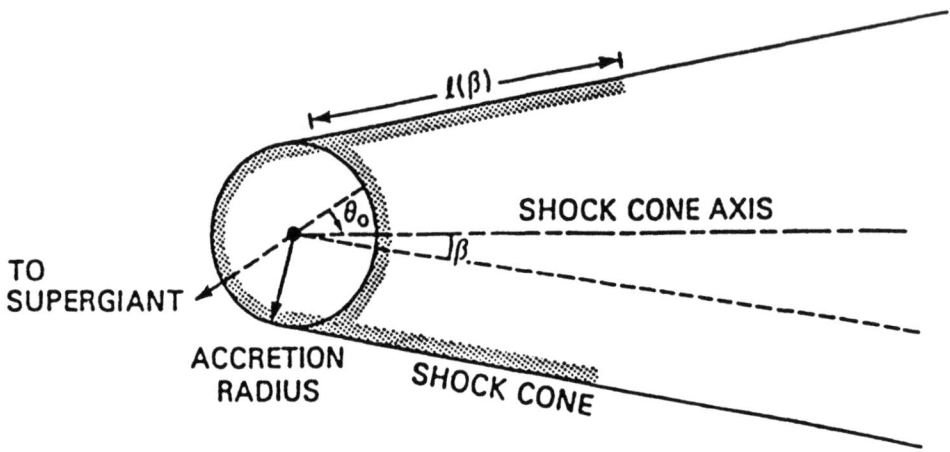

Fig. 1. Detailed sketch of refined ACK model for accretion onto the B dwarf in ζ Aur-type systems. β is the angle from the shock cone axis to the line of site. $l(\beta)$ is the effective length of the cone of shocked material. From Ahmad (1989).

take into account the more complex system 31 Cyg. The latter model is illustrated in Figure 1.

The interaction between the supergiant wind and the dwarf companion appears to be due to shock and accretion. Five years ago Ahmad and Chapman (1986) laid out an 'emerging picture' of these phenomena. We now reprise that picture in the light of subsequent research.

The shock cone is generally seen in the Si IV 1400 Å resonance doublet. In 22 Vul, ζ Aur, and 31 Cyg it may be seen in the hotter C IV 1550 Å doublet. In 22 Vul and 32 Cyg it is event visible in the N V 1240 Å doublet. The boundary layer may also be seen in some of all of these lines.

The reverse P Cygni profiles detected by Chapman (1981), Ahmad (1986) and Che-Bohnenstengle and Reimers (1986) is due to the boundary layer shock, a ring or disk of material falling onto the dwarf companion.

The accretion column is directly observable in the MgII 2800 Å doublet. These complex lines include a P Cygni profile, typical for an expanding stellar wind and a sharp interstellar profile at low radial velocity. The velocity of the interstellar lines is consistent with the Crutcher (1982) model of the local interstellar medium (Ahmad and Stencel 1988). During the obscuration phase the accretion column may be seen as an additional absorption feature shifted to the longward side by the velocity of its recession toward the dwarf.

The accretion column is best seen in MgII during the obscuration phase. The shock cone is best seen at the ingress and egress of the accretion phase. These results are the same for ζ Aur (Ahmad 1986), 32 Cyg (Ahmad 1986), and 22 Vul (Ahmad and Parsons 1986). In the case of 31 Cygni the situation is less clear. The shock cone absorption is visible at angles outside the narrow range expected from the simple Jackson (1975) theory. It has yet to be determined whether this is due to a low cooling rate (broadening the shock cone as suggested by Che-Bohnenstengle and Reimers 1986) or to the detailed geometry of the accretion shock (Ahmad 1989). This problem could be resolved by observing the system at phases at which cooling rate would require the presence of MgII from the accretion column.

A pertinent question is whether X-rays are present in any of these systems. Unfortunately X-rays have only been searched for in two cases. McCluskey and Kondo (1984) did detect X-rays in 31 Cyg, but not in ζ Aurigae. Ahmad and Chapman (1986), however, have noted that the latter system was observed when the accretion shock was eclipsed.

When HR 6902 is contrasted to the four systems discussed above, it appears that the dominant factor is spectral type of the secondary. *HR 6902 has no extended chromosphere.* Schröder (1990) has completed a curve of growth analysis of near eclipse phases of IUE spectra of HR 6902. Ahmad (1990) has examined all the high resolution IUE observations of HR 6902 available through the end of 1989 (including the spectrum at MgII 2800). There is no evidence for a stellar wind, nor atmospheric variability. The absence of MgII 2800 Å emission in the LWP observations (even during eclipse) is especially convincing in this regard. MgII resonance scattering is very strong in the other three systems and normally gives a P-Cygni profile indicating the presence of a wind.

These conclusions are supported by Reimers *et al.*'s (1990) detection of a rapid jump in temperature in HR 6902 to about 0.6×10^5 K within a stellar radius of the surface based on their study of the Fe II lines. Using a semi-empirical models of the stellar wind, Kuin and Ahmad (1989) have demonstrated that 22 Vul is close to the dividing line between coronae and cool stellar winds. Schröder (1990) and Ahmad (1990) have argued on empirical grounds that the coronal dividing line must fall near 22 Vul, which has been characterized as a hybrid system, but above the less luminous HR 6902.

5. A Vogt-Russell Theorem for Binary Stars

The Vogt-Russell theorem held that all the observables of a single star were uniquely determined by its age, mass, and initial composition. That theorem has lost its luster now the importance of rotation in determining observables has become appreciated. Nonetheless, it served an instructive purpose for many decades. An analogy for binary stars might be considered. It would be based on a matrix composed not only of age, mass, and initial compositions of the components, but also including their separation and initial eccentricity. (Period would play a derivative role.) As in the case of single stars, rotation denies this the status of a true theorem, but it can still serve an instructive purpose. For supergiant stars, at least, rotation cannot be very large. For the ζ Aurigae stars, at least, the parameters enumerated seem to correspond with the variations in observables which we have considered. How, and whether, more dramatic variations of these parameters would account for a transition from these systems to those in which binary interaction is more severe remains to be addressed.

6. Secular Variations

The existence of secular variability in these systems has been clear from the early optical studies (e.g., McKeller and Petrie (1952)). Recent research has confirmed that fact and suggested variability in the binary interactions as well (e.g., Ahmad 1986, Ahmad 1991). Evidence for transient interactions have also been demonstrated (e.g., Parson, Ake, and Hopkins 1984, Ahmad 1988). It is an area in which much work remains to be done.

POLARIZED RADIATION FROM CLOSE BINARIES

R. H. KOCH

1. Introduction

1.1. GENERAL

Comprehensive reviews of results from polarimetric observations of close binary stars include the brief survey by Koch (1988) and the compendium by Coyne et al. (1988). These have cutoff dates in the late 1980's. By now, almost 700 binary systems are known to have been observed polarimetrically and a total of 98 new observational series has been presented since 1985. Most of these are of visible-band, linear data but more than 30% contain circular component radio, IR, or visible-band measures. The organization of this chapter is complementary to that of Koch's earlier review and the reader can easily gauge the evolutionary states of close binaries which have recently seemed to be most rewarding for both observation and analysis.

1.2. OBSERVING PRACTICE

A small assortment of polarimeters have been described in some detail in recent literature. Some of these built around chromatic beam-folding optics permit simultaneous observing of more than one wavelength interval.

Most observing teams, however, give only meager detail with regard to their actual telescope practice so that a new worker commonly endures a trial-and-error effort before maturing techniques appropriate for his station and instrumentation.

One of the fundamental details of observing practice rests in the polarimetric analogue with UBV-standardized light curves rather than with the conventional magnitude-difference light curves obtained by frequent reference to a comparison star. For light curve accumulation the time needed for obtaining a comparison star measure is very short so that one commonly observes the program star several times between successive comparison star measures.

Because it takes so long to make a polarization measure, very frequent repetition of a standard or comparison star on one night is not common. If short-term transients in the instrumental system exist, they may remain unrecognized.

1.3. REDUCTION PROCEDURES

The linear Stokes parameters are inevitably developed from a modulation of electrical signals as a 2nd harmonic function of the azimuthal orientation around the beam direction. The fact that the functional relationship is known a priori simplifies the evaluation of the polarization parameters but cannot be used as a rationale for neglecting the evaluation of other harmonics of the signal. For instance, migration of the Ramsden disk on the cathode will impose a 1st harmonic onto the observations and higher (including 2nd) harmonics of measurable amplitude are not to be discounted unless proved to be negligible within the errors of observation. For this reason recourse to standard polarization stars is necessary.

The circular Stokes parameter is typically evaluated as a dc electrical signal. This does not mean that the signal is free of azimuthal dependence, perhaps due to crosstalk from the channels which carry the electrical forms of the linear parameters, and this possibility must be checked.

The number of stellar standards is not large. This circumstance arises from two causes. The first is founded in the belief that it is unrewarding to develop the information. Such an attitude is as misguided as a supposition that stellar photometry could have advanced to its present stage without the effort put into the validation of the UBV and uvby photometric systems. The second reason for the small number of standards is the attrition among the stars originally propounded as, for one or another reason, they have been found to be polarization variables. It is very disturbing that main sequence spotted stars and relatively long-period binaries are being found not to be useful as standards.

1.4. INTERSTELLAR POLARIZATION COMPONENTS

There exist many surveys of stellar polarization, most having been conducted for mapping of the galactic systematics and inferring the orientation of the local magnetic field. Of the thousands of such measures of the hundreds of stars, it inevitably happened that many of the stars are high luminosity ones of intrinsically variable polarization. Because repetitive observation of the survey stars has been uncommon, it is impossible to sort out exactly the constant from the variable polarization objects in order to determine the interstellar component at the 3-dimensional position of a program star. Furthermore, even though there do exist many field star measures and even though most close binaries are Milky Way disk population objects, the number of interstellar probe stars within a, say, 5° radius of a program star is typically small. It is not uncommon to have fewer than 15 stars within 75 square degrees distributed over 500 pc radially. For such a situation, the precision of the interstellar component to be attributed to a program star must inevitably be small. The situation is further exacerbated by the typical survey being a 1-filter or unfiltered one. While Serkowski, Mathewson, and Ford (1975) have developed the empirical evidence for the chromatic variation of the interstellar polarization, there is no such systematic known for the electric vector parameter. It remains an imponderable whether or not one should assume the interstellar vector to be of constant orientation or not for a given program star.

1.5. CLOSE BINARY INFORMATION TO BE GAINED

One aim of all stellar, and especially close binary, polarization investigations is to evaluate the number density and geometrical distribution of the scatterers and the changes, whether secular, non-monotonic, or phase-locked, in these representations. A second aim seeks to characterize the environment of the scatterers as a magnetic or non-magnetic one. Lastly, the distributions of sizes and shapes and the chemical identities of the scatterers can, in principle, be quantified. The Keplerian cycling of a binary imposes constraints on interpretation provided the orbital and bulk-stellar properties are themselves known independently.

It is naturally true that the foregoing parameters can be enmeshed in very intricate correlations and it is additionally true that the evaluations of any of the polarization parameters and their measures of precision are deeply model-dependent. Nonetheless, the appeal of being able to quantify the sense of orbital revolution for a point source and the orbital inclination with respect to the plane of the sky is very strong. This is particularly so, respectively, if a binary be part of a triple or multiple system and if the removal of the inclination projection factor is important to the knowledge of an otherwise-unknowable mass.

The following sections of this chapter are organized by advancing stage of binary evolution as far as this is known. It is certainly possible that some misinterpretations of evolutionary stage will become evident in time. However, the review is so arranged that only the last section depends upon but one single binary example. Each section is structured so that the expected properties of the scattering medium are described and then the evidence in favor of or against this expectation is summarized. A final judgment for each evolutionary stage completes each section. In every section conceptual advances and difficulties are emphasized and numerical illustration is eschewed since it is individual for each binary.

2. Pre-Main Sequence Close Binaries

2.1. GENERAL

The generalized model which presently exists for pre- main sequence stars and for the more complex case of a close binary in this evolutionary stage encourages belief that such close binaries ought to show a rich repertoire of polarization variability.

2.2. SPECIFIC BINARY SYSTEMS

BM Ori is the faintest member of the Orion Trapezium with a Keplerian period of about 6.5 days. In telescopes with small-to-moderate focal plane scales, it is notoriously difficult to observe radiometrically and polarimetrically because of the small angular separations of the stars and because of the non-uniform nebulous background or foreground. A modern interpretation of the system posits substantial intrinsic light variability arising from a disk-envelope structure possibly confined to the cooler, less massive member which is still contracting to the main sequence. Appenzeller (1966) measured a rather large polarization of about 0.4%, but with a mean error of ±0.2%. Wolf (1972) also did not succeed in making a 3σ linear measure in four wide and narrow bandpasses although only one night was used. Only in red did he detect a signal. He also failed to achieve a significant circular measure in blue and red on the same night. A similarly null, but much more precise, circular result has been published by Kemp, Wolstencroft, and Swedlund (1972), again on one night. A priori, BM Ori would be expected to demonstrate significant polarization if only from the scattering of the hotter, brighter star's light from the disk so these null results remain a puzzle. Evaluation and removal of the interstellar polarization are necessities and perhaps more extended monitoring may be useful.

A much more bizarre object is HD 44179, whose alias is The Red Rectangle. The most comprehensive study of this system is that of Cohen *et al.* (1975) wherein the bound nature of the visual binary is made clear and a classification of B9-A0 III is assigned to at least the one star embedded in an IR-radiating dust envelope. The likelihood of a pre-main sequence assignment for HD 44179 is not small but its characterization as a close binary is, at present, justifiable only in terms of its eventual interaction at the red giant stage. Coyne's polarimetry in the Cohen *et al.* paper removes a relatively small interstellar polarization and postulates that the residual polarization in the blue represents the contribution from Thomson scattering, although the seat of this scattering is not specified. The residual spectrum of the electric vector is essentially flat with respect to bandpass, the vector being nearly aligned with the N-S sides of the Rectangle. The residual intrinsic polarization rises monotonically to longer wavelengths without reaching a maximum at nearly 1 μm. This scattering is attributed to rather large grains in the systemic nebulosity but a distribution of the sizes of the grains is expected.

The coolest system discussed here is RY Tau containing a star classified as K1 IV-Ve accompanying a F8 V object. Several long term cyclical light variations are known but the Keplerian period is not yet convincingly evaluated. An evolutionary stage as advanced as that of, say, RS CVn seems unlikely and a slow-mass-exchange condition even more so. By default, a pre-ZAMS attribution appears possible.

2.3. CONCLUSION

The expectation that a wealth of polarization phenomena should be demonstrated by the structure and activity of pre-ZAMS binaries is not known to be true at present.

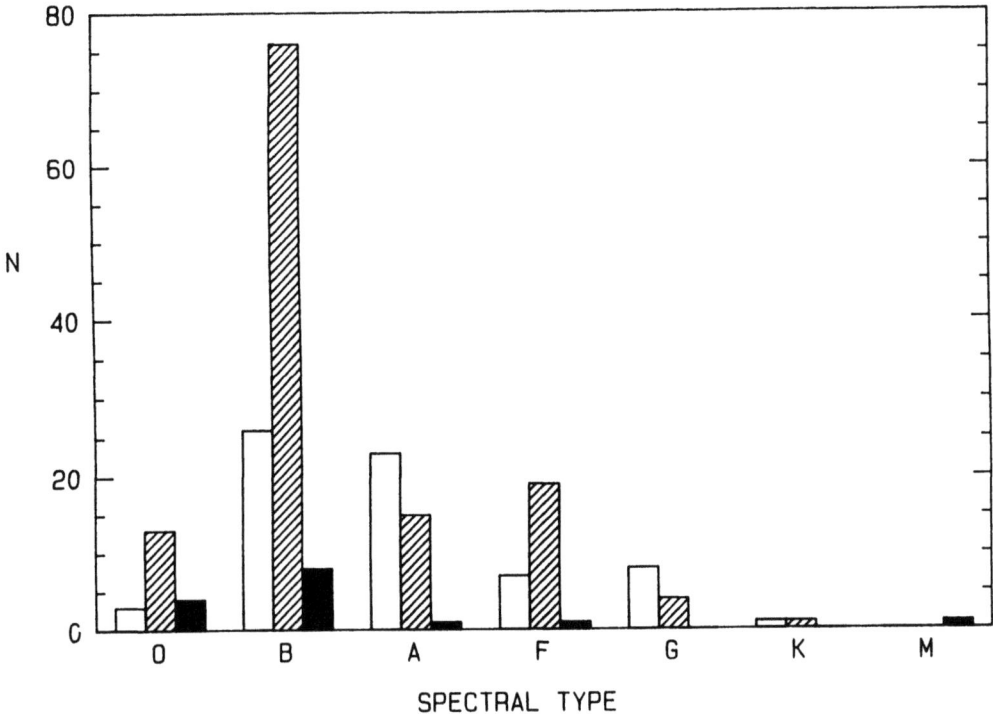

Fig. 1. Distribution of 209 close binaries.

3. Non-Contact Main Sequence Close Binary Stars

3.1. GENERAL

A literature search identified 209 close binaries, customarily classified as main sequence ones and not in the contact condition, which have been measured polarimetrically at least once. Their distribution by spectral type may be seen in Figure 1, wherein they are distinguished by measurement of a signal which is null within its error, or by detection of a non-null signal, or by evidence of polarimetric variability.

It is possible that the null and non-null detections could contain variable objects which have been insufficiently observed. To a small extent, this proposition may be examined for it is somewhat more likely that the non-null detections contain unrecognized variable-polarization stars than would be found in the null-signal sample. In order to examine this possibility, the measures for stars assigned spectral types B through K and not known to be binaries, were extracted at random from the Mathewson and Ford (1970) catalogue. These are inferred not to be polarimetrically variable. Within each spectral class, the number of stars randomly chosen is equal to the sum from the two categories for the known close binaries. For O-type stars, the Hiltner (1956) catalogue was used in order to find a control, single star sample. The distribution of the stars in the control sample is shown in Figure 2. For O-, B-,

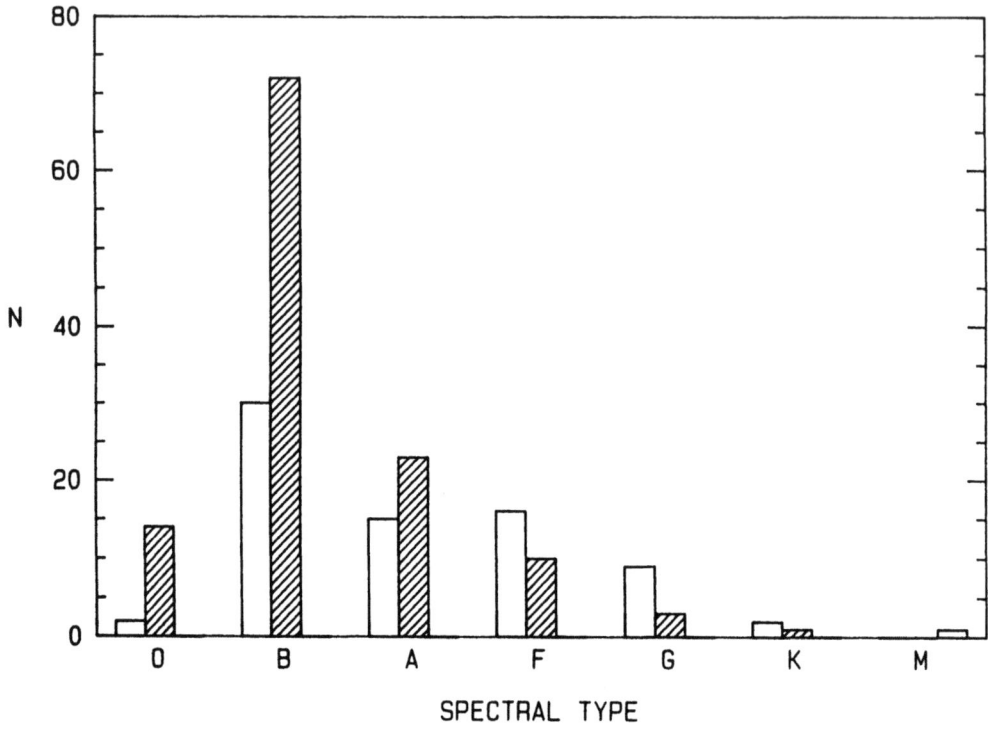

Fig. 2. Distribution of the stars in the control sample.

and G- type stars the ratios of non-null to null detections are evidently the same for the close binary and control, single star samples. The seeming differences between the binary and control samples of A- and F-type systems are not significant. Thus, there are unlikely to be many polarimetric variables among the binaries not yet known to be variable.

3.2. POSSIBLE MECHANISMS OF POLARIMETRIC VARIABILITY

The interesting objects, therefore, are the twelve polarimetrically-variable binaries. Several possible mechanisms may be imagined to function in main sequence binaries so as to create this effect. It cannot be denied that some stars, while classified as MK V for luminosity do actually possess shells which are themselves variable. Another, more subtle origin may be seated in the stellar winds which evidence themselves only in the spacecraft UV. Finally, there may be systems of such elegantly poised geometry that one may search for the Chandrasekhar (1946) Effect to be displayed by them during one or both eclipses.

3.3. O-TYPE CLOSE BINARIES

It is worth considering in detail the evidence of the four O-type systems SZ Cam, XZ Cep, θ^2 Ori, and σ Ori. This is important, not only for the conclusions which may be drawn, but equally usefully because of the great range of weight which exists for the assorted stars. It will be found that all categories of pairs offer a comparable mixture of documented and inferred conclusions based on greatly different data sets. For SZ Cam, Hall and Mikesell's (1950) two data differ by 0.9% with an error for each measure expected to be about 0.3% while Pfeiffer (1977) shows a dispersion of 0.25% among his five homogeneous measures. Since data are so few, it is impossible to know if there is any phase pattern to the measures. Saute and Martel (1979) show likely variability of both polarization parameters for XZ Cep. Since there is no evident phase locking, the variability must be sporadic. Several observers have detected the large polarization of θ^2 Ori but it is only in the extended series by Coyne and Gehrels (1967) that variability can be validated within the errors of measurement and yet the data remain too few to search for phase locking. Wolf found no significant circular component in his 1 observation. For the very curious object σ OriE, the situation is very different for Kemp and Herman (1977) collected abundant blue and ultraviolet data. The blue data fold onto the nearly 1.2-day period quite well with a double sinusoid but the violet measures accord not at all with the blue ones. Although it would be possible to model the blue data in terms of a systemic scattering envelope, these authors chose instead a model in which a thin disk partially embeds the faint member of the pair, contributing one polarization component from the light of the encircled star and the disk itself and a second polarization component in the light of the O-type star reflected from the disk. Clarke and McGale (1988) conceive of the object as a single magnetic Bp star (despite its customary classification of O9.5 V) with two atmospheric bulges on the magnetic equator which is not a great circle. Neither model has anything to say to the weak polarization pattern in the ultraviolet. It is not evident that any simple model can, in fact, accomodate both data sets.

3.4. B-TYPE CLOSE BINARIES

There are now known to be eight B-type systems which show intrinsic polarization. Koch and Pfeiffer (1989a,b) argue that the causes of the polarization for Y Cyg and U Oph are, in fact, the same: partly that due to the sporadic wind from each star of each pair and partly that due to the intersection of the winds from the paired components. These authors emphasize that evidence for the intrinsic nature of the polarization is indirect: very weak periodicities embedded in large-amplitude noise envelopes, the historical migration of the seasonal means for the electric vector angle for Y Cyg, and Coyne's (1970) independent, previously published evidence of variability for U Oph. Several authors have investigated the variability of ϕ Per but the series of Coyne and McLean (1975) led to the description of a completely ionized envelope with a non-negligible optical scattering depth in the shell. The emission in the lower Balmer lines is also partially polarized. In wide bandpasses Stokes *et al.* (1974) were unable to measure a significant circular component of the polarization. For HD 162679 Luna (1982) has published the data journal but remarks only that the short-term variability appears not to be periodic. Hayes's (1983) contented himself with establishing the low-level variability of the similar B2 Ve system, 66 Oph, and implying that it would be profitable to study simultaneous polarization and spectroscopic measures.

For the other B-type binaries known to have been investigated, τ^5 Eri, ζ Phe, AL Scl, variability is either asserted without published data or may be inferred from few data on the basis of published errors.

3.5. COOLER CLOSE BINARIES

Two A-type binaries have been studied. Nothing is known in detail of the declared variability for HD 6619. AR Aur continues to be monitored sporadically at Pennsylvania because of the remote possibility of detecting a Chandrasekhar Effect at each eclipse. No such phase-locked variation has yet been seen.

The few data taken for V1143 Cyg, an F-type pair, show no variability. ρ Tuc is the only other F-type system known to have been studied in detail. Variability is claimed for the system but no details are known.

Pfeiffer's (1977) several observations of YY Gem appear to show variability on the basis of their formal errors but this result is not conclusive.

3.6. CONCLUSIONS

The evidences for the polarimetric variability of main sequence binaries are therefore quite various but not strong. Rather weak winds from hot stars and presumably conventional Be-type envelopes appear to be implicated but the examples of these effects are not yet numerous. Presumably much work can be done productively for cooler main sequence systems.

4. Am- and Ap-Type Close Binaries

4.1. GENERAL

It is generally accepted that Am stars are all members of close binaries; no similar claim is made for Ap stars. It would be expected, therefore, that a significant number of them would have been studied polarimetrically. In fact, this is not the case. As may be seen in Figure 3, the number so observed is small and almost all of these have been observed but once or a few times. There is no significant difference in the samples which are believed to be unpolarized and those for which a non-zero signal has been detected. Presumably the latter sample simply shows interstellar polarization. Even though these two groups dominate the sample observed so

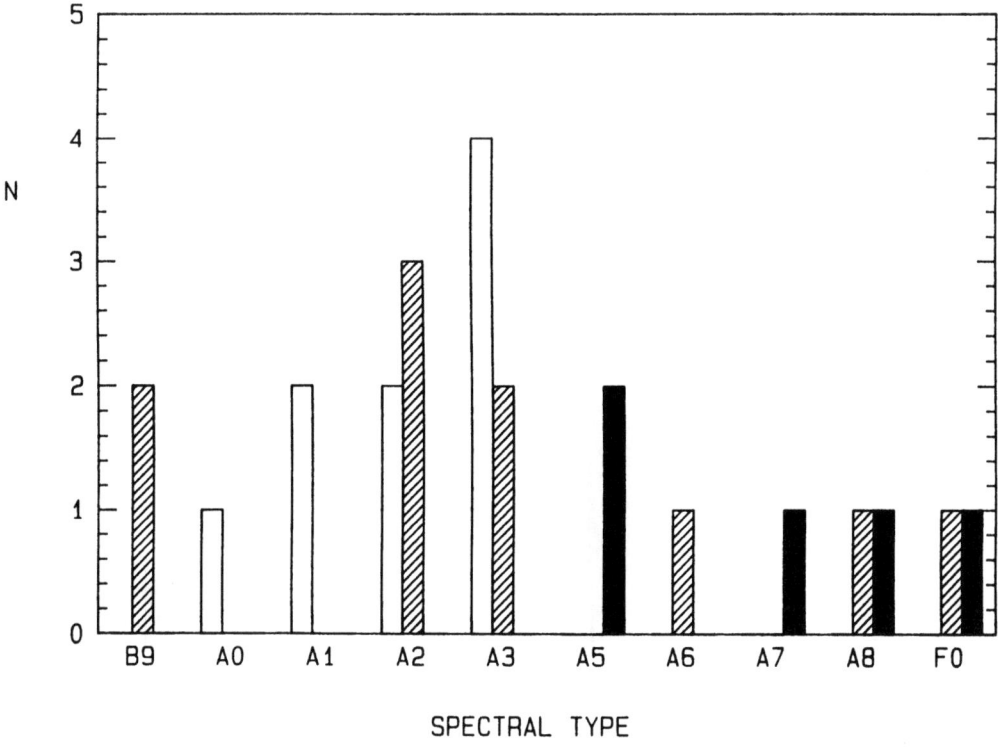

Fig. 3. Am stars.

far, intrinsic polarization is certainly a possibility in a magnetic environment and variability should definitely be sought.

4.2. POLARIMETRICALLY-VARIABLE AM BINARIES

Nothing is known in detail regarding the announced variability of ζ Eri. The few linear measures of β CrB by Serkowski and Chojnacki (1969) strongly indicate variability and possibly phase-locked variability with the 18.5-day short period (which is not the years-long Keplerian period) but nothing further is known about the system. The null circular observation by Kemp, Wolstencroft, and Swedlund does not encourage an attempt to find a magnetic explanation for the linear variability. Even though very little is known of the remaining three variables, there may eventually be validated a bias for polarimetrically-variable Am stars to occur preferentially among the later sub-types, as Figure 3 suggests.

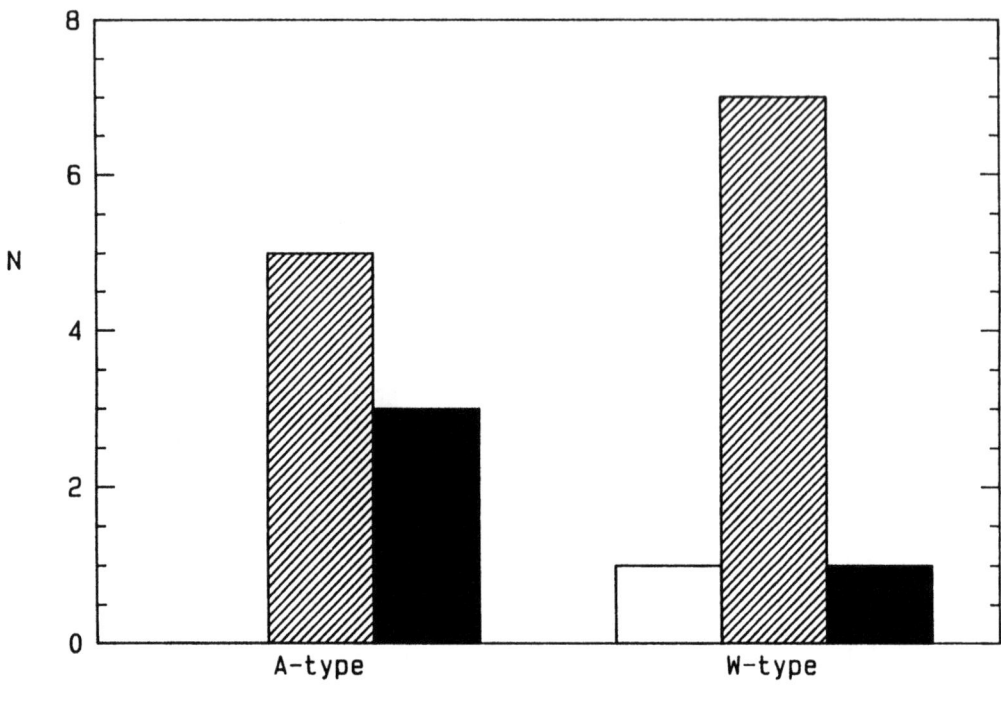

CONTACT BINARIES

Fig. 4. A- and W-types.

5. Near-Main Sequence Contact/Overcontact Close Binaries

5.1. GENERAL

It is true that the pioneering study by Buerger and Collins (1970) regarding phase-locked polarimetric variability would not be considered a completely modern one anymore. Nonetheless, contact pairs would seem to offer a very likely seat of intrinsic polarization because they depart so significantly from spherical symmetry, their envelopes are so active, and because circumbinary gas flow is so vigorous. Koch (1987) has reviewed the evidence from these systems. Reinterpretation of his conclusions, denying variability for S Ant and GK Cep and sorting by Binnendijk's (1970) A- and W-types, is shown in Figure 4. The single null result, that for 44 BooB, may be due entirely to the very small distance of this star from Sun or it may be due in part to light dilution from 44 BooA. Although the precision of many of these data sets is not high, it may be inferred that the detections of polarization for most of the systems are due simply to interstellar components.

5.2. POLARIMETRICALLY-VARIABLE CONTACT BINARIES

Only four UV data were actually presented for RR Cen by Serkowski (1970) but, by the criterion of his published errors, they display variability. Piirola (1977) found a small phase-locked variable signal for AW UMa and Koch confirmed his results but the signal is so small that little interpretation is possible. By far, the strongest evidence for intrinsic polarization is in Luna's (1980) series for ϵ CrA. A large phase-locked modulation is evident and the observations were interpreted modestly within the frame of the scattering-envelope model of Brown, McLean, and Emslie (1978). Koch showed the weak evidence for the intrinsic polarization of XY Leo. These data do not phase comparably well with the ephemeris for the short-period companion SB to the eclipsing pair discovered by Barden (1987). Nonetheless, it is possible that activity in the atmospheres of Barden's M-type pair could be the cause of the variability.

5.3. CONCLUSION

The results for contact close binaries are certainly disappointing for these would seem, on independent ground, to offer abundant opportunity for studying scattering in the envelopes of cooler stars. It is possible that polarization activity is high-frequency with respect to the duty cycle of obtaining observations but this seems unlikely at this time. Perhaps the simplest explanation rests in recognizing that ϵ CrA, the only object to show a convincing signal, is by far the brightest object of the class so that observational noise for it is smaller than for the other systems. Certainly, at this time there seems no useful distinction between the A- and W-types.

6. RS CVn-Type, Spotted Close Binaries

6.1. GENERAL

Numerous close binaries showing the RS CVn-type syndrome have been observed polarimetrically. Compared to the situation for other types of close binaries, data are fairly abundant due mostly to the intensive survey by Liu and Tan (1987). The entries in Figure 5, wherein systems are located by the MK class of their brightest members, show clearly the preference for cooler and brighter systems to be polarimetric variables.

6.2. THE CHARACTER OF THE VARIABILITY

All varieties of polarimetric variability are expressed by these systems. This may be seen, for example, in Liu and Tan's observations of AR Psc, V711 Tau, and ER Vul. In addition, the observations of II Peg by Scaltriti *et al.* (1992) suggest the electric vector to rotate significantly in the photometric period while the polarization changes hardly at all. On the evidence of Pfeiffer's (1979) results, RS CVn itself

(A)

	F-type	G-type	K-type
II–III		1z	1v
III		1d	3d / 1v
III–IV		1z / 1v	2d / 1v
IV	1d	1d	5v
IV–V	1d / 1v	1v	1d / 2v
V	2d / 1v	1z, 1d / 2v	2d

a: z: zero polarization within errors of measurement; d: polarization detected but may not be intrinsic to the binary; v: an intrinsically-polarized close binary.

(B)

	O5–O7	O8–O9	B0–B2	B3–B6	B7–B9
I	4d / 1v	1z 2d / 2v	10d / 3v	1d / 1v	3d
II=f	1z 3d / 2v	1z 5d	4d	4d	
III		4d / 2v	1z 13d / 2v	1z 9d / 3v	2z 8d
IV		2d	8z 14d / 3v	1z 11d	1z 2d

a: coding the same as in Table 1.

(C)

	A-type	F-type	G-type
I	3d	1z 2d	
II	1d		
III	3z 6d	3z 2d	
IV	6z 3d	12z 6d / 1v	1z 2d / 1v

a: coding the same as in Table 1.

(D)

	G-type	K-type	M-type
I	1z 4d	1z 2d	
II	2d / 1v	1d / 1v	1v
III	7z 6d	13z 19d	1z 6d / 1v
IV	1z 1d	1z	

a: coding the same as in Table 1.

Fig. 5. (A) Distribution of spotted close binaries by MK classification and polarimetric condition[a]. (B) Distribution of hat, massive, evolved close binaries by MK classification and polarimetric condition[a]. (C) Distribution of close binaries in the Hertzsprung Gap by MK classification and polarimetric condition[a]. (D) Distribution of giant close binaries by MK classification and polarimetric condition[a].

is inert much of the time but has shown sporadic seasonal variations and has even displayed a quasi-cyclical variation for about four months. This was interpreted as scattering from a cloud of very small mass ejected presumably from the K0 star. Spiralling or pushed away from the star, the cloud eventually diffused so that polarization diminished to an unmeasurably small level. This idea might now be tested further with the formalisms presented by Clarke and McGale (1986).

While it is possible to conceive of the cloud postulated for RS CVn to have been constrained by the field lines frozen into the stellar wind of the K star, there is no independent evidence for this concept. More emphatically, Vogt (1980) claimed no evidence for a field in his Zeeman measures of II Peg. It is, however, noteworthy that at spot phase $0.167P$ a conspicuous Hα detail exists in a spectrum of II Peg and this is neither interpreted nor explained away by Vogt. It is of great importance therefore that Kemp *et al.* (1987) were able to show the variability, and thus the intrinsic character, of the V- Stokes parameter for λ And. The reality of the magnetic environment is strongly supported by polarization, at least in this one case. It is almost equally important that there is no time-correlation between the V-parameter and broad-band light variations. Thus, the scattering locale associated with a magnetic loop is not spatially identical with the seat(s) of the light variations.

6.3. Conclusion

It is always hazardous to generalize from a few specific examples. For spotted binaries generalization is almost impossible at the present time. If there is one detail which looks promising, it is the possibility of studying the magnetic properties of spotted stars and their winds but much more survey work needs to be done in order to build upon the data base of Liu and Tan.

7. Hot, Massive Evolving Close Binaries

7.1. General

Because the pioneering surveys concentrated on hot stars in order to map the interstellar polarization, there are many observations of hot evolved close binaries. As may be seen in Figure 6, this effort and additional detailed monitoring have uncovered a significant number of binaries which are polarimetric variables. Randomized selection from the Mathewson and Ford catalogue so as to match the total numbers of III and IV B-type binaries with similar single stars uncovered essentially the same distributions of null measures and detections for the two samples. The single stars, however, commonly do not show polarization variability. For supergiants, polarization variability appears to be about equally common for single and binary stars.

7.2. POSSIBLE MECHANISMS OF POLARIZATION VARIABILITY

De Jager and Nieuwenhuizen (1977) have shown that the levels of stellar wind activity, parametrized as annual mass loss rates, increase significantly with evolution past the main sequence. In the steady state the original model by Brown, McLean, and Emslie may apply. Correction for the finite stellar sizes is, in principle, possible. Insofar as the mass flux would not be isotropic and steady, the arguments of Cassinelli, Nordsieck, and Murison (1987) offer a basis for understanding the observed polarization levels and variations. The same authors point to nonradial pulsations and magnetic fields as possible causes of the mass flux anisotropy. Since, however, close binaries of luminosity classes III and IV are more likely than comparable single stars to show polarization variability, it seems possible that the scattering and polarization for these binaries arise in the locales where the components' winds intersect. Rudy and Kemp (1977) have considered the detection of the polarized component of flux re-radiated from a stellar atmosphere after it had been absorbed from a close companion.

7.3. POLARIMETRICALLY-VARIABLE HOT, EVOLVED BINARIES

Most of these binaries known to be variable are so known from only a limited number of observations. The u Her system, however, has been observed abundantly by Rudy and Kemp (1977), who developed both reflection-effect and gas streaming models to represent their phase-locked measures. They favored the former model as the explanation for the dominant cause of the polarization variability. It is apparent, however, that some version of the scattering-envelope model of Brown, McLean, and Emslie will also do justice to the observations. In view of the variability of the hot component discovered by Eaton (1978) and the (probably real) scatter of the data, the stability implicit in the reflection-effect hypothesis seems unlikely. Since Brown, McLean, and Emslie also represented Rudy and Kemp's (1976) data for AO Cas reasonably with their model, the nearly-completely-ionized envelope for hot, evolved binaries appears to be well-founded. There are, however, ample evidences for non-static envelopes in Snow and Hayes's (1978) and Pfeiffer's (1976) data sets for δ Ori and AO Cas, respectively.

7.4. CONCLUSION

As far as is known at present, a conventional scattering envelope explanation for phase-locked polarization is sufficient to interpret most data. Presumably the envelopes are sustained by the stellar wind activity. The example of AO Cas's sporadic variability is most likely to be traced to a variation away from steady wind flow. Very much more can be done with these systems. For example, the polarization and the electric vector orientation are known to be variable for o Per and ζ Tau. This is undoubtedly to be explained by activity in the shells of one or both stars of these systems but there has never been an inquiry whether these variations require an intrinsic binary ingredient in their explanations.

8. Hertzsprung-Gap Binaries

At present, 43 close binaries in the Hertzsprung Gap (presumably for the first time) have polarimetric measures. Almost all of these have been observed only a few times and the results may be interpreted as either unpolarized within errors or as due to interstellar polarization. The two exceptions to this uniformity are ζ Hor (F4 IV) and HD 16589 (G0 IV). Neither of these is known as a spotted star. The data series for these systems are due to Luna (1981) but no detailed information is available yet.

9. Non Mass-Transferring Cool Giant Binaries

9.1. GENERAL

This is a far-from-homogeneous category, embodying systems with Be companions to cool giants as well as pairs with two cool-evolved stars. Although not the pioneering studies concerning polarimetric behavior of red giants, Shawl's (1975a) work shows clearly that intrinsic polarization is common among such stars. With respect to the distributions between the unpolarized condition and detected polarization which may be interstellar in part, it is again possible to find single stars which match well the close binary cool giants. At present, 73 cool giant binaries are known to have been measured polarimetrically.

9.2. POSSIBLE POLARIZING MECHANISMS

Shawl's (1975b) theoretical work leads one to expect scatterers of iron, silicate, and graphite grains of a restricted range of sizes, and a variety of possible geometries for the grains. Inhomogeneities and wind flow involving these particles are very likely.

9.3. INDIVIDUAL BINARIES

Four intrinsically polarized binaries are known at present. Undoubtedly, many other systems now known to show a polarized signal will be found to be variable when sufficient measures are accumulated. AX Mon has been known to have a polarized signal for a long time. The published data plus a certain number of unpublished measures show a peak-to-peak amplitude of about 0.7% in the long term and there is some meager evidence for a 0.2% peak-to-peak amplitude phase-locked to the Keplerian period. The red-yellow-blue polarization spectrum is itself variable. Wolf's measures of the circular component are not statistically significant. Deshpande, *et al.* (1987) have shown conspicuous variability for both linear polarization parameters of R Aqr with the curious circumstance that the amplitude of variability works in opposite sense with respect to wavelength for the two parameters. These authors present a rich interpretation involving a Mira dust shell, an accretion disk around the putative white dwarf companion to the Mira variable, and transient polar jets associated with the white dwarf. Gravina and Martel (1985) emphasize the intrinsic variability of the perhaps similar object CI Cyg whose polarization has changed tenfold and whose polarization spectrum is itself very variable on a long time scale. The variability of AG Peg (investing a W-R companion) can be judged only from Serkowski's short list of data.

9.4. Conclusion

For neither the amplitude nor the time scale of variability nor the polarization spectral variability is there much to distinguish the close binaries from the single stars in, say, Shawl's (1975a) list. Perhaps this is to due to relatively few data, for the interpretation of the binary character of R Aqr suggests that the variety of phenomena should be much greater for the binaries.

10. Hot Star Plus Cool Star, Mass-Transferring Close Binaries

10.1. General

Very little can be added to the content of the review by Koch et al. (1989) of the polarization and its variability for Algols and their direct antecedents and progeny. In this summary, the supergiant systems will be grouped with the aforesaid systems, for all these pairs are Roche-lobe filling binaries. The tally of these systems thus becomes binaries 20 through 34 of Koch, et al. plus ϵ Aur, ζ Aur, BM Cas, VV Cep, and 32 Cyg. Not all of these have been observed even with sketchy phase coverage but it is safe to say that all, and other objects in similar evolutionary stages, will be discovered to be polarization variables.

10.2. Possible Polarizing Mechanisms

Potentially the sources and seats of scattering and polarization are quite numerous: the Chandrasekhar Effect in the hot atmosphere at a deep primary eclipse; inhomogeneities in the wind from each star of a pair; the shock front at the intersection of the two winds if there are two winds; the gas flow and its inhomogeneities around the gainer star; structure in polar jets which might be associated with the accretion disk around the gainer; and a systemic envelope.

10.3. The Binaries Themselves

Only for β Per has evidence been claimed for the Chandrasekhar Effect (Kemp et al. 1983). A critique of this claim appears in Koch (1989) to which may be added another comment: the variations seen at primary eclipse in the extended seasonal monitoring of the binary at Oregon make this explanation quite unlikely. Another binary potentially favorable for the discovery of the effect is δ Lib. Desultory monitoring of this object at Pennsylvania has, so far, uncovered no evidence for the Effect.

Rich data sets, in the sense of multi-seasonal coverage, exist for ϵ Aur, ζ Aur, VV Cep, β Lyr, and β Per. These should be the best candidates to be examined for the existence of polarizations arising from other seats in the binary systems. It may, however, be remarked that Koch et al. show that the entire sample of systems which they studied, whether abundantly observed or not, demonstrate a correlation between the mass in the stellar wind and the mass of scatterers derived from intrinsic polarization interpreted by the envelope model of Simmons, Aspin, and Brown (1980). Therefore, there is ample support for orbital-plane scattering disks in this type of binary. Koch et al. also show that there is not a random distribution of the scatterers with respect to longitude in the orbital plane and that there is asymmetry of the scatterers with respect to the plane itself. All of these circumstances may be considered as validating the existence of inhomogeneties within the binary envelope and disk. Sorting out the contributions due to wind and Roche overflow causes has not yet been attempted with such weak data sets.

For the well-observed systems, a bit more information may be surmised. In the case of β Per changes in the scattering optical depth of the order of 50% emerge from the comparison of different seasons but the longitudinal location of the inhomogeneities remains fixed within the errors of determination. It might be supposed that a particular seat of enhanced wind flow from the B8 V star has remained variably active for several seasons but a simpler interpretation postulates variable overflow from the cool companion. Such an interpretation would also satisfy the high-frequency variability seen by Piirola (1981) for U Cep. In the case of β Lyr, all observers have documented high-frequency variability and this binary has almost the fastest mass-transfer rate known.

For the cool supergiant systems, ζ Aur and VV Cep, the situation is probably more complicated. In the case of ζ Aur, Koch and Pfeiffer (1982) localize the cause of the scattering to chromospheric inhomogeneities of the K-star. The implication of this conclusion is that overflow to the hot companion plays only a minor role at most. The much more complicated VV Cep-system has been interpreted by Pfeiffer and Koch (1987) to display scattering from the M-star expanding chromosphere, from its overflow to the O-type companion, and from the disk which embeds the hot star. For all these binaries except VV Cep, electron scattering is sufficient to explain the polarization spectra. The spectrum for VV Cep itself requires contributions from variable concentrations of grains whose identities, sizes, and geometries are not well-established. It may also be remarked that despite independent evidence of a magnetic field for β Lyr by Skul'skii (1985), there is no evidence for an intrinsic V-Stokes parameter of polarization for this system.

The astonishing richness in all types of observation of ϵ Aur set this object apart from almost all other close binaries. Most modern models now engross some sort of disk or toroid (embedding a hot star) although the details of the structure vary bewilderingly. The numerous historical polarization measures gave little hint of the cyclical behavior seen during the last atmospheric eclipse by Henson, Kemp, and Kraus (1984). Insofar as may be judged from the brief published details, the polarization spectrum looks white but large variations of the electric vector orientation take place. If there is a characteristic time scale for these phenomena, it is of the order of 100 days. These data do not stand alone for the polarization activity outside eclipse reported by Kemp *et al.* (1986) apparently is associated with a non-radial pulsation of the supergiant.

10.4. Conclusion

It would appear that all possible polarizing mechanisms operate in the entire ensemble of mass-transferring binary systems except for the Chandrasekhar Effect. This single deficit, which after all refers only to a static condition in the atmosphere, would be more than compensated by the verification of a new effect: significant polarization responding to non-radial pulsations.

11. Wolf-Rayet Stars

11.1. General

There are 22 Wolf-Rayet binaries which have been investigated for polarization signals. Of these, all but two are surely variable and thus intrinsically polarized. The interest in these objects rests in their advanced evolutionary states and the possibility of recovering the orbital inclination from the polarization curve. With one unknown fewer, the mass of the Wolf-Rayet object or limits to it and that of its companion can be evaluated.

11.2. Possible Polarizing Mechanisms

The high level of ionization evidenced in the spectra of these stars leads to an expectation of pervasive electron scattering in their envelopes. For some stars dust is known to form in the outer portions of the stellar wind flows.

11.3. Individual Binaries

The most recent comprehensive study of Wolf-Rayet stars is that by Schulte-Ladbeck and van der Hucht (1989). These authors conclude that phase-locked variability is the common pattern for both WC- and WN-types. Further, even for the incoherent variability of W-R 103, a preferred symmetry plane could be found. Variability on an assortment of time scales is possible for some of these systems as is indicated by the behavior of the curious object, EZ CMa, observed by Drissen *et al.* (1989). It is also of interest that the multi-bandpass polarimetry which does exist shows flat spectra. The multi-bandpass data for θ Mus by Serkowski yield a polarization spectrum which decreases monotonically from visual through blue to violet, but there is little additional information concerning non-electron scatterers. It is also of interest that Drissen *et al.* found no circular component to the polarization of EZ CMa.

11.4. Conclusion

The Wolf-Rayet stars are just now attracting the observational attention which they merit. Potentially, the polarization spectrum can tell much about the composition of the wind flow and, inferentially, about past nucleosynthesis of these stars. If they really are to emerge as white dwarfs or neutron stars, it will be very useful to extend the circular-component measures to as many of the sample as possible so as to look for evidence of the progenitor state of magnetized collapsed objects.

12. White Dwarfs in Binaries

12.1. General

After it was accepted that all or nearly all novae result from a thermonuclear event on or near the surface of a white dwarf, it was logical to see if polarized signals could be detected from these sources. Additionally, the surveys with X-ray satellites and follow-up optical searches rather quickly resulted in the identification of numerous X-ray sources with mass-transferring binaries in which the flow passes from a non-degenerate star to a white dwarf. Since many of these systems are faint, fairly large optical telescopes are necessary to study them in detail. This opportunity did eventuate but a number of the systems have not yet been observed in filtered mode. By now, 32 binaries containing such a white dwarf are known to have been polarimetrically observed and intrinsic polarization has been associated with all but five of these: T CrB, RS Oph, RR Pic, V471 Tau, and BE UMa. These systems are probably not really anomalies for they have not been observed repeatedly although it must be admitted that the entire variety of novae phenomena is probably not yet known. Quite apart from the motivation to understand the physics of the flow and accretion processes, there exists the possibility that the orbital inclination may be recovered from the polarimetry, this parameter being otherwise inaccessible except in an eclipsing system.

12.2. Polarizing Mechanisms

Although not all conceivable structures occur in a given system, the mass-transferring stream, the accretion disk and/or column(s), and the stellar or systemic envelope undoubtedly oblated by the fast Keplerian motion all offer asymmetric environments with sufficient populations of scatterers so as to yield an intrinsic and time-varying polarization. It may well be, however, that there are not enough photospheric photons scattering off the stream and disk to create a measurable signal. However, dust formation during the post-maximum interval of a nova explosion is very likely and image morphology makes it clear that the ejecta are not dispersed isotropically. These opportunities have been overshadowed, however, by the discovery of mega-gauss fields in white dwarfs creating a natural cyclotron-like environment. The motivation to accumulate both linear and circular polarization measures has, therefore, been strong.

12.3. INDIVIDUAL WHITE DWARF BINARIES

With regard to the evidence for dust formation, V1668 Cyg offers fairly clear evidence. The polarization spectrum by Blitzstein *et al.* (1980) from before the report of dust formation contrasts convincingly with that by Piirola and Korhonen (1979) taken after that epoch. A significant rotation of the electric vector also characterizes these data sets. Unfortunately, no near-IR measures exist so that no detailed interpretation of the grains may be attempted.

Insofar as the Blitzstein *et al.* data can have a interstellar polarization accurately removed from them, the residue from the subtraction could represent the polarized signal from the fraction of the disk and envelope that survived the nova event itself. This, however, is impossible to prove because the data are so few. Evidence on this matter is possibly provided by the non-null observations of RS Oph and V471 Tau all taken when the stars were quiescent. Comparison of Serkowski's and Schulte-Ladbeck's (1985) data for RS Oph indicate the possibility of a polarized signal derived from scattering in a stable disk and/or envelope and Kemp and Rudy's (1975) measures of V471 Tau show the same possibility.

The state of information as of 1985 with respect to polarization arising from magnetic white dwarfs was summarized by Koch (1988). Since that time, study of these objects has continued at a busy pace, 13 of them having been the subjects of at least one investigation in the intervening five years. One serendipitous consequence of the effort has been to increase the statistical weight of information regarding intermediate and strong polars. For instance, Morris *et al.* (1987) comment on the concentration of AM Her stars near the 115-min. period. Also, Berriman and Smith (1988) show that three eclipsing AM Her stars have now been discovered. With the natural experiment of eclipses running their courses, it is possible to sharpen models of these systems by a large factor. The polarization of AM Her itself, observed in the faint state by Bailey, Hough, and Wickramasinghe (1988), bears a certain spectral resemblance to the intermediate polar BG CMi as observed by West, Berriman, and Schmidt (1987). It is futile to try to recite all the polarization patterns which have been observed for these systems, but it is useful to point to the comprehensive observational base developed by Piirola, Reiz, and Coyne (1987) for EF Eri as the level of information which is desirable in order to attempt to model one of these binaries and the changes which occur in them.

A significant advance exists in the interpretation of V1500 Cyg by Stockman, Schmidt, and Lamb (1988). This ex-nova, historically very well documented, has been shown to modulate in circular polarization with a period significantly shorter than the photometric period, in distinction to the typically synchronously- rotating AM Her-type variable. Stockman, Schmidt, and Lamb develop a picture wherein the system was a normal AM Her variable before outburst, albeit with very large magnetic obliquity. By an intricate coupling during and after outburst among the white dwarf, the systemic envelope, and the companion star leading to angular momentum loss, the white dwarf developed its present fast spin. Obviously, recurrent novae can test this model in detail.

12.4. Conclusion

If it is really true that the AM Her-phenomenon is always associated with the pre- or post-outburst state of novae, there exist many opportunities to study magnetically-guided accretion and its consequences for magnetic white dwarfs.

13. Neutron Stars

13.1. General

There are now known 13 neutron star binaries which have been observed polarimetrically - an increase of just a few in five years. The linear polarization of all the objects is variable but so far only two of them have shown variable circular signals. These binaries are also distinguished from those of preceding types by having a certain number of X-ray and radio polarimetric measures. Just as for white dwarf systems, facing of the phase-locked polarization variations against a convincing model is an effective way to evaluate a system's orbital inclination.

13.2. Polarizing Mechanisms

The first principles of the understanding of neutron stars in close binaries lead to easy acceptance of scattering in accretion disks and in polar jets. The likelihood of magnetic environments is also very high and extended scattering envelopes around the B-type supergiant companions provide still another seat of possible polarization.

13.3. Individual Neutron Star Binaries

The few objects which are identified at present include some of the most famous (and thus strongest) sources in the X-ray sky: SS433, Cen X-3, Cyg X-2, Cyg X-3, Her X-1, Sco X-1, and Vela X-1. Since not all of these can be described in detail here, a few examples will be emphasized.

Consider, first of all, the results of Dolan and Tapia (1988) for V884 Sco and GP Vel. In the case of V884 Sco sporadic variability and the lack of a theoretical wave form for one of the Stokes parameters prevented very detailed discussion of the data. Nonetheless, it was possible to show that the dominant seat of the scatterers is in the stream transferring mass from the B-star to the neutron star and that in the stream recombination leads to Rayleigh scattering. For GP Vel not all the measures can be phase-locked onto one cycle showing that a Keplerian period is a characteristic time scale for changes in the scattering effect. The polarization spectrum is directly interpretable as due to Thomson scattering and the greatest concentration of scatterers lags (in the sense of Keplerian revolution) behind the neutron star. The dominant source of the scatterers is the B-star's stellar wind, rather than Roche lobe overflow and the sporadic variability is ascribed to inhomogeneities in the wind.

For the more complicated case of V1343 Aql, low-amplitude, phase-locked optical variability derives from a binary wherein an accretion disk and polar jets both precess with a period longer than the Keplerian one. The very detailed polarimetry of this object by Efimov, Piirola, and Shakhovskoy (1984) is encumbered by a large interstellar component. The dominant component of the intrinsic signal has been interpreted as arising not in the jets but from a concentration of scatterers flattened to the orbital plane. The lesser fraction of the signal, that of the second harmonic, likely derives from the accretion disk. Null (within their errors) circular measures forbid an interpretation based on a cyclotron effect associated with any magnetic field supporting and driving the jets.

13.4. Conclusion

It is clear that binaries with a neutron star member present a wealth of polarimetric detail not all of which can be understood at present. Models lack the fine structure to account for much of the observational information. There is no doubt that the motivation to work on such objects will continue unabated for a long time not only in order to eke out all latent information but also because the technique remains the most powerful way to evaluate the orbital inclination and free the projected mass and mass function values from its effect.

14. Black Hole Binaries

There must be many such systems but the case for Cyg X-1 = V1357 Cyg remains far the strongest. At this time, informed opinion is surely on the side that this system does contain a stellar black hole. There are a contentious few years of history concerning the orbital inclination of the binary for the obvious reason that the mass of the compact object hinges critically on this value. Not only has the inclination itself been argued fervently but the error limits on it have been the subject of just as much concern. The most recent polarimetric series, that of Dolan and Tapia (1989), leads to a rather modest inclination and assurance of a black hole structure but it must be admitted that phase coverage is not so detailed as one would wish.

Better phase coverage is, in fact, vested in the polarization curves of Kemp (1980) but the subtlety of the situation is well exemplified by Dolan and Tapia's conclusion that Kemp's analysis and the revisionist one by Simmons, Aspin, and Brown (1980) are both subject to the unavoidable limitations of observing opportunities and subsequent data handling procedures induced by the intrinsically-variable polarization. While all this may certainly be true, it is impossible to prove that the Dolan and Tapia series is not afflicted with variability running its course on time scales other than the Keplerian one.

15. Summary

From the foregoing, several generalizations may be drawn. Scattering and polarization effects in stellar envelopes have been understood in part only for the most extended layers of the largest close binary supergiants and have not even been measured for other systems. A considerable measure of interpretative success is associated with scattering from magnetically-guided mass flow in systems with compact components. Lesser observational efforts for non-degenerate binaries, for spotted-star systems, and for contact pairs have not yet yielded satisfaction at the desired level. No doubt, these binaries will be studied more intensively in the very near future. The polarimetric technique has almost nothing to say about stellar layers where optical depth is high since multiple scattering typically depolarizes the radiation. However, information concerning these layers is commonly a most important guide for interpreting the polarimetric data themselves. A closing remark may, therefore, be appropriate: any student of the polarimetric properties of close binaries must be completely informed on the other observational and analytical methods by which information is to be gained and has already been gained for these stars.

Acknowledgements

I am indebted to R. J. Pfeiffer for reading the manuscript before it was submitted to the Editors and to R. J. Mitchell for preparation of the Figures.

RADIO EMISSION FROM ACTIVE LATE-TYPE BINARIES

N. M. ELIAS II and R. L. MUTEL

1. Introduction

Radio emission is one of several indicators for enhanced stellar activity connected with strong magnetic fields. Recent advances in receiver and telescope technology have allowed radio detections of stars distributed nearly everywhere on the H-R diagram. However, the majority of stars with detectable radio emission are in close binary systems, usually with at least one component cooler than early F spectral type (e.g., RS CVn and Algol binaries). Except for an occasional reference to other stellar types for comparison, we will restrict our focus to these close binaries. For a more general discussion of stellar radio emission, see the excellent reviews by Dulk (1985) and Hjellming (1988).

In this chapter, we will address two categories of questions which the radio observations raise. The first category relates directly to the properties of the radio emission itself:

1. What are the physical mechanisms responsible for the observed radio emission? Do the emission mechanisms vary with spectral type?
2. Is radio luminosity correlated with other stellar properties, such as orbital period, spin period, rotational velocity, Rossby number, H+K line luminosity, C IV line luminosity, and X-ray continuum luminosity?
3. What do multiwavelength observations of close binaries tell us?
4. What types of close binaries have detectable radio emission? What is the probability of radio emission for each type of binary?

The second set of questions is more interpretative and/or speculative:

1. What role does binarity play in the radio emission process?
2. What do the observed properties of radio emission tell us about stellar magnetospheres?
3. What critical observations are needed to further our understanding of the physics of stellar radio emission?

This chapter is organized such that each section addresses one of the above questions in turn. For easy reference, we have summarized recent radio observations of individual stellar classes in an Appendix.

2. Emission Mechanisms

Late-type binaries emit cm radiation by several different mechanisms. For example, RS CVn and Algol systems exhibit sporadic, nonthermal synchrotron flares from their chromospheres and transition regions (TRs), supposedly directly above their photospheric active regions, namely starspots and faculae (Gary, Linsky, and Dulk 1983; Mutel *et al.* 1985). The flare scenario is described as follows. A magnetic coronal loop, which has its feet in the active regions of the photosphere, expands and "reconnects", ejecting electrons that produce the nonthermal synchrotron radiation. The spectrum of this radiation decreases sharply with increasing frequency. The radiation is also highly polarized (sometimes almost as much as 100%), and the brightness temperatures are usually greater than $T_B \sim 10^{10}$ K. As a matter of fact, flares have been observed at 70 and 20 cm with $T_B \sim 10^{16}$ K (Bastian *et al.* 1990). Flares tend to be infrequent, which means that gathering enough observations for good statistics is not easy.

RS CVn and Algol systems also produce quiescent radio emission within their TRs and extended coronae; some of these emitting regions have been imaged directly by VLBI (Mutel *et al.* 1984; Mutel *et al.* 1985; Lestrade 1988). Models for this type of emission have been proposed by Drake, Simon, and Linsky (hereafter 1989; DSL) and by Morris, Mutel, and Su (hereafter 1990; MMS). Brief descriptions of these models follow.

The DSL model is based on comparisons between the 6-cm luminosity of the active component (normalized to the bolometric luminosity) and the CIV line luminosity/X-ray luminosity (cf. Section 3). Because all of these luminosities appear to be correlated, DSL deduce that the emitting region is the same for all three wavelengths. They also state that the radio emission mechanism is thermal gyrosynchrotron radiation, with a brightness temperature $T_B \sim 5 \times 10^7$ K, an electron number density $n_e \sim 2 \times 10^8$ cm^{-3}, and a magnetic field strength $B \sim 200$ G.

The MMS model is not entirely consistent with the DSL model. MMS suggest that 1) the radio emission mechanism is nonthermal gyrosynchrotron radiation from power-law electrons (at cm wavelengths, the *nonthermal* gyrosynchrotron spectrum is not too different from the *thermal* gyrosynchrotron spectrum in the DSL model, so they are difficult to distinguish), 2) the radio emitting region is confined to a torus (by the dipolar magnetic field of the active star) *separated* from the X-ray emitting region which surrounds this torus, and 3) the emitting region is axisymmetric. This model was created in order to account for various properties of the radio emission, such as the lack of correlation between radio luminosity and orbital phase (Feldman *et al.* 1978), an increasing circular polarization with decreasing inclination angle (a result of the dipolar magnetic field; Mutel *et al.* 1987), a polarization reversal between 20 and 6 cm (due to the emission and absorption coefficients for each mode; MMS), and correct radio and X-ray luminosities (Majer *et al.* 1986). Additional details concerning this model are described in Section 7.

It should be pointed out that thermal/nonthermal gyrosynchrotron radiation has been observed in star systems other than RS CVns and Algols. These systems include "pre-Algols" (e.g., β Lyr, and the 'emission-line Algols' RZ Sct, W Ser, and V367 Cyg; Wade and Hjellming 1972; Elias 1990), Ap/Bp stars (Drake et al. 1987), weak-lined T Tau stars (Phillips, Lonsdale, and Feigelson 1991), and a few O-type stars (Abbott, Bieging, and Churchwell 1984, hereafter ABC; White 1985; Abbott et al. 1986; Persi et al. 1990). Actually, there has been some controversy concerning the cause of gyrosynchrotron emission in O-type stars. The models which have been put forward to explain this phenomenon include accretion onto an early-type, non-compact companion and turbulence/shocks in the stellar winds.

Lastly, for completeness, we mention another cm emission mechanism, thermal bremmstrahlung. It is produced by thermal electrons with temperatures ranging from 10^4 to 10^5 K. This emission mechanism is observed in the extended winds of early-type (O˙ and WR) single and binary stars (ABC; Abbott et al. 1986; Persi et al. 1990). It is also observed in the winds and mass transfer regions of symbiotic binaries (Seaquist and Taylor 1987; Kenny, Taylor, and Seaquist 1991).

3. Radio Luminosity - Stellar Correlations

Of all classes of late-type stellar radio emitters, only the RS CVn binaries have been studied enough such that correlations can be determined. These correlations are useful because they allow for the comparison of various physical phenomena as well as the relationships between them. Morris and Mutel (1988) observed 103 candidate stars, of which 53 were detected. They found a correlation between the radio and bolometric luminosities, but state that it may be the result of selection effects (distance, flux limits, etc.). They did not find a correlation between the rotational period of the active components for the entire sample of RS CVns, but for a subsample of binaries with periods less than 10 days, they found a weak correlation.

Morris and Mutel also sought correlations of radio luminosity with $B - V$ and Rossby number (the ratio of the rotational period of the active star to its convective turnover time). There appears to be a good correlation between the radio luminosity and $B - V$ for the active components (the stars with strong H+K emission) of the binaries while there was no correlation for the other components. This implies that the radio emission is associated only with the active members of these systems. There was no correlation with the Rossby number, which suggests that the stellar dynamos, which supposedly lead to the radio emission, are not well understood.

DSL also studied the relationships between radio emission and stellar parameters for RS CVn binaries. For example, they have determined that there is no correlation between the 6-cm luminosity and the orbital or rotational periods. They have found, however, that there is a weak correlation between the 6-cm luminosity and the rotational velocity (to the power 1.0 ± 0.3). They found much stronger correlations using the 6-cm luminosity normalized to the bolometric luminosity; it is strongly correlated with the orbital period (power -0.9 ± 0.2), the rotational velocity of the active component (power 1.4 ± 0.3), and the orbital period (for giant and subgiant systems only, power -1.1). These authors also note that the correlations with orbital period are the same for both giants and subgiants (the active components), but correlations with rotational period are *not* the same. All of these correlations (and lack thereof) imply that it is the rotation of the active star which is important for radio emission, and that the radio emission depends heavily on the conditions (internal and external) of the active component.

H+K emission lines (chromosphere), CIV emission lines (TR), and X-ray continuum (corona) have been suggested as indicators of stellar activity (Morris and Mutel 1988; Vaiana 1983; Basri, Laurent, and Walter 1983; Gilliland 1985), which implies that they may be reasonable indicators for radio emission as well. The rotationally driven dynamo is assumed to produce magnetic fields that transport energy to the solar atmosphere via MHD waves or other mechanisms (Spruit and Roberts 1983; Schüssler 1983; Gilman 1983). While single late-type stars show correlations between rotation and activity at many wavelengths, Majer *et al.* (1986) suggest that RS CVns do not follow these trends. However, Morris and Mutel (1988) have found correlations between cm emission with H+K emission, and DSL have also found strong correlations between the normalized 6-cm flux with C IV emission (power 1.4 ± 0.1) and X-ray continuum (power 2.2 ± 0.3). At present, it is not clear why all these results are contradictory. These correlations were used to create a model for the cm, ultraviolet line, and X-ray emitting regions, which were assumed to be the same (cf. Section 2).

4. Multiwavelength Observations

Multiwavelength observations are becoming more and more common in observational stellar astronomy because they allow study into the nature of stellar activity and the relationships between different physical phenomena. Several examples of multiwavelength observing programs follow.

Van den Oörd *et al.* (1988) and van den Oörd *et al.* (1989) observed Algol and σ CrB at both X-ray and radio wavelengths; the Algol observations will be described here. The X-ray observations were performed using EXOSAT, and the run lasted for 35 hours. Low-level variability was observed, but there was no evidence for variability due to primary eclipse, suggesting that the X-ray emitting region is comparable in size to that of the orbit. The temperature of the emission was about 2.5×10^7 K and the total luminosity was 5×10^{30} erg s^{-1} from 0.1 to 10 KeV. Also, a flare was observed which lasted about 8 hours; its temperature was 6×10^7 K and its luminosity was 1.4×10^{31} erg s^{-1}. The length of the flare suggested that radiation cooling dominated.

The radio observations were performed at 20 cm. Two hours of these cm observations were coincident with the quiescent X-ray emission, while four more hours were coincident with the first half of the flare. During the first observation, there was a flux of ~ 47 mJy and no detectable circular polarization. At the beginning of the flare, the flux was about 30 mJy and rose steadily to about 48 mJy and declined back to about 39 mJy. The circular polarization was about 10% and declined to about 3%. The authors modelled the quiescent radio emission as optically thick synchrotron radiation, and the flaring emission as synchrotron emission by a small population of electrons accelerated to MeV energies. The observed variability of the radio and X-ray radiation suggests that the emission regions were cospatial.

Linsky et al. (1989) observed V711 Tau at radio, optical, and ultraviolet wavelengths. The observed chromospheric and TR emission lines indicated that the flare occurred near the K subgiant. The electron number density of the flaring plasma was about 10^{11} cm^{-3} (an order of magnitude larger than normal) and the temperature was about 6×10^4 K. The emitting volume of the flare was approximately 2×10^{30} cm^3, about 200 times smaller than the quiescent corona. The footprint separation was about 5×10^{10} cm, which was about 1/5 the size of the corresponding starspot observed by visible-band photometry. The luminosity of the flare was of order 10^{32} erg s^{-1}. The temporal variability of the 4.7-cm emission was comparable to that of the ultraviolet emission, which implies that the emission regions were cospatial. The brightness temperature was assumed to be 2×10^{10} K, which corresponds to an emission region 1.4 times the surface area of the starspot, containing mildly relativistic electrons producing gyrosynchrotron radiation. Using this information, Linsky et al. created a model consisting of a starspot permeated with magnetic flux tubes. Near the footprints, the flux tubes are filled with $\sim 10^5$ K plasma, producing the ultraviolet emission. The rest of the volume contains 1.7 MeV electrons which produce the radio emission. Note that the size of this flaring region is smaller than that of Algol described in the last paragraph.

Ultraviolet and radio observations were performed by Lang and Willson (1988) on the RS CVn binaries UX Ari and HR 1099. They detected flare activity both in the ultraviolet and in the radio, but never simultaneously, suggesting that the ultraviolet and radio emission regions for these stars were not the same. These results are different from the V711 Tau results of Linsky et al. (1989), implying that there may be several different flare scenarios for RS CVn binaries.

5. Radio Emission by Binary Type

Late-type binaries (containing a G, K, M dwarf, subgiant, or giant) emit radio radiation via mechanisms involving electrons moving through magnetic fields. This radiation may be either quiescent or flaring. Detailed descriptions of these mechanisms may be found in Section 2.

As a group, late-type binaries have low to moderate detection rates. DSL compiled a list of all VLA observations of RS CVn objects and found that the probability of detecting these binaries at least once was about 50%. This percentage, however, does not take into account selection effects such as distance, but it is a useful number because it shows that there is a good chance for detecting additional stars in future radio surveys. Stewart *et al.* (1989) have published a survey of Algol-type binaries. They reconfirmed earlier results by Slee *et al.* (1987a) that 14 out of 47 binaries in the survey were radio sources. The Algols are divided into two catagories, EA1 and EA2. EA1 binaries with two early components; they emit nonthermal radiation. EA2 binaries are similar to Algol, with a hot and cool component (binaries referred to as "Algols" are by default EA2). The cm emission mechanism for EA2 binaries is similar to that of RS CVn binaries, and the detection rate for EA2 is about twice as high as for EA1 (neglecting selection effects).

Seaquist and Taylor (1990) have recently summarized the radio properties of symbiotics based on a large VLA survey of essentially all systems visible from the northern hemisphere; about 25% of the 107 systems were detected. The emission mechanism is believed to be thermal bremmstrahlung. Similarly, O-type stars, whether single or binary, emit thermal bremmstrahlung (sometimes nonthermal radiation as well), within their winds. As far as we know, no surveys of O-type stars have been published, so it is difficult to say what the detection rates for these stars are.

Although most Ap/Bp stars are either single or in wide binaries, we mention that Drake *et al.* (1987) published the results of a survey of 34 Ap/Bp stars; 5 of these stars were detected (all Bp, detection rate 14%). Again, this detection rate may not be accurate because the sample is small and it does not take into account selection effects. The cm-emission mechanism, however, is believed to be similar to the quiescent emission of late-type binaries.

6. The Role of Binarity in Quiescent Radio Emission

Up to and including 1980, about 700 stars were observed at radio wavelengths. Gibson (1980) took a subset (201 stars) of these observations (subject to criteria which reduced systematic biases) to classify the radio emission of these objects. He concluded that stars (single or binary) were much more likely to be radio emitters if they have at least one component later than F5 (mainly RS CVns and Algols). He attributed this result to stars that exhibit what he has termed "stellar magnetic activity". His analysis also suggests that binary stars are more likely to be radio sources than single stars, although this result is not so conclusive as the former.

Since 1980, many additional stars have been observed in the radio. Most of these observations are better than the earlier ones because they were performed with the VLA (better receivers, reduced source confusion, etc.). These additional observations include those of DSL and Morris and Mutel (1988) (RS CVns); Stewart *et al.* (1989) (Algols); Drake *et al.* (1991) and Elias and Dorren (1990) (single, active, late-type stars). The detection rates for the RS CVn and Algol binaries are fairly high (cf. Section 5). The single-star observations, however, have yielded no detections. These results seem to imply that binarity is an important condition for cm emission for late-type stars, consistent with the statements of

Gibson. It should be mentioned here that we have not listed FK Com stars (Hughes and McLean 1987) or T Tau stars (Phillips, Lonsdale, and Feigelson 1991) because these are extreme cases of single-star activity: FK Com stars are single, late-type stars which show much activity because of their very rapid rotation (they may, in reality, be the result of two stars merging), and T Tau stars are rapid rotators which have not reached the main sequence.

The magnetic activity of the so-called active, late-type stars is believed to be the result of the 'dynamo mechanism', which is very vigorous convection in the layers underlying the photosphere (due to rapid rotation), which in turn produces active regions (starspots and faculae) on the surface of one or both of the stars. RS CVn and Algol binaries are believed to have rapidly rotating components, presumably produced by tidal interactions. Young, late-type, single, main-sequence stars rotate quickly as well, because the effects of magnetic braking have yet not slowed them down (e.g., HD 129333; Dorren and Guinan 1991). Why then is binarity important? How do single stars and binary stars differ?

There are two main differences between single and binary late-type stars, either or both of which may be the correct answer to the above questions. Firstly, binaries experience tides due to the stars' mutual gravitational interaction. This interaction surely affects the dynamo mechanism, but it is not clear at this time what the effects are. Secondly, the amount of circumstellar material (most importantly, electrons for our purposes) in binaries is most likely greater than for single stars. The mass-loss rates of single, main-sequence stars are so small ($\sim 10^{-14} M_\odot$ yr^{-1}; calculated from values listed in Allen 1973), that they barely affect their evolution and add little to the circumstellar environment. The mass-loss and mass-transfer rates for Algols are typically larger because the cool component has filled or is near filling its Roche lobe. This assertion is confirmed by period studies (Koch and Guinan 1978) and visible-band polarimetry (Koch *et al.* 1989). Additional circumstellar material (mainly dust, but possibly electrons as well) is difficult to explain for RS CVn binaries because neither component has reached its Roche lobe, but there is evidence that it exists from visible-band polarimetry (Coyne *et al.* 1991) and far-infrared photometry (infrared excesses; Busso *et al.* 1987).

Radio observations of early-type stars (ABC), however, indicate that binarity is not so important. The reason for this is as follows. The radio emission from these objects is mostly thermal bremmstrahlung from expanding ionized winds, independent of any companion. The only exception to this rule may be the nonthermal radiation observed for some of these stars (ABC; Abbott *et al.* 1986; Persi *et al.* 1990). It has been suggested that this nonthermal radiation may be produced by shocks within the winds of a single star (White 1985), but it has also been suggested that accretion onto non-compact, less luminous companions (ABC) may produce it. Visible-band polarimetry of the luminous, close O-type binary DH Cep (Corcoran 1988) suggests that interactions between the winds of the two stars may be another source of nonthermal cm radiation.

7. Magnetospheric Models

Radio observations of active binaries provide valuable constraints on physical properties of the emitting region. This is because in most cases the emission mechanism is well understood, allowing reliable estimates of the magnetic field strength and electron number density from the observed spectrum, polarization, brightness temperature (with VLBI) of the radio emission. Any realistic model of the extended coronae of late-type binaries must satisfy the following observational constraints:

Overall size. For the RS CVn and Algol systems, extensive VLBI observations (e.g., Mutel *et al.* 1985; Lestrade 1988; Lestrade *et al.* 1988) show that the quiescent emission arises from a region comparable with the size of the binary separation ($a \sim 10^{11-12}$ cm). Furthermore, for eclipsing binaries there is no evidence for a reduction in the radio light curve during eclipse (Feldman *et al.* 1978; Doiron and Mutel 1984). This is also true of X-ray observations of the high-temperature ($T > 10^7$K) component of eclipsing binaries such as Algol (White *et al.* 1986).

Magnetic Geometry. For the RS CVn and Algol systems, it has been determined that the amount of circular polarization increases with decreasing orbital inclination, i (Mutel *et al.* 1987). This effect is produced by a large, time-invariant, axisymmetric, dipolar magnetic field of the active star. For a binary at $i = 0^{\deg}$, the magnetic field lines from only one pole are visible, which means that there will be a large circular polarization signal. If i is increased, however, the magnetic field lines from the other pole increasingly come into view, causing partial cancellation of the circular polarization signal until total cancellation at $i = 90^{\deg}$.

Polarization Reversal. For RS CVn systems, reversals in circular polarization have been observed between 20 and 6 cm (MMS). These are produced by changes in the emission and absorption coefficients for each mode (ordinary, o; extraordinary, e) with wavelength.

MMS have proposed a model in order to satisfy the above conditions. In this model, there are two distinct regions, a "dead zone" and a "wind zone". The dead zone is the inner toroidal region of both thermal and relativistic plasma; the plasma is trapped by the dipolar magnetic field lines ($B \sim 10$ to 100 G). The electron number density of the thermal plasma in the dead zone is of order 10^9 cm^{-3}, and the relativistic plasma is a small fraction of this number. The temperature of the thermal plasma is of order 10^7 K. The wind zone is the region where thermal plasma escapes from the system via radial magnetic field lines. The cm emission is believed to be produced by the relativistic plasma in the dead zone, *distinctly separated* from the thermal bremmstrahlung observed at X-ray wavelengths in the wind zone. The spectra of the DSL model and the MMS model are similar at the cm wavelengths of interest, but the MMS model appears to be better because it is consistent with the polarization observations, including polarization reversals between 20 and 6 cm.

Although the above discussion applies mainly to RS CVn and Algol binaries, it may apply to other systems as well. For example, Havnes and Goertz (1984) successfully used a simple axisymmetric "stretched" dipolar magnetospheric geometry to model the hydrogen line emission and infrared excesses from Ap/Bp stars. This same magnetosphere is believed to produce the observed radio emission of Ap/Bp stars (Drake *et al.* 1991). Also, similar conditions may exist in the emission-line Algol binaries (Elias 1990), all of which may have evolved, late-type components.

8. Suggestions for Future Observations

Surveys are the first step in the study of any class of objects, which includes, of course, radio stars (radio surveys are often called "detection experiments"). Most proposals for radio detection experiments of stellar objects use activity at other wavelengths as their scientific justification. Indicators of possible radio emission for late-type binaries include the manifestations of magnetic activity, such as starspots, flares observable at all wavelengths, strong H+K emission, strong ultraviolet emission (C IV, etc.), and above normal X-ray radiation. There are also indicators used for other types of stars. For instance, to search for quiescent radio emission from Algol and pre-Algol circumbinary environments (accretion disks and circumbinary envelopes) the criteria typically used are strong ultraviolet emission lines, strong hydrogen emission lines, active light curves, variable optical polarization, and period changes (which are produced by mass transfer and/or mass loss; Pringle 1985). The same criteria have been applied to other binaries, such as the so-called symbiotic binaries.

Many radio star detection experiments have been performed in the past. We present here a brief list:

RS CVn binaries. Morris and Mutel (1988); DSL.
Algol binaries. Stewart *et al.* (1989).
'Peculiar emission-line Algol' binaries. Elias (1990).
'Symbiotic' binaries. Seaquist and Taylor (1990).
BY Dra objects. Caillault, Drake, and Florkowski (1988).
Late-type dwarfs. Drake *et al.* (1991).
M dwarfs. Caillault (1989)
Flare stars. Bastian, Dulk, and Slee (1988).
DQ Her novae. Bookbinder and Lamb (1987).
X-ray binaries. Nelson and Spencer (1988).
Late-type giants. Drake, Linsky, and Elitzur (1987); Slee *et al.* (1989); Drake *et al.* (1991).

If the candidates for a survey are numerous and well chosen (to reduce biases due to distance and flux limitation), it may be possible to determine the true fraction of radio-loud systems. Also, period-luminosity relationships, characteristic luminosities, maximum possible distances for detection, etc. may be determined if the distances to the systems are known well enough.

Of the approximately 300 known RS CVn binaries, about 120 have been observed at radio wavelengths (DSL). Although most of the nearby RS CVns have already been observed, there is no reason that a detection experiment containing some of the presently unobserved stars could not be performed. More detections (or even a lack of detections) could yield useful information on the statistics of these objects. Also, there are many late-type dwarfs that have not been observed at cm wavelengths. If we want to study the question of binarity and its relationship to radio emission, this is a definite must. Similar questions may be answered by additional surveys for all the stellar types mentioned above.

Stars from detection experiments which are strong radio sources should reobserved for additional information ("strong" is a vague term, because it depends the type of observations desired). For example, multiple observations of a single source show short- and long- term variability; short-term variability may be a good indicator of source size and intrinsic variation (which includes flaring), and long-term variability may show orbital modulation or stellar cycles (not necessarily solar-type cycles). Radio spectra are also important in the study of radio stars, because they give clues to which emission mechanism(s) is (are) operating. Long-term monitoring of radio stars has shown that different mechanisms turn on and off with time (e.g., O-type stars; ABC). A good discussion on the spectra produced by different radiation mechanisms has been written by Dulk (1985). If a stellar radio source is stronger than about 1 mJy, VLBI observations can be performed to search for structure at the milliarcsecond (mas) level and reveal information about the brightness temperature of individual sources (*if* they are resolved; brightness temperatures also provide clues for radiation mechanisms). Lastly, multiwavelength observations can provide additional information about the properties of stars, including the physical relationships between the various emission mechanisms and emission regions (cf. Section 4).

What radio stars need further observations (for variability, cm spectra, multiwavelength campaigns, etc.)? The simple answer to this question is *all* of them, but we will go into some detail here. We are just beginning to understand both the quiescent and flaring emission of late-type binaries (RS CVns, Algols, and BY Dras). Further observations will help us create better models for the cm activity. Also, if any single, solar-type stars are detected at cm wavelengths, we may use the information to understand our own Sun and its evolution better.

The evolution of massive, close binaries is, at present, not well understood. One possible scenario is a β Lyr-type object losing mass to become a emission-line Algol object and finally an Algol-type object (Sahade 1987). Additional radio surveys and observations of confirmed radio sources, as well as observations at different wavelengths, will definitely help us make better models. Additional radio observations of symbiotic stars promise a wealth of additional information, especially in light of the AG Peg results of Kenny, Taylor, and Seaquist (1991).

Their cm observations using a variety of VLA array configurations found a multishell structure surrounding the star, allowing them to "look back in time". Also, recent X-ray results (Hjellming et al. 1990a; Hjellming et al. 1990b) have found anti-correlations between their radio and X-ray emissions (cf. Section 7 of the Appendix). Further observations will allow for better modelling. O-type stars (single and binary) have been observed in the radio. The scientific justification for radio observations of these stars include hydrogen emission and P Cygni lines from their winds, as well as variable polarization at optical wavelengths. Nonthermal emission has been observed for some of these stars, and several possible explanations have been suggested (see above). In order to choose between these alternatives, it is necessary to search for this phenomenon in other stars and continue observations of suspect stars for confirmation.

Appendix
Summary of Radio Observations by Class

CONTACT BINARIES

Contact binaries (e.g., W UMa systems) are mostly evolved systems which are sufficiently close so that both components exceed their Roche lobes. They have periods between 6 to 15 hours. There is well established evidence for an active, high temperature chromosphere and corona, starspots, intense photospheric and chromospheric lines, as well as enhanced radio and X-ray emission. The reader is referred to two excellent reviews by Rucinski (1985a, 1985b) which summarize the observational and theoretical properties of this class.

The first report of radio emission from contact binaries was a 6-cm VLA survey of 12 eclipsing systems by Hughes and McLean (1984). They detected two systems (VW Cep, V502 Oph), and obtained upper limits for ten others in the range $3 \times 10^{13} < L_R < 10^{16}$ erg s^{-1} Hz^{-1}. The detection of V502 Oph was marginal ($S = 0.4 \pm 0.13$ mJy, $L_R = 4 \times 10^{15}$ erg s^{-1} Hz^{-1} at 91 pc.) and only at a single epoch. Surprisingly, the source was double, with a separation of 2."6 (220 AU at 91 pc). The closer component was 0."6 from the expected stellar position. While either or both sources could be extragalactic, Hughes and McLean suggest the farther component is an as yet undetected companion which is accreting mass from the stellar wind of the primary. A similar model has been proposed for the double radio source α Sco by Hjellming and Newell (1983).

VW Cep ($P_{orb} = 6.^h5$, K0V) was detected three times with flux densities between $0.07 \leq S \leq 4.8$ mJy ($10^{13} \leq L_r \leq 4 \times 10^{14}$ erg s^{-1} Hz^{-1}). During the second epoch, the flux density of VW Cep rose gradually from less than 0.1 mJy to 4.8 mJy in $\sim 3^h$. Rucinski and Seaquist (1988) also studied VW Cep at the 2-, 6-, and 20-cm wavelengths, but had only a single marginal detection ($S = 0.11 \pm 0.04$ mJy) at 6-cm. Vilhu et al. (1988) reported simultaneous VLA 6-cm radio and EXOSAT X-ray measurements of the contact binaries VW Cep and XY Leo.

Both the X-ray and radio light curves for XY Leo showed no evidence for orbital modulation. Although the timescale for the VW Cep radio flare is similar to that observed in semi-detached systems such as RS CVns, the peak luminosity is only $\sim 1\%$ that of typical RS CVn systems (Mutel et al. 1987).

CLASSICAL ALGOL BINARIES

The classical Algol binaries (Budding 1986) are semi-detached systems consisting of a mass-gaining, early-type primary and either a late-type (type EA2, later than F5) or early type (EA1, F5 or earlier) secondary. Radio surveys of Algol binaries have been made by Umana et al. (1989) and Slee et al. (1987b). The survey of Umana et al. consisted of VLA observations of 15 Algols at 6 cm, of which 60% (9) were detected. The radio luminosities of the detected systems were in the range $(0.3 - 219) \times 10^{15}$ erg s^{-1} Hz^{-1}. Both the radio luminosity range and the fractional detection rate are comparable to the RS CVn systems (Morris and Mutel 1988; Drake, Simon, and Linsky 1989). The survey of Slee et al. included 47 Algols, of which 30% (15) were detected using the Parkes 70-meter telescope at 3.6 cm. Stewart et al. (1989) analyzed the Slee et al. observations and concluded that the emission is likely gyrosynchrotron emission from mildly relativistic electrons ($E \lesssim 1$ MeV) gyrating in magnetic fields of $B \sim 150$ to 300 G. They stress that at least five and posssibly seven of the detected systems were of type EA1, for which *both* primary and secondary are earlier than F5. The implied surface magnetic fields of these stars are then 1000 to 2000 G, assuming $B \propto r^{-3}$ and a source size of several stellar radii. The source size of the prototype system Algol (B8V + G8IV) was directly measured using VLBI (Mutel et al. 1985; Lestrade et al. 1988), and was consistent with the above values.

RS CVN BINARIES

The RS Canum Venaticorum (RS CVn) binaries (Hall 1976) are detached binaries with active late-type dwarfs, subgiants, or giant components. The radio emission from these systems has both "quiescent" and "flare" phases. The flare emission has two components: 1) broad-band, unpolarized emission, with a decay time of several hours, and 2) highly circularly polarized, narrow-band emission with a time scale of a few minutes. The former is probably synchrotron emission from relativistic electrons ($E \gtrsim 5$ MeV; Mutel et al. 1985), while the latter appears to be coherent emission, possibly an electron-cyclotron maser (Lestrade et al. 1988; Mutel et al. 1987).

The quiescent component is broad-band, moderately circularly polarized, and variable over a period of hours. VLBI observations have demonstrated that it arises from an extended region comparable with the expected size of the extended corona of the active star in RS CVn systems (Lestrade 1988). There is strong evidence that this emission is gyrosynchrotron radiation from mildly relativistic ($E \sim 1$ MeV) electrons spiraling in the enhanced coronal magnetic field of at least one of the two component stars (Owen, Jones, and Gibson 1976; Mutel *et al.* 1985; Klein and Chinderi-Drago 1987). If this is the case, the emergent flux should be directly related to the magnetic field strength (e.g., $S \propto B^2$). In addition, the polarization properties of this radiation should contain information on the geometry of the magnetic field. Radio observations of such systems provide the only direct probe of the strength and configuration of extended stellar magnetospheres. This empirical information can then be used to place constraints on models of dynamo-driven magnetic field enhancement and the magnetic heating and confinement thought to be responsible for the enhanced optical and X-ray emission observed from these systems.

'EMISSION-LINE ALGOL' BINARIES

The 'emission-line Algols' (Eggleton 1985; Plaveč 1989) comprise a subclass of Algol binaries. They are strongly interacting, quickly evolving binary systems consisting of a hot (typically B or A) primary and a cool (typically K) secondary with periods between 10 and 40 days. There is considerable mass transfer from the cooler to the hotter star resulting in a geometrically thick, high opacity accretion disk. The mass gainer (hot star) is often optically invisible because of obscuration by the accretion disk. The light curves tends to be noisy, probably because of variations in the accretion disk. Likewise, the periods are variable because of variations in the mass transfer/loss rates (Koch and Guinan 1978).

Elias (1990) has recently published a small radio survey of 6 emission-line Algols using the VLA at 6 and 3.6 cm. Only one source (V367 Cyg) was detected at 6 cm while 3 sources were detected at 3.6 cm (W Ser, RZ Sct, and V367 Cyg). The radio luminosity of V367 Cyg is typical of other semi-detached binaries, of order 10^{17} erg s^{-1} Hz^{-1} (within an order of magnitude of the luminosity of a related star, β Lyr; Wade and Hjellming 1972). The spectral index between 6 and 3.6 cm was marginally positive ($\alpha = 0.4 \pm 0.3$), just as the quiescent emission spectral index of RS CVns and Algols (Mutel *et al.* 1987). There was no evidence for circular polarization, but the low flux levels ($S \sim 0.1$ to 0.4 mJy) preclude a sensitive measurement. Elias suggested that the emission was gyrosynchrotron radiation due to electrons which were thermalized by collisional interactions with the circumbinary gas (Gibson 1980).

The distances to the emission-line Algols in this sample are greater than 500 pc, which is much farther than the typical RS CVn or Algol system in comparable radio survey. Since all detections are near the sensitivity limit of the VLA in "snapshot" mode, the preponderance of non-detections may simply be the result of the large distances to emission-line Algols rather than any fundamental difference in radio emission properties. Although Elias finds that the emission is consistent with Gibson's suggestion of thermalized gyrosychrotron emission, he suggests that the physical configurations of the systems cannot be the same, since their radio luminosities are about 10 times lower than that of β Lyr. For example, he cites evidence that the accretion disks may be larger in typical emission-line Algols than in β Lyr (another possible member of this "group"; Plaveč 1989). This implies that infalling material is less energetic because it does not fall so deeply into the potential well, ultimately producing weaker cm radiation.

DME STARS

The study of stellar radio emission from stars other than the sun began with the study of decimetric flares from nearby red dwarfs at Jodrell Bank in the 1960's. In recent years, both the VLA and Arecibo have been used extensively to study radio flares from over a dozen dMe stars, some of which are binaries. There appear to be at least two types of emission: a quiescent, unpolarized, slowly varying component attributed to gyroresonant emission from hot ($T \gtrsim 10^7$ K) thermal electrons (e.g., Gary, Linsky, and Dulk 1983), and an impulsive, circularly polarized flare component which displays rapid, narrow-band frequency drifts on dynamic spectra (Bastian and Bookbinder 1987). The flare emission is thought to be due to a coherent radiation process driven by a loss-cone anisotropy in the converging field lines of a magnetic loop. This results in either an electron-cyclotron maser or coherent plasma radiation, depending on whether the ratio of the electron plasma to gyrofrequency is less than or greater than three, respectively (Bastian et al. 1990).

The VLA has been used to spatially resolve several dMe radio sources which are binary systems. For example, both EQ Peg AB (Topka and Marsh 1982) and UV Ceti AB (Gary, Linsky, and Dulk 1983) have similar quiescent radio characteristics. Short-duration flares have been observed on a single component from each system. Since the radio characteristics of dMe stars are similar for the binary and single star sub-classes, it is unlikely that binary effects such as interacting magnetospheres or internal tidal effects play a significant role in the radio emission process.

CATACLYSMIC VARIABLES

Cataclysmic variables (cvs) are strongly interacting binary systems consisting of a white dwarf primary which is accreting mass from a (typically) late-type dwarf secondary that is overflowing its Roche lobe. They are short-period systems with periods between 1^h and 15^h, and include such sub-classes as DQ Her novae, dwarf novae, recurrent novae, ultrashort period AM CVn binaries, and AM Her binaries (Wade and Ward 1985).

The defining property of the DQ Her class of CVs is the presence of coherent pulsations in the optical light curve. It is believed that there is a strong magnetic field which disrupts the accretion disk for radii smaller than the Chapman-Ferraro radius (i.e., the effective magnetospheric radius).

Radio emission has been detected from AE Aqr (Bookbinder and Lamb 1987; Bastian, Dulk, and Chanmugam 1988), one of 10 known members of the DQ Her class. The system consists of a $0.94 M_\odot$ white dwarf and a $0.74 M_\odot$ K5 dwarf secondary in a nearly circular orbit ($P_{orb} = 9.^h9$) with an inclination of 58^{deg}. Bastian, Dulk, and Chanmugam found highly variable emission (1 to 12 mJy in several hours at 2 cm), with an average long-term spectral index $\alpha \sim 0.3$ to 0.4, but with no measurable circular polarization. This corresponds to a radio luminosity of $L_R \sim 10^{16-17}$ erg s^{-1} Hz^{-1} at an assumed distance of 84 pc. They suggest a model consisting of a superposition of discrete synchrotron emitting flares which are nearly continuous. The source geometry is not known, although it is argued that the emission is likely to be associated with the interaction between the thin accretion disk and the white dwarf magnetic field. The system could be a low power analog to powerful X-ray binaries such as Cygnus X-3 (cf. Section 9.7) which are similar except that the evolved object is a neutron star or black hole instead of a white dwarf.

The AM Her binaries have highly polarized optical line radiation indicating a magnetic field $B \sim 10^7$ G on the surface of the white dwarf. Both quiescent and highly circularly polarized flare radio emission have been detected from the prototype system AM Her (Chanmugam and Dulk 1982; Dulk, Bastian, and Chanmugam 1983), the only source detected in a 6 cm VLA survey of 6 AM Her binaries. The quiescent emission is thought to arise from the usual gyrosychrotron process, but with an interesting twist: the electrons are accelerated to mildly relativistic energies by a "unipolar induction" scheme. Since the red-dwarf secondary is imbedded in the magnetosphere of the neutron star, small departures from synchronous rotation can lead to a large potential difference (~ 400 KV) across the secondary, which accelerates electrons. The flare emission is probably an electron-cyclotron maser operating near the surface of the red dwarf. If this is correct, it implies a surface magnetic field for the secondary $B \sim 1000$ G, which could affect the mass transfer properties of the system.

X-RAY BINARIES

The radio emission from X-ray binaries comes from a region near, but not necessarily the same as, the X-ray emission. The radio and X-ray emissions are anticorrelated. For example, Cyg X-2 exhibited radio flares at the 2 to 5 mJy level during an episode of quiescent hard X-rays, and at another time was radio quiet from 0.3 to 1.5 mJy with softer X-rays which were more variable and flare-like (Hjellming *et al.* 1990a).

There are two possible explanations for this behavior, which depend on the rate of mass transfer within the binary. If the mass transfer rate increases during the episode of X-ray flaring, this means that there must be a reduction of plasma in the radio-emitting regions, and that the radio and X-ray regions cannot be entirely coincident. If the mass transfer rate decreases during an episode of X-ray flaring, on the other hand, this means that the radio and X-ray regions become more confined with enhanced electron energies. Further observations are necessary to determine which model is the best.

Similar results were obtained with Sco X-1 (Hjellming *et al.* 1990b), although the authors state that they have found two separate components to the radio emission. One component peaks at about 20 cm and the other at about 3.6 cm. The higher frequency component was found to be variable over timescales of 2 to 3 hours.

SYMBIOTIC STARS

Symbotic stars have spectra indicative of a late-type giant and a very hot star. They are thought to be binaries consisting of a mass-losing, late-type giant and a hot companion which ionizes some part of the stellar wind (Kenyon 1986; Allen 1988). They can be divided into two classes according to their infrared properties as follows: S-type symbiotics exhibit IR emission characteristic of a red giant stellar atmosphere, while D-types have an IR spectrum indicative of dust. Stars in the D-class most likely consist of a Mira-type variable and an accreting white dwarf.

Seaquist and Taylor (1990) have recently summarized the radio properties of symbiotics based on a large VLA survey of essentially all systems visible from the northern hemisphere. They detected about 25% of the 107 systems surveyed to a limiting flux level of 0.5 mJy, including nearly all of the D-type systems. The radio luminosity was correlated with spectral type in the sense that later spectral types are more radio-luminous. The spectral index of all measured systems is positive ($0 \lesssim \alpha \lesssim 1$ where $S \propto \nu^{\alpha}$) between 20 and 2 cm. No circular polarization has been detected. Although a few symbiotics are strongly variable radio sources (e.g., RX Pup), most have not varied by more than 30% over a period of several years.

The radio observations are consistent with optically thick thermal bremsstrahlung radiation from the ionized component of the red giant's outflowing wind. A large fraction of the symbiotics with spectral indices measured at three frequencies indicate the emission becomes optically thin near 1 cm (although this has not been measured directly). In the D-types, the radio emission probably originates within an extended volume surrounding the binary system (typical sizes 0.1 to 100 AU). For the larger S-type systems (semi-major axis 100 to 1000 AU) both the radio and Hβ line emission may originate in very small dense regions within the mass-loss wind region.

A recent paper by Kenny, Taylor, and Seaquist (1991) reports on multiple shells surrounding the symbiotic star AG Peg. These results have been obtained with several VLA observations using different configurations. In the center of the radio emission is an unresolved point source ($< 0.''1$) which is partially optically thick thermal bremsstrahlung and variable, probably in phase with the orbit. Next, there is a nebular shell about $2''$ in diameter which appears almost dipolar. It has been attributed to a He I flare in the wind of the giant star near the compact star circa 1950. If this is true, its expansion velocity now is approximately 60 km s^{-1}. Its mass is estimated to be about $4 \times 10^{-5} M_{\odot}$. The next shell outward is approximately $20''$ in diameter. It has been attributed to mass loss after an eruption of this star circa

1850. Its total mass is believed to be between 10^{-4} to 10^{-3} $M_\odot$. Lastly, there is a shell which is 1' in diameter. It is believed to be the remnant of the 1850 outburst. Its total mass is between 10^{-5} to 10^{-4} $M_\odot$. All the extended emission regions are believed to radiate via thermal bremsstrahlung.

NONTHERMAL RADIO EMISSION FROM SINGLE STARS

While most sources of nonthermal stellar radio emission are associated with binary systems, some single stars have also been detected. In this section, four classes of single stars are discussed whose nonthermal radio emission resembles many of the binary systems discussed earlier. This similiarity suggests that the physical mechanism responsible for strong nonthermal radio emission in close binaries does not depend entirely on binarity itself.

Rapidly Rotating Pre-Main-Sequence Stars

The first detection of radio emission from a rapidly rotating, pre-main-sequence (PMS) star was from T Tauri (Spencer and Schwartz 1974). Since then, many other PMS stars have been detected, sometimes with asymmetric or jet-like structures. For many PMS stars, the radio emission is consistent with thermal bremmstrahlung from ionized winds or accretion disks (e.g. T Tau(S); Schwartz, Simon, and Campbell 1986; André 1987). However, several nonthermal PMS stars have been detected recently. For example, during a VLA survey of 160 PMS objects in the ρ Ophiuchi dark cloud (André, Montmerle, and Feigelson 1988), two (DoAr21, ROX31) of the nine radio detections were highly variable sources (Stine, Feigelson, and André 1988). The radio spectra during flares were positive, while quiescent spectra were nearly flat between 6 and 2 cm. Flare luminosities ranged up to $L_{6\ cm} \sim 10^{17.7}$) erg s^{-1} Hz^{-1}. Both of these characteristics are quite similar to those of active binaries such as RS CVns.

At least one of the non-flaring radio sources (S1; André et al. 1988) also appears to be a nonthermal emitter, and circular polarization ($\sim 7\%$) was detected at 6 and 2 cm. André et al. suggest that S1 is a single, rapidly rotating magnetic B star surrounded by an equatorially elongated magnetosphere (cf. Section 7). They suggest that the emission mechanism is gyrosynchrotron and arises from an equatorial plasma sheet beyond the "dead-zone" or radiation belt regions proposed by MMS for RS CVn systems and Drake et al. (1987) for Ap/Bp stars. They also detected a large extended thermal halo (diameter $\sim 10^{17}$ cm) which they interpret as an optically thin compact HII shell ionized by the central B3-B5 star.

Massive OB and WR Stars

Since binarity is not necessary for radio emission from OB and WR stars, these stars are listed along with other single stars. About 50% of OB and Wolf-Rayet stars appear to be single. For most massive OB and Wolf-Rayet stars, the radio spectral index, angular size, and lack of polarization are consistent with thermal radiation from a hot ionized wind (e.g., Abbott *et al.* 1986). However, in a few cases clear evidence for nonthermal emission has been found. For example, Cygnus OB2 Nr. 9 has rapid time variability (ABC), significant circular polarization, and a high brightness temperature $T_B \sim 10^{8.5}$ K as determined by VLBI measurements (Phillips, Lonsdale, and Feigelson 1991). The physics of the emission is not well understood, although it could be the result of particle acceleration by shocks in the radiation-driven winds, which are also thought to carry magnetic flux outward.

Magnetic Ap/Bp Stars

Drake *et al.* (1987) reported the detection of 5 out of 34 chemically peculiar B- and A- type stars during a 6-cm survey at the VLA. The radio luminosities range from $10^{16.2} \leq L_R \leq 10^{17.9}$ erg s^{-1} Hz^{-1} with nearly flat spectral indices between 20 and 2 cm. Some of the stars are variable on a timescale of hours. No circular polarization was detected, although, since the flux densities were quite low (0.5 to 3 mJy), it would have impossible to detect modest ($\lesssim 10\%$) polarizations. The nonthermal nature of the emission from these stars has since been confirmed by VLBI observations (Phillips and Lestrade 1988), which found brightness temperatures $T_B \gtrsim 10^8$ K.

The emission model proposed by Drake *et al.* is gyrosynchrotron emission from a "radiation-belt" geometry with an outer radius two to ten times the stellar radius, and an energetic electron number density which increases with radius. This is very similar to the "dead-zone" magnetospheric model (MMS; cf. Section 7 in the main text) discussed earlier for RS CVn and Algol systems.

Single Late-Type Stars

There have been several radio detections of late-type stars with many of the characteristic features of active RS CVn systems (starspots, CaII H and K emission, intense X-rays, intense chromospheric lines), but which are *single* stars. One example is FK Com. It is a single star with a rotational period of $2.^d4$ and a spectral type between G0II and G8IV (Rucinski 1981). The light curve has modulations attributed to large starspots that cover 10 to 15% of the surface (Dorren and Guinan 1984). The radio emission was studied by Hughes and McLean (1987), who measured a radio luminosity $L_{6\ cm} = 5 \times 10^{17}$ erg s^{-1} Hz^{-1} with a concave spectrum that turns over near 6 cm. These data, combined with a large X-ray luminosity, led Hughes and McLean to conclude that the emission was nonthermal.

A similar example is HD 36705 (AB Dor), a spotted K2 dwarf (Rucinski 1985a, 1985b) with a period of $0.^d51$ and strong, highly variable nonthermal radio emission. Vihlu *et al.* (1988) suggest that this star is just arriving on the main sequence and is still contracting. This may explain its rapid rotation rate, since it has not yet suffered significant magnetic braking (Mestel and Spruit 1987).

Slee *et al.* (1989) recently completed an 3.6-cm Parkes survey of all 82 late-type (G, K, and M) single stars of luminosity class I-III within 30 parsecs and south of the celestial equator. They detected 24 stars, including 19 GK giants, two GK bright giants, and five M-type giants. They argue that the emission is "almost certainly" nonthermal (probably gyrosynchrotron), since the implied size for an optically thick thermal source ($\sim 10^{15}$ cm) is inconsistent with the rapid time variability observed. The maximum radio luminosity was two orders of magnitude lower than that for RS CVn binaries. These detections are very surprising, since unlike FK Com stars, these are *slow* rotators with periods of months or years. However, Slee *et al.* argue that these stars have radio luminosities consistent with the radio surface brightness-rotational velocity relation found by Slee and Stewart (1989) derived from a sample of active binaries.

Acknowledgements

RLM wishes to acknowledge support from National Science Foundation grants AST 88-42712 and 89-19386.

FORMATION OF BINARY STARS

A. P. BOSS

1. Introduction

Convincing evidence for the physical association of double stars was produced by William Herschel over 200 years ago. Starting in 1782, Herschel published catalogues of far more double stars than could be accounted for by a random distribution of single stars, and went on to confirm the concept of physical association by discovering the mutual orbital revolution of several visual binary star pairs (see Berry 1898). Laplace produced the first theoretical explanation for the origin of binary stars soon thereafter (in 1796), and the flow of suggestions for binary origins has continued to the present day. Because of the solid body of observational information about binary systems that is now available for use in discriminating between hypotheses of origin, and because of recent theoretical work on the mechanics of the hypotheses, we have made substantial progress in evaluating the possibilities. This chapter will detail the reasoning behind this optimistic claim.

While the emphasis of this volume is on the astrophysics of close binary stars, in this chapter we discuss theoretical efforts at explaining the origin of all types of Population I binary stars. There is no *a priori* reason to believe that close binary stars require an origin distinct from that of wider binaries, and so it appears wisest to consider the binary formation problem in its entirety. Previous reviews of theoretical work on binary star formation have been given by Tassoul (1978), Lucy (1981), Zinnecker (1984), Boss (1988), and Bodenheimer, Ruzmaikina and Mathieu (1990).

This chapter starts with a brief review of the fundamental properties of main-sequence binary stars, and of the results beginning to emerge from studies of pre-main-sequence binary stars. The three classical theories of binary formation are discussed next, followed by the three more recently proposed theories. Each theory is confronted both with its own dynamical problems and with the observational data in order to assess its likelihood of contributing to the binary star population.

2. Binary Star Properties

In this section we summarize the basic physical and dynamical properties of main-sequence and pre-main-sequence binary stars as determined by observations. These properties will then serve as benchmarks against which the various theories of binary origin can be judged.

2.1. MAIN-SEQUENCE STARS

Duquennoy and Mayor (1990; hereafter DM) have observed all F7-G9 dwarf stars within about 22 pc of the sun. The DM sample includes 166 primaries and 44 secondaries, for a total of 210 G dwarf stars that have been studied spectroscopically with precisions of 0.3 km s^{-1} or better. This parallax-limited sample from the Gliese catalogue should provide the most unbiased information on G-type stars and on main-sequence stars in general.

2.1.1. Frequency

DM found that 65% of their primary stars had a companion with mass ratio $q = M_2/M_1 \geq 0.1$. Of the remaining 35%, about 50% could have a very low mass companion in the range $q = 0.01 - 0.1$. Hence less than 25% of the sample may be true single stars. DM also found numerous multiple systems; their figure 4 depicts 38 binary, 9 triple, and 5 quadruple systems. These detections imply a frequency of binary and multiple systems comparable to the results of previous studies of G dwarfs (Abt 1983; 1987).

2.1.2. System Type

Multiple systems are primarily hierarchical, with greatly different periods for the closest and widest members of the system (Fekel 1981). Trapezium type systems also occur, where the stars are all roughly equidistant, but these systems are subject to rapid orbital evolution toward ejection or a more stable hierarchical configuration (e.g., Anosova 1989).

2.1.3. Period and Separation

The DM sample shows a roughly Gaussian distribution in orbital period, ranging from days to millions of years, with a peak around 300 years, corresponding to separations ranging from a few solar radii to ~ 0.1 pc, with a peak between 10 and 100 AU.

2.1.4. Eccentricity

Systems in the DM sample with periods less than 11 days have nearly zero eccentricity, consistent with the expected orbit circularization by tidal effects for a population as old as the galactic disk. The longest period (> 1000 days) systems show large eccentricities ($e \sim 0.2 - 1$), while intermediate period systems have a mean eccentricity of 0.35, a result that appears to hold for both halo and disk populations and for very low mass secondaries in the DM sample. The high eccentricity of these stellar systems contrasts with the small eccentricities of the planets in our solar system, implying a fundamental difference between stars and planets: stars form as highly eccentric systems, whereas planets do not.

2.1.5. Spin-Orbit Alignment

The correlation between spin rotational velocities and orbital inclinations for visual binaries implies that spin rotation axes tend to be aligned parallel to the orbital rotation axes (Weis 1974); the correlation is stronger for F stars than for A stars. Tidal evolution can remove components of the spin that are perpendicular to the orbital axis, but such tidal effects are negligible in visual binaries. Eclipsing close binary systems allow the simultaneous determination of the spin rotation direction and the orbital rotation direction; in every case, the spin direction is the same (prograde) as the orbital direction (Kopal 1978). Tidal effects are important for close binary systems, but tidal effects cannot reverse the sense of rotation, so the spin-orbit alignment must be primordial.

2.1.6. Orbit-Orbit Alignment

Weis (1974) found an inconclusive tendency toward coplanarity between spectroscopic and visual components of multiple systems. Fekel (1981) found that out of 20 close visual multiple systems, coplanarity (defined as inclinations within $15°$) was possible in about 2/3 of the systems.

2.1.7. Mass Ratio

Halbwachs (1987) studied a sample of 205 spectroscopic binaries, selected to exclude systems capable of undergoing mass exchange. Halbwachs (1987) found the mass ratio q ($= M_2/M_1$, where 1 designates the primary and 2 the secondary component) for these close binary systems to be very similar to that of visual binaries (Halbwachs 1986): a broad distribution from $q \sim 0.2$ to 1.0, with a peak around $q = 0.4$. This result is consistent with a common origin for both close and wide binary systems.

Trimble (1990) analyzed a sample of 164 spectroscopic binaries (mostly K giants) observed by R. F. Griffin and colleagues. As in the Halbwachs (1986, 1987) studies, the Griffin systems show a broad distribution, with a major peak around $q \sim 0.2$, but with a minor peak close to $q = 1$.

The DM sample yielded a distribution of mass ratios that rises slowly from a minor peak at $q = 1$ to a maximum around $q \sim 0.3$. For smaller values of q the situation is uncertain, but the distribution could be flat or even rising again toward $q = 0$. The DM distribution is well represented by an initial mass function derived from low mass field stars (Kroupa, Tout, and Gilmore 1990), except for around $q = 1$ and for $q < 0.1$, where the observed distributions may considerably exceed the prediction from the field star initial mass function.

While all three of these distributions appear to be satisfyingly similar, other distributions may occur in more restricted samples, such as common proper motion stars or very close binaries (Trimble 1990). For example, solar-type contact binaries appear to have started primarily as $q = 1$ systems (van't Veer 1981).

2.2. PRE-MAIN-SEQUENCE STARS

Observers are beginning to compile statistics on binary pre-main-sequence star properties (Reipurth 1988; Zinnecker 1989; Bodenheimer, Ruzmaikina, and Mathieu 1990), and while we generally do not yet have a large enough sample to draw firm conclusions, the systems found so far already yield important information. Observations are also beginning to appear of suspected binary *protostellar* objects (e.g., Wootten 1989; Sasselov and Rucinski 1990), and while the detection of a true solar-type protostar has not yet been agreed upon, and we can expect these types of observations eventually to be the final arbiters of formation questions.

2.2.1. Frequency

Mathieu, Walter, and Myers (1989) found 6 spectroscopic pre-main-sequence binaries by observing 'naked' (weak-lined) T Tauri stars, implying a frequency of occurrence of short period (< 100 days) systems of about 0.10, quite close to the main-sequence frequency (0.12) for solar-type stars with these periods (Abt 1983). M. Simon and W. P. Chen and their colleagues have used lunar occultations to detect 9 double systems in 31 pre-main-sequence stars, for a binary frequency of 0.29 (Chen *et al.* 1990). This frequency is also in agreement with the binary frequency (0.28 - 0.33) of main-sequence solar-type stars (Abt 1983) for the separations accessible by lunar occulations. If substantiated by further observations, these frequencies imply that essentially all binary stars are already formed by the time of the pre-main-sequence phase.

Triple pre-main-sequence systems have been found in surveys for visual binary systems (Zinnecker 1989), but the numbers involved are too small to assess the frequency of multiple pre-main sequence systems.

2.2.2. Age

Spectroscopic binary T Tauri systems have been found with inferred ages ranging from $\sim 10^6$ years (Mathieu, Walter, and Myers 1989) to as little as $\sim 10^5$ years (Bodenheimer, Ruzmaikina, and Mathieu 1990); in the latter cases, the stars are found very close to the birthline in the luminosity-effective temperature diagram (Stahler 1988). DF Tau is a very young, intermediate separation (~ 1 AU) binary which is also very close to the birthline (Chen *et al.* 1990). A young visual binary system has been found with an age on the order of 10^5 years (Zinnecker 1989), still deeply embedded in its placental NH_3 cloud clump. Evidently binary stars are present at the earliest possible epochs and binary star formation, in some cases at least, must occur coevally with formation of the stars themselves.

2.2.3. Separation

Cohen and Kuhi (1979) found 34 close pairs of T Tauri stars and argued on the basis of their similar extinctions and luminosities that the pairs are physical binaries. B. Reipurth and H. Zinnecker found 28 visual pre-main-sequence binaries with separations in the range 200 to 2400 AU (Reipurth 1988). Chen and Simon (1990, private communication) have used infrared array imaging to detect seven T Tauri binaries with separations on the order of 500 AU. Chen *et al.* (1990) previously used lunar occultations to discover 5 pre-main-sequence systems, with projected separations ranging from about 1 to 80 AU. Both classical and naked T Tauri stars have been found that are spectroscopic binaries with periods on the order of days to months (Bodenhemier, Ruzmaikina, and Mathieu 1990). The range of separations of pre-main-sequence binaries thus appears to cover the entire range occupied by their main-sequence counterparts.

2.2.4. Eccentricity

Pre-main-sequence binaries with periods less than 4 days have essentially zero eccentricities, while longer period systems are quite eccentric (Mathieu, Walter, and Myers 1989), with $e \sim 0.1 - 0.5$. The circular orbits of the shortest period systems presumably result from tidal evolution, which should be particularly rapid for convective pre-main-sequence stars. The cutoff at 4 days is consistent with these stars being a much younger ($\sim 10^6$ years) population than the main-sequence stars in the DM sample, where the cutoff occurs at a period of 11 days. Binaries thus appear to *form* with substantial eccentricities, regardless of their separation.

2.2.5. Spin-Orbit Alignment

Pre-main-sequence stars often show evidence of circumstellar disks and energetic bipolar flows. Circumstellar disks are likely to be aligned perpendicular to the stellar rotation axis, while bipolar flows are thought to be aligned with the rotation axis, holding out the possibility that future observations may be able to determine spin orientations for binary pre-main-sequence stars and disks (Zinnecker 1989).

2.2.6. Mass Ratio

An eclipsing pre-main-sequence double-lined binary system has not yet been found (excluding EK Cep, which has a main-sequence primary and a possible pre-main-sequence secondary; Popper 1987), so pre-main-sequence masses cannot be reliably calibrated. Mathieu, Walter, and Myers (1989) have used a non-eclipsing (to date) double-lined spectroscopic binary to place lower limits on the masses of the primary ($1.02 \pm 0.06 M_\odot$) and secondary ($0.94 \pm 0.05 M_\odot$), limits that appear to be *in*compatible with theoretical evolutionary models implying masses of 0.7 or 0.8 $M_\odot$. Chen *et al.* (1990) found the K band luminosities of their secondaries to range from 0.1 to 1.0 of the primaries, perhaps implying a range of mass ratios, ages, or circumstellar extinctions. Clearly substantial work remains to be done in the area of determining pre-main-sequence mass ratios.

3. Classical Theories of Binary Star Formation

In this section we discuss the classical theories for binary star formation, in the order in which they were first suggested.

3.1. SEPARATE NUCLEI

In 1796 Laplace suggested that Herschel's double stars could be explained as having arisen from the condensation of nuclei that began to orbit around each other for some reason (Tassoul 1978). As stated by Laplace, the hypothesis is extremely vague, and apparently no one has attempted to develop the separate nuclei idea more fully in the intervening centuries. However, the separate nuclei hypothesis deserves to be reborn in the light of what we currently know about molecular clouds and the dense cores found within, which appear to be frequent sites of star formation (e.g., Myers 1987). Through some as yet unknown process molecular clouds become highly structured and produce dense cloud cores. These cores typically have masses of $\sim 0.3 - 10 M_\odot$ and usually contain embedded infrared sources indicative of ongoing star formation. Because of the absence of evidence of large-scale collapse in molecular clouds, the evolution leading toward the formation of these cores is likely to be quasistatic and controlled by magnetic fields (e.g., Shu, Adams, and Lizano 1987).

3.1.1. Dynamics

While we may not fully understand the evolution that leads to isolated molecular cloud cores, it is clear observationally that the outcome of this process is production of stellar mass clouds ripe for star formation. Hence regardless of what dynamical or quasiequilibrium processes lead to this structure, we already know that they 'work', in contrast to the remaining hypotheses of binary formation, where observations of the process in action may be elusive.

3.1.2. Observations

Considering the paucity of predictions by the newly reborn hypothesis of separate nuclei, opportunities for comparison with the desired properties of binary stars are limited, but ultimately quite productive. On the positive side, binary molecular cloud cores would exist from the earliest phases, and so could explain the youngest binary systems. Also, a wide range of mass ratios would be expected, given the observed range of cloud core masses.

The negative evidence is considerably more compelling, however. First, all cloud cores mapped to date have been limited by radio telescope beam sizes to separations of ~ 0.2 pc or more. While it is conceivable that future interferometric observations will find quasistatic cores on smaller scales, at present it appears that cloud cores are too widely spaced to account for even the widest binaries (maximum $a \sim 0.1$ pc). Second, we do not know whether any of the dark cloud cores are gravitationally bound in pairs (or other system types) and hence are likely to yield physical rather than optical double stars. Third, spin-orbit and orbit-orbit alignments would not be expected unless the processes that create the cloud cores preferentially align them. Despite this largely negative assessment, the separate nuclei hypothesis as redefined here does serve as a necessary antecedent to the more promising 'cloud clump collisions' hypothesis discussed in section 4.2.

3.2. CAPTURE

In 1867 Stoney proposed the formation of binary stars through the capture of single stars (Tassoul 1978). In order to produce a bound orbit from an initially unbound pair, a source of energy dissipation must be identified. Several variants of the capture hypothesis exist, depending on the means of achieving this energy dissipation: three-body encounters, tidal dissipation, or gas disk interactions.

3.2.1. Three-Body Encounters

Mansbach (1970) has calculated the rate of formation of binary stars through three-body encounters. Three initially unbound single stars undergo an encounter where one of the stars absorbs sufficient kinetic energy to leave the remaining two stars gravitationally bound. Assuming a Maxwell-Boltzmann distribution in phase space, Mansbach (1970) finds that the formation rate varies as the stellar number density to the third power and is similarly directly dependent on powers of the final binary separation and the stellar masses. Three-body capture thus works best in regions of high stellar density, and should preferentially form wide, massive binary systems. The latter tendency is at odds with observations: wide binaries actually become *rarer* as the separation increases.

For the stellar number density of the galactic disk ($n \sim 0.1$ pc^{-3}), however, the formation rate is extremely small; less than one solar-type binary of 2000 AU separation is likely to have formed by this means over the age of the galaxy (Mansbach 1970). Captures during the main-sequence phase of course would occur too late in time to explain the frequency of pre-main-sequence binaries.

Because substantial present-day star formation occurs in clusters with higher number densities and lower relative velocities than the galactic disk, the three-body capture rate is considerably higher there and must be evaluated. Open clusters and

T Tauri star associations have $n \sim 10 - 10^2$ pc^{-3}, while the central core of ρ Ophiuchus and the Orion Trapezium have $n \sim 10^3$ pc^{-3} (Wilking and Lada 1983). The Lynds 1641 dark cloud contains a dense cluster with a suspected stellar density $n \sim 7 \times 10^3$ pc^{-3}, whereas Taurus-Auriga does not contain similarly dense clusters at all (Strom, Margulis, and Strom 1989). Applying Mansbach's (1970) formula to a very dense cluster of solar-type stars with $n = 10^4$ pc^{-3} and a mean relative velocity of 1 km s^{-1}, we find that the rate of formation of binaries with separations of ~ 200 AU is still so small that no binaries are likely to form through three-body capture over the $\sim 10^6$ year lifetime of a cluster.

3.2.2. Tidal Dissipation

Tidal friction during a close encounter of two stars can dissipate relative kinetic energy and potentially lead to capture. Because of the strong dependence of tidal dissipation on separation, tidal capture is only feasible for close encounters within roughly the Roche radii of the stars, i.e., the tidal capture cross section is ~ 10 times larger than the geometrical cross section. The kinetic theory of gas can be used to approximate the mean time (τ) between encounters for a given cross section σ. For solar-type stars in the galactic plane with $n = 1$ pc^{-3} and relative velocities v_r of 10 km s^{-1}, the mean time between tidal encounters per star is $\tau \sim (n\sigma v_r)^{-1} > 10^{19}$ years. Even in a very dense cluster with $n = 10^4$ pc^{-3} and $v_r = 1$ km s^{-1}, $\tau > 10^{16}$ years for any individual star. Hence capture through tidal forces on stellar atmospheres is completely negligible.

3.2.3. Gas Disk Interaction

The gaseous disks that accompany the formation of rotating protostars could produce dissipation through processes such as gas drag, gravitational drag, tidal forces, shock heating, and radiation, if two protostellar disks should undergo a close encounter. While the hydrodynamics and thermodynamics of these processes have not been elucidated as yet, we will assume that the interaction is highly dissipative, and that encounters where the two disks overlap substantially lead to a bound binary system. Note that this may be an overly optimistic outcome: collisions between rotating, spherical clouds produce a variety of outcomes, including mergers, dispersal of the overlapping regions, and induced gravitational collapse (Lattanzio and Henriksen 1988). The quasiequilibrium disk produced by the collapse of a cloud core to form a single protostar is on the order of 10 AU in size. Applying the kinetic theory of gas to a very dense ($n = 10^4$ pc^3) cluster of protostars with 10 AU disks and $v_r = 1$ km s^{-1} yields $\tau \sim 10^{10}$ years per star, a negligible rate considering the cluster lifetime.

Larson (1990) has argued that the binary formation rate could be quite large in very dense clusters, because of dynamical friction associated with passage of a star past a protostellar disk. Using the gravitational drag found numerically by Shima *et al.* (1985) for flow of gas past a gravitating sphere, Larson (1990) estimated that dynamical friction would increase the effective capture cross section radius to roughly 350 AU. The resulting rate could lead to the formation of binaries in a substantial fraction (~ 0.4) of stars in a very dense cluster with an age of 10^6 years (Larson 1990). However, the gravitational drag calculation of Shima *et al.* (1985) applies to the steady state response of gas initially distributed uniformly over a region much larger than the accretion radius $R_A = 2GM/v_r^2 \approx 2000$ AU; the steady state gas distribution that results in strong gravitational drag resembles a paraboloid of revolution of similar size. Thus the Shima *et al.* (1985) gravitational drag probably is an upper limit on the drag experienced by a star exciting a transient response by passing ~ 350 AU from the center of a thin disk with most of its mass inside a radius of ~ 40 AU; the actual drag may be considerably lower. Without detailed calculations of the interaction, estimates of the capture frequency will remain highly uncertain.

Binaries formed through gaseous disk capture in protoclusters might be consistent with the ages of pre-main-sequence binaries, but would be limited to separations comparable to protostellar disk sizes ($\sim 10 - 100$ AU); subsequent orbital decay caused by ongoing gravitational drag by the disk could produce closer systems (Larson 1990).

3.2.4. Observations

In addition to the specific comparisons already noted, several observational comparisons apply equally well to all three capture mechanisms. Capture should lead to a wide range of mass ratios and to highly eccentric orbits, as observed. However, the formation of multiple systems would seem to require multiple capture events, which would lower the total probability of occurrence to a very small value indeed. In general, single event capture probabilities are already quite low. Spin-orbit and orbit-orbit alignments would not generally be expected to result from capture. Capture thus appears to be limited from several points of view in the numbers of binaries it could have created: a few binaries formed in dense protostar clusters appear most plausible.

3.3. FISSION

In 1883 Kelvin and Tait proposed the formation of binary stars through fission (Tassoul 1978). Fission was envisioned to result from the contraction with conserved angular momentum of a rotating fluid body; such a body will eventually become rotationally unstable, and Kelvin and Tait hypothesized that the resulting instability would result in fission into two bodies. The fission hypothesis is based on analyses of the equilibrium and linear stability of idealized spheroidal and ellipsoidal bodies performed by a number of the great classical workers such as Poincaré, Darwin, Liapounoff, Jeans, and Cartan. Maclaurin spheroids were shown to be dynamically unstable to the growth of bar-shaped perturbations when the ratio of rotational to gravitational energy $T/|W|$ exceeded 0.274, and secularly unstable to deformation into Jacobi ellipsoids for $T/|W| > 0.138$. The Jacobi ellipsoids themselves become dynamically unstable to pear-shaped perturbations for $T/|W| > 0.163$. Fission scenarios generally envision quasistatic contraction through spheroidal and/or ellipsoidal forms, terminating in a phase of dynamical instability that leads directly to binary formation. The classical workers, however, were unable to follow the nonlinear evolution of the dynamical instability with analytical methods.

3.3.1. Polytropic Dynamics

The last decade has seen a serious examination of the fission hypothesis through numerical studies of the nonlinear evolution of a rotationally unstable body (e.g., Durisen and Tohline 1985). Hydrodynamics codes can calculate the fully three dimensional, nonlinear, time evolution of rotating, self-gravitating gaseous bodies. While obviously limited in spatial resolution, three dimensional codes do appear to have sufficient resolution to study the evolution of low order perturbations like the bar-mode. Furthermore, studies of the rotational instability of polytropic spheroids have been performed with several different numerical codes, using quite different numerical techniques, and have reached a common conclusion (Durisen et al. 1986). For differentially rotating, axisymmetric polytropes (n = 1.5) with initial $T/|W| \geq 0.33$, the growth of a bar-mode perturbation leads to trailing spiral arms that are efficient at transporting angular momentum outward through gravitational torques on the same dynamical time scale as perturbation growth. The outward angular momentum transport reduces the value of $T/|W|$ in the central regions and thus tends to stifle the rotational instability. Both finite difference (FD) and smoothed particle hydrodynamics (SPH) codes agree that the resulting race between fission and transport is won by transport: the central region evolves toward $T/|W| < 0.274$, while the trailing spiral arms windup into an ejected ring containing a small fraction of the initial mass and a large fraction of the angular momentum. One of the FD codes has been successfully compared with linearized analyses of perturbation growth (Tohline, Durisen, and McCollough 1985), lending further credence to the numerical results. The failure of the fission instability to lead to binary star formation was extended to polytropes with varied indices (n = 0.8 and 1.8) and lower $T/|W|$ (= 0.31) by Williams and Tohline (1988).

3.3.2. Realistic Dynamics

The polytropic fission calculations just described employed certain approximations that potentially could affect the outcome: (a) starting from an initial state well above the critical value of $T/|W|$ that actually would be approached from below during a quasistatic contraction, (b) starting from a spheroidal (axisymmetrical) configuration rather than the ellipsoidal configuration envisioned in most scenarios, (c) use of a polytropic equation of state rather than the complete thermodynamics of a radiating star, and (d) ignoring the likelihood of mass accretion during quasistatic contraction as a result of formation from a molecular cloud core. One of the FD codes from Durisen *et al.* (1986) has now been used to study the fission instability in a calculation that relaxes all four of these approximations (Boss 1989). The calculation involves the formation of the final protostellar core from a collapsing cloud core, including three dimensional radiative transfer and complete gas thermodynamics. [Because of the presence of a collapsing envelope, this model could be argued to involve 'fragmentation' rather than 'fission', as they are commonly defined, but since this model involves rotational instability in a strongly thermally supported, quasiequilibrium body, it is more closely related to fission.] The protostellar core forms with $T/|W| < 0.274$; the core is strongly rotationally flattened but still axisymmetric at this point. As the core contracts and further gas is accreted, $T/|W|$ reaches 0.274, and at that point nonaxisymmetry begins to grow exponentially. The evolution, however, is quite similar to that of the polytropic models: the core fissions into a tight binary system, but growing spiral arms remove the orbital angular momentum from the binary, and the binary merges into a single bar-shaped core again. Ongoing mass and angular momentum accretion soon drive the core to $T/|W| > 0.274$ again, but spiral arms again transport sufficient angular momentum to prevent the formation of a stable binary, in this cycle starting from an ellipsoidal core. The fission instability appears to be an efficient means of transporting angular momentum rather than of forming binaries.

3.3.3. Observations

Fission appears to lead to the ejection of a low mass ring or disk rather than formation of a binary system. Even if all of the gas in this ring could be gathered together into a single companion, the resulting system would have $q \ll 1$. The requirement of a quasiequilibrium initial state combined with our understanding of protostellar evolution means that any systems formed would be restricted to separations on the order of ~ 10 AU or several $R_\odot$. The eccentricity of any body formed out of the ring is likely to be small. Multiple system formation would require a few discrete events, which appears to be unlikely; the fission instability is likely to be highly cyclical and more nearly continuous. On the positive side, spin-orbit and orbit-orbit alignments would be expected and as well as early formation, but the negative aspects of fission are already overwhelming.

4. Recent Theories of Binary Star Formation

We now turn to consideration of the theories that have been proposed in this century. The hypothesis of binary formation through disintegration of a star cluster will not be explicitly considered, however, because cluster disintegration immediately raises the question of formation of the star cluster itself, a question that leads in a direction beyond the scope of this review. Small-scale cluster formation may be similar to multiple system formation, and to that extent cluster disintegration is considered implicitly.

4.1. FRAGMENTATION

Fragmentation, defined as breakup resulting from gravitational instability during the collapse phase of protostellar evolution, was suggested by Hoyle (1953) on the basis of a Jeans mass argument. Because the Jeans mass decreases as collapse proceeds at fixed temperature (as occurs during the isothermal collapse phase), Hoyle (1953) hypothesized that successively smaller mass entities would become self-gravitating and separate out of the overall cloud collapse. Hunter (1962) demonstrated analytically that a collapsing pressureless sphere would be unstable to the growth of perturbations that could fragment the cloud, and that the time scale for perturbation growth was shorter than that for the collapse of the overall cloud. However, Layzer (1963) argued that the perturbations would simply lead to turbulence rather than to self-gravitating entities. Fragmentation was revived through two dimensional calculations by Larson (1972), who found that rapidly rotating spherical clouds would collapse, flatten due to rotation, and form rings. Larson (1972) was quick to point out that these rings would be likely to fragment in a fully three dimensional calculation, and suggested fragmentation of rapidly rotating clouds as a means of forming binary and multiple protostars. Bodenheimer (1978) used his own two dimensional calculations to suggest that hierarchical systems might be formed through the collapse of clouds that undergo several phases of fragmentation and subfragmentation.

About this time fully three dimensional hydrodynamical codes were developed in order to attack the collapse and fragmentation problem (at least 8 different codes were developed; see Boss 1990). Boss and Bodenheimer (1979) showed that a rapidly rotating cloud with a large initial density perturbation would fragment directly into a binary protostar, without passing through a phase of ring formation. However, Tohline (1980) showed that rapidly rotating clouds starting from conditions close to Jeans equilibrium would damp even large initial perturbations and form rings that did not always readily fragment. Bodenheimer, Tohline, and Black (1980) surveyed a large part of the parameter space for isothermal clouds ($\rho < 10^{-13}$ g cm^{-3}, $T \sim 10$K) and concluded that binary formation would result from the collapse of clouds with a wide variety of initial conditions. Boss (1986) performed a similar survey for the nonisothermal regime ($\rho > 10^{-13}$ g cm^{-3}, $T > 10$ K) and found that while fragmentation could still occur, rising thermal support eventually stifles fragmentation. However, all of these calculations assumed a uniform density initial cloud; if the cloud is initially strongly centrally condensed (Shu 1977), the initial prejudice toward single protostar formation cannot be overcome (Boss 1987). Furthermore, the neglect of magnetic fields in these calculations limits their application to phases when magnetic fields are dynamically negligible, such as during the dynamic collapse phase of a magnetically-supported cloud following ambipolar diffusion (Lizano and Shu 1989; Tomisaka, Ikeuchi, and Nakamura 1990).

In the remainder of this section we will present a number of new three dimensional models of fragmentation, and use them to illustrate comparisons between fragmentation and the basic observational constraints on binary star formation.

4.1.1. New Numerical Models

Previous three dimensional FD models (e.g., Boss 1986, 1987) often were restricted to a symmetry that precluded the formation of unequal mass binaries; in general, only equal mass binaries were allowed. The new numerical models are intended largely to explore unequal mass binary formation. Compared to Boss (1986, 1987), the actual grid resolution in the crucial azimuthal (ϕ) direction has been increased from $N_\phi = 16$ to $N_\phi = 64$, and the assumption of symmetry through the rotational axis has been relaxed, allowing inclusion of both odd and even modes ($m = 1, 2, 3, ...16$). The increase in N_ϕ should also decrease the amount of numerical damping of perturbations, and thus provide a more realistic simulation of protostellar fragmentation.

The numerical code calculates explicit FD hydrodynamics including self-gravity and radiative transfer, on a radially contracting, spherical coordinate, Eulerian grid. Boss (1980) describes the modified donor-cell solution of the equations of hydrodynamics and the spherical harmonic solution of the Poisson equation for the gravitational potential. The hydrodynamical equations are solved in conservation law form, ensuring the global conservation of advected quantities. Implementation of the Eddington approximation for radiative transfer is described by Boss (1984).

4.1.2. Frequency

Star formation is believed to occur primarily through the collapse of dense molecular cloud cores. Previous numerical studies have shown that binary fragmentation can occur in a dense, collapsing cloud core provided that: [1] the magnetic field is dynamically insignificant; [2] the cloud contains several Jeans masses ($\alpha = E_{therm}/|E_{grav}| <\sim 1/2$); [3] appreciable rotation is present ($\beta = E_{rot}/|E_{grav}| >\sim 0.02$); and [4] collapse does not start from a power law ($\rho \propto r^{-1}$ or r^{-2}) initial density profile.

Unfortunately, it is not clear to what extent these restrictive conditions are met. [1] There is indirect observational evidence that magnetic fields become dynamically unimportant in dense ($n > 10^4$ cm^{-3}) cores (Heyer 1988), but direct measurements of magnetic field strengths in such dense cores are lacking. [2] Thermally supported clouds might be expected to begin collapse with only slightly more than a Jeans mass and thus to avoid fragmentation, but clouds supported by a combination of magnetic fields, rotation, and thermal pressure and evolving by ambipolar diffusion (Tomisaka, Ikeuchi, and Nakamura 1990) could begin collapse with $\alpha_i \sim 0.3$ (however, see [4] then). Alternatively, thermally supported clouds could begin collapse following rapid compression produced by shock waves or by cloud collisions (Lattanzio and Henriksen 1988); observations of high galactic latitude clouds have been interpreted as evidence of cloud collisions capable of inducing collapse (Keto and Lattanzio 1989). [3] Because of thermal line widths and complex velocity fields, evidence for cloud rotation is usually restricted to the most rapidly rotating ($\beta > \sim 0.1$) clouds (e.g., Goldsmith and Arquilla 1985); rotation rates of most clouds are unknown. [4] The centrally condensed structure of cloud cores containing embedded young stars is consistent with formation from 'singular isothermal spheres' (Shu 1977), with collapse starting at the center of the sphere and leading to accretion of the cloud envelope onto the central protostellar core. Such singular initial conditions may be the result of evolution of magnetically supported clouds by ambipolar diffusion (Lizano and Shu 1989). However, considering that a majority of dense cores contain detectable embedded infrared sources, it may be that all such cores already contain young stars, in which case the precursors to dense cores remain to be identified.

Without more detailed observational knowledge of the initial conditions for the collapse phase, the frequency of binary formation through fragmentation cannot be assessed, and distributions of properties such as mass ratios cannot be determined. However, ranges of properties can be estimated by assuming various possible initial conditions.

Restriction [2] is illustrated in Figure 1. The critical value of α_i for binary fragmentation (with $\beta_i = 0.21$ and a modest density perturbation) in the new models appears to lie around 0.42; clouds with higher α_i may be dominated by growth of the $m = 1$ mode, producing an off-axis protostellar object. For comparison, in a similar sequence of models (B1 through B6) Bodenheimer, Tohline, and Black (1980) found that for $\alpha_i = 0.6$ to 0.2, the clouds formed rings prior to fragmentation, but for $\alpha_i = 0.5$ to 0.4, the rings did not readily fragment. In another similar sequence, Miyama, Hayashi, and Narita (1984) found that for $\alpha_i = 1.0$, the cloud did not collapse, for $\alpha_i = 0.8$ the cloud collapsed but did not fragment, and for $\alpha_i \leq 0.6$ the clouds collapsed and fragmented; Monaghan and Lattanzio (1990) found the same behavior in their SPH models. The models shown in Figure 1 appear to be in qualitative agreement with the behavior found by Miyama, Hayashi, and Narita (1984) for $\alpha_i \leq 0.8$, though the critical α_i for binary fragmentation is somewhat lower.

4.1.3. 10^5 yr Age

Because fragmentation by definition occurs during the collapse process that produces stars, binary stars produced by fragmentation should be present by the time that protostars are first formed.

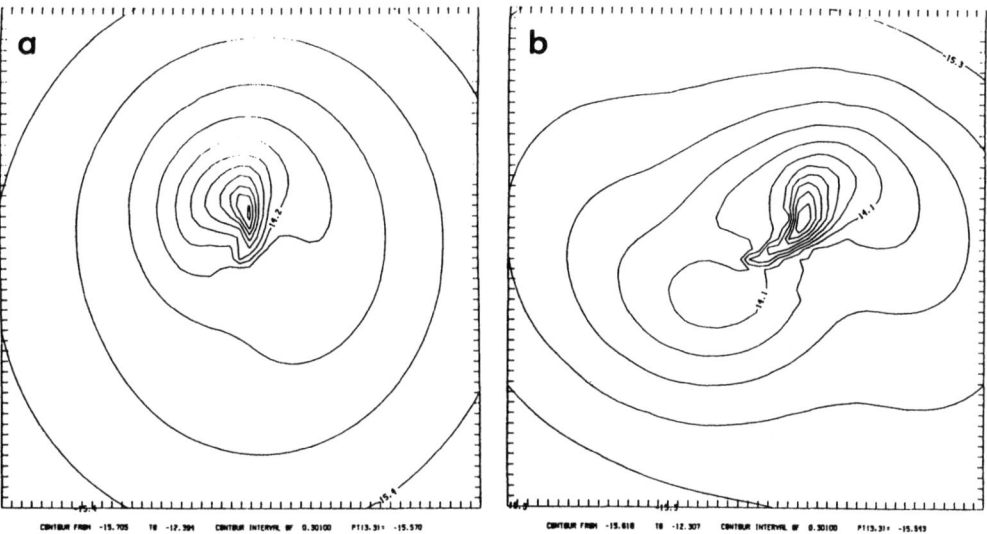

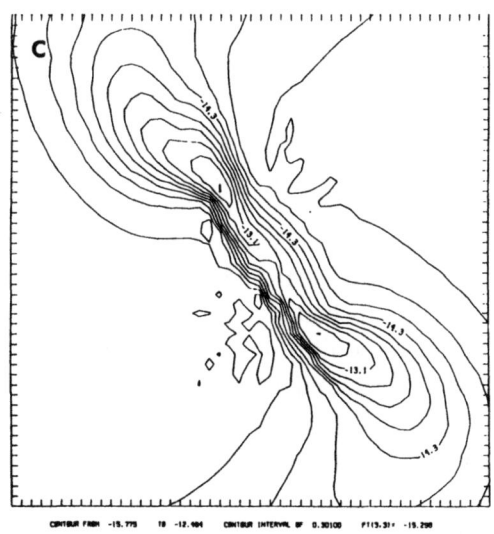

Fig. 1. Equatorial density contours for three models all starting with $\beta_i = 0.21$, $M_i = 1.0 M_\odot$, and $T_i = 10K$, but with varied $\alpha_i = 0.54$ (a), 0.42 (b), and 0.32 (c). In this contour plot and the others, each contour (labelled in g cm^{-3}) represents a change in density by a factor of 2; the rotation axis falls at the center of the plot, and the initial rotation is uniformly counterclockwise. Each model had an initial density perturbation consisting of a mixture of $m = 2, 3, 4, ...8$ modes (cos $m\phi$) with amplitudes of 0.1; in addition, the density perturbation in each cell was multiplied by a random number between 0 and 1. All three models are shown with a radius of about 4×10^{15}cm, but at different times: (a) $1.741 t_{ff}$; (b) $1.500 t_{ff}$; (c) $1.448 t_{ff}$, where t_{ff} is the free fall time of the initial cloud. As α_i is lowered to the range 0.42-0.32, binary fragmentation occurs.

4.1.4. System Type

Fragmentation directly into Trapezium type multiple systems has been obtained in clouds with small random density perturbations (Miyama, Hayashi, and Narita 1984) and in clouds with ordered perturbations (Bodenheimer, Tohline, and Black 1980); in the latter study, one strongly perturbed cloud encountered fragmentation into four bodies distributed in a line, while a cloud with a milder perturbation experienced ring formation and subsequent breakup into four bodies. Figure 2 shows a weakly perturbed cloud that formed an eccentric ring and fragmented into at least three bodies. Fragmentation through an intermediate ring configuration also has been seen in a SPH model calculated by Monaghan and Lattanzio (1990).

The most common outcome of clouds strongly perturbed with an $m = 2$ mode is formation of a binary system (Boss and Bodenheimer 1979). Because the binary fragments formed have low values of α_i and are subject to substantial tidal forces, these fragments are likely to subfragment during their collapse and so produce a hierarchical system (Bodenheimer 1978). Hierarchical fragmentation is hard to simulate in a single calculation because of the extreme demands on spatial resolution, but the model shown in Figure 3 appears to simulate the beginning of hierarchical fragmentation.

4.1.5. Separation

Rapidly rotating clouds can fragment to form binaries with separations only slightly smaller than the initial cloud radius (Boss and Bodenheimer 1979). Dense molecular cloud cores have radii up to about 0.2 pc (Myers 1987), so provided that a large cloud core is rapidly rotating, the widest binaries could be formed through fragmentation. Slower rotating clouds would fragment to form closer binaries, and smaller cloud cores that start collapse from higher initial densities would form even closer binaries. Fragmentation is halted soon after clouds enter the nonisothermal regime and form the first quasiequilibrium cores, yielding the smallest binary separations (Figure 4), as close as about 1 AU. Fragmentation does not appear to be possible during the second collapse phase leading to final protostellar core formation (Boss 1989). Hence, if formed through fragmentation, very close binaries with separations much less than ~ 1 AU must result from orbital decay of binaries with initially larger separations; rapid orbital decay accompanying fragmentation has been found numerically (Boss 1986) and actually poses a threat to the continued existence of nonisothermal binaries. The full range of binary separations thus appears to be plausible through fragmentation.

4.1.6. Eccentricity

Figure 5 displays the expected eccentricities and semimajor axes for 14 binaries formed in the new set of models. Eccentricities as low as 0.1 were found, but most systems had very large eccentricities ($\sim 0.8 - 1.0$). High eccentricities are

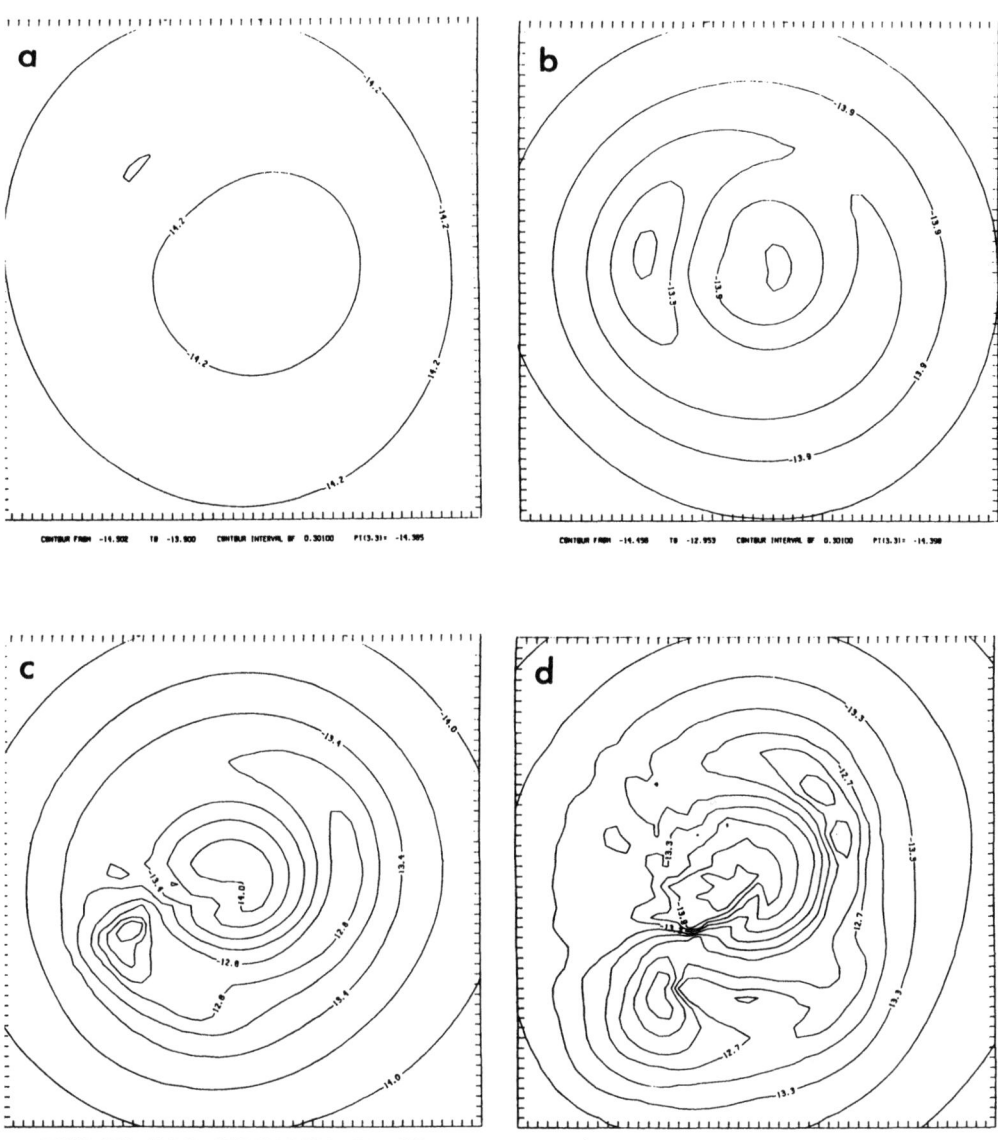

Fig. 2. Equatorial density contours for a model starting with $\alpha_i = 0.22$, $\beta_i = 0.21$, $M_i = 1.0 M_\odot$, and $T_i = 10$K. This model had a very small initial density perturbation, consisting of noise with maximum cell to cell variations of 10%. Radii and times for the sequence: (a) 3.0×10^{15} cm, $1.379 t_{ff}$; (b) 3.0×10^{15} cm, $1.423 t_{ff}$; (c) 2.3×10^{15} cm, $1.451 t_{ff}$; (d) 1.8×10^{15} cm, $1.463 t_{ff}$. Density minima lie at the center of each plot. An eccentric ring forms (a), which undergoes a 'banana mode' ($m = 1$) instability (b); eventually the banana fragments (d) into at least two more objects ($q \approx 0.44, 0.26$).

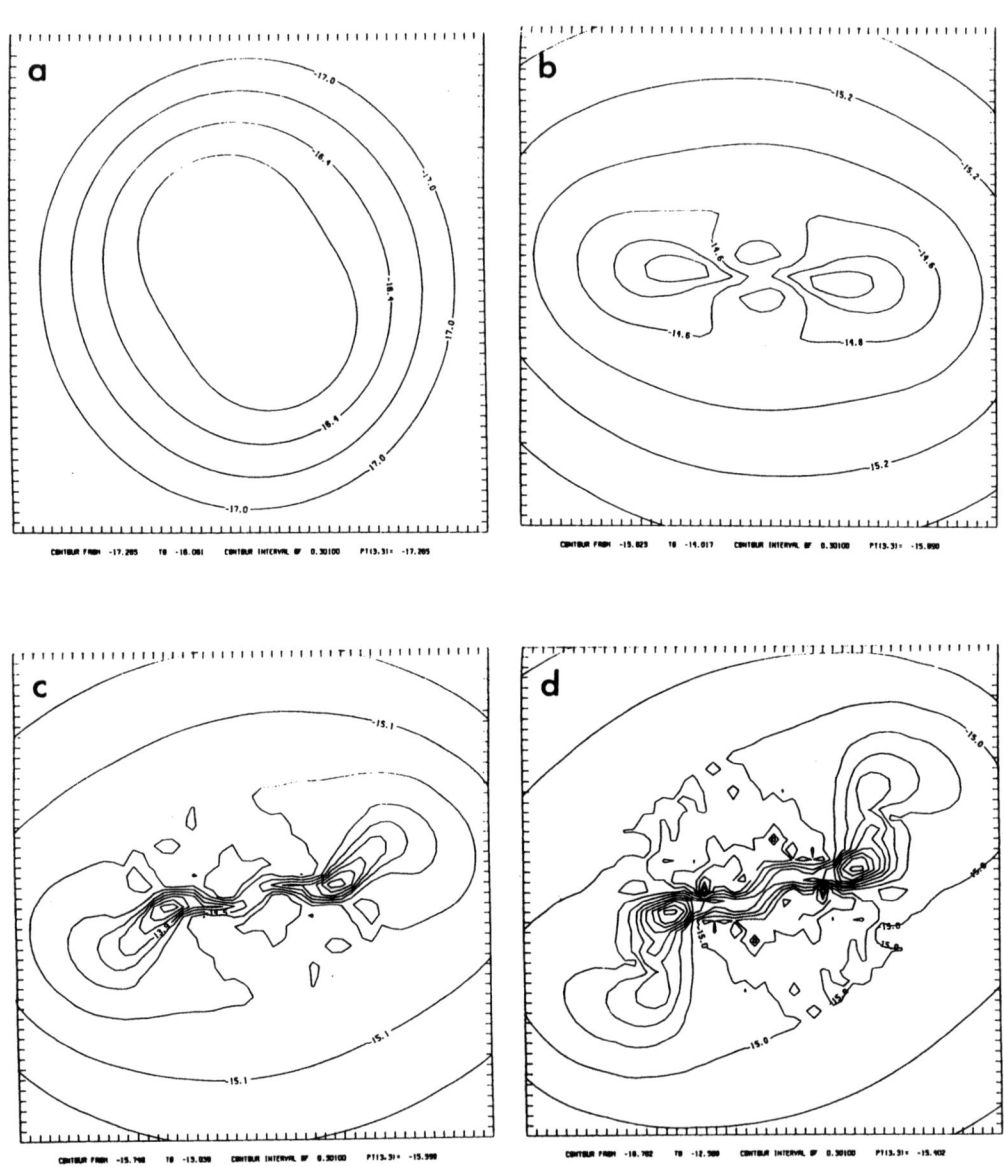

Fig. 3. Equatorial density contours for a model starting with $\alpha_i = 0.22$, $\beta_i = 0.21$, $M_i = 1.0 M_\odot$, and $T_i = 10K$. This model had a small initial $m = 2$ density perturbation, with an amplitude of 0.1. Radii and times for the sequence: (a) 3.2×10^{16} cm, $1.022 t_{ff}$; (b) 1.0×10^{16} cm, $1.277 t_{ff}$; (c) 1.0×10^{16} cm, $1.349 t_{ff}$; (d) 1.0×10^{16} cm, $1.402 t_{ff}$. A binary forms through fragmentation of the collapsing cloud (b), and then both members of the binary begin to subfragment into unequal mass binaries (d), producing a hierarchical multiple system with $q \approx 1$ and $q \approx 0.1$. The small diamond-shaped regions in (d) are density minima.

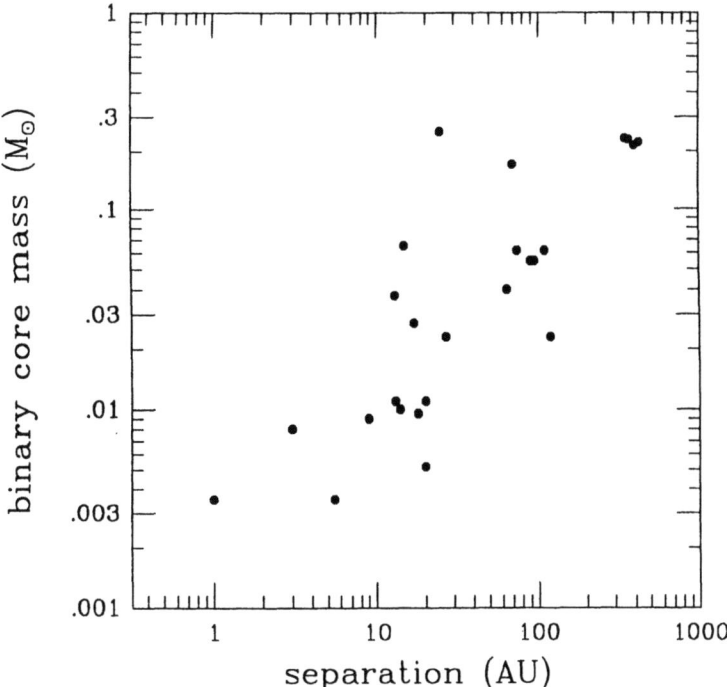

Fig. 4. Binary core masses and separations from the nonisothermal survey of Boss (1986). This survey was limited to low mass clouds; also, the binary cores should increase in mass through subsequent accretion. The separations are approximate measures of expected semimajor axes; note that because of the possibility of partial orbital decay, these should be considered upper bounds. A wide range of binary separations can be produced by fragmentation.

expected to result from fragmentation (Lucy 1981; Boss 1988) because collapsing clouds may start far from centrifugal balance, whereas binaries formed only after formation of a nearly Keplerian disk would be expected to be more nearly circular. Of course, the evaluation in Figure 5 was done at a very early phase of evolution, prior to formation of the final protostellar cores, and so the values appropriate for pre-main-sequence binaries are not certain. These high eccentricities are probably upper bounds to the pre-main-sequence values, because interactions with the rest of the cloud envelope should tend to dissipate energy and circularize orbits.

4.1.7. Spin-Orbit and Orbit-Orbit

Nearly all the FD calculations performed to date have assumed symmetry above and below the equatorial plane, ensuring that any fragments that form from a rotating cloud have spin and orbital angular momenta perfectly aligned — clearly

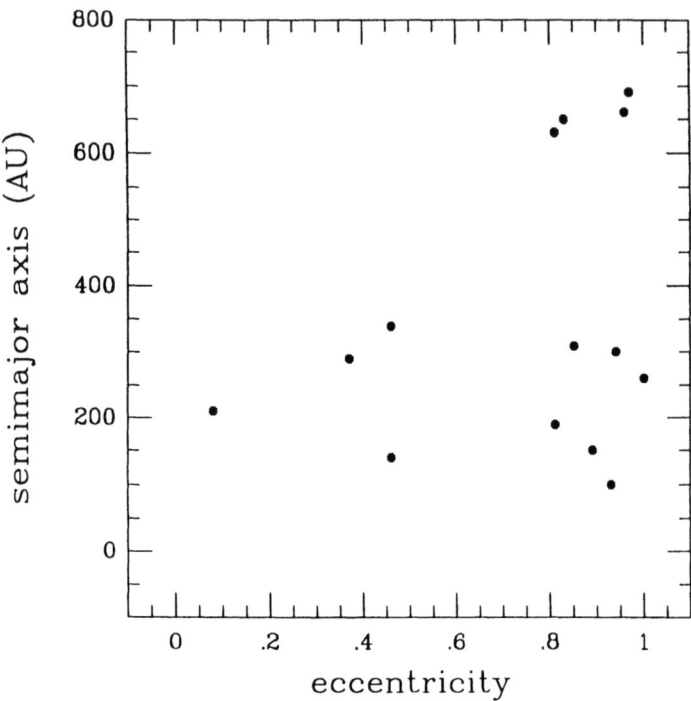

Fig. 5. Eccentricity versus semimajor axis for the new binary models. Orbital parameters are based on fragment masses and velocities and were derived from a solution given by Danby (1988). Evidently fragmentation tends to lead to binaries with initially very high eccentricity; subsequent evolution with the rest of the protostellar cloud may reduce this eccentricity. Binary separations are restricted here to the range $\sim 100 - 700$ AU because these models all started from the same initial mass, β, and temperature.

this overly simplistic alignment is a result of the symmetry imposed. The SPH calculations do not impose this constraint, but because models typically start from a cloud in rapid solid body rotation, the fragments produced in the SPH models to date probably have a fair degree of alignment. More complex initial velocity fields than solid body rotation should lead to some degree of misalignment between the various spin and orbital angular momentum vectors. Recent FD calculations by Sigalotti (1990) relax the equatorial symmetry assumption, and imply that even in the case of clouds initially in solid body rotation, substantial equatorial asymmetry may develop, provided $\alpha_i > 0.1$ or $\beta_i < 0.2$ holds.

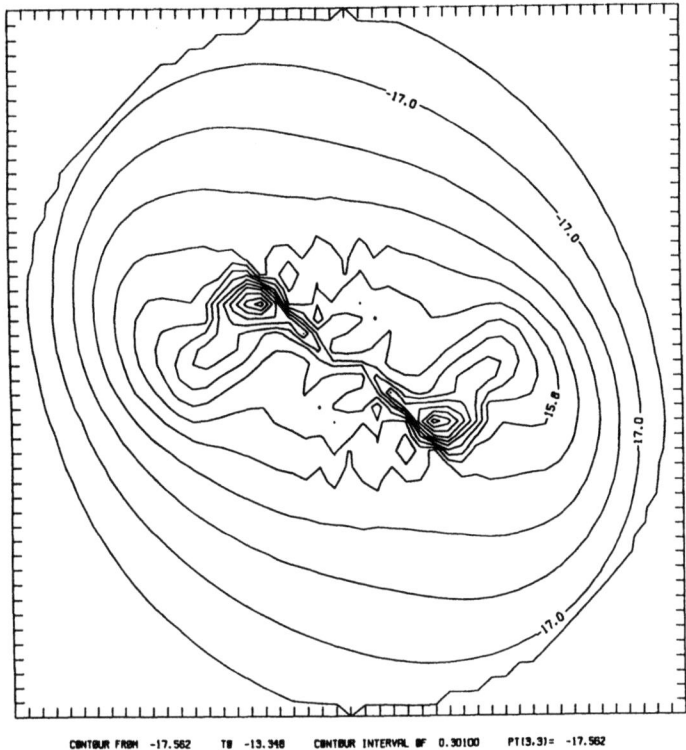

Fig. 6. Equatorial density contours for a model starting with $\alpha_i = 0.22$, $\beta_i = 0.21$, $M_i = 1.0 M_\odot$, and $T_i = 10K$. This model initially had a uniform density perturbed by a 50% $m = 2$ mode. Region shown is 3.2×10^{16} cm in radius and is at $1.363 t_{ff}$. A binary protostar has formed, with fragment properties (e.g., $\alpha_f \approx 0.05$, $\beta_f \approx 0.1$, $M_f \approx 0.16 M_\odot$, $q \approx 1$) very similar to those of the standard test models of Bodenheimer and Boss (1981). In spite of the retention of odd modes, the binary fragments have equal masses ($q = 1$); hints of possible subfragmentation are evident.

4.1.8. Mass Ratio

Previous FD surveys have preferentially formed $q = 1$ binaries because of symmetry assumptions, whereas SPH calculations have produced $q \neq 1$ binaries; in general, however, little attention has been paid to mass ratios. The new models show that suitable cloud cores (e.g., $\alpha_i = 0.22$, $\beta_i = 0.21$) may collapse and fragment into binary systems with mass ratios ranging from $q \sim 0.1$ to 1.0. Clouds with bar-like initial conditions tend to remember the strong initial perturbation and form $q \sim 1$ binaries (Figure 6); in some cases, such clouds appear to undergo subfragmentation into $q \sim 0.1$ systems (Figure 3). Clouds with less well-ordered initial density distributions may fragment directly into low q binaries or multiple systems (Figure 2). Without a more precise knowledge of the initial conditions for cloud collapse, however, the distribution of mass ratios produced by fragmentation cannot be specified.

4.2. CLOUD CLUMP COLLISIONS

Pringle (1989) has hypothesized that binary stars might be formed from collisions between the 'clumps' found in molecular clouds. The largest clumps are the dense cloud cores associated with embedded young stellar objects. Noting that the embedded objects usually do not coincide exactly with the center of the cloud cores, Pringle (1989) infers that relative motions on the order of 0.5 km s^{-1} exist between the embedded objects and the cloud envelopes, and suggests this motion is evidence of a past external influence on the cloud collapse process. One likely external perturbation is a collision with another clump, and provided that the clumps are primarily thermally rather than magnetically supported, clump collisions should lead to collapse (e.g., Lattanzio and Henriksen 1988) and possibly fragmentation. Pringle (1989) further suggests that the clumps may have quite long lifetimes, slowly accreting gas in their envelopes, and periodically forming stars in their centers following collisions.

4.2.1. Dynamics

The cloud clump collision hypothesis is largely a synthesis of the separate nuclei and fragmentation hypotheses — pre-existing cloud clumps, formed by processes poorly understood at present, undergo collisions that induce collapse, perhaps resulting in fragmentation into binary or multiple protostellar systems. Unfortunately, no theoretical models designed to explore this specific hypothesis are available at this time. Nevertheless, enough work has been done on nonmagnetic cloud collisions and fragmentation to make a preliminary assessment.

Pringle (1989) prefers collisions with relative velocities comparable to the sound speed c. Lattanzio and Henriksen (1988) have modeled head-on and off-center collisions at $v_r = c$ of spherical, equal mass, rotating clouds, and found that in each case a single, nearly spherical clump results and begins to collapse. However, these calculations do not attempt to follow the collapse and so the degree of subsequent fragmentation was not ascertained. Based on the results of the fragmentation models described in Section 4.1, we can expect that if collisions do in general lead to nearly spherical clouds at the beginning of collapse, then subsequent fragmentation may not occur unless the cloud's post-collision α has been reduced to a sufficiently low value. If collisions can lead to strongly asymmetric clouds at the start of collapse, clearly the prospects for fragmentation are enhanced, but a reduced value of α is still necessary.

4.2.2. Observations

The frequency of binary formation through cloud clump collisions depends on how frequently clumps collide and at what velocity, their degree of magnetic support, and their internal density and angular velocity fields. The situation is thus the same as for fragmentation: without detailed knowledge of the precursor clouds, the fate of colliding, collapsing clouds is uncertain.

Clump collisions certainly could explain youthful binary systems and presumably the formation of wide binaries; closer binaries would require orbital decay through disk interactions (Pringle 1989). If the colliding clouds do not completely disappear, then highly eccentric orbits and a wide range in mass ratios might be expected. Even if the clouds merge prior to collapse, subsequent fragmentation may produce these characteristics (Section 4.1), and possibly others as well, such as multiple systems and spin-orbit and orbit-orbit alignment. In the latter case, the clump collision model would become more of a means of inducing collapse leading to fragmentation rather than an independent hypothesis.

4.3. ECCENTRIC DISK INSTABILITY

Adams, Ruden, and Shu (1989) proposed the formation of binary stars through spiral gravitational instability in circumstellar disks. Adams, Ruden and Shu (1989) found that thin, massive, Keplerian disks would be unstable to the growth of $m = 1$ modes, equivalent to eccentric motions by fluid elements in the disk, with growth occurring on a time scale comparable to the rotational period at the outer edge of the disk. In order to preserve the center of mass, growth of the $m = 1$ mode in a disk generally requires motion (wobble) of the central star, and the resulting coupling may enhance the chances for growth of the $m = 1$ mode in an otherwise axisymmetrically stable disk.

4.3.1. Dynamics

Shu *et al.* (1990) more fully explored the physical basis of the eccentric instability, and showed that in some cases a marginally unstable disk would have a mass of about 0.24 $M_\odot$ for a central star with a mass of 0.76 $M_\odot$; gravitational instability would then require the disk to be more massive, nearly as massive as the central star. Shu *et al.* (1990) also stressed that the growth of the eccentric instability is critically dependent on positive feedback through wave reflections off a 'sharp' outer disk boundary, which may not exist in a real disk, but argued that disks might be 'self-sharpening', or that interactions with an eccentrically distorted edge might lead to continued growth.

The numerical analysis of Adams, Ruden, and Shu (1989) was based on linear perturbation theory and so was unable to follow the growth of the $m = 1$ modes into the nonlinear regime where the possibility of binary formation could be best assessed. Cassen *et al.* (1981) studied the nonlinear growth of gravitational instability in thin, massive disks, stabilized by central stars, but did not focus on $m = 1$ modes. Nonlinear growth of $m = 1$ modes has been demonstrated in rotating isothermal rings by Norman and Wilson (1978), and dominant $m = 1$ modes have been encountered in nonlinear collapse calculations as well [e.g., Figure 1(a,b) and

Figure 2(a,b)], so $m = 1$ growth well into the nonlinear regime may very well result. However, the nonlinear outcome need not necessarily be binary formation, if the outcome of the fission instability is an appropriate analogy. That is, a disk that gains enough mass to become gravitationally unstable may encounter substantial growth of the $m = 1$ mode, but because this growth occurs only on the time scale of the rotation period of the outer edge of the disk, and the binary companion is hypothesized to form at the $m = 1$ mode's corotation radius (located well inside the disk), angular momentum transport by trailing spiral arms outside this corotation radius may result in a relatively rapid disk evolution that prevents formation of a stable binary. Clearly a numerical treatment of the nonlinear evolution of these modes is required for further progress.

4.3.2. Observations

Without knowledge of the nonlinear fate of the eccentric modes, the frequency of binary star formation by this mechanism cannot be determined, nor can we assess the chances of forming multiple systems, highly eccentric systems, or diverse mass ratios. The eccentric disk instability probably could account for very young binaries, assuming that the instability can occur prior to the central star evolving very far along its pre-main-sequence track, and could account for spin-orbit and orbit-orbit alignments by virtue of formation in a prograde rotation disk. Binary separations would appear to be limited by circumstellar disk sizes to values on the order of 10 to 100 AU; wider binaries may not form through this mechanism.

5. Conclusions

Table I summarizes the comparisons made between the six theories of binary star formation and the observational constraints. The assignment of plus and minus signs is necessarily somewhat subjective, and other workers might very well differ on specific assignments. Nevertheless, it is gratifying at least to note that the question marks are in the minority; evidently we are making significant progress in our ability to discriminate between the competing hypotheses. The remaining question marks highlight the areas where future work is necessary to further our understanding of binary star formation.

Table I is intended to reveal if a single mechanism is capable of explaining all types of binary and multiple systems. Given our current knowledge, fragmentation appears to come the closest to success as a universal mechanism of binary formation. Further progress in assessing the importance of fragmentation appears to require improved observational knowledge of the initial conditions for protostellar collapse. Other promising mechanisms, such as capture through protostellar disk interactions, cloud clump collisions, and eccentric disk instability, primarily require sustained theoretical development. Of course, it must be admitted that a universal mechanism may not exist; perhaps several mechanisms contribute by producing

TABLE I
Summary of the adequacy of six theoretical hypotheses as *universal* mechanisms for explaining binary and multiple star formation. Plus signs denote agreement with the constraint, minus signs denote disagreement, and questions marks denote situations where too little is known to make even a tentative comparison.

Constraint	Separate Nuclei	Capture	Fission	Fragmentation	Clump Collisions	Eccentric Disk
Dynamics	?	+	–	+	?	?
Frequency	?	?	–	?	?	?
10^5 yr age	+	+/–	+	+	+	+
System type	?	–	–	+	?	?
Separation	–	+/–	+/–	+	+/–	+/–
Eccentricity	?	+	–	+	?	?
Spin–orbit	–	–	+	+	?	+
Orbit–orbit	–	–	+	+	?	+
Mass ratio	+	+	–	+	?	?

only certain types of binary stars, so we should not dismiss hypotheses capable only of producing a subset of binaries.

The higher spatial resolution calculations of fragmentation presented here have confirmed much of what was previously known about fragmentation, and have yielded new results as well. The formation of $q \neq 1$ systems has been shown to occur in fragmenting clouds with varied initial density perturbations. Highly eccentric binary systems were found to be the usual outcome of fragmentation. Certain clouds with ordered initial density perturbations were found to be likely to undergo hierarchical fragmentation; the binary fragments formed from the initial perturbation begin to undergo subfragmentation, implying formation of a hierarchical multiple system. These results have buttressed the case for fragmentation as the mechanism most capable of explaining the formation of the majority of binary and multiple stellar systems.

Acknowledgements

I thank Wen Ping Chen and Robert Mathieu for kindly providing preprints of their work, and Wen Ping Chen for his comments on the manuscript. The calculations were performed on the Apollo DN-10040 workstation of the Carnegie Institution of Washington. This research was partially supported by U. S. National Science Foundation grant AST 88-17334.

THE EVOLUTION OF INTERACTING BINARY SYSTEMS

Observational Aspects

G. E. MCCLUSKEY, JR. and Y. KONDO

1. Introduction

The very existence of interacting binaries was hardly perceived before the early nineteen forties. In 1991 the study of the evolution of interacting binaries is an important part of astrophysics and some of the most bizarre of stellar objects are to be found in these systems. Understanding the formation and evolution of interacting binaries is one of the principal goals of modern astronomical research. Interacting binary stars afford us the opportunity to acutely test the theory of evolution of stars, which in turn is one of the basic underpinnings of nearly all of astrophysics. A number of factors have been responsible for the rise to prominence of interacting binaries.

First was the recognition that such systems exist. The pioneering research of Struve (1941), Kuiper (1941), Gill (1941), and Greenstein and Page (1941), among others, led to the recognition that gas flow, mass exchange and mass loss play a major role in the evolution of close binaries. Second was the development of the theory of stellar evolution for single stars and its application to understanding the Hertzsprung-Russell diagram. By the mid-sixties the evolution of single stars was becoming sufficiently understood in order to apply it to interacting binaries. Third was the development of observational techniques and the application of computers to data analysis and modelling. The advances in observational methods include improvements of standard methods, e.g., photometric and spectroscopic techniques; the application of new methods, e.g., high speed photometry and charge coupled devices, and the opening up of unexplored or poorly explored regions of the electromagnetic spectrum, e.g., space astronomy. Finally, the discovery of the existence of exotic interacting binaries such as the cataclysmic variables, X-ray binaries, SS 433, and binary pulsars, has greatly stimulated interest in interacting binaries.

This chapter discusses some observational aspects of the evolution of interacting binaries. Many subjects of interest in this regard are discussed in other chapters of this book. The discussion in this section is restricted to interacting binaries in which neither component is degenerate. Low mass systems, e.g., W UMa stars, are discussed elsewhere in this volume and are not considered here. A few recent contributions to this subject are McCluskey (1982), De Loore (1984), Kopal (1984),

Eggleton (1985), Budding (1986), De Greve (1986), and McCluskey and Sahade (1987).

2. Pre-Main Sequence Binaries

The origin of close binaries is discussed elsewhere in this volume by Boss. Here the discussion is limited to two systems which are believed to have a pre-main-sequence secondary component. In fact, very few such systems are known and this stage in the evolution of close binaries is very poorly understood.

The strongly interacting binary HD 37021=BM Ori appears to consist of a B3 V zero-age main sequence star detached from its Roche critical equipotential surface and a less massive but larger A-type peculiar secondary star with a disk-like structure surrounding it (Popper and Plavec 1976). Since the primary does not seem to be losing mass at a significant rate it seems unlikely that the disk derives from that component. There is no indication of circumstellar material near the B-star indicating that the secondary star is not an evolved star losing mass by so-called Roche lobe overflow. The secondary star appears to be highly flattened although untangling the disk and star is very difficult. Arnold and Hall (1976) found primary eclipse to show significant time dependent variations in depth and shape. It certainly seems possible that the secondary star is in the last stages of final accretion in star formation. Unfortunately, this system has not been given the attention it deserves.

The detached system EK Cep was found to have a secondary component which is probably a pre-main sequence star. Popper (1987) finds it to have anomalous values of color, temperature and gravity for its mass. With the exception of a small number of systems which might have secondary components not yet quite on the zero-age main sequence, BM Ori and EK Cep seem to be the only close binaries known to be of this type.

3. Massive Interacting Binaries

Considerable observational and theoretical effort has been devoted to understanding the evolution of massive systems. In this discussion a massive binary is one in which at least one component initially had a minimum of 10–15 solar masses.

In interacting binary research the concept of the critical Roche lobes has had immense influence. Regrettably, the origin of this concept often seems to have been forgotten, particularly for massive interacting binaries. The critical Roche lobes are particular equipotential surfaces in the restricted three-body problem. They are meaningful only insofar as the conditions assumed for the classical restricted three-body problem are valid. One primary assumption is that the only force acting is gravity. The very existence of a stellar wind belies the existence of critical Roche lobes as defined above. The effects of radiation pressure on the critical Roche lobes has been discussed by Schuerman (1972), Kondo and McCluskey (1976), Kondo *et al.* (1976), Vanbeveren (1977), Zorec and Niemela (1980a, b), and Huang and Taam (1990), among others. Even if radiation pressure is too weak to completely counteract gravity, it distorts the lobes and if a wind exists critical Roche lobes do not.

The meaning of an equipotential surface in the context of the restricted three-body problem is itself very restricted. A surface exists only for non-interacting particles which satisfy a given set of initial conditions. This particular surface does not exist for particles with different initial conditions; this surface does not exist for gas streaming around the equator of a star; this surface does not exist if pressure plays any role whatsoever. Sahade and Wood (1978) pointed out the dangers of indiscriminate use of these equipotential surfaces and critical labels. Irrespective of all of this, the literature all too often still contains claims that the O-type component in some particular system is in contact with its critical Roche lobe. This despite the fact that this O-star is known to be losing mass at 10^{-6} solar mass per year via a powerful stellar wind.

This is not to say that the gravitational field, etc., of a close companion star has no effect. Quite the opposite. But the classical critical Roche lobes are often utterly inadequate in modelling these effects. It would require incredibly good fortune for the classical model to be at all accurate. Since massive stars almost always have winds, dimensions of massive binaries based on assumptions about critical Roche lobes must be considered as quite suspect. Here we simply use the term overflow for mass lost due to the binary nature of the system.

3.1. UNEVOLVED SYSTEMS

Unevolved systems are those in which no mass overflow has occurred and both components are in the H-core burning stage of their evolution. If masses, radii, and effective temperatures can be accurately determined, these systems provide relatively straight forward tests of stellar evolution theory. Hilditch and Bell (1987) compared observation and theory for a number of well-determined massive and moderate mass binaries. They concurred with a number of earlier theoretical and observational investigations (e.g., Humphreys and McElroy 1984; Doom, De Greve and De Loore 1985; Stothers and Chin 1985; Mermilliod and Maeder 1986) that classical evolutionary models of massive stars must be modified to account for convective core overshooting and stellar wind mass loss. Examples of massive unevolved systems are V1182 Aql, AH Cep, Y Cyg, V453 Cyg, V380 Cyg, CW Cep, and α Vir (Hilditch and Bell 1987). Other massive unevolved systems with well-determined parameters are EM Car (Andersen and Clausen 1989) and NY Cep (Holmgren *et al.* 1990).

3.2. EVOLVED SYSTEMS

A large effort has been expended to arrive at an understanding of the evolution of massive interacting binaries. The subject has been reviewed by Vanbeveren and De Loore (1980), De Loore (1984), De Loore and Sutantyo (1984), Chiosi and Maeder (1986), and Hilditch (1989).

Mass loss may proceed via a stellar wind and/or overflow. Vanbeveren (1982) found that the effects of stellar wind mass loss during H-core burning on stellar evolution was marginal for initial masses below 40 solar masses but that up to 20-30% of the initial mass could be lost by more massive stars. In post-H-core burning evolution, large amounts of mass can be lost by all massive stars via 'winds'. The O-type binary UW CMa is losing mass at a rate of about 3×10^{-6} solar mass per year via a 750 km/s stellar wind (McCluskey, Kondo, and Morton 1975). Unfortunately, it is invariably called a semi-detached or a contact system (Leung and Schneider 1978; Eaton 1978; Wilson and Rafert 1981) with at least the Of component at its Roche lobe. This is also the claim for many O-type binaries (Wilson and Rafert 1981). The O9.5 III + B0.5 III eclipsing binary LY Aur provides another example of the unwillingness of astronomers to believe that the classical Roche lobes are meaningless for most, if not all, O-type stars in interacting binaries. Eaton (1978) states that LY Aur could be a contact system while also stating that radiation pressure causes significant distortion of the system. Li and Leung (1985), without a word about the powerful stellar wind known to be present in LY Aur, call LY Aur a semi-detached system. Recently, Howarth and Prinja (1989) have surveyed the stellar winds of over 200 O-type stars. The stellar wind mass loss rate for LY Aur is 5×10^{-7} solar mass per year. If accurate dimensions are to be determined for systems like UW CMa or LY Aur, the classical Roche lobe idea must be scrapped and replaced by a more appropriate model.

It has become certain that mass overflow in massive interacting binaries causes much mass and angular momentum to be lost from the system. Vanbeveren and De Loore (1980) studied evolved systems where the mass ratio had been reversed, presumably by overflow, and concluded that at least 50% of the mass lost by the primary star leaves the system taking about 50% of the system orbital angular momentum with it. Huang and Taam (1990) discussed the non-conservative evolution of massive binaries. System mass loss leads to reduced orbital periods and smaller remnant masses for the original primary star as compared to conservative evolution. The period decrease is predominantly caused by loss of angular momentum.

The later stages in the evolution of massive interacting binaries have been discussed by De Loore and Sutantyo (1984), among others. The primary star loses much or all of its hydrogen rich envelope, becomes a Wolf-Rayet star and eventually undergoes a supernova explosion of Type II and becomes a neutron star or black hole. Since the original secondary star is now the more massive star, the supernova explosion is not likely to disrupt the system. If the original secondary has or develops a stellar wind, a massive X-ray binary results as the degenerate component accretes material from this wind. Eventually the original secondary overflows, loses its envelope and becomes a Wolf-Rayet star with a low mass companion. The original secondary becomes a supernova and if the binary system survives a binary pulsar system may result. If a double neutron star system is formed at sufficiently small separation, the more massive neutron star will tidally disrupt the less massive component as spiral-in occurs due to gravitational radiation. Accretion of this disrupted material may give birth to a millisecond pulsar. If either component in the original binary loses sufficient mass, a white dwarf may result. A number of variations of this scheme are possible. These late stages of interacting binary evolution are discussed more fully elsewhere in this volume by Trimble.

Two rapidly evolving massive binaries are BD $+40°$ 4220 and HDE 228766. Bohannan and Conti (1976) find minimum masses of 47 and 11 solar masses for the O7 and O6 components of BD $+40°$ 4220. The O6 star is almost as large as the O7 star and must already have lost considerable mass. Both are classified as Of supergiants. Some emission line characteristics of the O6 star are intermediate between ordinary Of and Wolf-Rayet stars. It is postulated that the O6 star is becoming a Wolf-Rayet star. Vreux (1985) claims that the O6 star is a Wolf-Rayet star but is accreting matter from a strong stellar wind or overflow of the O7f massive component. It is not made clear how a Wolf-Rayet star losing mass at a very high rate can accrete matter. This system warrents intensive study. Massey and Conti (1977) proposed that HDE 228766, usually considered a Wolf-Rayet binary, consists of two O-type supergiants of nearly equal mass, with one of these supergiants making the transition from an Of star to a Wolf-Rayet star.

The Wolf-Rayet stars provide some fascinating puzzles. Doom (1987) has given an extensive review of these stars. Approximately 30% of Wolf-Rayet stars have an O- or early B- type companion while 15–20% are binary with no spectroscopically or photometrically detectable companion. It is usually assumed in this latter case that the companion is a neutron star or black hole although there is no direct evidence for this. Consequently, it appears that about half of the Wolf-Rayet stars are binaries and half are single stars. It is presumed that overflow is the cause for the rapid mass loss, $1-15 \times 10^{-6}$ solar mass per year, in some of the binary Wolf-Rayet stars but since several have periods of months or years this is not likely to be the case for all of the binaries. The single Wolf-Rayet stars are losing mass via some unknown instability although it is possible that an enhancement of the radiative stellar wind mechanism operating in hot stars may provide an explanation. No significant difference between the single and binary Wolf-Rayet stars has yet been discovered and it is not at all clear whether or not the presence of a companion has any effect on the Wolf-Rayet phenomenon and its evolution.

The most important need in future research is to determine the physical parameters for massive binaries *without* using the classical Roche model.

4. Moderate Mass Interacting Binaries

A moderate mass binary is one in which the most massive star in the original system will not become a Type II supernova but is more massive than 1 or 2 solar masses. This requires an upper limit of about 10–11 solar masses. The majority of moderate mass interacting binaries are usually referred to as Algol-type binaries. The classical Algol-systems consist of a B -or early A-type main sequence star and a G- or K-type sub-giant or giant star which is presumed to be at its critical Roche lobe. In general, the Algol systems are a large subset of the semi-detached binaries in which one component is presumed to fill its critical lobe. The use of the classical Roche lobes for moderate mass binaries is perhaps not as questionable as its use for massive systems, since radiation pressure is of little or no importance so far as is known. However, the cool stars in Algol systems probably have significant magnetic fields with all of the phenomena which accompany them, e.g., starspots, flares, chromospheres and coronae. Since gas is observed to escape from these binary systems even during quiescent phases, stellar winds must be present (Kondo 1989). In addition, during rapid overflow the conditions assumed for the Roche model are invalid. Once again it is important to determine the physical parameters of these sytems without assuming the classical Roche model. Mass loss and several other aspects of Algol systems are discussed elsewhere in this volume by McCluskey and by Sahade so this discussion will concentrate on other aspects of their evolution.

It is convenient to divide the Algol systems into three types. The most active systems will be referred to as dynamic Algols. Examples are SX Cas or V367 Cyg. Moderately active systems such as U Cep, TT Hya, or SW Cyg will be referred to as active Algols. Finally, low activity Algols such as S Cnc, δ Lib, or U Sge are the least active systems.

The Algol systems are also subject to large amounts of loss of mass and angular momentum during overflow. De Greve and Vanbeveren (1980) found that more than 50% of the original primary's mass and a large percentage of orbital angular momentum are lost to the system. Popper and Tomkin (1984) give masses of 2.33 and 0.17 solar masses for the B9 V and G8 IV components of S Cnc. The G8 IV star must have lost at least 80–90% of its mass and much of this probably left the system.

4.1. PERIOD VARIATIONS

Mass transfer and/or loss in a binary system will generally result in a period increase or decrease depending on the specific circumstances. Many systems do show simple increases or decreases in period but more complex variations are observed for other systems. Often, non-linear terms are present in the period increase or decrease indicating variable mass and angular momentum loss. A particularly interesting period variation has been observed for a number of Algol binaries.

Abrupt fractional period changes on the order of 10^{-4} to 10^{-7} occur on time scales from months to decades. The duration of time over which these changes occur is not known but is probably much shorter than one year and could be on the order of a few orbital periods or less. The signs of these changes tend to alternate although a few successive changes may have the same sign. Heintze and Van Gent (1989) reported on seven period jumps in U CrB over 10–20 years. In 1980, a change of approximately 10^{-4} occurred twice, once positive and once negative. Olson (1985) reports on four abrupt period changes of U Cep in 11 years with alternating signs. Interestingly, these period variations do not seem to be related to the mass flow outbursts which occur on a timescale of 5–15 years. Similar abrupt changes have been observed in RS Cep (Olson and Stoehr 1986) and U Sge (Olson 1987). At least one abrupt period change has been observed in a number of interacting systems. These abrupt variations may be even more common than presently known since they are easily obscured by a non-linear term in the period variation if observed times of minima are separated by long periods of time.

Hall (1989) proposed that the quasi-periodic alternations of these abrupt period changes in classical Algol systems are related to a magnetic cycle occurring in the cool component. For 31 stars, periods of 7–109 years with a median of 50 years were found. This does not explain the abrupt variations occurring on shorter timescales and Marsh and Pringle (1990) find that the energy required for these quasi-periodic changes would take 10–100 times longer to generate than their observed timescale.

4.2. ROTATION

In the great majority of Algol systems, the hotter, main sequence component is accreting matter from the cool subgiant or giant component. The accreted matter brings orbital angular momentum with it and this is converted, at least in part, to rotational angular momentum of the accreting star. This rotational angular momentum creates an equatorial bulge or even a disk-like structure around the star. Rapid non-synchronous rotation of the hot star in Algols with orbital periods less than a few weeks is a strong indicator of rapid accretion. Van Hamme and Wilson (1990) have studied the rotational velocities of several dozen Algol primaries. For the most part, those primaries which are rapidly rotating are those in systems in which mass flow is taking place at a moderate to high rate. The primaries of SW Cyg, RW Per, and RZ Sct are rotating near the centrifugal limit and those of U Cep, RW Mon, and RY Per are rotating at 5–10 times the synchronous rate. All of these systems are known for their relatively high rates of mass flow. Polidan (1988) finds a rotational velocity of 10–11 times synchronous for the hotter component of V356 Sgr which puts it near the centrifugal limit. The system β Lyr, which by a stretching of the definition might be called a dynamic Algol binary, contains a hot, primary star which is either disk-like or is surrounded by an optically and geometrically thick disk. This structure is presumably rotating at or near the centrifugal limit although

4.3. LONG PERIOD SYSTEMS

In Algol systems with periods longer than 7–10 days, a relatively large volume of space is available for both evolutionary expansion of the stars and for mass flow. This leads to a rich variety of possibilities. The original primary star can evolve into a true giant or supergiant before overflow occurs if the orbital period is sufficiently long.

Ake, Parsons, and Kondo (1985) have analyzed ultraviolet spectra and ground-based observations of 22 Vul (G3 Ib-II + B9; P = 249 days). A dense wind from the cool supergiant envelopes the B-star although the radius of the supergiant is much smaller than that required for overflow.

Olson (1988) analyzed the light curve of KU Cyg (F4p + K5 III; P = 38.44 days). The cool star appears to be too small for classical overflow but clearly strong mass flow is occurring. The light curve is very complicated and a thick, slightly dusty disk surrounds the F-star. The K-star has a mass of only about 1/5 that of the F-star and may already have lost a large fraction of its original mass. The accretion rate of the F-star is time dependent with maximum values estimated at approximately 10^{-8} solar mass/year. The presence of dust in KU Cyg is indicative of dust formation which occurs in a number of long period systems with large mass flow and relatively cool stars.

Knee et al. (1986) analyzed RZ Oph (F5 II–III + K5 III; P=261.93 days) which appears to be well detached from the classical Roche lobes. The K-giant has a mass of 0.70 solar mass and radius of 41 solar radii while the F-star has 5.65 solar masses and 5.4 solar radii. The K-star appears to be highly evolved. There is also evidence for the existence of a disk around the F-star. Research into the tidal effects of a companion star on a giant or supergiant star's convective envelope would be very useful. We do not know whether the K-star has reached its present state via single star evolution or whether it has been significantly affected by its companion.

The only dynamic Algol systems known (RX Cas, SX Cas, W Cru, V 367 Cyg, β Lyr, and W Ser) have orbital periods between 12 and 37 days except for W Cru with a 198.53 day period. The shorter period system R Ara (P = 4.43 days) is quite active and may also be a dynamic Algol system. It is not clear that these systems form a coherent physically meaningful group but all are characterized by very distorted and variable light and velocity curves, extensive circumbinary and circumstellar gas, large period variations and partial or total obscuration of at least one component. It is not clear that these are semi-detached systems but that assumption is almost invariably made in order to estimate physical parameters for these systems. Whatever the cause, these systems are undergoing rapid mass loss.

4.4. MAGNETIC FIELDS

A summary of direct and indirect evidence for magnetic fields in cool stars is given by Gray (1989). In interacting binaries of moderate mass, direct detection of magnetic fields has proven difficult, with the exception of a few RS CVn binaries, e.g., λ And (Giampapa, Golub, and Worden 1983) which has a magnetic field of about 1.3 kilogauss over approximately one-half of its surface, due to the faintness of the cool components and the complexity of the binary spectrum. Indirect evidence is given by the existence of starspots, X-ray emission, radio emission and other chromospheric and coronal effects.

White and Marshall (1983) detected variable X-ray fluxes from 6 of 9 Algol systems observed with the Einstein Observatory. The X-ray flux was similar to that of normal single stars of the same spectral and luminosity class as the cool Algol components. McCluskey and Kondo (1984) detected several other Algol systems at similar X-ray luminosities. It appears that proximity to the Sun is one factor in determining which systems were detected.

The nearby low activity Algol β Per has been studied in the X-ray region of the spectrum in some detail. White *et al.* (1986) observed β Per for 35 hours continuously centered about secondary eclipse at 0.1–10 kev with EXOSAT. Quiescent emission from 1–10 kev with a temperature of 25×10^6 K and an X-ray flare were observed. The quiescent X-ray emission was not eclipsed by the hot star indicating a scale height equal to or greater than the radius of the cool star. The quiescent emission is presumed to arise in the corona of the K-star. The X-ray flare seen from 0.1–10 kev had a rise time of 1700 seconds and decayed in 7000 seconds. The peak luminosity was 1.4×10^{34} erg/s with a peak temperature of 60×10^6 K. This flare is believed to have arisen in a magnetic loop 0.1–0.2 K-star radii above the K-star's photosphere.

Slee *et al.* (1987) detected 6 of 20 Algol systems at 5.8 and/or 8.4 GHz with the 64-meter Parkes radio telescope. The mean brightness temperature of the Algols was approximately 1.1×10^{10} K, sufficiently high that a coherent process such as plasma or cyclotron masering might be present. Alternatively, the radio source might be much larger than the system. The low activity system δ Lib was the most active of the detected systems and exhibited an intense radio flare at 8.4 GHz. All sources were highly variable. Stewart *et al.*(1989) discussed 15 eclipsing binaries detected at 8.4 GHz. In addition to the additional detections of two of the six Algols detected by Slee *et al.* (1987), the Algol δ Cap was detected as well as six early type binaries. Several other binaries with various spectral classes were also detected. All were detected once or twice out of six to twenty-two attempts except for TZ Men (B9-A1 III + B9-late A) detected seven out of sixteen times, δ Lib (A0 IV/V + G2 IV) detected eighteen out of thirty-three times, V393 Sco (B3-9 III + F III?) detected seven out of thirteen times, and V505 Sgr (A2 V + F7-G7 IV) detected three out of fourteen times. The detection of the early-type binaries ζ Phe (B6 V + B8 V), δ Ori A (O9 III + B0 III), TZ Men, V760 Sco (B4–8 V + B8-A3?),

V1647 Sgr (A1 V + A2 V), and V822 Aql (B3 V + B9 V) indicates the presence of moderately strong coronal magnetic fields of the order of 100 Gauss between the stars which indicates a surface magnetic field of $\sim 10^3 - 10^4$ Gauss on one or both components. If this interpretation is correct, the gas dynamics in early-type systems may be quite different than heretofore imagined.

Elias (1990) has detected the dynamic Algol V367 Cyg at 3.6 and 6 cm and the dynamic Algol W Ser at 3.6 cm with the VLA. The active system RZ Sct (B2 II + A0 II-III) was also detected at 3.6 cm. This radio emission is attributed to thermalized gyrosynchrotron radiation which would again emphasize the importance of magnetohydrodynamics in at least some interacting binaries.

Lestrade *et al.* (1988) obtained dual-polarization and dual-frequency VLBI observations of β Per at 1.6, 2.3 and 8.4 GHZ. The system had brightness temperature varying from $3-50\times 10^8$ K. This is attributed to gyrosynchrontron emission in an active coronal region or regions by mildly relativistic electrons in a magnetic field with a strength of a few tens of Gauss. The radiating region was about three times the radius of the K-star in size and is associated with the K-star. The separation of the two stars is about six times the radius of the cool star. In addition, two radio outbursts were detected. A broad-band outburst with a brightness temperature of $1.5-3.0\times 10^{10}$K in a magnetic field of about 10G at approximately 0.7–3.0 K-star radii above the K-star was observed. The second short outburst detected at 1.66 GHz lasted about 15 minutes, had a brightness temperature greater than $1.5 \times 10^9 K$ and may have occurred in a 300 G field at the foot of a magnetic loop. These radio detections are similar to VLBI detections of a number of RS CVn systems where very strong evidence exists for the presence of active magnetic fields.

Along these lines, Hall (1989) has noted that the late-type secondaries in Algol binaries are often relatively rapidly rotating convective stars and should consequently be chromospherically active. For example, the secondaries of U Cep, AR Mon, and S Vel, among others, should show properties similar to those found in chromospherically active stars.

Strassmeier *et al.* (1990) have surveyed late-type stars, both single and those in binaries, and found that the dominant factor determining the level of chromospheric activity is the rate of rotation. Higher rotation rates yield higher chromospheric activity. The binarity per se plays no role in those systems studied. Naturally, synchronous rotation due to tidal forces in an interacting binary leads to rapid rotation for late-type stars. It seems quite likely that this rotation also leads to enhanced magnetic field activity.

Richards (1990) obtained infra-red (1.2 μ) light curves of β Per and detected time-dependent variations in the depths and phases of both eclipses, the asymmetric shape of the secondary minimum, cyclic variations of ±350K in the mean temperature of the secondary star on a time scale of 1–2 years and a wave distortion of the light curve. These effects are quite similar to those observed in the optical light curves of the RS CVn binaries in which they are attributed to magnetic activity, e.g., starspots on the chromospherically active K-type secondaries. It would certainly appear that the K2 IV secondary of β Per is magnetically active. Clearly, magnetic phenomena are occurring in many Algol systems. Magnetic fields can have significant effects on gas flow and on the long term evolution of the system by means of magnetic torques. Little to nothing is known of these processes in Algol systems.

4.5. EVOLUTION

The theory of the evolution of moderate mass interacting binaries has been developed in considerable detail by Iben and Tutukov (1985), and Iben (1985, 1986), among others.

In general, the final evolutionary state of these stars is a white dwarf. Iben (1985) finds that stars with initial masses between 2.3 and 10 solar masses become carbon-oxygen white dwarfs. Helium white dwarfs result from lower mass stars. If a pair of white dwarfs with a separation of less than approximately three solar radii results, then coalescence will occur via magnetic braking and/or gravitational radiation on a time-scale less than the Hubble-time. This will lead to a single white dwarf or a supernova, presumably of Type I. It is possible that stars with initial masses above 7–8 solar masses may become supernovae and become neutron stars (Iben 1986).

Eggleton and Tout (1989) discuss the possibility that a tidally enhanced stellar wind may reduce the original mass of the primary below the original mass of the secondary before overflow occurs. If the orbital period is fairly long, wind loss can reduce or prevent rapid overflow. If the period is shorter, but sufficiently long for the original primary to develop a fully convective envelope, overflow can be drastic and a common envelope system may result as a large amount of mass is lost in a short time. Magnetic braking could also reduce the separation of the components and enhance mass loss even further.

Pastetter and Ritter (1989) find that if the primary star becomes an asymptotic branch giant before overflow occurs, then its original mass can be reduced by more than 50% and, if overflow finally occurs, it will be slow and a large (500–1000 solar radii) cool disk can form around the companion star. Could this be ϵ Aur?

De Greve and Packet (1990) analyse the possibility that due to mass accretion, the original secondary star evolves more rapidly, begins to overflow and reverses the flow. Their discussion is severely limited by the assumption of conservative mass flow but this possibility has generally been neglected and must be examined under more realistic conditions.

It would be useful to find what Leung (1989) has called reverse Algols, systems in which the more massive star is overflowing so that the mass ratio has not yet been reversed. Leung (1989) lists seven candidates but ADS 9010 is a W UMa system as is probably the case for RT Scl. HD 199497, V 1425 Cyg, V790 Cen, and V3894 Sgr have such low inclinations (52°–71°) that light curve orbital elements are really indeterminate. That leaves only BF Aur (B5 V + B5 V) which is a borderline case. Reverse Algols are probably yet to be found.

It is conceivable, e.g., Sahade (1986,1987) and Wilson (1989), that some dynamic Algols evolve into active Algols which evolve into low activity Algols. This scheme probably has a number of ifs, ands and buts. In our present state of ignorance, it is a beginning.

4.6. A FEW PECULIAR SYSTEMS

The unique system β Lyr has figured very prominently in the binary star literature. Sahade (1980) has reviewed the system and only a few recent results will be discussed here.

Polidan (1989) has observed β Lyr from the Lyman limit to 3200 Å with the International Ultraviolet Explorer satellite and the Voyager spacecraft. A strong ultraviolet continuum is present down to 912 Å. Remarkably, no eclipses are detectable below 1200 Å as observed with Voyager. A rapid ultraviolet brightening of 50% centered on 955 Å occurred in 40 hours. The far-ultraviolet spectrum resembles that of a cataclysmic variable accretion disk. If the secondary object (the B8p star is photometrically the primary star) is a massive cataclysmic variable-like disk, the central object in the disk would have to have a radius equal to or less than that of the Sun in order to produce the observed far-ultraviolet continuum by accretion processes.

Harmanec (1990), following many others, has derived masses of 4.3 ± 1.1 and 14.1 ± 1.1 solar masses for the primary and secondary components, respectively but states that 2 and 12 solar masses cannot be ruled out.

The more we observe β Lyr, the more confused we become. Hopefully, β Lyr will continue keeping astronomers from becoming too sure of themselves.

Another intriguing system is ϵ Aur. The system was reviewed by Sahade and Wood (1978). With an orbital period of 27.1 years, the 1982–85 eclipse has led to a spate of publications.

Lissauer and Backman (1984) propose that the F0 supergiant is eclipsed by a disk, the usual model for the peculiar eclipsing object, which is about 10 AU in diameter and at least 1 AU in thickness assuming a radius of about 1.3 AU for the supergiant. The outer regions of the disk have a temperature of about 500K. The mass of the supergiant is about 20 solar masses and that of the disk, plus that of the companion star presumed to exist at its center, is approximately 16 solar masses. The semi-major axis of the system is then 30 AU.

On the other hand, Lambert and Sawyert (1986) propose that the supergiant is a low-mass ($\leq$ 3 solar masses) post-asymptotic giant branch star which is losing sufficient mass to create a large, cool accretion disk around a companion star of 3–6 solar masses.

Carroll *et al.* (1989) analysed the light curve of the recent eclipse and found it best reproduced by a tilted disk with a central hole and a large, semi-opaque central region. The central region could be formed by a hot star. The disk is 10 times the radius of the supergiant and is tilted by 2° to the orbital plane.

It does seem reasonable that a disk-like structure of considerable size is present, but the nature of any putative central object is still uncertain.

The interacting binary v Sgr (B8-A2 Ip + OB V: ; P = 137.943 days) is one of four known binaries (KS Per, LSS 4300, and CPD–58°2721 are the others) with very H-deficient primary stars. Only the supergiant is detectable in the optical spectrum.

Parthasarathy, Cornachin, and Hack (1986) have analysed the International Ultraviolet Explorer spectra of v Sgr. A stellar wind is present with maximum velocities from 300–800 kms^{-1}. They estimate the masses to be approximately one solar mass for the supergiant and three solar masses for the companion.

Morrison (1988) has made a detailed photometric analysis of v Sgr and arrived at an evolutionary scenario based partially on the theoretical results of Iben and Tutukov (1985) and Iben (1987). Morrison (1988) confirms the results of Malcolm and Bell (1986) that no eclipses are present and that quasi-periodic variations on a time scale of 20 days occur. Morrison (1988) finds that this 0.17 magnitude variation is due to radial pulsation of the supergiant in its fundamental mode. If the mass of the supergiant is assumed to be one solar mass, the lack of significant ellipsoidal variations seem to imply an orbital inclination below 30° and a companion mass of at least 12 solar masses. The ultraviolet flux is satisfied by a B2 V star of about 11 solar masses. The evolutionary history consists of two overflows of the supergiant which was originally the more massive star. The first occurred after H-core exhaustion as the star became a red giant. The initial mass of this star may have been 7–10 solar masses, while its companion began with 6–9 solar masses. The primordial orbital period would have been no more than approximately 600 days. The first mass loss stage led to a remnant of 1.3–2.0 solar masses. The second phase of overflow occurs when this star expands after exhausting helium in its core. The remnant is still losing mass slowly ($\sim 10^{-7}$ solar mass/year) near the end of this stage and has reduced its mass to about one solar mass. The secondary star accreted some mass but is probably still on the main sequence. The low mass supergiant will not ignite carbon in its core and will become a carbon-oxygen white dwarf. If the companion's mass is below about 10–11 solar masses, it will follow an evolutionary path similar to that of the original primary star and become a carbon-oxygen white dwarf or, if its mass is near the upper limit for avoiding a supernova, (10.6 solar masses according to Iben and Tutukov (1985)) it may become an oxygen-neon white dwarf. If the secondary's mass is too large it becomes a type I supernova with a neutron star remnant and possible system disruption. If the orbital period of the resulting double degenerate system is sufficiently short, spiral-in will occur in a Hubble time or less.

The evolution of these systems is discussed elsewhere in this volume by Rucinski (W Ursae Majoris Stars) and Nomoto and Starrfield (Binarity, Novae, and Supernovae Phenomena). White dwarfs are the end result in the evolution of these systems but getting there is most of the fun. Spiral-in may result in a single white dwarf or perhaps a "quiet" supernova yielding a neutron star. The reader is referred to the above chapters for details.

5. Chemical Composition

The products of nucleosynthesis in stellar interiors may appear at the stellar surface where they are directly detectable in two primary ways: Mixing and/or loss of the original unburned stellar envelope. In interacting binaries, overflow can contribute dramatically to envelope loss. The simplest abundance anomalies to be expected from nuclear burning of hydrogen, helium, or carbon would be under- or overabundances of H, He, C, O, Ne, and Mg. In addition, it is expected that carbon is initially severely depleted while nitrogen is enhanced during hydrogen-burning via the CNO cycle in stars above 2 solar masses. Other, more subtle anomalies are also expected and elements heavier than magnesium can be created during very late evolution nuclear burning in massive stars. Consequently, it is expected that abundance anomalies of various kinds should occur in interacting binaries where one or both stars have lost significant amounts of mass.

The Wolf-Rayet stars are usually interpreted as having overabundances of N (WN stars) due to CNO processing and mass loss or overabundances of carbon and oxygen with depletion of nitrogen (WC and WO stars) due to helium burning and mass loss. Due to the loss of much or all of the hydrogen-helium envelope and perhaps mixing, hydrogen appears to be underabundant. As discussed in an earlier section, it is not clear if or how single star mass loss or overflow in interacting binaries differ in the Wolf-Rayet stars they produce and these systems will not be further discussed here.

Walborn (1976) reviewed a group of massive stars known as OBN or OBC stars in which nitrogen or carbon, respectively, appear to be overabundant. Kudritzki and Hummer (1986) find that the ON stars have a helium abundance which is two times the solar value, that the N/C ratio is 30–80 times the solar value, and that the N/O ratio is about 20 times the solar value. The oxygen abundance is the same as that of the Sun. Levato et al. (1988) find that almost 100% of the OBN stars are binary but that less than 25% of the OBC stars are binary. At present there is no evidence that the binary nature of an OBN/OBC star is at all related to the apparent abundance anomalies.

The cool subgiant or giant in the classical Algol systems might be expected to show abundance anomalies since up 80% or 90% of the original mass may have been lost in some cases. Parthasarathy, Lambert, and Tomkin (1983) find that $[Fe/H] = 0.0 \pm 0.3$, $[C/Fe] = -0.5$, and $[N/Fe] = +0.5$ for the secondaries of U Cep and U Sge. In addition, $[C/Fe] = -0.5$ for the very low mass secondary of S Cnc. In comparison with average field giants, $[C/Fe] = 0.25 \pm 0.1$ and $[N/Fe] = +0.15 \pm 0.1$ for these systems. The observed overabundance of nitrogen is significantly less than expected from CNO nucleosynthesis while carbon should be 10–100 times less abundant than observed. It was proposed that after some mass loss from the original primary star, mixing between the core and envelope dilutes the enhancement of nitrogen and the corresponding carbon deficiency.

It is normally impossible to observe the secondary star's spectrum with sufficient resolution to allow accurate abundance analysis because the primary star dominates the spectrum. Since the primary star has presumably accreted matter from the secondary star, abundance anomalies might be expected in its atmosphere. Cugier and Hardorp (1988a, b) have investigated the [C/H] and [N/H] ratios in many B-type single stars and the [C/H] ratio in a number of Algol system primaries. They proposed that rapid rotation can cause large scale mixing and desired to test this hypothesis. Of 108 single stars, only 5 showed nitrogen anomalies but these were uncorrelated with carbon abundances. No clear cut dependence on projected rotational velocity was detected, although two of the most rapidly rotating stars, α Leo and ψ^2 Aqr, showed the lowest carbon abundances; 0.07 and 0.02 times the solar value, respectively. The Algols δ lib, RS Vul, and U Sge had solar carbon abundances as did the B-type binary U Her. The Algol systems TX UMa, β Per, λ Tau, and U CrB exhibit carbon underabundances of factors of 1.8–2.1 with respect to the sun. Non-LTE effects were investigated and would lead to [C/H] values larger than the LTE values by 0.10–0.15.

Tomkin and Lambert (1989) determined the chemical composition of the primary star in the low activity Algol system R CMa which has a very low mass secondary component. They found [C/H]= 0.0 ± 0.2, [N/H]= 0.4 ± 0.2, [O/H] = 0.3 ± 0.3, [S/H]= 0.1 ± 0.2, and [Fe/H]= 0.1 ± 0.1. It is concluded that the previous mass transfer in R CMa must have been non-conservative and/or that mixing occurred. Considering all of the assumptions involved, it is difficult to be certain whether or not any abundance anomalies have been detected in classical Algol systems.

In most Algol systems absorption lines of Si IV, C IV, and sometimes N V are detected outside of eclipse. These ions are often detected in emission during the totality phase of primary eclipse (Plavec 1989). In the dynamic Algols the emission lines are always detectable. Peters and Polidan (1984) analysed the C IV absorption in several classical Algol systems and concluded that carbon is about 10 times underabundant. However, C IV emission lines detected during totality in Algols are almost always the strongest emission lines (Sahade 1986). McCluskey and Sahade (1987) have suggested that C IV emission is filling in the C IV absorption, causing it to mimic underabundance. Gimenez and Claret (1989) study the irradiation of Algol secondaries by the hot primary star and find that the irradiated spectra can be very different from the normal spectrum and that some absorption lines can be filled-in by emission.

An interesting Algol system is V 356 Sgr (A2 II + B3/4 V; P = 8.896 days), which has been observed with the International Ultraviolet Explorer and Voyager. Polidan (1988) detected no C IV emission during the total eclipse of the B-star by the A-star and no carbon was detected in the A-star's ultraviolet spectrum. Si IV and N V emission are quite strong during totality. Perhaps the A-star really is carbon-deficient.

Balachandran et al. (1986) have found the B8 primary component in β Lyr to be overabundant in nitrogen by a factor of 20 with respect to the Sun. Carbon is found to be at least 400–500 times less abundant and oxygen at least 300 times abundant than in the Sun. The material at the surface of the B8 primary, which is the mass-losing star, is fully processed CNO cycle material if these abundances are correct.

Sahade (1988) discusses β Lyr and the H-deficient stars v Sgr and KS Per. He gives a table of H, He, C and O abundance determinations for β Lyr made between 1960 and 1986. The different determinations differ by factors of 10–1000 or more. Sahade (1988) concludes that abundance determinations in complex spectra are not likely to be of much value with our present state of knowledge.

A more subtle abundance anomaly is the existence of Ba II stars. McClure (1983) concluded that all Ba II stars are binaries. The Ba II component is a Pop I giant enhanced in s-process elements with strong Sr II and Ba II lines. The companion stars are probably white dwarfs of 0.2–0.6 solar mass. The average separation is 2 AU. Iben and Truran (1978) found that s-process elements are mixed to the surface in stars of several solar masses while they are giants and/or supergiants. But McClure (1983) notes that the Ba II stars have s-process abundances similar to that of low mass stars where no mixing is expected. It is concluded that either the companion caused mixing to occur or the companion contaminated the star with s-process elements which it produced. Bohm-Vitense, Nemec, and Proffitt (1984) prefer the latter since the orbital period seems much too long for tidal effects to cause mixing. They believe that the masses of the Ba II stars are too low for the creation of s-process elements.

McClure and Woodsworth (1990) further study Ba II stars and include a small group of stars in which the CH band is abnormally strong. They conclude that all of these stars are binaries with orbital periods from 80 days to longer than 10 years. The Ba II stars have about 1.5 solar masses while the CH stars have approximately 0.8 solar mass. All have white dwarf companions of about 0.6 solar mass. The CH stars are the Pop II analog of the Ba II stars.

6. Conclusions

Interacting binary star research, both observational and theoretical, continues apace. New observational techniques, the opening up of nearly the entire electromagnetic spectrum via observatories in space, and computer modelling permit the accomplishment of many things which were nigh impossible ten or twenty years ago. Yet a number of complex problems have hardly been addressed (or even admitted) and must be at the forefront of interacting binary research in the near future if further significant progress is to result.

The classical Roche model is completely inappropriate when radiation pressure, stellar winds, and/or rapid mass flow are present in interacting binaries. A much more complete theory of surface flow on a star at or near overflow is sorely needed. No one has yet justified the hose-like nature of gas flow at the L_1 point which is invariably assumed by some theorists to occur. Surface flow on an accreting star is also very poorly understood.

The analysis of the physical state of circumstellar and circumbinary gas in interacting binaries is in an extremely primitive and unreliable state. This gas will also be seriously affected by magnetic fields which are known to exist in some binaries and may well be ubiquitous. These magnetic effects have been completely neglected, with few exceptions, in all research.

If accurate comparisons are to be made between theoretical predictions of abundances and observed abundances, the assumption that all Pop I stars have the same chemical composition when on the zero age main sequence must be questioned. It is probably no more likely to be correct than the assumption made by many astronomers early in this century that the interstellar material was uniformly or simply distributed in the plane of our Galaxy.

Theoreticians and observers alike must find ways to begin to understand how these problems can be attacked and then to begin to solve them. Problems at the forefront of human knowledge are always difficult but that makes them all the more worthy of our dedication.

PENDING PROBLEMS AND FINAL COMMENTS

J. SAHADE, G. E. MCCLUSKEY, JR and Y. KONDO

In roughly two centuries astronomers have gone from an almost complete disbelief in the existence of binary stars to a near certain knowledge that more than one-half of the stars are members of binary or multiple systems. Analogously, in approximately one-half century the field of interacting binaries has changed from the study of a very few apparently unique systems to a very active multifaceted area of astrophysics.

A complete summary of our understanding of the formation of binary stars is given in the chapter by Boss. Of the various evolutionary stages which occur in interacting binaries, it is this stage which is least understood. To paraphrase Eddington: if binary stars did not exist, we would surely have an excellent theory to explain why. The primary problem is, as it is for single stars, that star formation occurs in regions of very high visual absorption and a firm observational basis which would limit theoretical speculation is absent. The development of infrared and radio observational techniques to probe star forming regions is beginning to alleviate this situation and future developments in observation and in three-dimensional numerical investigations hold great promise. At present it appears that fragmentation plays a major role.

Chapters by Lubow, McCluskey, McCluskey and Kondo, Meyer-Hofmeister and Ritter, and Sahade discuss evolution, mass flow and accretion phenomena in interacting binaries.

Sahade has proposed a very tentative evolutionary scheme for the gaseous structures present in intermediate mass binaries: R Arae $\rightarrow$ W Serpentis $\rightarrow$ GG Carinae $\rightarrow$ β Lyrae $\rightarrow$ V453 Scorpii $\rightarrow$ U Cephei $\rightarrow$ β Persei. Unfortunately, basic astrophysical properties, e.g., mass and radius, are very poorly known at best for the first four systems or any of the few similar systems known. The reasons for this are that these evolutionary stages incorporate rapid mass loss with a short time scale so that few systems are observed in this dynamic stage of evolution and those that are exhibit extremely complex photometric and spectroscopic phenomena. In the most dramatic cases at least one component may be so shrouded in gas and, in the case of cooler stars, dust that it is not directly observable spectrosopically. As pointed out by Sahade, the evolutionary scenario mentioned above probably does not include all physically distinct stages and may not be unique even if by some good fortune (often referred to as a brilliant idea) it happens to be applicable at times.

The W Ursae Majoris systems, discussed in the chapter by Rucinski, provide a few critical puzzles. We really do not know from whence they come and whither they are going. In addition, the transfer of large amounts of energy and mass from one component to the other seems required by the observations but still defies theoretical explanation. Will all of these systems fuse into single stars? Will any of the components become white dwarfs? Do pre-W UMa systems exist and if so what do they look like?

Much effort has been devoted to the evolution of massive stars, particularly since they result in the formation of neutron stars and, according to current astrophysical wisdom (hopefully not an oxymoron), black holes. The analysis of Wolf-Rayet stars convincingly shows that massive single stars lose significant fractions of their mass and it is unclear at present how, or if, a single Wolf-Rayet star differs in its structure and evolution from a Wolf-Rayet in a presumably interacting binary. Systems with compact components have been discussed in the chapter by Trimble while chapters by King and Lasota and by Starrfield discuss white dwarf phenomena in interacting binaries. One significant problem in these compact systems is an almost complete lack of even reasonably accurately determined masses for the low mass X-Ray binaries. This is due to the faintness, often to the point of undetectability, of the optical component and to rarity of pulsars in these systems. The origin and fate of many of these systems is also not well understood. In that most exotic system SS433, it is not certain whether a neutron star or black hole is involved. In addition, the complex nature of x-ray generation and degradation to longer wavelengths is a very difficult problem.

The chapter by Nomoto summarizes our theoretical understanding of SNeI but it must be noted that no progenitor of a SNI has ever been observed. King and Lasota discuss the polars, interacting binaries containing a magnetic white dwarf, in their chapter. These systems may prove to be excellent laboratories for the study of magnetic field interactions. Dozens more will probably be discovered by ROSAT.

Starrfield's chapter reviews the nova phenomenon. Nova theory has succeeded in explaining many aspects of the observational material and a reasonably good theory is apparently available. Many details remain to be clarified, e.g., dependence of the phenomena on white dwarf mass, chemical composition, magnetic properties, and on the nature of the cool stars.

Koch's chapter covers the effects of polarization in interacting binaries. This technique allows us to study the optically thin regions of gas flows and holds promise for analysing magnetic field effects and determining orbital spatial orientation with the appropriate technological development.

The chapter by Weekes discusses a new area of astrophysics; the putative detection of particle energies from 10^{12} to 10^{16} or more electron volts from a few X-Ray binaries such as Hercules X-1. The basic problems are quite primitive: are the detections real and if so are we indirectly detecting photons or some unknown particle or both? If these events are real, the extreme high energy gamma-ray spectrum and possibly brand new astrophysics will become available to us.

We conclude with a brief discussion of some of the fundamental problems facing us in the study of interacting binaries.

1. The concept of the Roche critical lobe is of limited value and is regularly used where it is not valid. Any star which is losing mass for any other reason than simple overflow, e.g., radiation pressure, magnetic effects, convective overshooting, pulsation, rotation or other poorly understood dynamic effects, cannot obey the Roche model. A more adequate model is badly needed in particular for hot stars and for stars undergoing rapid mass loss. The restricted three-body problem from which the Roche model derives neglects *all* forces other than the gravitational forces of these three bodies considered to be point masses!

2. It is becoming clearer and clearer that loss of mass from the system is very often more important than mass exchange. Both observational and theoretical research is required in order to quantify the effects of systemic mass loss on the evolution of interacting binaries.

3. The evolutionary path taken by an interacting binary will depend on many parameters but primarily on the initial masses, the initial mass ratio and the initial separation. A plethora of scenarios have been considered theoretically and a similar copius variety of types are observed. Yet a one-to-one relationship between theory and observation is not close. Uniqueness is probably not to be expected and all that can be done is to try to understand the observed systems as well as possible and to devise ways in which such systems might arise. Care must be taken to find physically meaningful quantities when postulating some evolutionary sequence or stage. For example, emission lines in themselves are *not* good indicators of anything except extensive low density gas. Interacting binaries as disparate as Wolf-Rayet binaries, β Lyrae, W Serpentis, AM Hercules and SS433 have emission lines and their interrelationships must be better understood before they become of significant value in a classification scheme.

4. Nearly all astrophysists use the term accretion disk to explain a wide variety of phenomenon in a wide variety of interacting binaries. In fact, strong evidence for truly disk-like structures is present only when the accreting star is compact, i.e., in catacysmic variables and low mass x-ray binaries. In most other cases the evidence is far weaker and the existence of actual disks should not be taken for granted. Certainly, circulating gas is present but the application of a CV-like disk to a system like W Serpentis or Algol is a large extrapolation and should be regarded with suspicion until more detailed observational research clarifies the nature of the circumstellar material. The chapter by Meyer-Hofmeister and Ritter discusses the observational evidence for accretion disks, particularly in *CVs*, and some of their properties.

5. Lubow discusses the theory of mass transfer in his chapter. Problem areas are in the details of circularization and synchronization, the effects of orbital eccentricity, the structure of the gas stream and the stream-disk, stream-star, and disk-star interactions.

6. Giménez and Guinan contributed a chapter on magnetic activity in close

binaries. Evidence is accumulating for the importance of magnetic field effects not only in cool components of interacting binaries but even in some B- and A-type stars. Advances have been made possible by x-ray, ultraviolet, and radio observations with very recent optical techniques beginning to contribute. X-Ray mapping, ultraviolet and visible Doppler imaging and use of the eclipses to map magnetic field structures are very promising techniques.

7. Many interacting binaries have detectable orbital period variations but these are rarely explainable by simple models of mass loss. Particularly puzzling are apparently abrupt (or occurring over a short period of time) period changes in which the period sometimes increases or decreases slightly in the same system. Magnetic field effects have been suggested as a cause but there appear to be major difficulties with this idea at present.

8. Nature apparently enjoys making things difficult for those who would understand it. This is clearly demonstrated in the comparison between observation and theory with regard to the composition of stars in general with the interacting binaries not to be excluded. For example, only if we invoke numerous nearly ad hoc assumptions such as convective overshooting, semiconvection, mixing, rotation, etc., can we "make" agreement between the observed abundance of carbon and nitrogen in Algol-type binaries and the theoretical predictions; and here we are presumably dealing with simple hydrogen burning via the carbon cycle.

9. The RS Canum Venaticorum systems have not received much attention in this volume, mainly because the great majority of them do not appear to be interacting in the usual sense of mass flow, etc. How or if they are related to interacting binaries is unknown. Their origin and future evolution is still a mystery.

10. Mutel and Elias have contributed a chapter on radio emission from active late-type binaries. This is a rapidly developing field and holds much promise. It appears that rapid rotation, a consequence of being a member of a short period binary star, is the primary factor involved. It is still not clear how binarity is involved but it does not seem likely that it would be of no consequence. For example, wind collisions and accretion effects are almost certainly involved.

11. It might be obvious but is well worth emphasizing the immense importance of observing at all possible wavelengths, from extremely high energy gamma rays to long wave radio waves. In addition, to observe simultaneously in as many wavelength regions as possible is even more beneficial but requires extensive planning, resources and luck.

The editors of this volume have always found the study of interacting binaries to be exciting, rewarding and of significant benefit to astrophysics in general. This will continue on into the future. Interacting binaries deserve our best efforts.

REFERENCES

THE DEVELOPMENT OF THE IDEA OF INTERACTING DOUBLE STARS
A. H. Batten and F. B. Wood

Andersen J.: 1991, *Astron. Astrophys. Rev.* **3**, 91.
Batten, A.H., Scarfe, C.D., Baldwin, B.W., and Fisher, W.A.: 1975, *Nature* **253**, 174.
Beer, A.: 1958, in Etoiles à raies d'émission, Mém. de la Soc. Roy. des Sciences de Liège, 4ème Série, XX, p. 387.
Budding, E. and Trodahl, H.J.: 1986, *Southern Stars* **32**, 19.
Carpenter, E.F.: 1930, *Astrophys. J.* **72**, 205.
Catalano, S. and Rodonò, M.: 1967, *Mem. Soc. astr. Ital.* **38**, 395.
Collier, A.C., Hearnshaw, J.B., and Austin, R.R.D.: 1981, *Mon. Not. Roy. Astron. Soc.* **197**, 769.
Crawford, J.A.: 1955, *Astrophys. J.* **121**, 71.
Dugan, R.S.: 1920, Contr. Princeton Obs. No. 5.
Fracastoro, M.G.: 1956, in Vistas Astron. (ed. A. Beer), Vol. 2, Pergamon Press, Oxford, p. 1198.
Fracastoro, M.G.: 1965, Kl. Veröff. Remeis-Sternw. Bamberg IV, Nr 40, p. 253.
Genet, R.M. and Hayes, D.S.: 1989, *Robotic Observatories*, Autoscope Corp., Mesa, Arizona.
Goodricke, J.: 1783, *Phil. Trans. Roy. Soc. London* **73**, 474.
Hall, D.S.: 1973, in *Extended Atmospheres and Circumstellar Matter in Spectroscopic Binary Systems* (Int. Astron. Un. Symp. No.51) ed. A.H. Batten, D. Reidel, Dordrecht, Holland, p.280.
Hall, D.S.: 1976, in Multiple Periodic Variable Stars (Int. Astron. Un. Coll. No.29) ed. W.S. Fitch, D. Reidel, Dordrecht, Holland, p. 287.
Hill, G.: 1979, *Publ. Dom. Astrophys. Obs.* **15**, 197.
Hoyle, F.: 1955, *Frontiers of Astronomy*, Wm. Heinemann, London, pp. 195–202.
Huang, S.-S.: 1963a, *Astrophys. J.* **138**, 342.
Huang, S.-S.: 1963b, *Astrophys. J.* **138**, 471.
Huang, S.-S.: 1966, *Ann. Astrophys.* **29**, 331.
Hubeny, I.: 1989, in *Algols* (Int. Astron. Un. Coll. No. 107, *Space Sci. Rev.* 50) ed. A.H. Batten, Kluwer Academic Publishers, Dordrecht, Holland, p. 117.
Hutchings, J.B. and Hill, G.: 1970, *Astrophys. J.* **162**, 265.
Joy, A.H.: 1942, *Publ. Astron. Soc. Pacific* **54**, 35.
Joy, A.H.: 1954, *Astrophys. J.* **120**, 377.
Kaitchuck, R.H. and Honeycutt, R.K.: 1982, *Pub. Astron. Soc. Pacific* **94**, 532.
Kippenhahn, R. and Weigert, A.: 1967, *Z. Astrophys.* **65**, 241.
Koch, R.H.: 1988 in *Trans. Int. Astron. Un.* XXA, ed. J.-P. Swings, Kluwer Academic Publishers, Dordrecht, Holland, p. 571.
Kopal, Z.: 1935, *Z. Astrophys.* **9**, 239.
Kopal, Z.: 1941, *Astrophys. J.* **94**, 145.
Kopal, Z.: 1955, *Ann. Astrophys.* **18**, 379.
Kopal, Z.: 1984, in *Double Stars, Physical Properties and Generic Relations* (Int. Astron. Un. coll. No. 80; *Astrophys. Space Sci.* 99) eds. B. Hidayat, Z. Kopal and J. Rahe, D. Reidel, Dordrecht, Holland, p. 3.
Kraft, R.P.: 1962, *Astrophys. J.* **135**, 408.
Kraft, R.P.: 1964, *Astrophys. J.* **139**, 457.
Kron, G.E.: 1946, *Astrophys. J.* **103**, 326.
Kron, G.E.: 1947, *Publ. Astron. Soc. Pacific* **59**, 261.
Kruszewski, A.: 1964, *Acta Astr.* **14**, 214.

Kuiper, G.P.: 1935, *Pub. Astron. Soc. Pacific* **47**, 15+121.
Kuiper, G.P.: 1941, *Astrophys. J.* **93**, 133.
Lamontagne, R., Moffat, A.F.J., and Seggewiss, W.: 1983, *Astrophys. J.* **269**, 596.
Lestrade, J.-F., Mutel, R.L., Preston, R.A., and Phillips, R.B.: 1988, *Astrophys. J.* **328**, 232.
Lucy, L.B.: 1967, *Astrophys. J.* **151**, 123.
Lucy, L.B.: 1968, *Astrophys. J.* **153**, 877.
Markowitz, W.: 1932, *Astrophys. J.* **75**, 69.
Massey, P., Conti, P.S., and Niemela, V.S.: 1981, *Astrophys. J.* **246**, 145.
McKellar, A. and Petrie, R.M.: 1957, *Publ. Dominion Astrophys. Obs.* **11**, 1.
McLaughlin, D.B.: 1952, *Pub. Astron. Soc. Pacific* **64**, 109.
McLean, B.J.: 1981, *Mon. Not. Roy. Astron. Soc.* **195**, 931.
McLean, B.J. and Hilditch, R.W.: 1983, *Mon. Not. Roy. Astron. Soc.* **203**, 1.
Mikolajewska, J., Friedjung, M., Kenyon, S.J., and Viotti, R. (eds): 1989, *The Symbiotic Phenomenon*, Kluwer Academic Publishers, Dordrecht, Holland.
Minchen, G.M.: 1895, *Proc. Roy. Soc. London* **58**, 142.
Moore, J.H.: 1935, in *The Binary Stars*, 2nd Edn, R.G. Ailken, McGraw-Hill, New York (also Dover Reprint), p. 125.
Morton, D.C.: 1960, *Astrophys. J.* **132**, 146.
Nha, I.-S. and Kang, Y.-W.: 1982, *Pub. Astron. Soc. Pacific* **94**, 496.
Oliver, J.P.: 1973, in *Extended Atmospheres and Circumstellar Matter in Spectroscopic Binary Systems* (Int. Astron. Un. Symp. No.51), ed. A.H. Batten, Reidel, Dordrecht, Holland, p. 270.
Olson, E.C.: 1980, *Astrophys. J.* **241**, 257.
Paczynski, B.: 1967, in *On the Evolution of Double Stars* (Int. Astron. Un. Colloq.) ed. J. Dommanget, Comm. Obs. Roy. Belgique, Ser. B, No. 17, p. 111.
Piotrowski, S.: 1947, *Astrophys. J.* **106**, 472.
Piotrowski, S.: 1948, *Astrophys. J.* **108**, 36.
Plavec, M.: 1973, in *Extended Atmospheres and Circumstellar Matter in Spectroscopic Binary Systems* (Int. Astron. Un. Symp. No.51) ed. A.H. Batten, Reidel, Dordrecht, Holland, p. 216.
Plavec, M.: 1983, *J. Roy. Astron. Soc. Canada* **77**, 283; 1983b, *Astrophys. J.* **275**, 251.
Popper, D.M.: 1970a, in *Spectroscopic Astrophysic* ed. G.H. Herbig, Berkeley: University of California Press, p. 141.
Popper, D.M.: 1970b, in *Mass Loss and Evolution in Close Binaries* (Int. Astron. Un. Coll. No. 6) eds. K. Gyldenkerne and R.M. West, Copenhagen Univ. Obs., p. 13.
Popper, D.M.: 1980, *Ann. Rev. Astron. Astrophys.* **18**, 115.
Prendergast, K.H. and Burbidge, G.R.: 1968, *Astrophys. J.* **151**, L83.
Pringle, J.E.: 1981, *Ann. Rev. Astron. Astrophys.* **19**, 137.
Ritter, H.: 1990, *Astron. Astrophys. Suppl.* **85**, 179.
Roach, F.E. and Wood, F.B.: 1952, *Ann. Astrophys.* **15**, 21.
Russell, H.N.: 1912, *Astrophys. J.* **35**, 315.
Russell, H.N.: 1948, in *Centennial Symposia* (Harvard Observatory Monographs No. 7) Harvard Observatory, p. 181.
Russell, H.N. and Merrill, J.E.: 1952, *Contr. Princeton Obs.* No. 26.
Russell, H.N. and Shapley, H.: 1912, *Astrophys. J.* **36**, 239, 385.
Sahade, J.: 1958, *Observatory* **78**, 79.
Sahade, J., Huang, S.-S., Struve, O., and Zebergs, V.: 1959, *Trans. Amer. Phil. Soc. New Series* **49**, Part 1.
Sahade, J. and Wood, F.B.: 1978, *Interacting Binary Stars* (Oxford: Pergamon Press) pp. 1-13.
Smak, J.: 1962, *Acta Astr.* **12**, 28.
Smak, J.: 1989 in *Algols* (Int. Astron. Un. Coll. No.107; *Space Sci. Rev.* 50) ed. A.H. Batten, Kluwer Academic Publishers, Dordrecht, Holland, p. 107.
Stebbins, J.: 1928, *Publ. Washburn Obs.* 15.
Struve, O.: 1941, *Astrophys. J.* **93**, 104.
Struve, O.: 1946, *Ann. Astrophys.* **9**, 1.
Struve, O.: 1948, in *Centennial Symposia* (Harvard Observatory Monographs No. 7) Harvard Observatory, p. 211.

Struve, O.: 1950, *Stellar Evolution*, Princeton University Press, pp. 233-4.
Struve, O.: 1955, *Sky Telesc.* **14**, 275.
Struve, O. and Sahade, J.: 1957, *Publ. Astron. Soc. Pacific* **69**, 41.
Taylor, M.: 1988, *Sky Telesc.* **76**, 351.
van Hamme, W. and Wilson, R.E.: 1984, *Astron. Astrophys.* **141**, 1.
Vogel, H.C.: 1890, *Astron. Nachr.* **123**, 289.
Vogt, S.S. and Penrod, G.D.: 1983, *Publ. Astron. Soc. Pacific* **95**, 565.
Walker, M.F.: 1954, *Publ. Astron. Soc. Pacific* **66**, 230.
Walter, K.: 1931, *Veröff. Konigsberg Sternw.* No. 2.
Walter, F.M. and Basri, G.S.: 1982, *Astrophys. J.* **260**, 735.
Wilson, O.C.: 1940, *Astrophys. J.* **91**, 379.
Wilson, R.E. and Devinney, E.J.: 1971, *Astrophys. J.* **166**, 605.
Wilson, R.E.: 1979, *Astrophys. J.* **234**, 1054.
Wilson, R.E.: 1989, in *Algols* (Int. Astron. Un. Coll. No. 107, *Space Sci. Rev.* 50) ed. A.H. Batten, Kluwer Academic Publishers, Dordrecht, Holland, p. 235.
Wilson, R.E.: 1991, in *Trans. Int. Astron. Union XXIA*, ed. D. McNally, Kluwer Academic Publishers, Dordrecht, Holland, p. 483.
Wood, F.B.: 1946, *Contr. Princeton Univ. Obs.* No. 21.
Wood, F.B.: 1950, *Astrophys. J.* **112**, 196.
Wood, F.B.: 1957, in *Non-Stable Stars* (Int. Astron. Un. Symp. No. 3) ed. G.H. Herbig, Cambridge University Press, p. 144.
Wright, K.O.: 1970, in *Vistas in Astron.* **12**, ed. A. Beer, Pergamon Press, Oxford, p. 147.
Wyse, A.B.: 1934, *Lick Obs. Bull.* **17**, 42.

THE INTERACTING BINARY ZOO

J. Sahade, G. E. McCluskey, Jr., and Y. Kondo

Abbles, J.G., McConnell, D., Jacka, C.F., McCulloch, P.M., Hall, P.J. and Hamilton, P.A.: 1989, *Nature* **342**, 158.
Batten, A.H., Fletcher, J.M. and MacCarthy, d.G.: 1989, *Eighth Catalogue of Orbital Elements of Spectroscopic Binary Systems*, Dominion Astroph. Obs. Publ., 17.
De Cuiper, J.-P.: 1985, in W. Boland and H. van Woerden, ed(s)., *Birth and Evolution of Massive Stars and Stellar Groups*, Reidel, Dordrecht, 207.
Eggleton, P.P.: 1986, in J. Truemper, W.H.G. Lewin and W. Brinkmann, ed(s)., *The Evolution of Galactic X-Ray Binaries*, Reidel, Dordrecht, 87.
Faulkner, J.: 1971, *Astroph. J. Letters* **170**, L99.
Kondo, Y.: 1988, in K.-C. Leung, ed(s)., *Critical Observations versus Physical Models for Close Binary Systems*, Gordon and Breach, 261.
Kraft, R.P.: 1962, *Astroph.J.* **135**, 448.
Kraft, R.P., Mathews, J., Greenstein, J.L.: 1962, *Astroph. J.* **136**, 312.
Krzeminski, W. and Kraft, R.P.: 1964, *Astroph. J.* **140**, 921.
Kumar, S.S.: 1969, in S.S. Kumar, ed(s)., *Low-Luminosity Stars*, Gordon and Breach, 255.
Ludendorff, H.: 1903, *Astr. Nachr.* **164**, 81.
Méndez, R.H.: 1989, in S. Torres-Peimbert, ed(s)., *Planetary Nebulae, IAU Symp. 131*, Kluwer Academic Publishers, Dordrecht, 261.
Nemoto, K. and Kondo, Y.: 1990, *preprint* , .
Paczynsky, B.: 1967, *Acta Astron.* **17**, 287.
Popper, D.M. and Tomkin, J.: 1984, *Astroph. J.* **285**, 208.
Rappaport, S., Joss, P.C., and Webbink, R.F.: 1982, *Astroph. J.* **245**, 616.
Sahade, J.: 1959, in , ed(s)., *Modèles d'étoiles et Evolution Stellaire*, Liège Coll., 76.
Sahade, J.: 1986, in E.J. Rolfe, ed(s)., *New Insights in Astronomy*, ESA SP-263, 267.
Sahade, J.: 1987, *Comments in Astroph.* **12**, 13.
Sahade, J.: 1988a, *J. Space Astron. Res.* **5**, 1.

Sahade, J.: 1988b, in K. Nomoto, ed(s)., *Atmospheric Diagnostics of Stellar Evolution, IAU Coll. 108*, Springer, 199.
Sahade, J.: 1988c, in V.M. Blanco and M.M. Phillips, ed(s)., *Progress and Opportunities in Southern Hemisphere, A.S.P. Conf. Series, vol. 1,*, 84.
Smak, J.: 1965, *Acta Astron.* **15**, 327.
Tomkin, J.: 1985, *Astroph.J.* **297**, 250.
Van der Hucht, K.A., Conti, P.S., Lundström, I. and Stenholm, B.: 1981, *Space Sci. Rev.* **28**, 227.

MASS LOSS IN INTERACTING BINARY STARS

J. Sahade

Aydin, C., Brandi, E., Engin, S., Ferrer, O.E., Hack, M., Sahade, J., Solivella, G. and Yilmaz, N.: 1988, *Astron. and Astroph.* **193**, 202.
Barr, J.M.: 1908, *J.R.A.S. Canada* **2**, 70.
Batten, A.H.: 1980, *P.A.S.P.* **82**, 574.
Batten, A.H.: 1973a, in *Extended Atmospheres and Circumstellar Matter in Spectroscopic Binary Systems*, IAU Symp No. 51, ed(s)., *A.H. Batten*, Reidel, Dordrecht, 1.
Batten, A.H.: 1973b, *Binary and Multiple Systems of Stars*, Pergamon, 196.
Batten, A.H. and Sahade, J.: 1973, *P.A.S.P.* **85**, 899.
Bolton, C.T.: 1989, in *Algols*: IAU Coll. No. 107, ed(s)., *A.H. Batten*, Kluwer, Dordrecht, 311.
Brandi, E., Ferrer, O.E. and Sahade, J.: 1989, *Astroph. J.* **340**, 1091.
Chen, H., Ringuelet, A.E., Sahade, J. and Kondo, Y.: 1989, *Astroph. J.* **347**, 1082.
Crawford, R.C. *Ph.D. Thesis*, U. of California, Los Angeles (1981), (quoted in Kaitchuck, Honeycutt and Faulkner 1989).
De Loore, C.: 1984, *Ap. Space Sc.* **99**, 199.
Dugan, R.S. and Wright, F.W. 1937 *Contr. Princeton Obs.* No. 19.
Elias, N.M., II: 1990, *Astroph. J.* **352**, 300.
Gies, D.R. and Wiggs, M.S. (1990), preprint.
Gillet, D., Mouchet, M. and North, P.: 1989, *Astron. and Astroph.* **219**, 219.
Huang, S.-S.: 1963, *Astroph. J.* **138**, 471.
Joy, A.H.: 1942, *P.A.S.P.* **54**, 35.
Kaitchuck, R.H. and Honeycutt, R.K.: 1982a, *P.A.S.P.* **94**, 532.
Kaitchuck, R.H. and Honeycutt, R.K.: 1982b, *Astroph. J.* **258**, 224.
Kaitchuck, R.H. and Park, E.A.: 1988, *Astroph. J.* **325**, 225.
Kaitchuck, R.H., Honeycutt, R.K. and Faulkner, D.R.: 1989, *Astroph. J.* **339**, 421.
Kaitchuck, R.H., Honeycutt, R.K. and Schlegel, E.M.: 1985, *P.A.S.P.* **97**, 1178.
Kondo, Y.: 1988, in *Critical Observations versus Physical Models for Close Binary Systems*, ed(s)., *K.-C. Leung*, Gordon and Breach, 261.
Kondo, Y, McCluskey, G.E., Jr. and Harvel, C.A.: 1981, *Astroph. J.* **247**, 202.
Kuiper, G.P.: 1941, *Astroph. J.* **93**, 133.
Lubow, S.H. and Shu, F.H.: 1975, *Astroph. J.* **198**, 383.
Lubow, S.H. and Shu, F.H.: 1976, *Astroph. J.* **207**, L53.
Olson, E.C.: 1980, *Astroph. J.* **241**, 257.
Olson, E.C.: 1987, *Astron. J.* **94**, 1043.
Olson, E.C.: 1989, in *Algols*, I.A.U. Coll. No. 107, ed(s)., *A.H. Batten*, Kluwer, Dordrecht, 23.
Olson, E.C. and Bell, P.J.: 1989, *P.A.S.P.* **101**, 907.
Peters, G.J.: 1989, in *Algols*: I.A.U. Coll. No. 107, ed(s)., *A.H. Batten*, Kluwer, Dordrecht, 9.
Plavec, M.: 1973, in *Extended Atmospheres and Circumstellar Matter in Spectroscopic Binary Systems*, IAU Symp. No. 51, ed(s)., *A.H. Batten*, Reidel, Dordrecht, 216.
Popper, D.M.: 1964, *Astroph. J.* **139**, 143.
Popper, D.M.: 1973, in *Extended Atmospheres and Circumstellar Matter in Spectroscopic Binary Systems*, IAU Symp. No. 51, ed(s)., *A.H. Batten*, Reidel, Dordrecht, 58.
Sahade, J.: 1952, *Astroph. J.* **116**, 35.

Sahade, J. 1958, Ètoiles á raies d'émission, Liége Coll. p. 401.
Sahade, J.: 1959, *P.A.S.P.* **71**, 151.
Sahade, J.: 1960, in *Stellar Atmospheres*, ed(s)., J.L. Greenstein, U of Chicago Press, 466.
Sahade, J.: 1973, in *Extended Atmospheres and Circumstellar Matter in Spectroscopic Binary Systems*, IAU Symp. No 51, ed(s)., A.H. Batten, Reidel, Dordrecht, 286.
Sahade, J.: 1980, *Space Sc. Rev.* **26**, 349.
Sahade, J.: 1986, in *New Insights in Astrophysics*, ESA SP-263, ed(s)., E. Rolfe, , 267.
Sahade, J.: 1987, *Comm. in Ap.* **12**, 11.
Sahade, J.: 1988a, in *Atmospheric Diagnostics of Stellar Evolution*, I.A.U. Coll. No. 108, ed(s)., K. Nomoto, Springer, 199.
Sahade, J.: 1988b, in *Progress and Opportunities in Southern Hemisphere Optical Astronomy*, A.S.P. Conference Series, vol. 1, ed(s)., V.M. Blanco and M.M. Phillips, , 84.
Sahade, J. and Brandi, E. (1990), preprint.
Sahade, J. and Cesco, C.U.: 1945, *Astroph. J.* **101**, 235.
Sahade, J. and Wood, F.B.: 1978a, *Interacting Binary Stars*, Pergamon Press, 1,69.
Sahade, J. and Wood, F.B.: 1978b, *Interacting Binary Stars*, Pergamon Press, 40.
Sahade, J. and Wood, F.B.: 1978c, *Interacting Binary Stars*, Pergamon Press, 130.
Sahade, J. and Wood, F.B.: 1978d, *Interacting Binary Stars*, Pergamon Press, 101.
Sahade, J., Kondo, Y., and McCluskey, G.E., Jr.: 1984, *Astroph. J.* **276**, 281.
Shore, S.N. and Brown, D.N.: 1988, *Astroph. J.* **334**, 1021.
Smak, J.: 1973, in *Extended Atmospheres and Circumstellar Matter in Spectroscopic Binary Systems*, I.A.U. Symp. No. 51, ed(s)., A.H. Batten, Reidel, Dordrecht, 57.
Stewart, R.T., Slee, O.B., White, G.L., Budding, E., Coates, D.H., Thompson, K.T. and Burton, J.D.: 1989, *Astroph. J.* **342**, 463.
Strickland, D.J., Lloyd, C., Pike, C.D., Willis, A.J. and Howarth, I.D.: 1986, in *New Insights in Astrophysics*, ESA-SP-263, ed(s)., E. Rolfe, , 487.
Struve, O.: 1941, *Astroph. J.* **93**, 104.
Struve, O.: 1944a, *Astroph. J.* **99**, 22.
Struve, O.: 1944b, *Astroph. J.* **99**, 89.
Struve, O.: 1947, *Astroph. J.* **106**, 255.
Struve, O.: 1948, *Ann. d'Ap* **11**, 117.
Struve, O. and Sahade, J.: 1951, *P.A.S.P.* **69**, 41.
Struve, O., Sahade, J. and Huang, S.-S.: 1958, *Astroph. J.* **127**, 148.
Tout, A. and Eggleton, P.P.: 1988, *Monthly Not. of the Roy. Astron. Soc.* **213**, 823.
Wallerstein, G., Willson, L.A., Salzer, J. and Brugel E.: 1984, *Astron. and Astroph.* **133**, 137.
Warner, B. and Nather, R.E.: 1971, *Monthly Not. of the Roy. Astron. Soc.* **152**, 219.
White, N.E., Culhane, J.L., Parmar, A.N., Kellet, B.J., Kahn, S., van der Oord and Kuipers, J.: 1986, *Astroph. J.* **301**, 262.
Willson, L.A., Wallerstein, G., Brugel, E.W. and Stencel, R.E.: 1984, *Astron. and Astroph.* **133**, 154.
Wood, F.B.: 1950, *Astroph. J.* **112**, 196.
Wood, F.B.: 1969, in *Mass Loss from Stars*, ed(s)., M. Hack, Reidel, Dordrecht, 149.

SOME ISSUES IN THE THEORY OF MASS TRANSFER

S. H. Lubow

Bahcall, J.N.: 1978, *Ann. Rev. Astr. and Astroph.* **16**, 241.
Campbell, C.G. and Papaloizou, J.: 1983, *Monthly Not. of the Roy. Astr. Soc.* **204**, 433.
Davidson, K. and Ostriker, J.P.: 1973, *Astroph. J.* **179**, 585.
Davies, R.E. and Pringle, J.E.: 1980, *M.N.R.A.S.* **191**, 599.
De Greve, J.P.: 1986, *Space Sci. Rev.* **43**, 139.
Edwards, D.A. and Pringle, J.E.: 1987, *M.N.R.A.S.* **229**, 383.
Frank, J., King, A.R., and Lasota, J.-P.: 1987, *Astr. and Astroph.* **178**, 137.
Giuricin, G., Mardirossian, F., and Mezzetti, M.: 1984a, *Astr. and Astroph.* **131**, 152.

Giuricin, G., Mardirossian, F., and Mezzetti, M.: 1984b, *Astr. and Astroph.* **134**, 365.
Giuricin, G., Mardirossian, F., and Mezzetti, M.: 1984c, *Astr. and Astroph.* **135**, 393.
Goldreich, P. and Keeley D.A.: 1977, *Astroph. J.* **211**, 934.
Goldreich, P. and Nicholson, P.D.: 1989a, *Astroph. J.* **342**, 1075.
Goldreich, P. and Nicholson, P.D.: 1989b, *Astroph. J.* **342**, 1079.
Goldreich, P. and Tremaine S.: 1978, *Icarus* **34**, 340.
Hellier, C. and Mason, K.O.: 1989, *Monthly Not. of the Roy. Astr. Soc.* **239**, 715.
Houghton, J.T.: 1977, *The Physics of Atmospheres*, Cambridge University Press, 129.
Hut, P. and Paczynski, B.: 1984, *Astroph. J.* **284**, 675.
Kovetz A., Prialnik D., and Shara M.M.: 1988, *Astroph. J.* **325**, 828.
Livio, M.: 1990, in F. Giovannelli, ed(s)., *Frontier Objects in Astrophysics and Particle Physics*, Vulcano Workshop, .
Livio, M., Soker, N., de Kool, M., and Savonije, G.J.: 1986a, *M.N.R.A.S.* **218**, 593.
Livio, M., Soker, N., de Kool, M., and Savonije, G.J.: 1986b, *M.N.R.A.S.* **222**, 235.
Livio, M., Soker, N., Matsuda, T., and Anzer, U.: 1992, *M.N.R.A.S.* **253**, 633.
Lubow, S.H. and Shu, F.H.: 1975, *Astroph. J.* **198**, 383.
Lubow, S.H. and Shu, F.H.: 1975, *Astroph. J.. Lett.* **207**, L53.
Lubow, S.H.: 1979, *Astroph. J.* **229**, 1008.
Lubow, S.H.: 1989, *Astroph. J.* **340**, 1064.
Mathieu, R.D. and Mazeh, T.: 1988, *Ap.J.* **326**, 256.
Mathieu, R.D.: 1992, in *Binaries and Tracers of Star Formation*, eds. Duquennoy and Mayor, Cambridge U. Press.
Matsuda, T., Inoue, M., and Sawada, K.: 1987, *M.N.R.A.S.* **226**, 781.
Matsuda, T., Sekino, N., Sawada, K., Shima, E., Livio, M., Anzer, U., and Bormer, G.: 1991, *Astr. Astrophys.* **248**, 301.
Matsuda, T., Ishii, T., Sekino, N., Sawada, K., Shima, E., Livio, M., and Anzer, U.: 1992, *M.N.R.A.S.*, in press.
Naylor, T.: 1989, *Monthly Not. of the Roy. Astr. Soc.* **238**, 587.
Paczynski, B.: 1971, *Ann. Rev. Astr. and Astroph.* **9**, 183.
Paczynski, B. and Sienkiewicz: 1972, *Acta Astr.* **22**, 73.
Plavec, M.: 1973, *Proc. IAU Symp.* **51**, 216.
Plavec, M.: 1990, in C.Ibanoglu, ed(s)., *Active Close Binaries*, Kluwer Academic Publishers, Dordrecht, 37.
Parmar, A.N., White, N.E., Giommi, P., Gottwald, M.: 1986, *Astroph. J.* **308**, 199.
Papaloizou, J.C.B and Pringle, J.E.: 1979, *Monthly Not. of the Roy. Astr. Soc.* **189**, 293.
Savonije, G.J. and Papaloizou, J.C.B.: 1983a, *Monthly Not. of the Roy. Astr. Soc.* **203**, 581.
Savonije, G.J. and Papaloizou, J.C.B.: 1983b, in P.P. Eggleton and J.E.Pringle, ed(s)., *Interacting Binaries*, Reidel, Dordrecht, 83.
Scharlemann, E.T.: 1981, *Astroph. J.* **246**, 292.
Scharlemann, E.T.: 1982, *Astroph. J.* **253**, 298.
Shu, F.H., Milione, V., and Roberts, W.W.: 1973, *Astroph. J.* **183**, 819.
Shu, F.H. and Lubow, S.H.: 1981, *Ann. Rev. Astr. and Astroph.* **19**, 277.
Spruit, H.C., Nordlund, A., and Title, A.M.: 1990, *Ann. Rev. Astr. Astroph.* **28**, 263.
Stahler, S.W., Shu, F.H., and Taam, R.E.: 1980, *Ap. J.* **242**, 226.
Stahler, S.W.: 1988, *Ap. J.* **332**, 804.
Stein, R.F. and Nordlund, A.: 1989, *Ap. J.* **342**, L95.
Stover, R.J.: 1981, *Astroph. J.* **249**, 673.
Taam, R.E. and Fryxell, B.A.: 1988, *Astroph. J.. Lett.* **327**, L73.
Tassoul, J.-L.: 1978, *Theory of Rotating Stars*, Princeton University Press, 409.
Starr, V.: 1968, *Physics of Negative Viscosity Phenomena*, McGraw Hill, .
White, N.E. and Mason, K.O.: 1985, *Space Sci. Rev.* **40**, 167.
White, N.E.: 1989, *Astr. and Astroph. Rev* **1**, 1, 85.
Zahn, J-P.: 1966a, *Ann. Ap.* **29**, 313.
Zahn, J-P.: 1966b, *Ann. Ap.* **29**, 489.
Zahn, J-P.: 1966c, *Ann. Ap.* **29**, 565.

Zahn, J-P.: 1975, *Astr. and Astroph.* **41**, 329.
Zahn, J-P.: 1977, *Astr. and Astroph.* **57**, 383.
Zahn, J.-P.: 1989, *Astron. Astrophys.* **220**, 112.
Zahn, J.-P. and Bouchet, L.: 1989, *Astron. Astrophys.* **223**, 112.

THE ALGOL-TYPE INTERACTING BINARIES

G. E. McCluskey, Jr.

Andersen, J., Pavlovski, K., and Piirola, V.: 1989, *Astron. Astrophys.* **215**, 272.
Andersen, J., Nordstrom, B., Mayor, M., and Polidan, R.S.: 1988, *Astron. Astrophys.* **207**, 37.
Balachandran, S., Lambert, D.L., Tomkin, J., and Parthasarathy, M.: 1986, *Monthly Notices Roy. Astron. Soc.* **219**, 479.
Batten, A.H.: 1974, *Publ. Dominion Astrophys. Obs.* **14**, 191.
Crawford, J.A.: 1955, *Astrophys. J.* **121**, 71.
Cugier, H. and Hardorp, J.: 1988, *Astron. Astrophys.* **202**, 101.
De Greve, J.P.: 1986, *Space Sci. Rev.* **43**, 139.
De Loore, C.: 1984, *Astrophys. Space Sci.* **99**, 199.
Dobias, J.J. and Plavec, M.J.: 1985, *Publ. Astron. Soc. Pacific* **97**, 138.
Elias, N., II. : 1990a, Ph.D. Dissertation.
Elias, N. II.: 1990b, *Astrophys. J.* **352**, 300.
Guinan, E.F.: 1989, *Space Sci. Rev.* **50**, 35.
Hack, M., Hutchings, J.B., Kondo, Y., McCluskey, G.E., Jr., Plavec, M., and Polidan, R.S.: 1975, *Astrophys.J.* **198**, 453.
Hall, D.S.: 1989, *Space Sci. Rev.* **50**, 219.
Kahn, M.A. and Budding, E.: 1986, *Astrophys. Space Sci.* **125**, 219.
Kaitchuck, R.H. and Honeycutt, R.K.: 1982, *Astrophys. J.* **258**, 224.
Kaitchuck, R.H. and Park, E.A.: 1988, *Astrophys. J.* **325**, 225.
Kaitchuck, R.H., Honeycutt, R.K., and Faulkner, D.R.: 1989, *Astrophys. J.* **339**, 420.
Kaitchuck, R.H., Honeycutt, R.K., and Schlegel, E.M.: 1985, *Publ. Astron. Soc. Pacific* **97**, 1178.
Koch, R.H.: 1989, *Space Sci. Rev.* **50**, 331.
Kondo, Y., Boggess, A., and Maran, S.P.: 1989, *Ann. Rev. Astron. Astrophys.* **27**, 397.
Kondo, Y., McCluskey, G.E., Jr., and Stencel, R.E.: 1979, *Astrophys. J.* **233**, 906.
Kopal, Z.: 1955, *Ann. Astrophys.* **18**, 379.
Kopal, Z.: 1984, *Astrophys. Space Sci.* **99**, 3.
Lestrade, J.-F., Mutel, R., Preston, R.A., and Phillips, R.B.: 1988, *Astrophys. J.* **328**, 232.
Li, Y.-F. and Leung, K.-C.: 1987, *Astrophys. J.* **313**, 801.
McCluskey, G.E., Jr.: 1982, in Y. Kondo, J.M. Mead, and R.D. Chapman, ed(s)., *Advances in Ultraviolet Astronomy*, NASA CP-2238, 102.
McCluskey, G.E., Jr. and Kondo, Y.: 1983, *Astrophys. J.* **266**, 755.
McCluskey, G.E., Jr. and Kondo, Y.: 1984, *Publ. Astron. Soc. Pacific* **96**, 817.
McCluskey, G.E., Jr. and Sahade, J.: 1987, in Y. Kondo, ed(s)., *Exploring the Universe with the IUE Satellite*, D. Reidel Publ. Co., Dordrecht, 1019.
McCluskey, G.E., Jr., McCluskey, C.P.S., and Kondo, Y.: 1988, in E.J. Rolfe ed(s)., *A Decade of UV Astronomy with the IUE Satellite*, ESA SP-281, 201.
McCluskey, G.E., Jr., McCluskey, C.P.S., and Kondo, Y.: 1991, *Astrophys. J.* **in press**, .
Olson, E.C.: 1976, *Astrophys. J.* **204**, 141.
Olson, E.C.: 1978, *Astrophys. J.* **220**, 251.
Olson, E.C.: 1980, *Astrophys. J.* **241**, 257.
Olson, E.C.: 1981, *Astrophys. J.* **250**, 704.
Olson, E.C.: 1985, in P.P. Eggleton and J.E. Pringle, ed(s)., *Interacting Binaries*, D. Reidel, Dordrecht, 137.
Olson, E.C.: 1986, *Astron. J.* **91**, 1421.
Olson, E.C.: 1987, *Astron. J.* **94**, 1043.

Olson, E.C.: 1989, *Astron. J.* **97**, 505.
Olson E.C. and Stoehr, C.A.: 1986, *Astron. J.* **91**, 1418.
Parthasarathy, M., Lambert, D.L., and Tomkin, J.: 1983, *Monthly Notices Roy. Astron. Soc.* **203**, 1063.
Peters, G.J.: 1989, *Space Sci. Rev.* **50**, 9.
Peters, G.J. and Polidan, R.S.: 1984, *Astrophys. J.* **283**, 745.
Plavec, M.J.: 1982, in Z. Kopal and J. Rahe, ed(s)., *Binaries and Multiple Stars as Tracers of Stellar Evolution*, D. Reidel, Dordrecht, 137.
Plavec, M.J.: 1983, *Astrophys. J.* **275**, 251.
Plavec, M.J.: 1985, in P.P. Eggleton and J.E. Pringle, ed(s)., *Interacting Binaries*, D. Reidel, Dordrecht, 155.
Plavec, M.J.: 1989, *Space Sci. Rev.* **50**, 95.
Polidan, R.S.: 1988, in E.J. Rolfe ed(s)., *A Decade of UV Astronomy with the IUE Satellite*, ESA SP-281, 205.
Polidan, R.S.: 1989, *Space Sci. Rev.* **50**, 85.
Popper, D.M. and Tomkin, J.: 1984, *Astrophys. J.* **285**, 208.
Rahe, J.: 1984, in J.M. Mead, R.D. Chapman, and Y. Kondo, ed(s)., *Future of Ultraviolet Astronomy Based on Six Years of IUE Research*, NASA CP-2349, 51.
Richards, M.T.: 1990, *Astrophys. J.* **350**, 372.
Sahade, J.: 1980, *Space Sci. Rev.* **26**, 349.
Sahade, J.: 1986, in E.J. Rolfe, ed(s)., *New Insights in Astrophysics*, ESA SP-263, 267.
Shore, S.N.: 1988, in E.J. Rolfe, ed(s)., *A Decade of UV Astronomy with the IUE Satellite*, ESA SP-281, 67.
Shore, S.N. and King, A.R.: 1986, *Astron. Astrophys.* **154**, 263.
Slee, O.B., Nelson, G.J., Stewart, R.T., Wright, A.E., Innin, J.L., Ryan, S.G., and Vaughn, A.E.: 1987, *Monthly Notices Roy. Astron. Soc.* **229**, 659.
Stewart, R.T., Slee, O.B., White, G.L., Budding, E., Coates, D.W., Thompson, K., and Bunton, J.D.: 1989, *Astrophys. J.* **342**, 463.
Tomkin, J.: 1978, *Astrophys. J.* **221**, 608.
Tomkin, J.: 1979, *Astrophys. J.* **231**, 495.
Tomkin, J.: 1981, *Astrophys. J.* **244**, 546.
Tomkin, J.: 1985, *Astrophys. J.* **297**, 250.
Tomkin, J. and Lambert, D.L.: 1989, *Monthly Notices Roy. Astron. Soc.* **241**, 777.
White, N.E. and Marshall, F.E.: 1983, *Astrophys. J. Letters* **268**, L117.
White, N.E., Culhane, J.L., Parmar, A.N., Kellett, B.J., Kahn, S., Van Den Oord, G.H.J., and Kuipers, J.: 1986, *Astrophys. J.* **301**, 262.
Wilson, R.E. and Plavec, M.J.: 1988, *Astron. J.* **95**, 1928.
Young, A. and Snyder, J.A.: 1982, *Astrophys. J.* **262**, 269.

MAGNETIC ACTIVITY IN CLOSE BINARIES

E.F. Guinan and A. Giménez

Allen, C. W. : 1981, *Astrophysical Quantities*, 3rd ed., Athlone, London.
Altamore, A., *et al.* : 1981, *Astroph. J.*, **245**, 630.
Applegate, J. H., and Patterson, J. : 1987, *Astroph. J.Lett.*, **322**, L99.
Babcock, H. W. : 1961, *A J.*, **133**, 572.
Baliunas, S. L., *et al.* : 1985, *Astroph. J.*, **294**, 310.
Baliunas, S. L., Guinan, E. F., and Dupree, A. K. : 1984, *Astroph. J.*, **282**, 733.
Baliunas, S. L., Loeser, J. G., Raymond, J. C., Guinan, E. F., and Dorren, J. D. : 1986, in *New Insights in Astrophysics: 8 Years of UV Astronomy with IUE*, E. Rolfe (ed.), ESA SP-263, 185.
Baliunas, S. L. and Vaughan, A. H. : 1985, *Ann. Rev. Astron. Astrophys.* **23**, 379.
Basri, G., Wilcots, E., and Stout, N. : 1989, *P.A.S.P.*, **101**, 528.
Bianchini, A. : 1990, *Astron. J.*, **99**, 1941.
Binnendijk, L. : 1965, *Astron. J.*, **70**, 209.

Binnendijk, L. : 1970, in *Vistas of Astronomy*, **12**, 217.
Bois, B., Lanning, H. H. and Mochnacki, S. W. : 1988, *Astron. J.*, **96**, 157.
Bopp, B. W. : 1983, in *IAU Colloq. 71, Activity in Red Dwarf Stars*, ed. P. B. Byrne and M. Rodonó (Reidel: Dordrecht) 363.
Bopp, B. W. and Fekel, F. C. : 1977, *Astron. J.*, **82**, 490.
Bopp, B. W. and Stencel, R. E. : 1981, *Astroph. J.(Let.)*, **247**, L131.
Bradstreet, D. H. : 1985, *Astroph. J.Suppl.*, **58**, 413.
Bradstreet, D. H. and Guinan, E. F. : 1988, in *A Decade of UV Astronomy with the IUE Satellite*, Rolfe (ed.), ESA SP-281, 303.
Bradstreet, D. H. and Guinan, E. F. : 1990, in *Active Close Binaries*, ed. C. Ibanoğlu (Kluwer: Dordrecht), 467.
Brandenburg, A., Moss, D., Rödiger, G. and Tuominen, I. : 1990, *Solar Phys.*, **128**, 243.
Bruning, D. H. : 1981, *Astroph. J.*, **248**, 274.
Busso, M., Scaltriti, R., Blanco, C., Catalano, S., Marilli, E., Pazzani, V., Rodonó, M. : 1984, *Astron. Astrophys.*, **135**, 255.
Busso, M. Scaltriti, F. and Cellino, A. : 1985, *Astron. Astrophys.*, **148**, 29.
Byrne, P. B. : 1989, in *Solar and Stellar Flares*, ed. B. M. Haisch and M. Rodonó (Kluwer: Dordrecht), 61.
Cannizzo, J. K. and Kenyon, S. J. : 1986, *Astroph. J.Lett.*, **309**, L43.
Clark, B. G., Kellermann, K. I. and Shaffer, D. : 1975, *Astroph. J.Lett.*, **198**, L123.
Clark, T. A.,*et al.* : 1976, *Astroph. J.Lett.*, **206**, L107.
Crawford, D. L. : 1975, *Astron. J.*, **80**, 955.
Cruddace, R. F. and Dupree, A. K. : 1984, *Astroph. J.*, **277**, 263.
Demircan, O. : 1990, in *Active Close Binaries*, ed. C. Ibanoğlu (Kluwer: Dordrecht), 431.
Donati, J.-F., Semel, M., Rees, D. E., Taylor, K. and Robinson, R. D. : 1990, *Astron. Ap.*, **232**, L1.
Dorren, J. D. and Guinan, E. F. : 1982a, *Astron. J.*, **252**, 296.
Dorren, J. D. and Guinan, E. F. : 1982b, *Astroph. J.*, **87**, 1546.
Dorren, J. D. and Guinan, E. F. : 1984, in *Lecture Notes in Physics, No. 193*, ed. S. L. Baliunas and L. Hartmann (Springer-Verlag: N.Y.), 259.
Dorren, J. D. and Guinan, E. F. : 1990, *Astroph. J.*, **348**, 703.
Dorren, J. D., Guinan, E. F. and Siah, M. J. : 1983, *Info. Bull Var. Stars*, No. 2305.
Dorren, J. D., Siah, M. J., Guinan, E. F., McCook, G. P. : 1981, *Astron. J.*, **86**, 572.
Dorren, J. D., Guinan, E. F. and Sion, E. M. : 1982, in *Advances in UV Astronomy: Four Years of IUE Research*, NASA Conf. Publ. 2238, 517.
Drake, S. A. : 1991, private communication.
Drake, S. A., Simon, T. and Linsky, J. L. : 1986, *Astron. J.*, **91**, 1229.
Drake, S. A., Simon, T. and Linsky, J. L. : 1989, *Astroph. J.Suppl.*, **71**, 905.
Drake, S. A., Linsky, J. L., Judge, P. G., Elitzur, M. : 1991, *Astron. J.,S Ap.Sp.* **101**, 230.
Dulk, G. A. : 1985, *Ann. Revs. Astr. Ap.*, **23**, 169.
Dworetsky, M. M., Lanning, H. H., Etzel, P. B., Patenaude, D. J. : 1977, *Monthly Not. Roy. Astr. Soc.*, **181**, 13.
Eaton, J. A. : 1983, *Astroph. J.*, **268**, 800.
Eaton, J. A. and Hall, D. S. : 1979, *Astroph. J.*, **227**, 907.
Eggen, O. J. and Iben, I. : 1989, *Astron. J.*, **97**, 431.
Elias, N. M. and Dorren, J. D. : 1990, *Info. Bull. Var. Stars*, No. 3541.
Etzel, P. B. : 1991, private communication.
Evren, S. : 1990, in *Active Close Binaries*, ed. C. Ibanoğlu (Kluwer: Dordrecht), 561.
Foukal, P. and Lean, J. : 1986, *Astroph. J.*, **302**, 826.
Foukal, P. and Lean, J. : 1988, *Astroph. J.*, **328**, 347.
Giampapa, M. S., Worden, S. P. and Gillman, L. B. : 1979, *Astroph. J.*, **229**, 1143.
Gilman, P. A. and DeLuca, E. E. : 1986, in *Lecture Notes in Physics, Vol. 254: Cool Stars*, 163.
Giménez, A., Reglero, V., De Castro, E., Fernandez-Figueroa, M. J. : 1986, *Adv. Sp. Res.*, **6**, 191.
Giménez, A., Reglero, V., De Castro, E., Fernandez-Figueroa, M. J. : 1991, *Astron. Astrophys.*, (in press).
Giménez, A., *et al.* : 1990, in *Evolution in Astrophysics: IUE Astronomy in the Era of New Space*

Missions, ed. E. J. Rolfe, ESA SP-310, 383.
Gray, D. F. : 1977, *Astroph. J.,* **211**, 198.
Guinan, E. F. : 1990, in *Evolution in Astrophysics: IUE Astronomy in the Era of New Space Missions,* ed. E. J. Rolfe, ESA SP-310, 73.
Guinan, E. F. and Bradstreet, D. H. : 1988, in *Formation and Evolution of Low Mass Stars,* ed. A. K. Dupree and M. T. Lago (Reidel: Dordrecht), 345.
Guinan, E. F. and McCook, G. P. : 1991, private communication.
Guinan, E. F., Robinson, C. R. and Wacker, S. W. : 1986b, in *Lecture Notes in Physics, Vol.* **254**, Springer-Verlag: Berlin, 304.
Guinan, E. F., Wacker, S. W., Baliunas, S. L., Loeser, J. G. and Raymond, J. C. : 1968a, in *New Insights in Astrophysics,* ESA SP-263, 197.
Haisch. B. M. : 1989, in *Solar and Stellar Flares,* ed. B. M. Haisch and M. Rodonó (Kluwer: Dordrecht), 3.
Hall, D. S. : 1972, *P.A.S.P.,* **84**, 323.
Hall, D. S. : 1976, in *Multiple Periodic Variable Stars,* ed. W. S. Fitch (Reidel: Dordrecht), 287.
Hall, D. S. : 1987, *Publ. Astr. Inst. Czech.,* **70**, 77.
Hall, D. S. : 1990, in *Active Close Binaries,* ed. C. Ibanoğlu (Kluwer: Dordrecht), 95.
Hall, D. S. and Kreiner, J. M. : 1980, *Acta. Astr.,* **30**, 387.
Hill, G. : 1989, *Astron. Astrophys.,* **218**, 141.
Hjellming, R. M. : 1988, in *Galactic and Extragalactic Radio Astronomy,* ed. G. L. Verschuur and K. I. Kellermann, 2nd edition, in press.
Horne, K. and Saar, S. H. : 1991, *Astroph. J.(Lett.),* in press.
Huang, S.-S. : 1966, *Astroph. J.,* **150**, 229.
Huenemoerder, D. P. and Ramsey, L. W. : 1987, *Astroph. J.,* **319**, 392.
Huenemoerder, D. P., Ramsey, L. W. and Buzasi, D. L. : 1989, *Astron. J.,* **98**, 1398.
Ibanoğlu, C. : 1990, in *Active Close Binaries,* ed. C. Ibanoğlu (Kluwer: Dordrecht), 515.
Jordan, C. : 1986, in *New Insights in Astrophysics,* Proc. Joint NASA/ESA/ SERC Confer- ence, ESA SP-263, 17.
Jordan, C. and Linsky, J. L. : 1987, in *Exploring the Universe with the IUE Satellite,* ed. Y. Kondo *et al.* (Reidel: Dordrecht), 259.
Kaluzny, J. and Shara, M. M. : 1987, *Astroph. J.,* **314**, 585.
Kawaler, S. D. : 1988, *Astroph. J.,* **333**, 236.
Kenyon, S. J. : 1986, *The Symbiotic Stars,* Cambridge Univ. Press, Cambridge.
Kron, G. E. : 1947, *P.A.S.P.,* **59**, 261.
Kuhl, L. V. : 1964, *P.A.S.P.,* **76**, 430.
Kuijpers, J. : 1989, in *Solar and Stellar Flares,* ed. B. M. Haisch and M. Rodonó (Kluwer: Dordrecht), 163.
Kwee, K. K. : 1966, *Bull. Astron. Inst, Netherlands Supple.,* **1**, 245; 265.
Lemen, J. R., Mewe, R., Schrijer, C. J. and Fuldra, A. : 1989, *Astroph. J.,* **341**, 474.
Lestrade, J.-F. : 1988, in *The Impact of VLBI on Astrophysics and Geophysics,* ed. M. J. Reid and M. J. Moran (IAU Publ.), 265.
Lestrade, J.-F., Mutel, R. L., Preston, R. A., Phillips, R. B. : 1988, *Astroph. J.,* **328**, 232.
Lestrade, J.-F., Phillips, R. B., Hodge, M. W. and Preston, R. A. : 1992, *Astroph. J.,* (in press).
Linsky, J. L. : 1984, in *Lecture Notes in Physics, Vol. 193: Cool Stars, Stellar Systems, and the Sun,* ed. S. L. Baliunas and L. Hartmann (Springer-Verlag: Berlin), 244.
Linsky, J. L. : 1988, in *Multiwavelength Astrophysics,* ed. F. Cordova (Cambridge Univ. Press: Cambridge), 49.
Linsky, J. L. : 1990, in *Active Close Binaries,* ed. C. Ibanoğlu (Kluwer: Dordrecht), 747.
Maceroni, C., Bianchini, A., Rodonó, M., Van't Veer, F. and Vio, R. : 1990, *Astron. Astrophys.,* **237**, 395.
Majer, P., Schmitt, J. H. M. M., Golub, L., Harnden, F. R. and Rosner, R. : 1986, *Astroph. J.,* **300**, NASA Conf. Publ. No. 2238, 554.
Mestel, L. : 1968, *Monthly Not. Roy. Astr. Soc.,* **138**, 359.
Mestel, L. : 1984, in *Lecture Notes in Physics, Vol. 193: Cool Stars, Stellar Systems, and the Sun,* ed. S. L. Baliunas and L. Hartmann (Springer-Verlag: Berlin), 49.

Mochnacki, S. W. : 1981, *Astroph. J.*, **245**, 650.
Montesinos, B., Giménez, A., Fernandez-Figueroa, M. L. : 1988, *Monthly Not. Roy. Astr. Soc.*, **232**, 261.
Morris, D. H. and Mutel, R. L. : 1988, *Astron. J.*, **95**, 204.
Mutel, R. L., Morris, D. H., Doiron, D. J. and Lestrade, J.-F. : 1987, *Astron. J.*, **289**, 262.
Neff, J. E., Walter, F. M., Rodonó, M. and Linsky, J. L. : 1989, *Astron. Astrophys.*, **215**, 79.
Nelson, B. and Young, A. : 1970, *P.A.S.P.*, **82**, 699.
Newton, H. W. and Nunn, M. L. : 1951, *Monthly Not. Roy. Astr. Soc.*, **111**, 413.
Noyes, R. W., Hartmann, L., Baliunas, S. L., Duncan, D. K. and Vaughan, A. H. : 1984, *Astroph. J.*, **279**, 763.
Olah, K. : 1990, in *Active Close Binaries,* ed. C. Ibanoğlu (Kluwer: Dordrecht), 545.
Olson, E. C. : 1985, in *Interacting Binaries,* ed. P. P. Eggleton and J. E. Pringle (Reidel: Dordrecht), 127.
Olson, E. C. : 1987, *Astron. J.*, **94**, 1043.
Olsen, E. H. : 1984, *Astron. Astrophys. Suppl.*, **57**, 443.
Olsen, E. H. : 1988, *Astron. Astrophys.*, **189**, 173.
Pallavicini, R., Golub, L., Rosner, R., Vaiana, G. S., Ayres, T., and Linsky, J. L. : 1981, *Astroph. J.*, **248**, 279.
Parker, E. N. : 1955, *Astroph. J.*, **122**, 393.
Parker, E. N. : 1981, *Astroph. J.*, **244**, 631.
Parker, E. N. : 1986, in *Lecture Notes in Physics, No. 254, Cool Stars, Stellar Systems, and the Sun,* ed. M. Zeilik and D. M. Gibson (Springer-Verlag: New York), 341.
Pettersen, B. R. : 1983, in *Activity in Red Dwarf Stars,* ed. P. D. Byrne and M. Rodonó (Reidel: Dordrecht), 17.
Pettersen, B. R. : 1989, *Solar Phys.*, **121**, 299.
Pfeiffer, R. J. : 1979, *Astroph. J.*, **232**, 181.
Pinsonneault, M. H., Kawaler, S. D., Sofia, S. and Demarque, P. : 1989, *Astroph. J.*, **338**, 424.
Plavec, M. J. : 1985, *Astroph. J.*, **275**, 251.
Popper, D. M. : 1988a, *Astron. J.*, **95**, 1242.
Popper, D. M. : 1988b, *Astron. J.*, **96**, 1040.
Popper, D. M. : 1990, *Astron. J.*, **100**, 247.
Popper, D. M. : 1991, *Astron. J.*, **101**, 220.
Ramsey, L. W. : 1990, in *6th Cambridge Workshop on Cool Stars, Stellar Systems, and the Sun,* ed. G. Wallerstein (ASP Conf. Series, 9), 171.
Ramsey, L. W. and Nations, H. L. : 1990, *Astroph. J.(Lett.)*, **239**, L121.
Reglero, V., Giménez, A., De Castro, E., Fernandez-Figueroa, M. J. : 1987, *Astron. Ap. Suppl.*, **71**, 421.
Reglero, V., Giménez, A. and Estela, A. : 1990, *Astron. Ap.*, **231**, 375.
Richards, M. T. : 1990, *Astroph. J.*, **350**, 372.
Rodonó, M. : 1981, in *Photometric and Spectroscopic Binary Systems,* ed. E. B. Carling and Z. Kopal (Reidel: Dordrecht), 285.
Rodonó, M. : 1986a, in *Lecture Notes in Physics, Vol. 254: Cool Stars, Stellar Systems, and the Sun,* ed. M. Zeilik and D. M. Gibson (Springer-Verlag: Berlin), 475.
Rodonó, M. : 1986b, *Highlights of Astronomy*, **7**, 429.
Rodonó, M. : 1988, in *A Decade of UV Astronomy with IUE,* ed. E. J. Rolfe, ESA SP-281, 45.
Rodonó, M. et al. : 1986, *Astron. Astrophys.*, **165**, 135.
Rosner, R., Tucker, W. H. and Vaiana, G. S. : 1978, *Astroph. J.*, **220**, 643.
Rucinski, S. M. : 1985, in *Interacting Binary Stars,* ed. J. E. Pringle and R. A. Wade (Cambridge Press: Cambridge), 85.
Rucinski, S. M. : 1992, this volume.
Rucinski, S. M. and Seaquist, E. R. : 1988, *Astron. J.*, **95**, 1837.
Rucinski, S. M. and Vilhu, O. : 1983, *Monthly Not. Roy. Astr. Soc.*, **202**, 1221.
Saar, S. H. : 1991, in *The Proceedings of IAU Colloq. 130: The Sun and Cool Stars - Activities,* No. 3228.
Saar, S. H. and Neff, J. E. : 1990, in *6th Cambridge Workshop on: Cool Stars, Stellar Systems, and*

the Sun, ed. G. Wallerstein (ASP Conf. Series 9), 171.
Scaltriti, F. : 1990, in *Active Close Binaries,* ed. C. Ibanoğlu (Kluwer: Dordrecht), 493.
Scaltriti, F., Piirola, V., Coyne, G. V., Koch, R. H., Elias, N. M., and Holenstein, B. D. : 1992, *Astron. Ap.,* in press.
Schatzman, E. : 1962, *Ann. Astrophys.,* **25,** 18.
Schnopper, H. W., Delvaille, J. P., Epstein, A., Heimken, H., Murray, S. S., Clark, G., Jernigan, G. and Doxsey, R. : 1976, *Astroph. J.(Lett.),* **210,** L75.
Shaw, J. S. : 1990, in *Active Close Binaries,* ed. C. Ibanoğlu (Kluwer: Dordrecht), 241.
Simon, T., Herbig, G. and Boesgaard, A. M. : 1985, *Astroph. J.,* **293,** 551.
Skumanich, A. : 1972, *Astroph. J.,* **171,** 565.
Soderblom, D. R. : 1983, *Astroph. J.Suppl.,* **53,** 1.
Soderblom, D. R. : 1988, in *Formation and Evolution of Low Mass Stars,* (NATO ASI Series), ed. A. K. Dupree and M.T.V.T. Lago (Kluwer: Dordrecht), 389.
Stepien, K. : 1990, *Acta. Astron.,* **30,** 315.
Stern, R. A., Haisch, B. M., Nagase, F., Uchida, Y. and Tsuneta, S. : 1990, in *6th Cambridge 9,* 224.
Strassmeier, K. G., Hall, D. S., Zeilik, M., Nelson, E., Eker, Z., and Fekel, F. C. : 1988, *Astr. Ap. Suppl.,* **72,** 291.
Strassmeier, K. G. : 1990, *Astroph. J.,* **348,** 682.
Strassmeier, K. G. : 1991, in *Robotic Telescopes in the 1990's,* ed. A. Filippenko (ASP Conf. Series), in press.
Struve, O. : 1950, *Stellar Evolution,* (Princeton Univ. Press: Princeton), 175.
Swank, J. H., White, N. E., Holt, S. S. and Becker, R. H. : 1981, *Astroph. J.,* **246,** 208.
Tang, F. : 1981, *Solar Phys.,* **69,** 399.
Tout, C. A. and Eggleton, P. P. : 1988, *Monthly Not. Roy. Astr. Soc.,* **231,** 823.
Tutukov, A. V. and Yungelson, L. R. : 1987, *Comments Astrophys.,* **12,** No. 2, 51.
Uchida, Y. and Sakurai, T. : 1983, in *IAU Colloq 71: Activity in Red Dwarf Stars,* ed. P. B. Bryne and M. Rodonó (Reidel: Dordrecht), 629.
Umana, G., Catalano, S. and Rodonó, M. : 1991, *Astron. Astrophys.,* (in press).
Van't Veer, F. : 1979, *Astron. Astrophys.,* **80,** 287.
Van't Veer, F. : 1991, in *Stades Avancés dans l'Evolution des Binaries Seriées,* CNRS preprint No. 347.
Verma, R. P., Iyengar, K. V. V. and Rengarajan, T. N. : 1987, *Astron. Astrophys.* **177,** 346.
Vilhu, O. : 1982, *Astron. Astrophys.,* **109,** 17.
Vilhu, O. : 1984, *Proc. 4th European IUE Conf.,* (ESA SP-218), 239.
Vilhu, O. : 1987, in *Lecture Notes in Physics Vol. 291: Cool Stars, Stellar Systems, and the Sun,* ed. J. L. Linsky and R. E. Stencel (Springer-Verlag: Berlin), 110.
Vilhu, O. and Walter, F. M. : 1987, *Astroph. J.,* **321,** 922.
Vilhu, O., Ambruster, C. W., Neff, J. E., Linsky, J. L., Brandenburg, A., Ilyin, I. V. and Shakjovskaya, N. : 1989, *Astron. Astrophys.,* **222,** 179.
Vogt, S. S. : 1988, in *The Impact of Very High S/N Spectroscopy on Stellar Physics,* ed. G. C. De Strobel and M. Spite (Kluwer: Dordrecht), 253.
Vogt, S. S. and Penrod. G. D. : 1983, *P.A.S.P.,* **95,** 565.
Wade, C. M. and Hjellming, R. M. : 1972, *Nature,* **235,** 270.
Walter, F. M, Charles, P. and Bowyer, S. : 1978, *Astroph. J.,* **350,** L119.
Walter, F. M., Neff, J. E., Gibson, D. M., Linsky, J. L., Rodonó, M. Gary, D. E. and Butler, C. J. : 1987, *Astron. Astrophys.,* **186,** 241.
Warner, B. : 1988, *Nature,* **336,** 129.
Webbink, R. F. : 1976, *Astroph. J.,* **209,** 829.
Webbink, R. F. : 1985, in *Interacting Binary Stars,* ed. J. E. Pringle and R. A. Wade (Cambridge Univ. Press: Cambridge), 39.
White, N. E., Culhane, J. L., Parmar, A. N., Kellett, B. J., Kahn, S., Van Den Oord, G. H. J. and Kuijpers, J. : 1986, *Astroph. J.,* **301,** 262.
White, N. E., Shafer, R. A., Horne, K., Parmar, A. N. and Culhane, J. L. : 1990, *Astroph. J.,* **350,** 776.
Yamasaki, A. : 1982, *Ap. Sp. Sci.,* **85,** 43.
Young, A., Skumanich, A., and Paylor, V. : 1988, *Astroph. J.,* **334,** 397.

Zwaan, C. : 1986, in *Lecture Notes in Physics, Vol. 254: Cool Stars, Stellar Systems, and the Sun*, ed. M. Zeilik and D. M. Gibson (Springer-Verlag: Berlin), 19.
Zwaan, C. : 1991, in *Mechanisms of Chromospheric and Coronal Heating*, ed. P. Ulmschneider, E. Priest, and R. Rosner (Springer-Verlag: Berlin), in press.

CONTACT BINARIES OF THE W UMA TYPE

S.M. Rucinski

Abt, H. A.: 1983, *Ann. Rev. Astron. Astrophys.* **21**, 343.
Allen, C. W.: 1973, *Astrophysical Quantities, 3rd Ed.*, The Athlone Press), 243.
Anderson, L. and Shu, F. H.: 1977, *Astrophys. J.* **214**, 798.
Baliunas, S. L. and Guinan, E. F.: 1985, *Astrophys. J.* **294**, 207.
Barden, S. C.: 1985, *Astrophys. J.* **295**, 162.
Batten, A. H. and Lu, W.: 1985, *Publ. Astr. Soc. Pacific* **98**, 92.
Bell, S. A., Hilditch, R. W. and King, D. J.: 1984, *Mon. Not. Roy. Astr. Soc.* **208**, 123.
Binnendijk, L. 1965, Kleine Veröff. Remeis- Sternw. Bamberg, **4**, No. 4, 36.
Binnendijk, L.: 1970, *Vistas in Astron.* **12**, 217.
Binnendijk, L.: 1977, *Vistas in Astron.* **21**, 359.
Bopp, B. W. and Rucinski, S. M.: 1981, in *Fundamental Problems in the Theory of Stellar Evolution*, ed(s)., D. Sugimoto *et al.*, Reidel Publ., Dordrecht, 177.
Bopp, B. W. and Stencel, R. E.: 1981, *Astrophys. J. Lett.* **247**, L 131.
Bradstreet, D. H.: 1985, *Astrophys. J. Suppl.* **58**, 413.
Bradstreet, D. H. and Guinan, E. F.: 1988, *A Decade of UV Astronomy with IUE, ESA SP-281*, Vol.1, 303.
Budding, E.: 1982, in *Binary and Multiple Stars as Tracers of Stellar Evolution*, ed(s)., Z. Kopal and J. Rahe, Reidel Publ., Dordrecht, 351.
Drake, S. A., Simon, T. and Linsky, J. L.: 1986, *Astron. J.* **91**, 1229.
Duerbeck, H. W.: 1984, *Astrophys. Space Sci.* **99**, 363.
Eaton, J. A., Wu, C.-C. and Rucinski, S. M.: 1980, *Astrophys. J.* **239**, 919.
Efremov, Y. N., Kholopov, P. N., Kukarkin, B. V., and Sharov, A. S. 1964, Inf. Bull. Var. Stars, No. 75.
Eggen, O. J.: 1961, *Roy. Obs. Bull.* No. 31, .
Eggen, O. J.: 1967, *Mem. Roy. Astron. Soc.* **70**, 111.
Flannery, B. P.: 1976, *Astrophys. J.* **205**, 217.
Gibson, D. M.: 1985, in *Radio Stars*, ed(s)., R. M. Hjellming and D. M. Gibson, Reidel Publ., Dordrecht, 213.
Gilliland, R. L., Brown, T. M., Duncan, D. K., Suntzeff, N. B., Lockwood, G. W., Thompson, D. T., Schild, R. E., Jeffrey, W. A. and Penprase, B. E.: 1991, *Astron. J.* **101**, 541.
Guinan, E. F. and Bradstreet, D. H.: 1988, in *Formation and Evolution of Low Mass Stars*, ed(s)., A.K. Dupree and M.T. Lago, Kluwer Acad. Publ., Dordrecht, 345.
Hall, D. S.: 1991, in *Active Close Binaries*, ed(s)., C. Ibanoglu, Kluwer Acad. Publ., Dordrecht, 1991.
Harris, H. C. and McClure, R. D.: 1985, *Publ. Astron. Soc. Pacific* **97**, 261.
Hilditch, R. W.: 1981, *Mon. Not. Roy. Astron. Soc.* **196**, 305.
Hilditch, R. W.: 1989, *Space Sci. Rev.* **50**, 289
   (= *Algols*, ed. A. H. Batten, (Kluwer Acad. Publ.), p.289).
Hilditch, R. W. and King, D. J.: 1986, *Mon. Not. Roy. Astr. Soc.* **223**, 581.
Hilditch, R. W., King, D. J. and McFarlane, T. M.: 1988, *Mon. Not. Roy. Astron. Soc.* **231**, 341.
Hill, G.: 1979, *Publ. D. A. O.* **15**, 297 (No.6).
Hill, G.: 1989, *Astron. Astrophys.* **218**, 141.
Hill, G., Fischer, W. A. and Holmgren, D.: 1989a, *Astron. Astrophys.* **211**, 81.
Hill, G., Fischer, W. A. and Holmgren, D.: 1989b, *Astron. Astrophys.* **218**, 152.
Hoffmann, M. 1978, Inf. Bull. Var. Stars, No.1487.
Hrivnak, B. J.: 1989, *Astrophys. J.* **340**, 458.

Hughes, V. A. and McLean, B. J.: 1984, *Astrophys. J.* **278**, 716.
Irwin, M. J. and Trimble, V.: 1984, *Astron. J.* **89**, 83.
Jameson, R. F. and Akinci, R.: 1979, *Mon. Not. Roy. Astron. Soc.* **188**, 421.
Kaluzny, J.: 1985, *Acta Astron.* **35**, 313.
Kaluzny, J.: 1990, *Astron. J.* **99**, 1207.
Kaluzny, J. and Rucinski, S. M.: 1986, *Astron. J.* **92**, 666.
Kaluzny, J. and Shara, M. M.: 1987, *Astrophys. J.* **314**, 585.
Kaluzny, J. and Shara, M. M.: 1988, *Astron. J.* **95**, 785.
Kholopov, P. N.: 1985, *General Catalogue of Variable Stars, 4th Ed.*, Nauka Publ. House, Moscow, .
Kreiner, J. M.: 1977, *The Interaction of Variable Stars with their Environment*, R. Kippenhahn *et al.* ed(s)., Veröff. Remeis-Sternw. Bamberg, 11, 393.
Kuhi, L. V.: 1964, *Publ. Astron. Soc. Pacific* **76**, 430.
Leonard, P. J. T.: 1989, *Astron. J.* **98**, 217.
Leung, K. C. and Jurkevich, I.: 1969, *Bull. Amer. Astron. Soc.* **1**, 251.
Linnell, A. P.: 1986, *Astrophys. J.* **300**, 304.
Lucy, L. B.: 1967, *Zeit. für Astroph.* **65**, 89.
Lucy, L. B.: 1968a, *Astrophys. J.* **151**, 1123.
Lucy, L. B.: 1968b, *Astrophys. J.* **153**, 877.
Lucy, L. B.: 1976, *Astrophys. J.* **205**, 208.
Maceroni, C., Milano, L. and Russo, G.: 1985, *Mon. Not. Roy. Astron. Soc.* **217**, 843.
Mateo, M., Harris, H., Nemec, J., Olszewski, E. and Schombert, J.: 1988, *Bull. Amer. Astron. Soc.* **20**, 717.
Mateo, M. Harris, H. C., Nemec, J. and Olszewski, E. W.: 1990, *Astron. J.* **100**, 469.
Mateo, M. and Krzeminski, W.: 1990, *Bull. Amer. Astr. Soc.* **22**, 1284.
Mauder, H.: 1972, *Astron. Astrophys.* **17**, 1.
McLean, B. J.: 1981, *Mon. Not. Roy. Astr. Soc.* **195**, 931.
McLean, B. J.: 1983, *Mon. Not. Roy. Astr. Soc.* **204**, 817.
Mochnacki, S. W.: 1981, *Astrophys. J.* **245**, 650.
Mochnacki, S. W.: 1985, in *Interacting Binary Stars*, ed(s)., J. E. Pringle and R. A. Wade, Cambridge Univ. Press, 51.
Mochnacki, S. W. and Doughty, N. A.: 1972a, *Mon. Not. Roy. Astron. Soc.* **156**, 51.
Mochnacki, S. W. and Doughty, N. A.: 1972b, *Mon. Not. Roy. Astron. Soc.* **156**, 243.
Mochnacki, S. W. and Whelan, J. A. J.: 1973, *Astron. Astrophys.* **25**, 249.
Moss, D.: 1985, *Astron. Astrophys.* **150**, 343.
Noyes, R. W., Hartmann, L. W., Baliunas, S. L., Duncan, D. K. and Vaughan, A. H.: 1984, *Astrophys. J.* **279**, 763.
Paczynski, B.: 1967, *Acta Astr.* **17**, 287.
Popper, D. M.: 1982, *Astrophys. J.* **262**, 641.
Rahunen, T.: 1981, *Astron. Astrophys.* **102**, 81.
Rigterink, P. V.: 1972, *Astron. J.* **77**, 230.
Robertson, J. A. and Eggleton, P. P.: 1977, *Mon. Not. Roy. Astron. Soc.* **179**, 359.
Rucinski, S. M.: 1968, *Acta Astron.* **18**, 49.
Rucinski, S. M.: 1974, *Acta Astron.* **24**, 119.
Rucinski, S. M.: 1976, *Acta Astron.* **26**, 227.
Rucinski, S. M.: 1980, *Acta Astron.* **30**, 373.
Rucinski, S. M.: 1985a, in *Interacting Binary Stars*, ed(s)., J. E. Pringle and R. A. Wade, Cambridge Univ. Press, 85 and 113.
Rucinski, S. M.: 1985b, in *Interacting Binaries*, ed(s)., P. P. Eggleton, J. E. Pringle, Reidel Publ., 13.
Rucinski, S. M.: 1985c, *Mon. Not. Roy. Astr. Soc.* **215**, 615.
Rucinski, S. M.: 1986, in *Instrumentation and Research Programmes for Small Telescopes*, ed(s)., J. B. Hearnshaw and P. L. Cottrell, Reidel Publ., 421.
Rucinski, S. M.: 1988, *Astron. J.* **95**, 1895.
Rucinski, S. M.: 1989, *Comm. on Astrophys.* **14**, 79.
Rucinski, S. M.: 1992, *Astron. J.*, in press, March 1992.
Rucinski, S. M. and Seaquist, E. R.: 1988, *Astron. J.* **95**, 1837.

Rucinski, S. M., Whelan, J. A. J. and Worden, S. P.: 1977, *Publ. Astron. Soc. Pacific* **89**, 684.
Samec, R. G. and Bookmyer, B. B.: 1987, *Publ. Astron. Soc. Pacific* **99**, 1298.
Samec, R. G., Van Hamme, W. and Boomyer, B. B.: 1989, *Astron. J.* **98**, 2287.
Sarna, M. J.: 1989, *Astron. Astrophys.* **224**, 98.
Shapley, H. 1948, Harvard Obs. Monograph, No. 7, 249.
Shenavrin, V. I. and Zhukov, G. V. 1984, Inf. Bull. Var. Stars, No. 2541.
Smith, R. C.: 1984, *Quart. J. Roy. Astron. Soc.* **25**, 405.
Stepien, K.: 1980, *Acta Astron.* **30**, 315.
Struve, O.: 1950, *Stellar Evolution*, Princeton Univ. Press, 175.
van't Veer, F.: 1975a, *Astron. Astrophys.* **40**, 167.
van't Veer, F.: 1975b, *Astron. Astrophys.* **44**, 437.
van't Veer, F.: 1978, *Astron. Astrophys.* **70**, 91.
van't Veer, F.: 1980, *Acta Astron.* **30**, 381.
van't Veer, F.: 1982, in *Binary and Multiple Stars as Tracers of Stellar Evolution*, ed(s)., Z. Kopal and J. Rahe, Reidel Publ., 279.
Vilhu, O.: 1982, *Astron. Astrophys.* **109**, 17.
Vilhu, O., Ambruster, C. W., Neff, J. E., Linsky, J. L., Brandenburg, A., Ilyin, I. V. and Shakhovskaya, N. I.: 1989, *Astron. Astrophys.* **222**, 179.
Vilhu, O., Caillault, J.-P. and Heise, J.: 1988, *Astrophys. J.* **330**, 922.
Vilhu, O. and Walter, F. M.: 1987, *Astrophys. J.* **321**, 958.
Walter, K.: 1979, *Astron. Astrophys.* **80**, 27.
Walter, K.: 1983, *Astron. Astrophys.* **128**, 391.
Walter, K. and Duerbeck, H. W.: 1988, *Astron. Astrophys.* **189**, 89.
Webbink, R. F.: 1976, *Astrophys. J.* **209**, 829.
Webbink, R. F.: 1979, *Astrophys. J.* **227**, 178.
Whelan, J. A. J.: 1972, *Mon. Not. Roy. Astron. Soc.* **156**, 115.
Whelan, J. A. J., Worden, S. P. and Mochnacki, S. W.: 1973, *Astrophys. J.* **183**, 133.
Whelan, J. A. J., Worden, S. P., Rucinski, S. M. and Romanishin, W.: 1979, *Mon. Not. Roy. Astr. Soc.* **186**, 729.
Williams, P. S. and Roxburgh, I. W.: 1976, *Mon. Not. Roy. Astron. Soc.* **176**, 81.
Wilson, R. E. and Biermann, P.: 1976, *Astron. Astrophys.* **48**, 349.
Wilson, R. E. and Devinney, E. J.: 1973, *Astrophys. J.* **182**, 539.
Yamasaki, A., Okazaki, A., Nakamura, Y., Lin, Q., Yang, Y., Zhang, Z. and Zhang, Y.: 1990, *Astron. J.* **99**, 1218.
Zhai, D. S. and Lu. W. X.: 1989, *Acta Astroph. Sinica* **9**, 208.

ACCRETION DISKS IN CLOSE BINARIES

E. Meyer-Hofmeister and H. Ritter

Abramowicz. M.A., Czerny, B., Lasota, J.P., Szuszkiewicz, E.: 1988, *Astrophys. J.* **332**, 646.
Adam, J., Störzer, H., Shaviv, G., Wehrse, R.: 1988, *Astron. Astrophys.* **193**, L1.
Bailey, J., Ward, M.: 1981, *Mon. Not. R. Astr. Soc.* **196**, 425.
Bath, G.T.: 1969, *Astrophys. J.* **158**, 571.
Bath, G.T.: 1972, *Astrophys. J.* **173**, 121.
Bath, G.T.: 1975, *Mon. Not. R. Astr. Soc.* **171**, 311.
Bath, G.T., Pringle, J.E.: 1981, *Mon. Not. R. Astr. Soc.* **194**, 967.
Bath, G.T., Edwards, A.C., Mantle, V.J.: 1983, *Mon. Not. R. Astr. Soc.* **205**, 171.
Batten, A., (ed.): 1988, *Algols*, IAU Coll. No. 107; *Space Sci. Rev.* **50**.
Cannizzo, J.K., Wheeler, J.C.: 1982, *Astrophys. J. Suppl.* **55**, 367.
Cannizzo, J.K., Kenyon, S.J.: 1986, *Astrophs. J.* **309**, L43.
Cannizzo, J.K., Kenyon, S.J.: 1987, *Astrophys. J.* **320**, 319.
Cannizzo, J.K., Cameron, A.G.W.: 1988, *Astrophys. J.* **330**, 327.
Canuto, V.M., Goldman, I., Chasnov, J.: 1988, *Astron. Astrophys.* **200**, 291.

Drury, L. O'C.: 1977, *Ph.D. Thesis*, Cambridge University, UK, .
Duschl, W.J.: 1986a, *Astron. Astrophys.* **163**, 56.
Duschl, W.J.: 1986b, *Astron. Astrophys.* **163**, 61.
Duschl, W.J.: 1988, in *The Symbiotic Phenomenon*, ed(s)., J. Mikolajewska *et al.*, Kluwer Academic Publishers, 137.
Duschl, W.J.: 1989, in Proc. 23rd ESLAB Symp. on Two Topics in X-Ray Astronomy, Bologna, Italy, ed(s)., J. Hunt and B. Battrich, ESA SP-296, 393.
Duschl, W.J.: 1991, *Astron. Astrophys.* **247**, 393.
Eardly, D.M., Lightman, A.P: 1975, *Astrophys. J.* **200**, 187.
Edwards, D.A.: 1985, *Mon. Not. R. Astr. Soc.* **212**, 623.
Edwards, D.A.: 1987, *Mon. Not. R. Astr. Soc.* **226**, 95.
Edwards, D.A.: 1988, *Mon. Not. R. Astr. Soc.* **231**, 25.
Frank, J., King, A.R., Raine, D.J.: 1985, *Accretion Power in Astrophysics*, Cambridge University Press.
Faulkner, J., Lin, D.N.C., Papaloizou, J.: 1983, *Mon. Not. R. Astr. Soc.* **205**, 359.
Flannery, B.P.: 1975, *Mon. Not. R. Astr. Soc.* **170**, 325.
Gilliland, R.L.: 1985, *Astrophys. J.* **292**, 522.
Hall, D.S.: 1988, in *Algols*, IAU Coll. No. 107, ed(s)., A. Batten, *Space Sci. Rev.*, 50 219.
Hassall, B.J.M.: 1984, in Proc. Fourth European IUE Conference, ed(s)., E. Rolfe & B. Battrick, ESA SP-218, 385.
Hirose, M., Osaki, Y.: 1989, in *Theory of Accretion Disks*, NATO ARW, ed(s)., F. Meyer *et al.*, Kluwer Academic Publishers, 207.
Hirose, M., Osaki, Y.: 1990, *Publ. Astron. Soc. Japan* **42**, 135.
Horne, K.: 1985, *Mon. Not. R. Astr. Soc.* **213**, 129.
Horne, K., Cook, M.C.: 1985, *Mon. Not. R. Astr. Soc.* **214**, 307.
Horne, K., Stiening, R.F.: 1985, *Mon. Not. R. Astr. Soc.* **216**, 933.
Hubeny, I.: 1989, in *Algols*, IAU Coll. No. 107, ed(s)., A. Batten, *Space Sci. Rev.*, 50 117.
Kato, S., Honma, F., Matsumoto, R.: 1989, *Publ. Astron. Soc. Japan* **40**, 709.
Kenyon, S.J.: 1986, *The Symbiotic Stars*, Cambridge University Press, .
King, A.R.: 1988, *Quart. J. R. Astr. Soc.* **29**, 1.
Kopal, Z.: 1959, *Close Binary Systems*, John Wiley and Sons Inc.
Kraft, R.P.: 1963, *Adv. Astron. Astrophys.* **2**, 43.
Křiž, S., Hubeny, I.: 1986, *Bull. Astr. Inst. Czech.* **37**, 129.
Kurucz, R.L.: 1979, *Astrophys. J. Suppl.* **40**, 1.
la Dous, C.: 1989, *Astron. Astrophys.* **211**, 131.
Lightman, A.P.: 1974, *Astrophys. J.* **194**, 429.
Lin, D.N.C., Pringle, J.E.: 1976, in Structure and Evolution of Close Binary Systems, IAU Symp. No. 73, ed(s)., P. Eggleton, S. Mitton, J. Whelan, D. Reidel, 237.
Lynden-Bell, D., Pringle, J.E.: 1974, *Mon. Not. R. Astr. Soc.* **168**, 603.
Marsh, T.R., Horne, K.: 1988, *Mon. Not. R. Astr. Soc.* **235**, 169.
Marsh, T.R., Horne, K.: 1990, *Astrophys. J.* **349**, 593.
Marsh, T.R., Horne, K., Schlegel, E.M., Honeycutt, R.K., Kaitchuck, R.H.: 1990, *Astrophys. J.* **364**, 637.
Matsuda, T., Sekino, N., Shima, E., Sawada, K., Spruit, H.: 1989, in Theory of Accretion Disks, NATO ARW, ed(s)., *F. Meyer et al.*, Kluwer Academic Publishers, 355.
Meyer, F.: 1984, *Astron. Astrophys.* **131**, 303.
Meyer F.: 1990, in Reviews in Modern Astronomy Vol. 3, ed(s)., *G. Klare*, Springer Verlag, 1.
Meyer, F., Meyer-Hofmeister: 1982, *Astron. Astrophys.* **106**, 34.
Meyer, F., Meyer-Hofmeister: 1983a, *Astron. Astrophys.* **121**, 29.
Meyer, F., Meyer-Hofmeister: 1983b, *Astron. Astrophys.* **128**, 420.
Meyer, F., Meyer-Hofmeister: 1984, *Astron. Astrophys.* **132**, 143.
Meyer, F., Meyer-Hofmeister: 1988, *Astron. Astrophys.* **194**, 135.
Meyer, F., Meyer-Hofmeister: 1990, *Astron. Astrophys.* **239**, 214.

Mineshige, S., Osaki, Y.: 1983, *Publ. Astron. Soc. Japan* **35**, 377.
Mineshige, S., Tuchman Y., Wheeler, J.C.: 1990, *Astrophys. J.* **359**, 176.
Mineshige, S., Wood, J.H.: 1990, *Mon. Not. R. Astr. Soc.* **247**, 43.
Mochnacki, S.W.: 1984, *Astrophys. J. Suppl.* **55**, 551.
Narayan, R., Goldreich, P., Goodman, J.: 1987, *Mon. Not. R. Astr. Soc.* **228**, 1.
O'Donoghue, D.: 1986, *Mon. Not. R. Astr. Soc.* **220**, 23p.
O'Donoghue, D.: 1990, *Mon. Not. R. Astr. Soc.* **246**, 29.
O'Donoghue, D., Fairall, A.P., Warner, B.: 1987, *Mon. Not. R. Astr. Soc.* **225**, 43.
Osaki, Y.: 1989, *Publ. Astron. Soc. Japan* **41**, 1005.
Osaki, Y., Hirose, M.: 1990, (private communication).
Paczyński, B.: 1971, *Ann. Rev. Astron. Astrophys.* **9**, 183.
Paczyński, B.: 1977, *Astrophys. J.* **216**, 822.
Paczyński, B., Rudak, B.: 1980, *Astron. Astrophys.* **82**, 349.
Papaloizou, J.C.B., Bath, G.T.: 1975, *Mon. Not. R. Astr. Soc.* **172**, 339.
Papaloizou, J.C.B., Pringle, J.E.: 1977, *Mon. Not. R. Astr. Soc.* **181**, 441.
Papaloizou, J.C.B., Faulkner, J., Lin, D.N.C.: 1983, *Mon. Not. R. Astr. Soc.* **205**, 487.
Papaloizou, J.C.B., Pringle, J.E.: 1984, *Mon. Not. R. Astr. Soc.* **208**, 721.
Papaloizou, J.C.B., Pringle, J.E.: 1985, *Mon. Not. R. Astr. Soc.* **225**, 267.
Pringle, J.E.: 1981, *Ann. Rev. Astron. Astrophys.* **19**, 137.
Pringle, J.E., Verbunt, F., Wade, R.A.: 1986, *Mon. Not. R. Astr. Soc.* **221**, 169.
Pringle, J.E., Wade, R.A.: 1985, *Close Binary Stars*, Cambridge University Press, .
Ritter, H.: 1986, in The Evolution of Galactic X-ray Binaries, NATO ASI Series C, Vol. 167, ed(s)., *J. Trümper, W.H.G. Lewin and W. Brinkmann*, D. Reidel, 271.
Ritter, H.: 1990, *Astron. Astrophys. Suppl. Ser.* **85**, 1179.
Rózyczka, M.: 1985, *Astron. Astrophys.* **143**, 59.
Rózyczka, M., Spruit, H.C.: 1989, in Theory of Accretion Disks, NATO ARW, ed(s)., *F. Meyer et al.*, Kluwer Academic Publishers, 341.
Sawada, K., Matsuda, T., Hachisu, I.: 1986, *Mon. Not. R. Astr. Soc.* **219**, 75.
Schoembs, R., Dreier, H., Barwig, H.: 1987, *Astron. Astrophys.* **181**, 50.
Schwarzenberg-Czerny, A.: 1981, *Acta Astron.* **31**, 241.
Schwarzenberg-Czerny, A., Rózyczka, M.: 1988, *Acta Astron.* **38**, 189.
Scott, B.D.: 1990, *Astrophys. J.* **357**, L53.
Shakura, N.I., Sunyaev, R.A.: 1973, *Astron. Astrophys.* **24**, 337.
Shaviv, G., Wehrse, R.: 1989, in *Theory of Accretion Disks*, NATO ARW, ed(s)., F. Meyer *et al.*, Kluwer Academic Publishers, 419.
Smak, J.: 1982, *Acta Astron.* **32**, 199.
Smak, J.: 1983, *Astrophys. J.* **272**, 234.
Smak, J.: 1984a, *Acta Astron.* **34**, 93.
Smak, J.: 1984b, *Publ. Astr. Soc. Pacific* **96**, 5.
Smak, J.: 1984c, *Acta Astron.* **34**, 161.
Smak, J.: 1989, *Space Sc. Rev.* **50**, 107.
Spruit, H.C.: 1991, in *Reviews in Modern Astronomy Vol. 4*, ed. G. Klare, Springer-Verlag, 197.
Tuchman, Y., Mineshige, S., Wheeler, J.C.: 1990, *Astrophys. J.* **359**, 164.
Tylenda, R.: 1981, *Acta Astron.* **31**, 127.
Verbunt, F.: 1987, *Astron. Astrophys. Suppl. Ser.* **71**, 339.
Vogt, N.: 1980, *Astron. Astrophys.* **88**, 66.
Vogt, N.: 1983, *Astron. Astrophys.* **128**, 29.
Wade, R.A.: 1984, *Mon. Not. R. Astr. Soc.* **208**, 381.
Wade, R.A.: 1988, *Astroph. J.* **335**, 394.
Warner, B.: 1974, *Mon. Not. R. Astr. Soc.* **168**, 235.
Warner, B.: 1988, *High Speed Astronomical Photometry*, Cambridge University Press, Cambridge.
Warner, B., O'Donoghue, D.: 1988, *Mon. Not. R. Astr. Soc.* **233**, 705.
Webbink, R.F.: 1985 in *Close Binary Stars*, ed(s). J. Pringle and R.A. Wade, Cambridge University Press, 39.
White, N.E.: 1989, *Astron. Astrophys. Rev.* **1**, 85.

Whitehurst, R.: 1988, *Mon. Not. R. Astr. Soc.* **232**, 35.
Williams, R.E.: 1980, *Astrophys. J.* **235**, 939.
Williams, R.E., Shipman, H.L.: 1988, *Astrophys. J.* **326**, 738.
Wood, J., Horne, K., Berriman, G., *et al.*: 1986, *Mon. Not. R. Astr. Soc.* **219**, 629.
Wood, J.H., Horne, K., Berriman, G., Wade, R.A.: 1989a, *Astrophys. J.* **341**, 974.
Wood, J.H., Marsh, T.R., Robinson, E.L., *et al.*: 1989b, *Mon. Not. R. Astr. Soc.* **239**, 809.
Wood, P.R.: 1977, *Astrophys. J.* **217**, 530.
Zhang, E.-H., Robinson, E.L.: 1987, *Astrophys. J.* **321**, 813.
Zola, S.: 1989, *Acta Astron.* **39**, 45.

MAGNETIC CATACLYSMIC VARIABLES

A. R. King and J. P. Lasota

Barrett, P., O'Donoghue, D., and Warner, B.: 1988, *Mon. Not. R. Astr. Soc.* **233**, 759.
Beuermann, K.: 1988, in *Circumstellar Polarisation*, Proceedings of the Vatican Conference, ed(s)., Coyne, G.V. *et al.*, Vatican Observatory, 125.
Beuermann, K. and Thomas, H.C., 1987. *IAU Circ.* No. 4486.
Beuermann, K., Thomas, H.C. and Schwope, A.: 1988, *Astron. Astrophys.* **195**, L15.
Campbell, C.G.: 1985, *Mon. Not. R. Astr. Soc.* **215**, 509.
Chanmugam, G., Frank, J., King, A.R. and Lasota, J.P.: 1990, *Astrophys. J. Lett.* **350**, L16.
Chanmugam, G. and Ray: 1984, *Astrophys. J.* **285**, 252.
Cropper, M.S.: 1986, *Mon. Not. R. Astr. Soc.* **222**, 225.
Cropper, M.S.: 1990, *Space Sci. Rev.* **54**, 195.
Ferrario, L., Wickramasinghe, D.T., Bailey, J.B., Tuohy, I.R. and Hough, J.H.: 1989, *Astrophys. J.* **337**, 832.
Frank, J., King, A.R. and Lasota, J.P.: 1987, *Astron. Astrophys.* **178**, 137.
Frank, J., King, A.R. and Lasota, J.P.: 1988, *Astron. Astrophys.* **193**, 113.
Frank, J., King, A.R. and Raine, D.J.: 1991, *Accretion Power in Astrophysics*, Cambridge University Press, in press.
Ghosh, P. and Lamb, F.K.: 1979, *Astrophys. J.* **232**, 259.
Giommi, P., Angelini, L., Osborne, J., Stella, L., and Tagliaferri, G., 1987 *IAU Circ.* No. 4486.
Hameury, J.M. and King A.R.: 1988, *Mon. Not. Roy. Astr. Soc.* **235**, 433.
Hameury, J.M., King, A.R. and Lasota, J.P.: 1986, *Mon. Not. Roy. Astr. Soc.* **218**, 695.
Hameury, J.M., King, A.R. and Lasota, J.P.: 1988, *Astron. Astrophys.* **195**, L12.
Hameury, J.M., King, A.R. and Lasota, J.P.: 1989a, *Mon. Not. Roy. Astr. Soc.* **237**, 39.
Hameury, J.M., King, A.R. and Lasota, J.P.: 1989b, *Mon. Not. Roy. Astr. Soc.* **237**, 845.
Hameury, J.M., King, A.R. and Lasota, J.P.: 1990, *Mon. Not. Roy. Astr. Soc.* **242**, 141.
Hameury, J.M., King, A.R., Lasota, J.P. and Livio, M.: 1989, *Mon. Not. Roy. Astr. Soc.* **237**, 835.
Hameury, J.M., King, A.R., Lasota, J.P. and Ritter, H.: 1987, *Astrophys. J.* **316**, 275.
Hameury, J.M., King, A.R., Lasota, J.P. and Ritter, H.: 1988, *Mon. Not. Roy. Astr. Soc.* **231**, 535.
Hellier, C.: 1991, *Mon. Not. Roy. Astr. Soc.* **in press**, .
Hellier, C., Mason, K.O., Rosen, S.R. and Córdova, F.A.: 1987, *Mon. Not. Roy. Astr. Soc.* **228**, 463.
Hellier, C., Mason, K.O., Smale, A.P., Corbet, R.H.D., O'Donoghue, D., Barrett, P.E. and Warner, B.: 1989, *Mon. Not. Roy. Astr. Soc.* **238**, 1107.
Horne, K. and Marsh, T.: in *The Physics of Accretion onto Compact Objects*, ed(s)., Mason, K.O, Watson, M.G. & White, N.E., Springer-Verlag, Berlin, 1.
King, A.R.: 1985, *Mon. Not. Roy. Astr. Soc.* **217**, 23P.
King, A.R.: , in *The Physics of Accretion onto Compact Objects*, ed(s)., Mason, K.O., Watson, M.G. & White, N.E., Springer-Verlag, Berlin, 137.
King, A.R.: 1988, *Q. Jl. R. Astr. Soc.* **29**, 1.
King, A.R.: 1989, *Mon. Not. Roy. Astr. Soc.* **241**, 365.
King, A.R., Frank, J. and Ritter, H.: 1985, *Mon. Not. Roy. Astr. Soc.* **213**, 185.
King, A.R., Frank, J. and Whitehurst, R.: 1990, *Mon. Not. Roy. Astr. Soc.* **244**, 731.

King, A.R. and Lasota, J.P.: 1979, *Mon. Not. Roy. Astr. Soc.* **181**, 653.
King, A.R. and Lasota, J.P.: 1990a, *Mon. Not. Roy. Astr. Soc.* **247**, 214.
King, A.R. and Lasota, J.P.: 1990b, *Astrophys. J.* **in press**, .
King, A.R. and Shaviv, G.: 1984, *Mon. Not. Roy. Astr. Soc.* **211**, 883.
King, A.R., Mouchet, M. and Lasota, J.P.: 1991, in *Structure and Emission Properties of Accretion Discs*, Proceedings of the 6th IAP/IAU Coll. No. 129, ed(s)., C. Bertout, S. Collin–Souffrin, J.P. Lasota and J. Trân Than Vân., Editions Frontiéres, 213.
Kuipers, J. and Pringle, J.P.: 1982, *Astron. Astrophys.* **114**, L4.
Lamb, D.Q. and Masters, A.R.: 1979, *Astrophys. J. Lett* **234**, L117.
Lamb, D.Q., and Melia, F.: 1987, *Ap. Space. Sci.* **131**, 511.
Liebert, J. and Stockman, H.S.: 1985, in Proc. 7th North American Workshop on CVs and LMXRBs, ed(s)., Lamb, D.Q. and Patterson, J., Reidel, Dordrecht, 151.
Lubow, S.H.: 1989, *Astrophys. J.* **340**, 1064.
Marcy, G.W.: 1984, *Astrophys. J.* **276**, 286.
Marsh, T.R., Horne, K.: 1990, *Astrophys. J.* **349**, 593.
Mason, K.O., Rosen, S.R. and Hellier, C.: 1988, in *The Physics of Compact Objects*, Adv. in Space Research, Vol. 8(2), ed(s)., White, N.S. and Filipov, L.G., Pergamon Press, Oxford, 293.
Meyer-Hofmeister, E., 1992, this volume.
Norton, A.J. and Watson, M.G.: 1989a, *Mon. Not. Roy. Astr. Soc.* **237**, 715.
Norton, A.J. and Watson, M.G.: 1989b, *Mon. Not. Roy. Astr. Soc.* **237**, 853.
Norton, A.J., Watson, M.G., King, A.R., McHardy, I.M. and Lehto, H.: 1990, in Proc. 11th North American Workshop on CVs and LMXRBs, ed(s)., C.W. Mauche, Cambridge Univ. Press, 209.
Osborne, J.: 1988, *Mem. S. A. It.* **59**, 117.
Patterson. J.: 1984, *Astrophys. J. Suppl. Ser.* **54**, 443.
Penning, W.R., Schmidt, G.D. and Liebert, J.: 1986, *Astrophys. J.* **301**, 881.
Plavec, M. and Kratochvil, P.: 1964, *Bull. Astron. Inst. Czech.* **15**, 165.
Rappaport, S.A., Verbunt, F. and Joss, P.C.: 1983, *Astrophys. J.* **275**, 713.
Ritter, H.: 1987, *Astron. Astrophys. Supp. Ser.* **70**, 335.
Rosen, S.R., Mason, K.O. and Córdova, F.A.: 1988, *Mon. Not. Roy. Astr. Soc.* **231**, 549.
Spruit, H.C. and Ritter, H.: 1983, *Astron. Astrophys.* **124**, 267.
Wang, Y.M.: 1987, *Astron. Astrophys.* **183**, 257.
Warner, B.: 1983, in *Cataclysmic Variables and Related Objects*, ed(s)., Livio, M. and Shaviv, G., Reidel, Dordrecht, 155.
Warner, B. and Wickramasinghe, D.T.: 1991, *Mon. Not. Roy. Astr. Soc.* **248**, 370.
Watson, M.G.: 1987, in *The Physics of Accretion onto Compact Objects*, ed(s)., Mason, K.O., Watson, M.G. and White, N.E., Springer-Verlag, Berlin, 97.
Watson, M.G., King, A.R. and Osborne, J.: 1985, *Mon. Not. Roy. Astr. Soc.* **212**, 917.
Whitehurst, R.: 1989, in *Theory of Accretion Discs*, ed(s)., F. Meyer, Duschl, W.J., Frank, J., and E. Meyer–Hofmeister, Kluwer, Dordrecht, 213.
Wickramasinghe, D.T., Wu, K. and Ferrario, L.: 1990, *Mon. Not. Roy. Astr. Soc.* **249**, 460.

SYMBIOTIC STARS

A. A. Boyarchuk

Allen, D.A.: 1979, in *Changing Trends in Variable Star Research*, ed(s)., Bateson *et al.*, U. Waikato Press, Hamilton, N.A, 125.
Allen, D.A.: 1983, *Mon. Not. Roy. Astr. Soc.* **204**, 113.
Baldwin, I.R., Frogel, I.A. and Persson, S.E.: 1973, *Astrophys. J.* **184**, 427.
Bath, T., Pringle, J.E.: 1982, *Mon. Not. Roy. Astro. Soc.* **201**, 345.
Belyakina, T.S.: 1979, *Izv. Crimean Astroph. Obs.* **59**, 133.
Belyakina, T.S.: 1983, *Izv. Crimean Astroph. Obs.* **68**, 108.
Belyakina, T.S., 1990 (private communication).
Blair, W.P. Stencel, R.E., Fiebelman, W.A., and Michalitsianos, A.G.: 1983, *Astrophys. J. Suppl.* **53**, 573.

Boyarchuk, A.A.: 1966, *Astron. Zh.* **43**, 976.
Boyarchuk, A.A.: 1967a, *Astron. Zh.* **44**, 122.
Boyarchuk, A.A.: 1967b, *Izv. Crimean Astroph. Obs.* **38**, 155.
Boyarchuk, A.A.: 1968, *Astrofiz* **4**, 291.
Boyarchuk, A.A.: 1969, *Non-Periodic Phenomena in Variable Stars*, Budapest, 395.
Boyarchuk, A.A.: 1970, in *Eruptive Stars*, ed(s)., Boyarchuk, A.A. and Gershberg, R.E., Nauka, Moscow (in Russian), 136.
Boyarchuk, A.A.: 1975, in *Variable Stars and Stellar Evolution*, ed(s)., Sherwood and Plant, , 377.
Boyarchuk, A.A.: 1984, *Sov. Sci. Rev. Ser. Astrophys. Phys.* **3**, 123.
Boyarchuk, A.A.: 1985, in Proc ESA Workshop: Recent Results on Cataclysmic Variables, ed(s)., ESA SP, 236, 97.
Boyarchuk, A.A.: 1986, *The Irish Astr. J.* **17**, 394.
Duschl, W.E.: 1988, in J. Mikolajewska et al., ed(s)., *The Symbiotic Phenomenon*, Kluwer Academic Publishers, Dordrecht, 137.
Feast, M.W., Robertson, B.S.C. and Catchpole, R.M.: 1977, *Mon. Not. Roy. Astr. Soc.* **179**, 499.
Feast M.W., Catchpole, R.M., Whitelock, P.A., Carter, B.S. and Roberts, G.: 1983, *Mon. Not. Roy. Astr. Soc.* **203**, 373.
Garcia, M.P. and Kenyon, S.J.: 1988, in J. Mikolajewska et al., ed(s)., *The Symbiotic Phenomenon*, Kluwer Acad Publ., Dordrecht, 27.
Glass, I.S. and Webster, B.L.: 1973, *Mon. Not. Roy. Astr. Soc.* **165**, 77.
Herman J., Burger, G.H. and Penninx, W.H.: 1986, *Astron. Astrophys.* **167**, 247.
Hummer, D.G.: 1963, *Mon. Not. Roy. Astr. Soc.* **165**, 461.
Iben, I. Jr.: 1982, *Astrophys. J.* **259**, 244.
Iijima, T.: 1981, in *Photometric and Spectroscopic Binary Systems*, ed(s)., B. Carling and Z. Kopal, Reidel Publ. Comp., Dordrecht, 517.
Kenyon, S.J., 1983, *Ph.D. Thesis*, Univ. Illinois.
Kenyon, S.J. and Gallagher, J.S.: 1983, *Astron. J.* **88**, 666.
Kenyon, S.J. and Webbink, R.F.: 1984, *Astrophys. J.* **279**, 252.
Kenyon, S.J, Fernandez-Castro, T. and Stencel, R.F.: 1986, *Astron. J.* **92**, 118.
Kenyon S.J.: 1986, *The Symbiotic Stars*, Cambridge Univ. Press, .
Kenyon S.J. and Fernandez-Castro T.: 1987, *Astron. J.* **93**, 938.
Kenyon S.J., Fernandez-Castro T. and Stencel, R.E.: 1988, *Astron. J.* **95**, 1817.
Kenyon, S.J.: 1988a, *Astron. J.* **96**, 337.
Kenyon, S.J.: 1988b, in *The Symbiotic Phenomenon*, ed(s)., J. Mikolajewska *et al.*, Kluwer Acad. Publ., Dordrecht, 27.
Knapp G.R. and Wilcots E.M.: 1987, in *The Later Stages of Stellar Evolution*, ed(s)., S. Kwok and R. Pottasch, Reidel, Dordrecht, 171.
Leibowitz, E.M. and Formiggini, L.: 1988, in J. Mikolajewska et al., ed(s)., *The Symbiotic Phenomenon*, Kluwer Acad. Publ., Dordrecht, 33.
Merrill, P.W.: 1929, *Astrophys. J.* **69**, 330.
Nussbaumer, H.: 1988, in *The Symbiotic Phenomenon*, ed(s)., J. Mikolajewska *et al*, Kluwer Acad. Publ., Dordrecht, 107.
Paczynski, B., Rudak, B.: 1980, *Astron. Astrophys.* **82**, 349.
Penston, M.V. and Allen, D.A.: 1985, *Mon. Not. Roy. Astr. Soc.* **212**, 939.
Purton, C.R., Feldman, P.A., and March, K.A.: 1973, *Nature Phys. Sci.* **245**, 5.
Sahade J., 1976, Mem. Soc. Roy. Sci. Liege, 6eme serie, tome IX, page 303.
Seaquist, E.R., Taylor, A.R. and Button, S.: 1984, *Astrophys. J.* **284**, 202.
Seaquist, E.R. and Taylor, A.R.: 1987, *Astrophys. J.* **312**, 813.
Seaquist, E.R.: 1988, in J. Mikolajewska et al., ed(s)., *The Symbiotic Phenomenon*, Kluwer Acad. Publ., Dordrecht, 69.
Slovak, M.H. and Code A.D.: 1988, in J. Mikolajewska et al., ed(s)., *The Symbiotic Phenomenon*, Kluwer Acad. Publ., Dordrecht, 27.
Swings, J.P. and Allen, D.A.: 1972, *Publ. Astr. Soc. Pacific* **84**, 523.
Szkody, P.: 1977, *Astrophys. J.* **217**, 140.
Taranova O.B. and Yudin, B.F.: 1981, *Soviet Astron.* **25**, 710.

Taranova, O.B. and Yudin, B.F.: 1982, *Soviet Astron.* **26**, 57.
Taylor A.R.: 1988, in J. Mikolajewska et al., ed(s)., *The Symbiotic Phenomenon*, Kluwer Acad. Publ., Dordrecht, 77.
Taylor A.R. and Seaquist, E.R.: 1984, *Astrophys. J.* **286**, 263.
Thronson H.A., Jr., Latter W.B., Black J.H., Bally, J., and Hacking, P.: 1987, in *The Later Stages of Stellar Evolution*, ed(s)., S. Kwok and S.R. Pottasch, Reidel, Dordrecht, 131.
Tutukov, A.V., Yungelson, L.R.: 1976, *Astrofiz.* **12**, 521.
Vogel M.: 1988, in J. Mikolajewska et al., ed(s)., *The Symbiotic Phenomenon*, Kluwer Acad. Publ., Dordrecht, 119.
Webster, B.L. and Allen, D.A.: 1975, *Mon. Not. Roy. Astr. Soc.* **171**, 171.
Whitelock, P.A.: 1987, *Publ. Astr. Soc. Pacific* **99**, 573.
Whitelock, P.A.: 1988, in J. Mikolajewska et al., ed(s)., *The Symbiotic Phenomenon*, Kluwer Acad. Publ., Dordrecht, 47.
Wright, A.E. and Allen D.A.: 1978, *Mon. Not. Roy. Astr. Soc.* **184**, 893.
Wright, A.E. and Barlow, M.J.: 1975, *Mon. Not. Roy. Astr. Soc.* **170**, 41.

THEORY AND OBSERVATIONS OF CLASSICAL NOVAE IN OUTBURST

S. Starrfield

Andrea, J, Drechsel, H., Snijders, M.A.J., and Cassatella, A.: 1990, *A. & A.* **244**, 111.
Arnett, W.D. and Truran, J.W.: 1969, *Ap. J.* **157**, 339.
Austin, S., Starrfield, S., Saizar, P., Shore, S. N., and Sonneborn, G.: 1990, in Evolution in Astrophysics: IUE Astronomy in the Era of New Space Missions, ed(s)., *E. Rolfe*, ESA SP-310; Noordwijk, 367.
Bath, G.T.: 1978, *M.N.R.A.S.* **182**, 35.
Bath, G.T. and Shaviv, G.: 1976, *M.N.R.A.S.* **183**, 515.
Benjamin, R.A., and Dinerstein, H.L.: 1990, *Astr. J.* **100**, 1588.
Bode, M.F.: 1986, *RS Oph (1985) and the Recurrent Nova Phenomenon*, VNU Science Press, Utrecht
Bode, M.F., and Evans, A.N.: 1989, *The Classical Nova*, Wiley, New York
Bode, M.F., and Kahn, F.D.: 1985, *M.N.R.A.S.* **217**, 205.
Boyarchuk, A.A., Galkina, T.S., Krasnobabtsev, V.I., Rachkovskaya, T.M., and Shakhovskaya, N.I.: 1977, *Sov. A. J.* **21**, 257.
Bruch, A.: 1986, *RS Oph(1985) and the Recurrent Nova Phenomenon*, VNU Science Press, Utrecht, 13.
Cassatella, A., and Viotti, R.: 1991, *The Physics of Classical Nova*, Springer-Verlag, Berlin
Castor, J, Abbott, D., and Klein, R.: 1975, *Ap. J.* **195**, 157.
Cohen, J.G.: 1985, *Ap. J.* **292**, 90.
Cohen, J.G., and Rosenthal, A.J.: 1983, *Ap. J.* **268**, 689.
Cordova, F.A., and Howarth, I.D.: 1987, in *Exploring the Universe with the IUE Satellite*, ed(s)., Y. Kondo, Kluwer, Dordrecht, 395.
Drechsel, H., Kondo, Y., and Rahe, J.: 1987, *Cataclysmic Variables. Recent Multifrequency Observations and Theoretical Developments*, Kluwer, Dordrecht
Ennis, D., Becklin, E.E., Beckwith, S., Elias, J., Gately, I., Matthews, K., Neugebauer, G., and Willner, S.P.: 1977, *Ap. J.* **214**, 33.
Ferland, G.J., Lambert, D.L., and Woodman, J.H.: 1986a, *Ap. J. Supp.* **60**, 375.
Ferland, G.J., Lambert, D.L., and Woodman, J.H.: 1986b, *Ap. J. Supp.* **62**, 939.
Ferland, G.J. and Shields, G.A.: 1978, *Ap. J.* **226**, 172.
Ford, H.C.: 1978, *Ap. J.* **219**, 595.
Friedjung, M.: 1977, *Novae and Related Stars*, Reidel, Dordrecht
Gallagher, J.S., and Ney, E.P.: 1976, *Ap. J. (Lett.)* **204**, L35.
Gallagher, J.S. and Starrfield, S.G.: 1976, *M.N.R.A.S.* **176**, 53.
Gallagher, J.S., and Starrfield, S.: 1978, *Ann. Rev. Astr. Ap.* **16**, 171.
Gehrz, R.D.: 1988, *Ann. Rev. Astr. Ap.* **26**, 377.

Gehrz, R.D.: 1991, in *The Physics of Classical Novae*, ed(s)., A. Cassatella and R. Viotti, Springer-Verlag, Berlin, 138.
Gehrz, R.D., Grasdalen, G.L., Greenhouse, M., Hackwell, J.A., Hayward,T., and Bentley, A.F.: 1986, *Ap. J. (Lett.)* **308**, L63.
Gehrz, R.D., Hackwell, J.A., Grasdalen, G.L., Ney, E.P., Neugebauer, G., and Sellgren, K.: 1980b, *Ap. J.* **239**, 570.
Gehrz, R.D., Grasdalen, G.L., and Hackwell, J.A.: 1985, *Ap. J. (Lett.)* **298**, L163.
Gehrz, R.D., Grasdalen, G.L., and Hackwell, J.A., and Ney, E.P.: 1980a, *Ap. J.* **237**, 855.
Gehrz, R.D., Hackwell, J.A., and Jones, T.W.: 1974, *Ap. J.* **191**, 675.
Gehrz, R.D., Truran, J.W., and Williams, R.E.: 1992, *Protostars and Planets III*, , in press.
Greenhouse, M.A., Grasdalen, G.L, Hayward, T.L., Gehrz, R.D., and Jones, T.J.: 1988, *Astron. J.* **78**, 172
Greenhouse, M.A., Grasdalen, G.L., Woodward, C.E., Benson, J., Gehrz, R.D., Rosenthal, E. and Skrutskie, M.F.: 1990, *Ap. J.* **352**, 307.
Hauschildt, P.H., Shaviv, G., and Wehrse, R.: 1989, *A. & A.* **210**, 262.
Hillebrandt, W. and Thielemann, F.-K.: 1982, *Ap. J.* **255**, 617.
Hjellming, R.M.: 1991, in *The Physics of Classical Novae*, ed(s)., A. Cassatella, Springer-Verlag, Berlin, 169.
King, A.R.: 1989, in Classical Novae, ed(s)., *N. Evans and M. Bode*, Wiley, New York, 17.
Kraft, R.P.: 1964, *Ap. J.* **139**, 457.
Krautter, J., and Williams, R.E.: 1989, *Ap. J.* **341**, 968.
Kutter, G.S. and Sparks, W.M.: 1987, *Ap. J.* **321**, 386.
Lance, C.M., McCall, M.M., and Uomoto A.K.: 1988, *Ap. J. Supp.* **66**, 151.
Liller, W. 1990, IAU Circular No. 4964.
Livio, M., Shankar, A., Burkert, A., and Truran, J.W.: 1990, *Ap. J.* **356**, 250.
MacDonald, J.: 1980, *M.N.R.A.S.* **191**, 933.
MacDonald, J.: 1983, *Ap. J.* **267**, 732.
MacDonald, J., Fujimoto, M.Y., and Truran, J.W.: 1985, *Ap. J.* **294**, 263.
Martin, P.G.: 1988, in *The Classical Nova*, ed(s)., M. Bode and N. Evans, Wiley, New York, 113.
Mason, K.O., Cordova, F.A., Bode, M.F., and Barr, P.: 1986, in *RS Oph and the Recurrent Nova Phenomenon*, ed(s)., M.F. Bode, VNU Science Press, Utrecht, 167.
McLaughlin, D.B.: 1960, in *Stellar Atmospheres: Stars and Stellar Systems VI*, ed(s)., J.S. Greenstein, University of Chicago Press, Chicago, 585.
Ney, E. and Hatfield, B.F.: 1978, *Ap. J. (Lett.)* **219**, L111.
Nofar, I., Shaviv, G., and Starrfield, S.: 1991, *Ap. J.* **369**, 440.
Ögelman, H., Beuermann, K., and Krautter, J.: 1984, *Ap. J. Lett.* **287**, L31.
Ögelman, H., Krautter, J., and Beuermann, K.: 1987, *A. & A.* **177**, 110.
O'Brien, Kahn, F.D., and Bode, M.F.: 1986, *RS Oph and the Recurrent Nova Phenomenon*, M.F. Bode, VNU Press, Utrecht, 177.
Paczynski, B.: 1971, *Acta Astron.* **21**, 417.
Payne-Gaposchkin, C.: 1957, *The Galactic Novae*, Dover, New York
Prialnik, D., Livio, M., Shaviv, G., and Kovetz, A.: 1982, *Ap. J.* **257**, 312.
Pringle, J.E.: 1981, *Ann. Rev. Astr. Ap.* **19**, 137.
Robinson, E.L.: 1975, *Astron. J.* **80**, 515.
Saizar, P. et al.: 1991, *Ap. J.* **367**, 310.
Schaefer, B.: 1990, *Ap. J. (Lett.)* **355**, L39.
Shara, M.M.: 1989, *Pub. Ast. Soc. Pac.* **101**, 5.
Shaviv, G., and Starrfield, S.: 1987a, *Ap. J. (Lett.)* **321**, L51.
Shaviv, G., and Starrfield, S.: 1987b, in *The Second Conference on Faint Blue Stars*, ed(s)., A. Philip, D.S. Hayes, and J. Liebert, L.D. Press, Schenectady, 427.
Shaviv, G., and Starrfield, S.: 1988, *Ap. J.* **335**, 383.
Shore, S.N, Sonneborn, G., and Starrfield, S.: 1990, in *Evolution in Astrophysics: IUE Astronomy in the Era of New Space Missions*, ed(s)., E. Rolfe, ESA SP-310; Noordwijk, 447.
Shore, S.N., Sonneborn, G., Starrfield, S., Hamuy, M., Williams, R.E., Cassatella, A., and Drechsel, H.: 1991, *Ap. J.* **370**, 193.

Sievers, J., 1970, Inter. Bull. Var. Stars No. 448.
Snijders, M.A.J.: 1986, in *RS Oph(1985) and the Recurrent Nova Phenomenon*, ed(s)., M. Bode, VNU press, Utrecht, 51.
Snijders, M.A.J., Batt, T.J., Seaton, M.J., Blades, J.C., and Morton, D.C.: 1984, *M.N.R.A.S.* **211**, 7.
Snijders, M.A.J., Batt, T.J., Roche, P.F., Seaton, M.J., Morton, D.C., Spoelstra, T.A.T., and Blades, J.C.: 1987, *M.N.R.A.S.* **228**, 329.
Sonneborn, G., Shore, S.N., and Starrfield, S.: 1990, in *Evolution in Astrophysics: IUE Astronomy in the Era of New Space Missions*, ed(s)., E. Rolfe, ESA SP-310; Noordwijk, 439.
Sparks, W.M. and Kutter, G.S.: 1987, *Ap. J.* **321**, 394.
Sparks, W.M., Starrfield, S., and Truran, J.W.: 1986, in *RS Oph(1985) and the Recurrent Nova Phenomenon*, ed(s)., M.F. Bode, VNU Press, Utrecht, 39.
Starrfield, S.: 1979, in *White Dwarfs and Variable Degenerate Stars*, ed(s)., H.M. Van Horn and V. Weidemann, University of Rochester, Rochester, 274.
Starrfield, S.: 1986, in *Radiation Hydrodynamics*, ed(s)., D. Mihalas and K.-H. Winkler, Reidel, Dordrecht, 225.
Starrfield, S.: 1987, in *New Insights in Astrophysics: Eight Years of Ultraviolet Astronomy with IUE*, ed(s)., E. Rolfe, ESA SP-263, Noordwijk, 239.
Starrfield, S.: 1988, in *Multiwavelength Astrophysics*, ed(s)., F.A. Cordova, Cambridge University Press, Cambridge, 159.
Starrfield, S.: 1989, in *Classical Novae*, ed(s)., N. Evans and M. Bode, Wiley, New York, 123.
Starrfield, S.: 1990, in *Evolution in Astrophysics: IUE Astronomy in the Era of New Space Missions*, ed(s)., E. Rolfe, ESA SP-310, Noordwijk, 101.
Starrfield, S.: 1991, in *The Physics of Classical Novae*, ed(s)., A. Cassatella, Springer-Verlag, Berlin, 127.
Starrfield, S. and Snijders, M.A.J.: 1987, in *Scientific Accomplishments of the IUE Satellite*, ed(s)., Y. Kondo et al., Reidel, Dordrecht, 377.
Starrfield, S., et al.: 1988b, in *A Decade of UV Astronomy with the IUE Satellite*, ed(s)., E. Rolfe, Estec Publ., Noordwijk, 163.
Starrfield, S., Sparks, W.M., and Shaviv, G.: 1988, *Ap. J.* **325**, L35.
Starrfield, S., Sparks, W.M., and Truran, J.W.: 1985, *Ap. J.* **291**, 136.
Starrfield, S., Sparks, W.M., and Truran, J.W.: 1986, *Ap. J. Lett.* **303**, L5.
Starrfield, S., Truran, J.W., and Sparks, W.M.: 1978, *Ap. J.* **226**, 186.
Starrfield, S., Truran, J.W., Sparks, W.M., and Krautter, J.: 1991a, in *EUV Astronomy*, ed(s)., R.A. Malina, 168,
Starrfield, S., Truran, J.W., Sparks, W.M., Krautter, J., and MacDonald, J.: 1991b, in *Physics of Classical Novae*, ed(s)., A. Cassatella and R. Viotti, Springer-Verlag, Heidelberg, 306.
Starrfield, S., and Wagner, R.M.: 1990, *Nature* **343**, 215.
Stickland, D., Penn, C.J., Seaton, M.J., Snijders, M.A.J., Storey, P.J., Kitchin, C.R.: 1981, *M.N.R.A.S.* **197**, 271.
Stockman, H.S., Schmidt, G., and Lamb, D.Q.: 1988, *Ap. J.* **332**, 282.
Stryker, L.L. et al.: 1988, in *A Decade of UV Astronomy with the IUE Satellite*, ed(s)., E. Rolfe, Estec Publications, Noordwijk, 149.
Truran, J.W.: 1982, in *Essays in Nuclear Physics*, ed(s)., C.A. Barnes, D.D. Clayton and D.N. Schramm, Cambridge University Press, Cambridge, 467.
Truran, J.W.: 1990, in *The Physics of Classical Novae*, ed(s)., A. Cassatella, Springer-Verlag, Berlin, 373.
Truran, J.W. and Livio, M.: 1986, *Ap. J.* **308**, 721.
Vogt, N.: 1990, *Ap. J.* **356**, 609.
Weiss, A. and Truran, J.W.: 1990, *A. & A.* **238**, 178.
Wehrse, R., Hauschildt, P.H., Shaviv, G., and Starrfield, S.: 1990a, in *Physics of Classical Novae*, ed(s)., A. Cassatella and R. Viotti, Springer-Verlag, Heidelberg, 264.
Wehrse, R., Hauschildt, P.H., Shaviv, G., and Starrfield, S.: 1990b, in *Evolution in Astrophysics: IUE Astronomy in the Era of New Space Missions*, ed(s)., E. Rolfe, ESA SP-310; Noordwijk, 461.
Wiescher, M., Gorres, J., Thielemann, F.-K., and Ritter, H.: 1986, *A. & A.* **160**, 56.
Williams, R.E.: 1990, in *The Physics of Classical Novae*, ed(s)., A. Cassatella, Springer-Verlag,

Berlin, 215.
Williams, R.E., Sparks, W.M., Gallagher, J.S., Ney, E.P., Starrfield, S.G., and Truran, J.W.: 1981, *Ap. J.* **251**, 221.
Williams, R.E., Ney, E.P., Sparks, W.M., Starrfield, S., Truran, J.W.: 1985, *M.N.R.A.S.* **212**, 753.
Wu, C.-C. and Kester, D.: 1977, *A. & A.* **58**, 331.
Young, P. and Schneider, D.P.: 1980, *Ap. J.* **238**, 955.

INTERACTING BINARIES AND TYPE I SUPERNOVAE

K. Nomoto

Axelrod, T.: 1980, in J.C. Wheeler, ed(s)., *Type I Supernovae*, Univ. of Texas:Austin, 80.
Benz, W., Bowers, R.L., Cameron, A.G.W., and Press, W.: 1990, *Ap. J.* **348**, 647.
Branch, D., Doggett, J.B., Nomoto, K., and Thielemann, F.-K.: 1985, *Ap. J.* **294**, 619.
Branch, D., Nomoto, K. and Filippenko, A.V.: 1991, *Comments on Astrophysics* **15**, 221.
Burrows, A., and Woosley, S.E.: 1986, *Ap. J.* **308**, 680.
Canal, R., Isern, J., and Labay, J.: 1980, *Ap. J. Letters* **241**, L33.
Canal, R., Garcia, D., Isern, J., and Labay, J.: 1990, *Ap. J. Letters* **356**, L51.
Ensman, L., and Woosley, S.E.: 1988, *Ap. J.* **333**, 754.
Filippenko, A.V.: 1988, *A. J.* **96**, 1941.
Filippenko, A.V.: 1992, *Ap. J. Letters* **384**, L37.
Filippenko, A.V. et al.: 1992, *Ap. J. Letters* **384**, L15.
Filippenko, A.V., Porter, A.C., and Sargent, W.L.W.: 1990, *A.J.* **100**, 1575.
Graham, J.R.: 1987, *Ap. J.* **315**, 588.
Habets, G.M.H.J.: 1986, *Astr. Ap.* **167**, 61.
Hachisu, I., Matsuda, T., Nomoto, K., and Shigeyama, T.: 1990, *Ap. J. Letters* **358**, L57.
Hachisu, I., Matsuda, T., Nomoto, K., and Shigeyama, T.: 1991, *Ap. J. Letters* **368**, L27.
Harkness, R.P.: 1991, in I.J. Danziger and K. Kjäer, ed(s)., *SN 1987A and Other Supernovae*, ESO, Garching, 447.
Harkness, R.P., and Wheeler, J.C.: 1990, in A. Petschek, ed(s)., *Supernovae*, Springer-Verlag: Heidelberg, 1.
Hashimoto, M., Nomoto, K., Arai, K., and Kaminisi, K.: 1986, *Ap. J.* **307**, 687.
Hashimoto, M., Nomoto, K., and Shigeyama, T.: 1989, *Astr. Ap.* **210**, L5.
Hoyle, F., and Fowler, W.A.: 1960, *Ap. J.* **132**, 565.
Iben, I. Jr., and Tutukov, A.: 1984, *Ap. J. Suppl.* **55**, 335.
Isern, J., Labay, J., Hernanz, M., and Canal, R.: 1983, *Ap. J.* **273**, 320.
Jeffery, D., Branch, D., Filippenko, A.V., and Nomoto, K.: 1991, *Ap. J. Letters* **377**, L89.
Kato, M., and Hachisu, I.: 1989, *Ap. J.* **346**, 424.
Kohkhlov, A.M.: 1991a, *Astr. Ap.* **245**, 114.
Kumagai, S., Shigeyama, T., Nomoto, K., Itoh, M., Nishimura, J. and Tsuruta, S.: 1989, *Ap. J.* **345**, 412.
Langer, N.: 1989, *Astr. Ap.* **220**, 135.
Leibundgut, B., Kirshner R.P., Filippenko, A.V., Shields, J.C., Foltz, C.B., Phillips, M.M., and Sonneborn, G.: 1991, *Ap. J. Letters* **371**, L23.
Meurs, E.J.A., and van den Heuvel, E.P.J.: 1989, *Astr. Ap.* **226**, 88.
Miyaji, S., and Nomoto, K.: 1987, *Ap. J.* **318**, 307.
Miyaji, S., Nomoto, K., Yokoi, K., and Sugimoto, D.: 1980, *Pub. Astr. Soc. Japan* **32**, 303.
Mochkovitch, R., and Livio, M.: 1990, *Astr. Ap.* **236**, 378.
Muller, R.A., Newberg, H.J.M., Pennypacker, C.R., Perlmutter, S., Sasseen, T.P., and Smith, C.K.: 1992, *Ap. J. Letters* **384**, L9.
Nariai, K., and Nomoto, K.: 1979, in H.M. Van Horn and V. Weidemann, ed(s)., *White Dwarfs and Variable Degenerate Stars, IAU Colloquium 53*, Univ. of Rochester, Rochester, 525.
Nariai, K., Nomoto, K., and Sugimoto, D.: 1980, *Pub. Astr. Soc. Japan* **32**, 473.
Nomoto, H.: 1989, *Exploring SN 1987A (in Japanese)*, Kodansha, Tokyo, .

Nomoto, K.: 1982a, *Ap. J.* **253**, 798.
Nomoto, K.: 1982b, *Ap. J.* **257**, 780.
Nomoto, K.: 1984, *Ap. J.* **277**, 791.
Nomoto, K.: 1986a, *Ann. NY Acad. Sci.* **470**, 294.
Nomoto, K.: 1986b, *Prog. Part. Nucl. Phys.* **17**, 249.
Nomoto, K.: 1987, *Ap. J.* **322**, 206.
Nomoto, K., Filippenko, A.V., and Shigeyama, T.: 1990, *Astr. Ap.* **240**, L1.
Nomoto, K., and Hashimoto, M.: 1988, *Physics Reports* **163**, 13.
Nomoto, K., and Kondo, Y.: 1991, *Ap. J. Letters* **367**, L19.
Nomoto, K., Miyaji, S., Sugimoto, D., and Yokoi, K.: 1979a, in H.M. Van Horn and V. Weidemann, ed(s)., *White Dwarfs and Variable Degenerate Stars, IAU Colloquium 53*, Univ. of Rochester, Rochester, 6.
Nomoto, K., Nariai, K., and Sugimoto, D.: 1979b, *Pub. Astr. Soc. Japan* **31**, 287.
Nomoto, K., and Sugimoto, D.: 1977, *Pub. Astr. Soc. Japan* **29**, 765.
Nomoto, K., Sugimoto, D., and Neo, S.: 1976, *Ap. Space Sci.* **39**, L37.
Nomoto, K., Thielemann, F.-K., and Yokoi, K.: 1984, *Ap. J.* **286**, 644.
Nomoto, K., Yamaoka, H., Shigeyama, T., Kumagai, S., and Tsujimoto, T.: 1991, in J. Ventura and D. Pines, ed(s)., *NATO ASI, Neutron Stars: Theory and Observation*, Kluwer, Dordrecht, 143.
Nomoto, K., Yamaoka, H., Shigeyama, T., Kumagai, S., and Tsujimoto, T.: 1992, in J. Audouze *et al.*, ed(s)., *Supernovae* (Les Houches, session LIV), Elsevier Sci. Publ., in press.
Panagia, N.: 1987, in F. Pacini, ed(s)., *High Energy Phenomena Around Collapsed Stars*, D. Reidel, Dordrecht, 33.
Ruiz-Lapuente, P., Cappellaro, E., Turatto, M., Gouiffes, C., Danziger, I.J., Della Valle, M., and Lucy, L.B.: 1992, *Ap. J. Letters*, in press.
Saio, H. and Nomoto, K.: 1985, *Astr. Ap.* **150**, L21.
Shigeyama, T., and Nomoto, K.: 1990, *Ap. J.* **360**, 242.
Shigeyama, T., Nomoto, K., and Hashimoto, M.: 1988, *Astr. Ap.* **196**, 141.
Shigeyama, T., Nomoto, K., Tsujimoto, T., and Hashimoto, M.: 1990, *Ap. J. Letters* **361**, L23.
Shigeyama, T., Nomoto, K., Yamaoka, H., and Thielemann, F.-K.: 1992, *Ap. J. Letters* **386**, L13.
Swartz, D.A., Wheeler, J.C., and Harkness, R.P.: 1991, *Ap. J.* **374**, 266.
Taylor, J.: 1992, in E.P.J. van den Heuvel, ed(s)., *NATO ARW, X-Ray Binaries and the Formation of Binary and Millisecond Radio Pulsars*, Kluwer, Dordrecht, 87.
Thielemann, F.-K., Hashimoto, M., and Nomoto, K.: 1990, *Ap. J.* **349**, 222.
Thielemann, F.-K., Nomoto, K., and Yokoi, K.: 1986, *Astr. Ap.* **158**, 17.
Tsvetkov, D.Yu.: 1985, *Sov. Astr.* **29**, 211.
Uomoto, A.: 1986, *Ap. J. Letters* **310**, L35.
Van den Heuvel, E.P.J.: 1991, in J. Ventura and D. Pines, ed(s)., *NATO ASI, Neutron Stars: Theory and Observation*, Kluwer, Dordrecht, 171.
Webbink, R.: 1984, *Ap. J.* **277**, 355.
Wheeler, J.C., and Harkness, R.: 1990, *Rep. Prog. Phys.* **53**, 1467.
Wijers, R.A.M.J.: 1991, PhD thesis, Univ. of Amsterdam.
Williams, F.A.: 1985, *Combustion Theory* (Menlo Park: Benjamin/Cummings).
Wolszczan, A.: 1991, *Nature* **350**, 688.
Woosley, S.E., Taam, R.E., and Weaver, T.A.: 1986, *Astr. Ap.* **301**, 601.
Woosley, S.E., and Weaver, T.A.: 1986a, *Lecture Notes in Physics* **255**, 91.
Woosley, S.E., and Weaver, T.A.: 1986b, *Ann. Rev. Astr. Ap.* **24**, 205.
Woosley, S.E., and Weaver, T.A.: 1992, in J. Audouze et al, ed(s)., *Supernovae* (Les Houches, Session LIV), Elsevier Sci. Publ., in press.
Yamaoka, H., and Nomoto, K.: 1991, in I.J. Danziger and K. Kjäer, ed(s)., *SN 1987A and Other Supernovae*, ESO, Garching, 193.
Yamaoka, H., Nomoto, K., Shigeyama, T., and Thielemann, F.-K.: 1992, *Ap. J. Letters* **393**, L55.
Yamaoka, H., Shigeyama, T., and Nomoto, K. 1992, *Astr. Ap.*, in press.

BINARY SYSTEMS WITH COMPACT COMPONENTS
V. Trimble

Achterberg, A. 1989, *Nature* **342**, 51.
Alexander, S.G. 1989, *Astroph. J.* **342**, 928.
Alexander, S.G. and Meszaros, P. 1989, *Astroph. J.* **344**, L1.
Alpar, M.A. and Shaham, J. 1985, *Nature* **326**, 239.
Alpar, M.A. et al. 1982, *Nature* **300**, 728.
Amnuel, P.R. et al. 1988, *Sov. Astron.* **32**, 520.
Anderson, S.G. et al. 1989, *Bull. Amer. Astron. Soc.* **21**, 1204.
Angelini, L., Stella, L. and Parmar, A.N. 1989, *Astroph. J.* **346**, 906.
Antokhina, E.A. and Cherepashchuk, A.M. 1987, *Sov. Astron.* **31**, 295.
Aurière, M. et al. 1989, *Astron. Astrophys.* **214**, 113.
Backer, D.C. and Hellings, R.W. 1986, *Ann. Rev. Astron. Astrophys.* **24**, 537.
Backer, D.C. et al. 1982, *Nature* **300**, 615.
Bailes, M. 1989, *Astroph. J.* **342**, 919.
Bailyn, C.D. 1988, *Nature* **332**, 330.
Bailyn, C.D. and Grindlay, J.E. 1990, *Astroph. J.* **353**, 159.
Bailyn, C.D. et al. 1989, *Astroph. J.* **344**, 787.
Balucinska, M. and Hasinger, G. 1989, in ESLAB 1989.
Band, D.L. 1989, *Astroph. J.* **336**, 937.
Basani, L. et al. 1989, *Astroph. J.* **343**, 313.
Beer, P. 1981, *Vistas in Astronomy* **25**, Nos. 1 & 2.
Berezhinski, V.S. 1988, *Nature* **334**, 506.
Bignami, G. 1987, in *High Energy Phenomena around Collapsed Stars*, ed. F. Pacini (Reidel), p. 297.
Bignami, G. et al. 1988, *Astron. Astrophys.* **212**, L1.
Bisnovatyi-Kogan, G.S. et al. 1989, *Astron. Astrophys.* **221**, L7.
Blaes, O.M. and Hawley, J.F. 1988, *Astroph. J.* **326**, 277.
Blandford, R.D., Applegate, J.H. and Hernquist, L. 1983, Mon. Not. Royal Astron. Soc. 204, 1025.
Bodenheimer, P. and Taam, R.E. 1986, in High Energy Phenomena around Collapsed Stars, ed. F. Pacini (Reidel), p. 13.
Boer, M. 1988, *Astron. Astrophys.* **202**, 117.
Bradt, H.V. and McClintock, J.E. 1983, *Ann. Rev. Astron. Astrophys.* **21**, 13.
Brainerd, J.J. 1989, *Astroph. J.* **338**, 281.
Brinkman, W. et al. 1989, *Astron. Astrophys.* **218**, L13.
Buchkarev, N.G. et al. 1988, *Sov. Astron.* **32**, 405.
Callanan, P.J. 1989, *Mon. Not. Royal Astron. Soc.* **238**, 25p.
Callanan, P.J. et al. 1989, *MNRAS* **241**, 27p.
Callanan, P.J. et al. 1989a, *MNRAS* **240**, 31p.
Caraveo, P. et al. 1989, *Astroph. J.* **338**, 338.
Carlini, A. and Treves, A. 1989, *Astron. Astrophys.* **215**, 283.
Carramiñana, A. et al. 1989, *Astroph. J.* **346**, 967.
Cassiday, G.L. et al. 1989a, *Phys. Rev. Lett.* **62**, 383.
Cassiday, G.L. et al. 1989b, *Phys. Rev. Lett.* **63**, 2329.
Chanmugam, G. and Brecher, K. 1989, *Nature* **327**, 696.
Chardin, G. and Gerbier, G. 1989, *Astron. Astrophys.* **210**, 52.
Cheng, A.F. 1989, *Astroph. J.* **339**, 29.
Cheng, K.S. and Ruderman, M. 1989, *Astroph. J.* **337**, L77.
Chevalier, C. 1989, *Astron. Astrophys.* **210**, 114.
Chlebowski, T. et al. 1989, *Astroph. J.* **341**, 427.
Ciampa, D., Clay, R.W. and Edwards, P.G. 1989, *Astroph. J.* **346**, 151.
Clark, G.W. 1975, *Astroph. J.* **199**, L143.
Corbet, R.H.D. 1987, p. 63 of Truemper *et al.* 1987.
Cordova, F.A. and Mason, K.O. 1984, in Lewin and Van den Heuvel (1983), p. 147.
Cote, J. and Pylyset, E.H.P. 1989, *Astron. Astrophys.* **218**, 131.

Cowley, A.P. et al. 1988, *Astroph. J.* **333**, 906.
Cudworth, K.M. 1988, *Astron. J.* **96**, 105.
Czerny, M. and King, A.P. 1988, *Mon. Not. Royal Astron. Soc.* **235**, 35p.
Damen, E. 1989, *MNRAS* **237**, 523.
Davies, S.R. et al. 1989, *MNRAS* **237**, 973.
Dennerl, K. 1989, in ESLAB 1989.
Dermer, C.D. 1989, *Astroph. J.* **347**, L13.
Dingus, B.L. et al. 1988a, *Phys. Rev. Lett.* **61**, 1906.
Dingus, B.L. et al. 1988b, *Phys. Rev. Lett.* **60**, 1785.
Djorgovski, S. and Evans, C.R. 1988, *Astroph. J.* **335**, L61.
Dolan, J.F. and Tapia, S. 1989, *Astroph. J.* **344**, 830.
Dotani, T. et al. 1989, *Publ. Astron. Soc. Japan* **41**, 427 & 577.
Ebisawa, K. et al. 1989, *PAS Japan* **41**, 519.
Eichler, D. and Ko, K. 1988, *Astroph. J.* **328**, 179.
Eichler, D. and Levinson, A. 1988, *Astroph. J.* **335**, L67.
Eichler, D. et al. 1989, *Nature* **340**, 126.
Emmering, R.T. and Chevalier, R.A. 1989, *Astroph. J.* **345**, 931.
Epstein, R.I. and Hartman, D. 1989, *Astroph. J.* **346**, 960.
Epstein, R.I. and Hurley, K. 1989, *Astrophys. Lett. Comm.* **27**, 220.
ESLAB 1989, 23rd Eslab Symposium. ESA Special Publication (in press for 1990).
Fabian, A.C. 1985, in *Interacting Binary Systems*, eds. J.E. Pringle and R.A. Wade (Cambridge U. Press), p. 71.
Fabian, A.C., Pringle, J.E. and Rees, M.J. 1975, *Mon. Not. Royal Astron. Soc.* **172**, 15p.
Fabian, A.C. et al. 1989, *MNRAS* **238**, 729.
Fabian, A.C., Pringle, J.E. and Whelan, J.A. 1974, *Nature* **247**, 351.
Fabian, G.C. and Gregory P.C. 1981, *Nature* **293**, 202.
Fedorova, A.V. and Ergma, E.V. 1989, *Astrophys. Space Sci.* **151**, 121.
Filippenko, A.V. 1988, *Astron. J.* **96**, 242.
Fortner, B. et al. 1989, *Nature* **342**, 775.
Frail, D.A. et al. 1987, *Astron. J.* **93**, 1506.
Fruchter, A.S., Taylor, J.H. and Stinebring, D.E. 1988, *Nature* **333**, 237.
Fruchter, A.S. et al. 1988, *Nature* **334**, 684.
Fujimoto, M. et al. 1988, *Astron. Astrophys.* **199**, L9.
Garcia, M.R. et al. 1989, *Astroph. J.* **341**, L75.
Geldzahler, B.J. et al. 1989, *Astroph. J.* **342**, 1123.
Ghosh, P. and Lamb, F.K. 1978, *Astroph. J.* **223**, L83.
Giovanelli, F. and Mannochi, G., eds. 1987, *High Energy and Ultra High Energy Behavior of Accreting X-Ray Sources*. Ital. Phys. Soc. Conf. Proc. Vol. 8.
Goodman, J. and Hut, P., eds. 1985, IAU Symp. 113, *Dynamics of Star Clusters*, Reidel.
Gottwald, M. et al. 1989, *Astroph. J.* **339**, 1044.
Gottwald, M. et al. 1987, *Astroph. J.* **323**, 525.
Gregory, P.C. and Fahlman, G. 1983, in *Supernova Remnants and Their X-Ray Emission*, IAU Symp. 101, eds. J. Danziger and P. Gorenstein (Reidel), p. 429.
Gregory, P.C. et al. 1989, *Astroph. J.* **339**, 1059.
Greiner, J. et al. 1987, *Astron. Nach.* **308**, 121.
Grindlay, J.E. 1986, in Truemper *et al.* 1986, p. 25.
Grindlay, J.E. 1988, in *Globular Cluster Systems in Galaxies*, IAU Symp. 126, eds. J.E. Grindlay and A.G.D. Philip, (Kluwer), p. 347.
Grindlay, J.E. 1989, in ESLAB 1989.
Grindlay, J.E. et al. 1983, in *Supernova Remnants and Their X-Ray Emission*, IAU Symp. 101, eds. J. Danziger and P. Gorenstein (Reidel), p. 459.
Gruber, D.E. 1988, *Astroph. J.* **328**, 265.
Haberl, F. 1987, *Astroph. J.* **314**, 266.
Habets, G.M.H.J. 1987, *Astron. Astrophys.* **184**, 109.
Haines, T. et al. 1990, in 14th Texas Symp. Relativistic Astrophys. (NY Acad. Sci., in press).

Halperin, J.P. and Tytler, D. 1988, *Astroph. J.* **330**, 201.
Hammeury, J.M. and Lasota, J.P. 1989, *Astron. Astrophys.* **210**, L15.
Harding, A.K. and Preece, R.D. 1989, *Astroph. J.* **338**, L21.
Hartmann, D. and Blumenthal, G.R. 1989, *Astroph. J.* **342**, 521.
Hanami, H. 1988, *Mon. Not. Royal Astron. Soc.* **233**, 423.
Hasinger, G. and Klis, M. van der 1989, *Astron. Astrophys.* **225**, 79.
Hasinger, G. et al. 1989, *Astroph. J.* **337**, 843.
Hasinger, G. et al. 1990, *Astron. Astrophys.* **235**, 131.
Helfand, D.S. and Huang, J.-H., eds. 1987, *Origin and Evolution of Neutron Stars*, IAU Symp. 125 (Reidel).
Hellier, C. and Mason, K.O. 1989, *Mon. Not. Royal Astron. Soc.* **231**, 715.
Henrichs, H.F. 1984, in Lewin and Van den heuvel (1984), p. 393.
Heuvel, E.P.J. van den 1981, in *Fundamental Problems in Theory of Stellar Evolution*, IAU Symp. 93, eds. D. Sugimoto et al. (Reidel), p. 155.
Heuvel, E.P.J. van den 1986, in J. Truemper et al. (1986), p. 107.
Heuvel, E.P.J. van den 1988a, in *The Physics of Compact Objects*, eds. N.E. White and C.G. Filipov (Pergamon), p. 355.
Heuvel, E.P.J. van den 1988, *Adv. Space Res.* **8**, 355.
Heuvel, E.P.J. van den and Habets, G.M.H.J. 1985, in *Supernovae, Their Progenitors, and Their Remnants*, eds. G. Srinivasan and V. Radhakrishnan (Bangalore Indian Academy of Sciences), p. 129.
Heuvel, E.P.J. van den and Paradijs, J. van 1988, *Nature* **334**, 227.
Heuvel, E.P.J. van den and Taam, R.E. 1984, *Nature* **309**, 234.
Hirano, T. et al. 1987, *Publ. Astron. Soc. Japan* **39**, 19.
Hjellming, R.M. and Johnston, K.J. 1988, *Astroph. J.* **328**, 600.
Hjellming, R.M. et al. 1990, *Astron. Astrophys.* **235**, 147.
Ho, C. and Epstein, R.I. 1990, *Astroph. J.* **343**, 277.
Hoshi, R. and Inoue, H. 1988, *Publ. Astron. Soc. Japan* **40**, 421.
Huang, M. and Wheeler, J.C. 1989, *Astroph. J.* **343**, 229.
Hulse, R.A. and Taylor, J.H. 1975, *Astroph. J.* **195**, L51.
Hurley, K. 1987, in *High Energy Activity around Collapsed Stars*, ed. F. Pacini (Reidel), p. 317.
Ilovaisky, S.A. 1988, *Mem. Ital. Astron. Soc.* **59**, 221.
Ilovaisky, S.A. et al. 1987, *Astron. Astrophys.* **179**, L1.
Janot-Pacheco, E. et al. 1988, *Astron. Astrophys.* **202**, 81.
Jenning, M.C. 1988, *Astroph. J.* **333**, 700.
Johnson, H.M., Kulkarni, S. and Oke, J.B. 1989, *Astrophys. J.* **345**, 492.
Jones, P.B. 1988, *Mon. Not. Royal Astron. Soc.* **233**, 875.
Joss, P.C. and Rappaport, S.A. 1979, *Astron. Astrophys.* **71**, 217.
Kallman, T.R. 1989, in ESLAB 1989.
Kallman, T.R. and White, N.E. 1989, *Astroph. J.* **341**, 995.
Kaluzny, J. 1988, *Acta Astron.* **38**, 207.
Kaminker, A.D. et al. 1988, *Sov. Astron.* **32**, 392.
Katz, J.I. and Smith, I.A. 1988, *Astroph. J.* **326**, 733.
Kawai, N. et al. 1989, *Publ. Astron. Soc. Japan* **41**, 491.
Kelley, R.L. 1986, in Truemper et al. 1986, p. 75.
Khruzina, T.S. and Cherepashchuk, A.M. 1986, *Sov. Astron.* **30**, 422.
Khruzina, T.S. and Cherepashchuk, A.M. 1988, *Sov. Astron. Lett.* **14**, 427.
Kiraly, P. and Meszaros, P. 1988, *Astroph. J.* **333**, 719.
Kirk, J.G. and Truemper, J.E. 1984, in Lewin and Van den Heuvel (1983), p. 261.
Kluzman, W. et al. 1988, *Nature* **334**, 225.
Kool, M. de et al. 1987, *Astron. Astrophys.* **183**, 47.
Koyama, K. et al. 1989, *Publ. Astron. Soc. Japan* **41**, 483.
Koyama, K. et al. 1989a, *Publ. Astron. Soc. Japan* **41**, 461.
Kraus, U. 1989, *Astron. Astrophys.* **223**, 246.
Kudryavtsev, M.I. and Svertilov, S.I. 1988, *Sov. Astron. Lett.* **14**, 90.

Kuiper, L. et al. 1988, *Astron. Astrophys.* **203**, 79.
Kulkarni, S. 1986, *Astroph. J.* **306**, L85.
Kulkarni, S. and Narayan, R. 1987, *Astroph. J.* **335**, 755.
Kulkarni, S., Narayan, R. and Romani, R. 1990, *Astrophys. J.* **356**, 174.
Kulkarni, S., Djorgovski, S. and Fruchter, A.S. 1988, *Nature* **334**, 524.
Kuznetsov, A.V. et al. 1987, *Sov. Astron. Lett.* **13**, 445.
Lamb, D.Q. 1984, *Ann. NY Acad. Sci.* **422**, 237.
Lamb, D.Q. and Petterson, J., eds. 1983, *Cataclysmic Variables and Low Mass X-Ray Binaries* (Reidel).
Lamb, R.C. and Weekes, T.C. 1987, *Science* **238**, 1528.
Lamb, R.C. et al. 1988, *Astroph. J.* **328**, L13.
Lawrence, M.A. et al. 1989, *Phys. Rev. Lett.* **63**, 1120.
Lehto, H. et al. 1989, *IAU* Circ. 4925.
Levine, A. et al. 1988, *Astroph. J.* **327**, 753.
Leonard, P.J.T. and Duncan, M.J. 1989, *Astron. J.* **99**, 608.
Lewin, W.H.G. and Heuvel, E.P.J. van den, eds. 1983, *Accretion Driven X-Ray Sources* (Cambridge U. Press).
Lewin W.H.G. and Joss, P.C. 1983, in Lewin and Van den Heuvel (1984), p. 41.
Liang, E. 1988, *Astroph. J.* **334**, 339.
Lochner, J.C. et al. 1989, *Astroph. J.* **337**, 827.
Long, K.S. and Speybroeck, L.P. van 1983, in Lewin and Van den Heuvel (1983), p. 117.
Lynden-Bell, D. 1969, *Nature* **223**, 690.
Lyubarskii, Yu.E. and Sunyaev, R.A. 1988, *Sov. Astron. Lett.* **14**, 390.
Magnier, E. et al. 1989, *Mon. Not. Royal Astron. Soc.* **237**, 729.
Makishima, K. et al. 1988, *Nature* **373**, 746.
Makishima, K. et al. 1989, *Publ. Astron. Soc. Japan* **41**, 531.
Makishima, K. et al. 1989a, *Publ. Astron. Soc. Japan* **41**, 697.
Margon, B. 1984, in Lewin and Van den Heuvel (1983), p. 287.
Margon, B. and Anderson, S.F. 1989, *Astroph. J.* **347**, 448.
Margon, B. and Bolte, M. 1987, *Astroph. J.* **321**, L61.
Margon, B. and Cannon, R. 1989, *Observatory* **109**, 82.
Mason, K.O. et al. eds. 1988, *The Physics of Accretion onto Compact Objects*, Lecture Notes in Phys. Vol. 266 (Springer).
Melia, F. 1989, *Astron. Astrophys.* **233**, L9.
Mendelson, H. and Mazeh, T. 1989, *Mon. Not. Royal Astron. Soc.* **239**, 733.
Meszaros, P. et al. 1989, *Astroph. J.* **337**, L29.
Michel, F.C. 1989, *Nature* **337**, 236.
Milgrom, M. 1978, *Astron. Astrophys.* **67**, L25.
Mineshiga, S. and Wheeler, J.C. 1989, *Astroph. J.* **343**, 241.
Mineshiga, S. and Wood, J.H. 1989, *Mon. Not. Royal Astron. Soc.* **241**, 259.
Mitsuda, K. and Dotani, T. 1989, *Pub. Astron. Soc. Japan* **41**, 537.
Mitsuda, K. and Tanaka, Y. 1986, in Truemper et al. (1986), p. 195.
Mitsuda, K. et al. 1989, *Pub. Astron. Soc. Japan* **541**, 97.
Molnar, L. 1988, *Astroph. J.* **331**, L25.
Molnar, L. 1989, *Mon. Not. Royal Astron. Soc.* **231**, 715.
Molnar, L.A. et al. 1988, *Astroph. J.* **331**, 494.
Motch, C. et al. 1989, *Astron. Astrophys.* **219**, 158.
Murakami, T. et al. 1988, *Nature* **335**, 234.
Nagase, F. 1989, *Publ. Astron. Soc. Japan* **41**, 1.
Nakamura, N. et al. 1988, *Publ. Astron. Soc. Japan* **40**, 209.
Nakamura, N. et al. 1989, *Publ. Astron. Soc. Japan* **41**, 617.
Narayan, R. 1987, *Astroph. J.* **319**, 162.
Narayan, R. and Popham, R. 1989, *Astroph. J.* **341**, L25.
Naylor, T. and Charles, P.A. 1989, *Mon. Not. Royal Astron. Soc.* **236**, 1p.
Oijen, J.G.J. van 1989, *Astron. Astrophys.* **217**, 115.

Paczyński, B. 1987, *Nature* **327**, 300.
Paczyński, B. 1987a, *Astroph. J.* **317**, L51.
Paczyński, B. 1988, *Astroph. J.* **335**, 525.
Paczyński, B. and Long, K.W. 1988, *Astroph. J.* **333**, 69.
Papaloizou, J.C.B. and Lin, D.N. 1989, *Astroph. J.* **344**, 645.
Paradijs, J. van 1984, in Lewin and Van den Heuvel (1983), p. 189.
Paradijs, J. van 1989, *Mon. Not. Royal Astron. Soc.* **238**, 45p.
Paradijs, J. van and Isaacman, R. 1989, *Astron. Astrophys.* **222**, 129.
Paradijs, J. van, Pennix, J.W. and Lewin, W.H.G. 1988, *Mon. Not. Royal Astron. Soc.* **233**, 437.
Paradijs, J. van et al. 1988, *Nature* **334**, 684.
Paradijs, J. van et al. 1989, *Astron. Astrophys. Suppl.* **76**, 185.
Paradijs, J. van et al. 1990, *Astron. Astrophys.* **235**, 156.
Parmar, A.N. and White, N.E. 1988, *Mem. Soc. Astron. Ital.* **59**, 147.
Parmar, A.N. et al. 1989, *Astroph. J.* **338**, 373.
Pennix, W. 1989, *Astron. Astrophys.* **208**, 146.
Peres, G. et al. 1989, *Astroph. J.* **336**, 140.
Petterson, J.A. 1984, in Lewin and Van den Heuvel (1983), p. 367.
Phinney, E.S. et al. 1988, *Nature* **333**, 832.
Prendergast, K.H. and Burbidge, G.R. 1968, *Astroph. J.* **Lett. 151**, L83.
Priedhorsky, W.C. and Holt, S.S. 1987, *Space Sci. Rev.* **45**, 291.
Pringle, J.E. 1981, *Ann. Rev. Astron. Astrophys.* **19**, 137.
Radhakrishnan, V. and Shukre, C.S. 1985, in *Supernovae, Their Progenitors, and Their Remnants*, eds. G. Srinivasan and V. Radhakrshnan (Bangalore Indian Academy of Sciences), p. 155.
Rankin, J.M. and Gil, J.A. 1989, *Comments on Astrophys.* **14**, 1.
Rappaport, S.J. et al. 1989, *Astroph. J.* **345**, 203.
Rasio, F.A. et al. 1989, *Astroph. J.* **342**, 934.
Raubenheimer, B.C. et al. 1989, *Astroph. J.* **336**, 394.
Rawley, L.A. et al. 1987, *Science* **238**, 761.
Romani, R. et al. 1987, *Nature* **329**, 309.
Ruderman, J.A., Shaham, J. and Tavani, M. 1989a, *Astrophys. J.* **336**, 507.
Ruderman, J.A., Shaham, J. and Tavani, M. 1989b, *Astrophys. J.* **343**, 272.
Sansom, A.E. et al. 1981, *Publ. Astron. Soc. Japan* **41**, 591.
Schaefer, B. et al. 1987, *Sky and Telescope* **73**, 604.
Schlickeiser, R. 1989, *Astron. Astrophys.* **213**, L23.
Schmidt, M. et al. 1988, *Astroph. J.* **329**, L85.
Schulz, N.S. et al. 1989, *Astron. Astrophys.* **225**, 48.
Shakura, N.I. and Sunyaev, R.A. 1973, *Astron. Astrophys.* **24**, 429.
Shapiro, M.M. and Wefel, J.P., eds. 1989, *Cosmic Gamma Rays, Neutrinos, and Related Astrophysics* (Kluwer).
Shara, M. et al. 1988, *Astroph. J.* **328**, 594.
Sheffer, E.K. 1988, *Sov. Astron. Lett.* **13**, 82.
Shibazaki, N. et al. 1989, *Nature* **342**, 656.
Smak, J. 1971, *Acta Astron.* **21**, 15.
Smale, A.P. and Charles, P.A. 1989, *Mon. Not. Royal Astron. Soc.* **238**, 595.
Smale, A.P. et al. 1988, *Mon. Not. Royal Astron. Soc.* **232**, 642.
Sokolov, V.V. 1988, *Sov. Astron.* **31**, 47.
Srinivasan, G. et al. 1990, *Current Science* **59**, No. 1.
Stephenson, C.B. and Sanduleak, N. 1977, *Astroph. J.* **Suppl. 33**, 459.
Stevens, I.R. 1989, *Mon. Not. Royal Astron. Soc.* **232**, 199.
Stollman, G.B. and Kuperman, M. 1988, *Astron. Astrophys.* **203**, 104.
Strom, R.G. et al. 1989, *Nature* **337**, 234.
Taam, R.E. 1986, in Truemper et al. (1986), p. 207.
Taam, R.E. and Fryxell, B.A. 1988, *Astroph. J.* **327**, L73.
Taam, R.E. and Fryxell, B.A. 1989, *Astroph. J.* **339**, 297.
Taam, R.E., Fryxell, B.A. and Brown, D.A. 1988, *Astrophys. J.* **331**, L117.

Taam, R.E. and Heuvel, E.P.J. van den 1986, *Astroph. J.* **305**, 325.
Tavani, M. 1989, *Bull. Amer. Astron. Soc.* **21**, 1204.
Tavani, M. et al. 1989, *Astroph. J.* **342**, L31.
Taylor, J.H. 1987, in Helfand and Huang (1987).
Taylor, J.H. and Dewey, R.J. 1988, *Astroph. J.* **332**, 771.
Taylor, J.H. and Weisberg, J.M. 1989, *Astroph. J.* **345**, 434.
Tawara, Y. et al. 1989, *Publ. Astron. Soc. Japan* **41**, 473.
Thorstensen, J.R. et al. 1988, *Astroph. J.* **334**, 430.
Tonwar, S.C. et al. 1988, *Astroph. J.* **330**, L11.
Treves, A., Maraschi, L. and Abramowicz, M., eds. 1989, *Accretion.* (Singapore: World Scientific.)
Trimble, V. 1991, *Contemporary Physics* **32**, 103.
Trimble, V. 1992, in Y. Kondo, R. Sistero & R. Polidan (eds.), IAU Symp. 151 (Kluwer).
Trimble, V. and Rees, M.J. 1971, *Astroph. J.* **166**, L85.
Trimble, V., Rose, W.K. and Weber, J. 1973, *Mon. Not. Royal Astron. Soc.* **162**, 1p.
Truemper, J. 1986, in *High Energy Activity around Collapsed Stars*, ed. F. Pacini (Reidel), p. 367.
Truemper, J., Lewin, W. and Brinkman, W., eds. 1986, *The Evolution of Galactic X-Ray Binaries* (Reidel).
Tsunemi, H. et al. 1989, *Astroph. J.* **337**, L81.
Tutukov, A.V. et al. 1988, *Sov. Astron. Lett.* **13**, 328.
Usov, V.V. 1988, *Astrophys. Space Sci.* **140**, 39.
Vanbeveren, D. 1989, *Astron. Astrophys.* **224**, 93.
Velusamy, T. and Subrahmanya, C.R. 1989, *Mon. Not. Royal Astron. Soc.* **239**, 281.
Verbunt, F. and Meylan, G. 1988, *Astron. Astrophys.* **203**, 295.
Verbunt, R. et al. 1987, *Nature* **329**, 312.
Verbunt, F., Lewin, W. and Paradijs, J. van 1989, *Mon. Not. Royal Astron. Soc.* **241**, 51.
Vrtilek, S.K. et al. 1986, *Astroph. J.* **307**, 698.
Vrtilek, S.K. et al. 1990, *Astron. Astrophys.* **235**, 162.
Wade, R.A. 1985, in *Interacting Binaries*, eds. P.P. Eggleton and J.E. Pringle (Reidel), p. 289.
Walker, M.A. and Meszaros, P. 1989, *Astroph. J.* **346**, 844.
Wang, J.C.L. et al. 1989, *Phys. Rev. Lett.* **63**, 1551.
Wasserman, I.M. and Cordes, J.M. 1988, *Astroph. J.* **333**, L91.
Weekes, T.C. et al. 1989, *Astroph. J.* **342**, 371.
Weiler, K.W. and Sramek, R.A. 1988, *Ann. Rev. Astron. Astrophys.* **26**, 295.
White, N.E. 1988, *Astroph. J.* **324**, 367.
White, N.E. 1989, *Astron. Astrophys. Rev.* **1**, 85.
White, N.E. and Filipov, L.G., eds. 1986, *Adv. Space Res.* **8**, No. 2-3.
Whitlock, L. et al. 1989, *Astroph. J.* **338**, 381.
Wijers, R.A.M.J. 1989, *Astron. Astrophys.* **209**, L1.
Wolszczan, A. 1989, *Nature* **337**, 531.
Zwitter, T. and Calvani, M. 1989, *Mon. Not. Royal Astron. Soc.* **326**, 581.
Zwitter, T. et al. 1989, *Fund. Cos. Phys.* **13**, 309.

TeV RADIATION FROM BINARY STARS

T. C. Weekes

Nagle, D.E., T.K.Gaisser, and R.J.Protheroe: 1988, *Ann. Rev. Nuclear and Particle Sci.* **38**, 609.
Weekes, T.C.: 1987, *Physics Reports* **160**, 1.
Vladimirsky, B.M., A.A. Stepanian, and V.P. Fomin, 1973, in Proc. 13th Intern. Cosmic Ray Conf., Denver, 1, 456.
Ramana Murthy, P.V.: 1989, *Nuclear Physics B* **14A**, 73.
Lamb, R.C. and T.C.Weekes: 1987, *Science* **238**, 1538.
Bonnet-Bidaud,J.M. and Chardin, G.: 1988, *Physics Reports* **170**, 326.
Cassiday, G.L. et al: 1988, *Phys. Rev. Lettr.* **62**, 383.

Chadwick, P.M. et al: 1985, *Nature* **318**, 642.
Zyskin, Yu. et al: 1987, *Astron. Circular, Bur. of Astron. Info. of U.S.S.R. Acad. of Sci.* **1508**, 1.
Weekes, T.C.: 1989, *Nuclear Physics B* **10B**, 41.
Resvanis, L.K. et al: 1988, *Astrophys. J.* **328**, L9.
Lamb, R.C. et al: 1988, *Astrophys. J.* **328**, L13.
Dingus, B.L. et al: 1988, *Phys. Rev. Lett.* **61(17)**, 1906.
Protheroe, R.J., 1987, in Proceedings of 20th International Cosmic Ray Conference, Moscow, August 2-15, 1987, 8, 21.
Hillas, A.M., Proc. 1986 Workshop on HE-UHE Behaviour of Accreting X-ray Sources, Vulcano, 1986.
Harding, A.K.: 1989, *Nuclear Physics B* **14A**, 3.
Weekes, T.C. et al: 1989, *Astrophys. J.* **342**, 379.

ζ AURIGAE ATMOSTPHERIC ECLIPSING BINARIES

I. A. Ahmad

Ahmad, I. A. 1986, *Astroph. J.* **301**, 275.
Ahmad, I. A. 1987, in *Cool Stars, Stellar Systems, and the Sun*, ed. by J. Linsky and R. E. Stencel, (New York: Springer-Verlag), 326.
Ahmad, I. A. 1988, in *A Decade of UV Astronomy with the IUE Satellite* (Proceedings of a Celebratory Symposium, Goddard Space Flight Center, Greenbelt, MD, ESA SP-281), vol 1, ed. by E. J. Rolfe, p. 217.
Ahmad, I. A. 1989, *Astroph. J.* **338**, 1011.
Ahmad, I. A. 1990, submitted to *Astroph. J.*
Ahmad, I. A. 1991, in preparation.
Ahmad, I. A., and Chapman, R. D. 1986 in *New Insights in Astrophysics* (Publications of 1986 International IUE Symposium, University College, London, ESA SP-263), ed. by E. J. Rolfe, p. 439.
Ahmad, I. A., Chapman, R. D., and Kondo, Y. 1983, *Astr. & Ap.*, **126**, L5 (ACK).
Ahmad, I. A., and Parsons, S. B. 1985, *Astroph. J. (Letters)*, **299**, L33.
Ahmad, I. A., and Parsons, S. B. 1986, *Bull. Am. Astron. Soc.*, **18**, 696.
Ahmad, I. A., and Stencel, R. E. 1988, *Astroph. J.*, **329**, 797.
Ake, T., Parsons, S. B., and Kondo, Y. 1985, *Astroph. J.* **298**, 772.
Bidelman, W. P. 1954, *Astroph. J. Suppl.*, **1**, 218.
Chapman, R. D. 1981, *Astroph. J.*, **248**, 1043.
Che, A., Hempe, K., and Reimers, D. 1983, *Astron. and Astroph.* **126**, 225.
Che-Bohnenstengle, A., and Reimers, D. 1986, *Astron. and Astroph.* **156**, 172.
Cowley, A. P. 1969, *P. A. S. Pac*, **81**, 297.
Crutcher, R. M. 1982, *Astroph. J.*, **254**, 82.
Dupree, A. K., and Reimers, D. 1987, in *Exploring the Universe with the IUE Satellite*, Y. Kondo, ed., (Dordrecht: D. Reidel) p. 321.
Eaton, J. 1988, *Astroph. J.* **333**, 288.
Griffin, R. E. M. 1988, *J. Roy. A. S. Canada*, **82**, 49.
Griffin, R., and Griffin, R. 1986, *Astron. and Astroph.*, 7, 195.
Griffin, R. E. M., Griffin, R. F., Schröder, K.-P., and Reimers, D. 1990, *Astron. and Astroph.*, **234**, 284.
Hack, M., and Stickland, D. 1987, in *Exploring the Universe with the IUE Satellite*, Y. Kondo, ed., (Dordrecht: D. Reidel) p. 445.
Hjellming, R. and Newell, R. 1983, *Astroph. J.* **275**, 704.
Jackson, J. C. 1975, *Monthly Not. of the Roy. Astr. Soc.*, **172**, 483.
Kiyokawa, M. 1967. *P.A.S. Japan* **19**, 72.
Kuin, N. P. M., and I. A. Ahmad 1989, *Astroph. J.*, **344**, 856.
McCluskey, Jr, G. E., and Kondo Y., 1984, *P.A.S. Pac.*, **96**, 817.

McKeller, A., and Petrie, R. M. 1952, *Monthly Not. of the Roy. Astr. Soc.g*, **112**, 641.
Parsons, S. B., Ake, T. B., and Hopkins, J. L. 1984, *P.A.S. Pac.*, **97**, 725.
Reimers, D., and Che-Bohnenstengle, A. 1986, *Astron. and Astroph.*, **166**, 252.
Saito, M. 1970, *P.A.S. Japan* **22**, 45.
Sahade, J., and Wood, F. B. 1978, in *Interacting Binary Stars* (Pergamon: Oxford), p. 121.
Schröder, K.-P. 1985, *Astron. and Astroph.*, **147**, 103.
Schröder, K.-P. 1986, *Astron. and Astroph.*, **170**, 70.
Schröder, K.-P. 1990, *Astron. and Astroph.*, **236**, 165.
Schröder, K.-P., Griffin, R. E. M., and Griffin, R. F. 1990, *Astr. and Ap.*, **234**, 299.
Stencel, R. E., Kondo, Y., Bernat, A. P., and McCluskey, G. E. 1979, *Astroph. J.*, **233**, 621.
Wilson, O. C. 1960, in *Stellar Atmospheres,* ed. J. L. Greenstein (U. of Chicago Press: Chicago), p. 436.
Wright, K. O. 1970. *Vistas in Astronomy* (ed. A. Beer) **12**, 147.

POLARIZED RADIATION FROM CLOSE BINARIES

R. H. Koch

Appenzeller, I. 1966, *Zeit. Astroph.*, **64**, 269.
Bailey, J., Hough, J. H., and Wickramasinghe, D. T. 1988, *Monthly Not. Roy. Astr. Soc.*, **233**, 395.
Barden, S. C. 1987, *Astroph. J.*, **317**, 333.
Berriman, G., and Smith P. S. 1988, *Astroph. J.*, **329**, L97.
Binnendijk, L. 1970, *Vistas in Astr.*, **12**, 217.
Blitzstein, W., Bradstreet, D. H., Hrivnak, et al., 1980, *Publ. Astr. Soc. Pac.*, **92**, 338.
Brown, J. C., McLean, I. S., and Emslie, A. G. 1978, *Astron. Astroph.*, **68**, 415.
Buerger, P. F., and Collins, G. W. 1970, *Astroph. J.*, **161**, 1025.
Cassinelli, J. P., Nordsieck, K. H., and Murison, M. A. 1987, *Astroph. J.*, **317**, 290.
Chandrasekhar, S. 1946, *Astroph. J.*, **103**, 351.
Clarke, D., and McGale, P. A. 1986, *Astron. Astroph.*, **169**, 251.
Clarke, D., and McGale, P. A. 1988, *Astron. Astroph.*, **205**, 207.
Cohen, M., Anderson, C. M., Cowley, A., et al., et al., 1975, *Astroph. J.*, **196**, 179.
Coyne, G. V. 1970, *Spec. Vat. Rich. Astr.*, **8**, 105.
Coyne, G. V., and Gehrels, T. 1967, *Astron. J.*, **72**, 887.
Coyne, G. V., Magalhaes, A. M., Moffat, A. F. J., et al., 1988, *Polarized Radiation of Circumstellar Origin*, U. Arizona Press, Tucson, U.S.
Coyne, G. V., and McLean, I. S. 1975, *Astron. J.*, **80**, 702.
DeJager, C., and Nieuwenhuizen, H. 1987, *Astron. Astroph.*, **177**, 217.
Deshpande, M. R., Joshi, U. C., Kulshrestha, A. K., et al., 1987, *Publ. Am. Soc. Pacific*, **99**, 62.
Dolan, J. F., and Tapia, S. 1988, *Astron. Astroph.*, **202**, 124.
Dolan, J. F., and Tapia, S. 1989, *Astroph. J.*, **344**, 830.
Drissen, L. Robert C., Lamontagne, R., Moffat, A. F. J., et al., 1989, *Astroph. J.*, **343**, 426.
Eaton, J. A. 1978, *Acta. Astr.*, **28**, 601.
Efimov, Y. S., Piirola, V., and Shakhovskoy, N. M. 1984, *Astron. Astroph.*, **138**, 62.
Gravina, R., and Martel, M. T. 1985, *Comm. 27 I.A.U. Inf. Bull. Var. Stars No. 2812*.
Hall, J. S., and Mikesell, A. H. 1950, *Publ. U.S. Nav. Obs.*, **17**, 1.
Hayes, D. S. 1983, *Comm. 27 I.A.U. Inf. Bull. Var. Stars No. 2350*.
Henson, G. D., Kemp, J. C., and Kraus, D. J. 1984, *Bull. A.A.S.*, **16**, 1013.
Hiltner, W. A. 1956, *Astroph. J. Suppl.*, **2**, 389.
Kemp, J. C. 1980, *Astron. Astroph.*, **91**, 108.
Kemp, J. C., Henson, G. D., Barbour, M. S., et al., 1983, *Astroph. J.*, **273**, L85.
Kemp, J. C., Henson, G. D., Kraus, D. J., et al., 1986, *Astroph. J.*, **300**, L11.
Kemp, J. C., Henson, G. D., Kraus, D. J., et al., 1987, *Astroph. J.*, **317**, L29.
Kemp, J. C., and Herman, L. C. 1977, *Astroph. J.*, **218**, 770.
Kemp, J. C., and Rudy, R. J. 1975, *Publ. Astr. Soc. Pacific*, **87**, 301.

Kemp, J. C., Wolstencroft, R. D., and Swedlund, J. B., 1972, *Astroph. J.*, **177**, 177.
Koch, R. H. 1987, *Comm. Astroph.*, **12**, 1.
Koch, R. H. 1988, in K.-C. Leung (ed.), *Critical Observations vs. Physical Models for Close Binary Systems*, Gordon and Breach, Montreux, p. 151.
Koch, R. H., Elias, N. M., Corcoran, M. F., et al., 1989, *Space Sci. Rev.*, **50**, 63.
Koch, R. H., and Pfeiffer, R. J. 1982, *Astron. J.*, **87**, 1409.
Koch, R. H., and Pfeiffer, R. J. 1989a, *Publ. Am. Soc. Pacific*, **101**, 279.
Koch, R. H., and Pfeiffer, R. J. 1989b, NATO ASI, Ser. C., No. 319, 81.
Liu, X., and Tan, H. 1987, *Acta Astr. Sinica*, **28**, 139.
Luna, H. G. 1980, *Publ. Am. Soc. Pacific*, **92**, 79.
Luna, H. G. 1981, *Rev. Mex.*, **6**, 111.
Luna, H. G. 1982, *Publ. Am. Soc. Pacific*, **94**, 692.
Mathewson, D. S., and Ford, V. L. 1970, *Mem. R.A.S.*, **74**, 139.
Morris, S. L., Schmidt, G. D., Liebert, J., et al., 1987, *Astroph. J.*, **314**, 641.
Nurmanova, U. A. 1982, *Perem. Zv.*, **21**, 655.
Pfeiffer, R. J. 1976, *Astron. J.*, **81**, 1000.
Pfeiffer, R. J. 1977, *Astron. J.*, **82**, 734.
Pfeiffer, R. J. 1979, *Astroph. J.*, **232**, 181.
Pfeiffer, R. J., and Koch, R. H. 1987, *Astron. J.*, **94**, 484.
Piirola, V. 1977, *Astron. Astroph.*, **56**, 105.
Piirola, V. 1981, *Astron. Astroph. Suppl.*, **44**, 461.
Piirola, V., and Korhonen, T. 1979, *Astron. Astroph.*, **79**, 254.
Piirola, V. Reiz, A., and Coyne, G. V. 1987, *Astron. Astroph.*, **186**, 120.
Rudy, R. J., and Kemp, J. C. 1976, *Astroph. J.*, **207**, L125.
Rudy, R. J., and Kemp, J. C. 1977, *Astroph. J.*, **216**, 767.
Saute, M., and Martel, M. T. 1979, *Comm. 27 I.A.U. Inf. Bull. Var. Stars No. 1681*.
Scaltriti, F., Piirola, V., Coyne, G. V., et al., 1992, Astron. Astrophys. (submitted).
Schulte-Ladbeck, R. 1985, *Astron. Astroph.*, **142**, 333.
Schulte-Ladbeck, R. E., and van der Hucht, K. A. 1989, *Astroph. J.*, **337**, 872.
Serkowski, K. 1970, *Astroph. J.*, **160**, 1083.
Serkowski, K., and Chojnacki, W. 1969, *Astron. Astroph.*, **1**, 442.
Shawl, S. J. 1975a, *Astron. J.*, **80**, 602.
Shawl, S. J. 1975b, *Astron. J.*, **80**, 595.
Simmons, J. F. L., Aspin, C., and Brown, J. C. 1980, *Astron. Astroph.*, **91**, 97.
Skul'skii, M. Yu. 1985, *Sov. Astr. Lett.*, **11**, 21.
Snow, T. P., and Hayes, D. P. 1978, *Astroph. J.*, **226**, 897.
Stockman, H. S., Schmidt, G. D., and Lamb, D. Q. 1988, *Astroph. J.*, **332**, 282.
Stokes, R. A., Swedlund, J. B., Avery, R. W., and Michalsky, J. J. 1974, *Astron. J.*, **79**, 678.
Vogt, S. S. 1980, *Astroph. J.*, **240**, 567.
West, S. C., Berriman, G., and Schmidt, G. D. 1987, *Astroph. J.*, **322**, L35.
Wolf, G. W. 1972, *Astron. J.*, **77**, 576

RADIO EMISSION FROM ACTIVE LATE-TYPE BINARIES

N. M. Elias II and R. L. Mutel

Abbott, D. C., Bieging, J. H., and Churchwell, E. : 1984, *Astrophys. J.*, **280**, 671.
Abbott, D.C., Bieging, J.H., Churchwell, E., and Torres, A.V. : 1986, *Astrophys. J.*, **303**, 239.
Allen, C. W. : 1973, *Astrophysical Quantities* (London: Athlone Press).
Allen, D.: 1988, in *IAU Colloqium No. 103, The Symbiotic Phenomenon*, ed. J. Mikolejewska, M. Friedjung, S. J. Kenyon, and R. Viotti (Dordrecht: Kluwer), p. 3.
André, P.: 1987, in *Protostars and Molecular Clouds*, ed. T. Montmerle and C. Bertout (Saclay: CENS), p. 143.
André, P., Montmerle, T., and Feigelson, E.D.: 1988, *Astron. J.*, 93, 1182.

André, P., Montmerle, T., Feigelson, E.D., Stine, P.C., and Klein, K.-L.: 1988, *Astrophys. J.*, **335**, 940.
Basri, G., Laurent, R., and Walter, F. : 1983, in IAU Symposium No. 102, *Solar and Stellar Magnetic Fields: Origins and Coronal Effects*, ed. J. O. Stenflo, (Dordrecht: Reidel), p. 199.
Bastian, T.S. and Bookbinder, J.A.: 1987, *Nature*, 326, 678.
Bastian, T. S., Dulk, G. A., and Slee, O. B. : 1988, *Astron. J.*, **95**, 794.
Bastian, T.S., Dulk, G.A., and Chanmugam, G.: 1988, *Astrophys. J.*, **324**, 431.
Bastian, T.S., Bookbinder, J., Dulk, G.A., and Davis, M.: 1990, *Astrophys. J.*, **353**, 265.
Bookbinder, J.A. and Lamb, D.Q.: 1987, *Astrophys. J. (Letters)*, **323**, L131.
Budding, E.: 1986, *Astr. Sp. Sc.*, 118, 241.
Busso, M., Persi, P., Robberto, M., Scaltriti, F., and Silvestro, G. 1987, in *IAU Symposium No. 122, Circumstellar Matter*, ed. I. Appenzeller and C. Jordan (Dordrecht: Reidel), pg. 113.
Caillault, J-P., Drake, S. A., and Florkowski, D. R. : 1988, *Astron. J.*, **95**, 887.
Caillault, J-P.: 1989, *Astron. J.*, **97**, 163.
Chanmugan, G. and Dulk, G.A.: 1982, *Astrophys. J. (Letters)* **255**, L107.
Corcoran, M. F.: 1988, *Ph.D. thesis, University of Pennsylvania*.
Coyne, G. V., Elias, N. M., Holenstein, B. D., Koch, R. H., Piirola, V., and Scaltriti, F. : 1991, in *IAU Colloquium 130, The Sun and Cool Stars: Activity, Magnetism, Dynamos, in press*.
Doiron, D.J. and Mutel, R.L.: 1984, *Astron. J.*, 89, 430.
Dorren, J.D. and Guinan, E.F.: 1984, in *Proceedings of the Third Cambridge Workshop on Cool Stars, Stellar Systems, and the Sun*, ed. S.L. Baliumas and L.W. Hartmann (Berlin: Springer-Verlag), p. 259.
Dorren, J. D. and Guinan, E. F. : 1991, *in press*.
Drake, S.A., Abbott, D. C., Bastian, T. S., Bieging, J. H., Churchwell, E., Dulk, G., and Linsky, J. L. : 1987, *Astrophys. J.*, **322**, 902.
Drake, S. A., Linsky, J. L., and Elitzur, M. : 1987, *Astron. J.*, **94**, 1280.
Drake, S.A., Simon, T., and Linsky, J.L.: 1989, *Astrophys. J. Suppl.*, **71**, 905.
Drake, S. A., Linsky, J. L., Judge, P. G., and Elitzur, M. : 1991, *Astron. J.*, **101**, 230.
Dulk, G.A., Bastian, T.S., and Chanmugam, G.: 1983, *Astrophys. J.*, **273**, 249.
Dulk, G.A.: 1985, *Ann. Rev. Astron. Astroph.*, 23,169.
Eggleton, P.P.: 1985, in *Interacting Binary Stars*, ed. J. E. Pringle and R. A. Wade (Cambridge: Cambridge University Press), p. 21.
Elias, N. M.: 1990, *Astrophys. J.*, **352**, 300.
Elias, N. M. and Dorren, J. D. : 1990, *Inf. Bull. Var. Stars*, No. 3541.
Feldman, P. A., Taylor, A. R., Gregory, P. C., Seaquist, E. R., Balonek, T. J., and Cohen, N. L. : 1978, *Astron. J.*, **83**, 1471.
Gary, D.E., Linsky, J.L., and Dulk, G.A.: 1983, in *Solar and Stellar Magnetic Fields: Origins and Coronal Effects*, ed. J.O. Stenflo (Dordrecht: Reidel), p. 387.
Gibson, D.M.: 1980, in IAU Symposium No. 88, *Close Binary Stars, Observations and Interpretation*, ed. M. J. Plaveč, D. M. Popper, and R. K. Ulrich (Dordrecht: Reidel), p. 31.
Gilliland, R. L. : 1985, *Astrophys. J.*, **299**, 286.
Gilman, P. A. : 1983, in IAU Symposium No. 102, *Solar and Stellar Magnetic Fields: Origins and Coronal Effects*, ed. J. O. Stenflo (Dordrecht: Reidel), p. 247.
Hall, D. S. : 1976, in *IAU Colloqium No. 29, Multiply Periodic Variable Stars*, ed. W. S. Fitch (Dordrecht: Reidel), p. 287.
Havnes, O. and Goertz, C.K.: 1984, *Astron. Astroph.*, 138,421.
Hjellming, R.M.: 1988, in *Galactic and Extragalactic Radio Astronomy*, ed. G. L. Verschuur and K. I. Kellermann (New York: Springer-Verlag), p. 381.
Hjellming, R. M. and Newell, R.T.: 1983, *Astrophys. J.*, **275**, 704.
Hjellming, R. M., Han, X. H., Córdova, F. A., and Hasinger, G. : 1990a, *Astron. Astrophys.*, **235**, 147.
Hjellming, R. M., Stewart, R. T., White, G. L., Strom, R., Lewin, W. H. G., Hertz, P., Wood, K. S., Norris, J. P., Mitsuda, K., Penninx, W., and van Paradijs, J. : 1990b, *Astrophys. J.*, **365**, 681.
Hughes, V.A. and McLean, B.J.: 1984, *Astrophys. J.*, **278**, 716.
Hughes, V.A. and McLean, B.J.: 1987, *Astrophys. J.*, **313**, 263.
Kenny, H. T., Taylor, A. R., and Seaquist, E. R. : 1991, *Astrophys. J.*, **366**, 549.

Kenyon, S.J. : 1986, *The Symbiotic Stars* (Cambridge: Cambridge University Press).
Klein, K-L. and Chiuderi-Drago, F. : 1987, *Astron. Astrophys.*, **175**, 179.
Koch, R.H. and Guinan, E.F.: 1978, *Inf. Bull. Var. Stars*, No. 1483.
Koch, R. H., Elias, N. M., Corcoran, M. F., and Holenstein, B. D. : 1989, *Space Sci. Rev.*, **50**, 63.
Lang, K. R. and Willson, R. F. : 1988, *Astrophys. J.*, **328**, 610.
Lestrade, J-F.: 1988, *IAU Symp. 129: VLBI in Astrophysics and Geophysics*, eds. J.M. Moran and M. Reid (Dordrecht: Kluwer), p. 265.
Lestrade, J-F., Mutel, R.L., Preston, R.A., and Phillips, R.B.: 1988, *Astrophys. J.*, **328**, 232.
Linsky, J. L., Neff, J. E., Brown, A., Gross, B. D., Simon, T., Andrews, A. D., Rodonó, M., and Feldman, P. A. : 1989, *Astron. Astrophys.*, **211**, 173.
Majer, P., Schmitt, J. H. M. M., Golub, L., Harnden, F. R., and Rosner, R. : 1986, *Astrophys. J.*, **300**, 360.
Mestel, L. and Spruit, H.C.: 1987, *Monthly Not. Roy. Astr. Soc.*, 226,57.
Morris, D.H. and Mutel, R.L.: 1988, *Astron. J.*, 95,204.
Morris, D.H., Mutel, R.L., and Su, B.: 1990, *Astrophys. J.*, **362**, 299.
Mutel, R. L., Doiron, D. J., Lestrade, J-F., and Phillips, R. B. : 1984, *Astrophys. J.*, **278**, 220.
Mutel, R.L., Lestrade, J-F., Preston, R.A., and Phillips, R.B.: 1985, *Astrophys. J.*, **289**, 262.
Mutel, R.L., Morris, D.H., Doiron, D.J., and Lestrade, J-F. : 1987 *Astron. J.*, 93,1220.
Nelson, R. F. and Spencer, R. E. : 1988, *Monthly Notices Roy. Astron. Soc.*, **234**, 1105.
Owen, F. N., Jones, T.W., Gibson, D. M. : 1976, *Astrophys. J. (Letters)*, **210**, L27.
Persi, P., Tapia, M., Rodríguez, L. F., Ferrari-Toniolo, M., and Roth, M. : 1990, *Astron. Astrophys.*, **240**, 93.
Phillips, R.B. and Lestrade, J-F.: 1988, *Nature*, 334, 329.
Phillips, R. B., Lonsdale, C. J., and Feigelson, E. D. : 1991, *submitted*.
Plavec, M.J.: 1989, *Space Sc. Rev.*, 50, 95.
Pringle, J. E. : 1985, *Interacting Binary Stars*, ed. J. E. Pringle and R. A. Wade (Cambridge: Cambridge University Press), p. 1.
Rucinski, S.M.: 1981, *Astron. Astroph.*, 104, 260.
Rucinski, S.M.: 1985a, in *Interacting Binary Stars*, ed. J. E. Pringle and R. A. Wade (Cambridge: Cambridge University Press), p. 113.
Rucinski, S.M.: 1985b, in *Interacting Binary Stars*, ed. J. E. Pringle and R. A. Wade (Cambridge: Cambridge University Press), p. 129.
Rucinski, S.M. and Seaquist, E.R.: 1988, *Astron. J.*, 95,1837.
Sahade, J.: 1987, *Comments on Astrophysics*, **12**, 13.
Schüssler, M.: 1983, in IAU Symposium No. 102, *Solar and Stellar Magnetic Fields: Origins and Coronal Effects*, ed. J. O. Stenflo (Dordrecht: Reidel), p. 213.
Schwartz, P.R., Simon, T., and Campbell, R.: 1986, *Astrophys. J.*, **303**, 233.
Seaquist, E. R. and Taylor, A. R. : 1987, *Astrophys. J.*, **312**, 813.
Seaquist, E.R. and Taylor, A.R.: 1990, *Astrophys. J.*, **349**, 313.
Slee, O. B. *et al.* : 1987a, *Proc. Astron. Soc. Australia*, 7, 55.
Slee, O.B., Nelson, G.J., Stewart, R.T., Wright, A.E., Innis, J.L., Ryan, S.G., and Vaughn, A.E. : 1987b, *Monthly Not. Roy. Astr. Soc.*, 229, 659.
Slee, O. B. and Stewart, R. T. : 1989, *Monthly Notices Roy. Astron. Soc.*, **236**, 129.
Slee, O.B., Stewart, R.T., Bunton, J.D., Beasley, A.J., Carter, B.D., and Nelson, G.J. : 1989, *Monthly Not. Roy. Astr. Soc.*, 239, 913.
Spencer, J.H. and Schwartz, P.R.: 1974, *Astrophys. J. (Letters)*, **188**, L105.
Spruit, H. C. and Roberts, B. : 1983, *Nature*, **304**, 401.
Stewart, R.T., Slee, O.B., White, G.L., Budding, E., Coates, D.W., Thompson, K., and Bunton, J.D.: 1989, *Astrophys. J.*, **342**, 463.
Stine, P.C., Feigelson, E.D., André, P., and Montemerle, T.: 1988, *Astron. J.*, 96, 1394.
Topka, K. and Marsh, K.A.: 1982, *Astrophys. J.*, **254**, 641.
Unama, G., Catalano, S., Rodonó, M., and Gibson, D.M.: 1989, *Space Sc. Rev.*, **50**, 370.
Vaiana, G. S.: 1983, in IAU Symposium No. 102, *Solar and Stellar Magnetic Fields: Origins and Coronal Effects*, ed. J. O. Stenflo (Dordrecht: Reidel), p. 165.
Van den Oörd, G. H. J., Mewe, R., and Brinkman, A. C. : 1988, *Astron. Astrophys.*, **205**, 181.

Van den Oörd, G. H. J., Kuijpers, J., White, N. E., van der Hulst, J. M., and Culhane, J. L. : 1989, *Astron. Astrophys.*, **209**, 296.
Vilhu, O., Caillault, J-P., Neff, J., and Heise, J.: 1988, in *Activity in Cool Stars*, ed. O. Havnes, B.R. Petterson, J.H.M.M. Schmidt, and J.E. Solheins (Dordrecht: Kluwer), p. 179.
Wade, C.M. and Hjellming, R.M.: 1972, *Nature*, 235, 270.
Wade, R.A. and Ward, M.J.: 1985, in *Interacting Binary Stars*, ed. J. E. Pringle and R. A. Wade (Cambridge: Cambridge University Press), p. 129.
White, N.E., Culhane, J.L., Parmar, A.N., Kellett, B.J., Kahn, S., van den Oörd, G.H.J., and Kuijpers, J.: 1986, *Astrophys. J.*, **301**, 262.
White, R. L. : 1985, *Astrophys. J.*, **289**, 698.

FORMATION OF BINARY STARS

A. P. Boss

Abt, H.A.: 1983, *Ann. Rev. Astr. Astrophys.* **21**, 343.
Abt, H.A.: 1987, *Astroph. J.* **317**, 353.
Adams, F.C., Ruden, S.P. and Shu, F.H.: 1989, *Astroph. J.* **347**, 959.
Anosova, J.P.: 1989, *Comments Astrophys.* **14**, 17.
Berry, A.: 1898, *A Short History of Astronomy*, Dover, New York
Bodenheimer, P.: 1978, *Astroph. J.* **224**, 488.
Bodenheimer, P., Tohline, J.E. and Black, D.C.: 1980, *Astroph. J.* **242**, 209.
Bodenheimer, P. and Boss, A.P.: 1981, *Month. Not. of the Roy. Astr. Soc.* **197**, 477.
Bodenheimer, P., Ruzmaikina, T. and Mathieu, R.D.: 1990, in *Protostars & Planets III*, ed(s)., E.H. Levy, J.I. Lunine, and M. S. Matthews, University of Arizona, Tucson, in press.
Boss, A.P. and Bodenheimer, P.: 1979, *Astroph. J.* **234**, 289.
Boss, A.P.: 1980, *Astroph. J.* **236**, 619.
Boss, A.P.: 1984, *Astroph. J.* **277**, 768.
Boss, A.P.: 1986, *Ap. J. Suppl.* **62**, 519.
Boss, A.P.: 1987, *Astroph. J.* **319**, 149.
Boss, A.P.: 1988, *Comments Astrophys.* **12**, 169.
Boss, A.P.: 1989, *Astroph. J.* **346**, 336.
Boss, A.P.: 1990, in *Physical Processes in Fragmentation and Star Formation*, ed(s)., R. Capuzzo-Dolcetta, C. Chiosi, and A. DiFazio, Kluwer, Dordrecht, 279.
Cassen, P.M., Smith, B.F., Miller, R.H. and Reynolds, R.T.: 1981, *Icarus* **48**, 377.
Chen, W.P., Simon, M., Longmore, A.J., Howell, R.R. and Benson, J.A.: 1990, *Astroph. J.* **357**, 224.
Cohen, M. and Kuhi, L.V.: 1979, *Ap. J. Suppl.* **41**, 743.
Danby, J.M.A.: 1988, *Fundamentals of Celestial Mechanics*, Willman-Bell, Richmond, Virginia
Duquennoy, A. and Mayor, M.: 1990, in Proceedings of the XI European Regional Astronomy Meeting of the IAU, ed(s)., M. Vazquez, Cambridge University Press, Cambridge, in press.
Durisen, R.H. and Tohline, J.E.: 1985, in *Protostars and Planets II*, ed(s)., D.C. Black, and M.S. Matthews, University of Arizona, Tucson, 534.
Durisen, R.H., Gingold, R.A., Tohline, J.E. and Boss, A.P.: 1986, *Astroph. J.* **305**, 281.
Fekel, F.C.: 1981, *Astroph. J.* **246**, 879.
Goldsmith, P.F. and Arquilla, R.: 1985, in *Protostars and Planets II*, ed(s)., D.C. Black, and M.S. Matthews, University of Arizona, Tucson, 137.
Halbwachs, J.L.: 1986, *Astron. and Astroph.* **168**, 161.
Halbwachs, J.L.: 1987, *Astron. and Astroph.* **183**, 234.
Heyer, M.H.: 1988, *Astroph. J.* **324**, 311.
Hoyle, F.: 1953, *Astroph. J.* **118**, 513.
Hunter, C.: 1962, *Astroph. J.* **135**, 594.
Keto, E.R. and Lattanzio, J.C.: 1989, *Astroph. J.* **346**, 184.
Kopal, Z.: 1978, *Dynamics of Close Binary Systems*, Reidel, Dordrecht
Kroupa, P., Tout, C.A. and Gilmore, G.: 1990, *Month. Not. of the Roy. Astr. Soc.* **244**, 76.

Larson, R.B.: 1972, *M. N. R. A. S.* **156**, 437.
Larson, R.B.: 1990, in *Physical Processes in Fragmentation and Star Formation*, ed(s)., R. Capuzzo-Dolcetta, C. Chiosi, and A. DiFazio, Kluwer, Dordrecht, 389-400.
Lattanzio, J.C. and Henriksen, R.N.: 1988, *Month. Not. of the Roy. Astr. Soc.* **232**, 565.
Layzer, D.: 1963, *Astroph. J.* **137**, 351.
Lizano, S. and Shu, F.H.: 1989, *Astroph. J.* **342**, 834.
Lucy, L.B.: 1981, in *Fundamental Problems in the Theory of Stellar Evolution*, ed(s)., D. Sugimoto, D.Q. Lamb, and D.N. Schramm, Reidel, Dordrecht, 75.
Mansbach, P.: 1970, *Astroph. J.* **160**, 135.
Mathieu, R.D., Walter, F.M. and Myers, P.C.: 1989, *Astron. J.* **98**, 987.
Miyama, S.M., Hayashi, C. and Narita, S.: 1984, *Astroph. J.* **279**, 621.
Monaghan, J.J. and Lattanzio, J.C.: 1990, *Astroph. J.* **in press**,
Myers, P.C.: 1987, in *Interstellar Processes*, ed(s)., D.J. Hollenbach and H.A. Thronson Jr., Reidel, Dordrecht, 71.
Norman, M.L. and Wilson, J.R.: 1978, *Astroph. J.* **224**, 497.
Popper, D.M.: 1987, *Astroph. J.* **313**, L81.
Pringle, J.E.: 1989, *Month. Not. of the Roy. Astr. Soc.* **239**, 361.
Reipurth, B.: 1988, in *Formation and Evolution of Low Mass Stars*, ed(s)., A.K. Dupree and M.T.V.T. Lago, Kluwer, Dordrecht, 305.
Sasselov, D.D. and Rucinski, S.M.: 1990, *Astroph. J.* **351**, 578.
Shima, E., Matsuda, T., Takeda, H. and Sawada, K.: 1985, *Month. Not. of the Roy. Astr. Soc.* **217**, 367.
Shu, F.H.: 1977, *Astroph. J.* **214**, 488.
Shu, F.H., Adams, F.C. and Lizano, S.: 1987, *Ann. Rev. Astron. Astrophys.* **25**, 23.
Shu, F.H., Tremaine, S., Adams, F.C. and Ruden, S.P.: 1990, *Astroph. J.* **358**, 495.
Sigalotti, L.D.G.: 1990, *Month. Not. of the Roy. Astr. Soc.* **in press**,
Stahler, S.W.: 1988, *Astroph. J.* **332**, 804.
Strom, K.M., Margulis, M. and Strom, S.E.: 1989, *Astroph. J.* **346**, L33.
Tassoul, J.-L.: 1978, *Theory of Rotating Stars*, Princeton University Press, Princeton
Tohline, J.E.: 1980, *Astroph. J.* **235**, 866.
Tohline, J.E., Durisen, R.H. and McCollough, M.: 1985, *Astroph. J.* **298**, 220.
Tomisaka, K., Ikeuchi, S. and Nakamura, T.: 1990, *Astroph. J.* **in press**,
Trimble, V.: 1990, *Month. Not. of the Roy. Astr. Soc.* **242**, 79.
van't Veer, F.: 1981, *Astron. and Astroph.* **98**, 213.
Weis, E.W.: 1974, *Astroph. J.* **190**, 331.
Wilking, B.A. and Lada, C.J.: 1983, *Astroph. J.* **274**, 698.
Williams, H.A. and Tohline, J.E.: 1988, *Astroph. J.* **334**, 449,
Wootten, A.: 1989, *Astroph. J.* **337**, 858.
Zinnecker, H.: 1984, *Astrophys. Space Sci.* **99**, 41.
Zinnecker, H.: 1989, in *Low Mass Star Formation and Pre-Main Sequence Objects*, ed(s)., B. Reipurth, European Southern Observatory, Garching, 447.

THE EVOLUTION OF INTERACTING BINARY SYSTEMS

G. E. McCluskey, Jr. and Y. Kondo

Ake, T.B., Parsons, S.B., and Kondo, Y.: 1985, *Astrophys. J.* **298**, 772.
Andersen, J. and Clausen, J.V.: 1989, *Astron. Astrophys.* **213**, 183.
Arnold, C.N. and Hall, D.S.: 1976, *Acta Astron.* **26**, 91.
Balachandran, S., Lambert, D.L, Tomkin, J., and Parthasarathy, M.: 1986, *Monthly Notices Roy. Astron. Soc.* **219**, 479.
Bohannan, B. and Conti, P.S.: 1976, *Astrophys. J.* **204**, 797.
Bohm-Vitense, E., Nemec, J., and Proffitt, C.: 1984, *Astrophys. J.* **278**, 726.
Budding, E.: 1986, *Astrophys. Space Sci.* **118**, 241.

REFERENCES

Carroll, S., Guinan, E., McCook, G., and Donahue, R.: 1989, *Space Sci. Rev.* **50**, 336.
Chiosi, C. and Maeder, A.: 1986, *Ann. Rev. Astron. Astrophys.* **24**, 329.
Cugier, H. and Hardorp, J.: 1988a, *Astron. Astrophys.* **197**, 163.
Cugier, H. and Hardorp, J.: 1988b, *Astron. Astrophys.* **202**, 101.
De Greve, J.P.: 1986, *Sci. Rev.* **43**, 139.
De Greve, J.P. and Packet, W.: 1990, *Astron. Astrophys.* **230**, 97.
De Greve, J.P. and Vanbeveren, D.: 1980, *Astrophys. Space Sci.* **68**, 433.
De Loore, C.: 1984, *Astrophys. Space Sci.* **99**, 199.
De Loore C. and Sutantyo, W.: 1984, *Astrophys. Space Sci.* **99**, 335.
Doom, C.: 1987, *Reports Phys.* **50**, 1491.
Doom C., De Greve, J.P. and De Loore, C.: 1985, in *Birth and Evolution of Massive Stars and Stellar Groups*, ed(s)., H. Van Woerden and W. Boland, D. Reidel Publ. Co., Dordrecht, 185.
Eaton, J.A.: 1978, *Astrophys. J.* **220**, 582.
Eggleton, P.P.: 1985, in *Interacting Binary Stars*, ed(s)., J.E. Pringle and R.A. Wade, Cambridge University Press, Cambridge, 21.
Eggleton, P.P. and Tout, C.A.: 1989, *Space Sci. Rev.* **50**, 165.
Elias, N.M., II: 1990, *Astrophys. J.* **352**, 300.
Giampapa, M.S., Golub, L. and Worden, S.P.: 1983, *Astrophys. J. Letters* **268**, L121.
Gill, J.R.: 1941, *Astrophys. J.* **93**, 118.
Gimenez, A. and Claret, A.: 1989, *Space Sci. Rev.* **50**, 343.
Gray, D.F.: 1989, in L.E. Cram and L.V. Kuhi, ed(s)., *FGK Stars and T Tauri Stars*, NASA SP-502, 7.
Greenstein, J.L. and Page, T.L.: 1941, *Astrophys. J.* **93**, 128.
Hall, D.S.: 1989, *Space Sci. Rev.* **50**, 219.
Harmanec, P.: 1990, *Astron. Astrophys.* **237**, 91.
Heintze, J.R.W. and Van Gent, R.H.: 1989, *Space Sci. Rev.* **50**, 257.
Hilditch, R.W.: 1989, *Astrophys. Space Sci.* **50**, 289.
Hilditch, R.W. and Bell, S.A.: 1987, *Monthly Notices Roy. Astron. Soc.* **229**, 529.
Holmgren, D.E., Hill, G., Fisher, W. and Scarfe, C.D.: 1990, *Astron. Astrophys.* **231**, 89.
Howarth, I.D. and Prinja, R.K.: 1989, *Astrophys. J. Suppl.* **69**, 527.
Huang, R.Q. and Taam, R.E.: 1990, *Astron. Astrophys.* **236**, 107.
Humphreys, R.M. and McElroy, D.B.: 1984, *Astrophys. J.* **284**, 565.
Iben, I., Jr.: 1985, *Quart. J. Roy. Astron. Soc.* **26**, 1.
Iben, I., Jr.: 1986, *Astrophys. J.* **304**, 201.
Iben, I., Jr.: 1987, *Astrophys. J.* **313**, 727.
Iben, I., Jr. and Truran, J.W.: 1978, *Astrophys. J.* **220**, 980.
Iben I., Jr. and Tutukov, A.V.: 1985, *Astrophys. J. Suppl.* **58**, 661.
Knee, L.B.G., Scarfe, C.D., Mayor, M., Baldwin, B.W., and Meatheringham, S.J.: 1986, *Astron. Astrophys.* **168**, 72.
Kondo, Y.: 1989, *Space Sci. Rev.* **50**, 79.
Kondo, Y. and McCluskey, G.E., Jr.: 1973, in *Structure and Evolution of Close Binaries*, IAU Symp. 73, ed(s)., P.P. Eggleton, S. Mitton, and J. Whelan, Reidel, Dordrecht, 277.
Kondo, Y., McCluskey, G.E. Jr., and Gulden, S.L.: 1976, *Proc. Symp. on X-ray Binaries, NASA SP-389*, , 531.
Kopal, Z.: 1984, *Astrophys. Space Sci.* **99**, 3.
Kuiper, G.P.: 1941, *Astrophys. J.* **93**, 133.
Kudritzki, R.P. and Hummer, D.G.: 1986, in *Luminous Stars and Associations in Galaxies*, ed(s)., W.H. DeLoore, A.J. Willis, and P. Laskarides, D. Reidel Publ. Co., Dordrecht, 3.
Lambert, D.L. and Sawyer, S.R.: 1986, *Publ. Astron. Soc. Pacific* **96**, 88.
Lestrade, J.-F., Mutel, R., Preston, R.A., and Phillips, R.B.: 1988, *Astrophys. J.* **328**, 232.
Leung, K.-C.: 1989, *Space Sci. Rev.* **50**, 279.
Leung K.-C.: 1978, *Astrophys. J.* **222**, 924.
Levato, H., Morrell, N., Garcia, B., and Malaroda, S.: 1988, *Astrophys. J. Supp.* **68**, 319.
Li, Y.-F. and Leung, K.-C.: 1985, *Astrophys. J.* **298**, 345.
Lissauer, J.L. and Backman, D.E.: 1984, *Astrophys. J. Letters* **286**, L39.
Malcolm, G.J. and Bell, S.A.: 1986, *Monthly Notices Roy. Astron. Soc.* **222**, 543.
Marsh, T.R. and Pringle, J.E.: 1990, *Astrophys. J.* **365**, 677.

Massey, P. and Conti, P.S.: 1977, *Astrophys. J.* **218**, 431.
McClure, R.D.: 1983, *Astrophys. J.* **268**, 264.
McClure, R.D. and Woodsworth, A.W.: 1990, *Astrophys. J.* **352**, 709.
McCluskey, G.E., Jr.: 1982, in *Advances in Ultraviolet Astronomy*, NASA CP-2238, ed(s)., Y. Kondo, J.M. Mead, and R.D. Chapman, , 102.
McCluskey, G.E., Jr. and Kondo, Y.: 1984, *Publ. Astron. Soc. Pacific* **96**, 817.
McCluskey, G.E., Jr., Kondo, Y., and Morton, D.C.: 1975, *Astrophys. J.* **201**, 607.
McCluskey, G.E., Jr. and Sahade, J.: 1987, in *Exploring the Universe with the IUE Satellite*, ed(s)., Y. Kondo, D. Reidel Publ. Co., Dordrecht, 427.
Mermilliod, J.-C. and Maeder, A.: 1986, *Astron. Astrophys.* **158**, 45.
Morrison, K.: 1988, *Monthly Notices Roy. Astron. Soc.* **233**, 621.
Olson, E.C.: 1985, in *Interacting Binaries*, ed(s)., P.P. Eggleton and J.E. Pringle, Cambridge University Press, Cambridge, 137.
Olson, E.C.: 1987, *Astron. J.* **94**, 1043.
Olson, E.C.: 1988, *Astron. J.* **96**, 1439.
Olson, E.C. and Stoehr, C.A.: 1986, *Astron. J.* **91**, 1418.
Parthasarathy, M., Cornachin, M., and Hack, M.: 1986, *Astron. Astrophys.* **166**, 237.
Parthasarathy, M., Lambert, D.L., and Tomkin, J.: 1983, *Monthly Notices Roy. Astron. Soc.* **203**, 1063.
Pastetter, L. and Ritter, H.: 1989, *Astron. Astrophys.* **214**, 186.
Peters, G.J. and Polidan, R.S.: 1984, *Astrophys. J.* **283**, 745.
Plavec, M.J.: 1989, *Space Sci. Rev.* **50**, 95.
Polidan, R.S.: 1988, in *A Decade of UV Astronomy with the IUE Satellite*, ESA SP-281, ed(s)., E.J. Rolfe, , 205.
Polidan. R.S.: 1989, *Space Sci. Rev.* **50**, 85.
Popper, D.M.: 1987, *Astrophys. J. Letters* **313**, L81.
Popper, D.M. and Plavec, M.J.: 1976, *Astrophys. J.* **205**, 462.
Popper, D.M. and Tomkin, J.: 1984, *Astrophys. J.* **285**, 208.
Richards, M.T.: 1990, *Astrophys. J.* **350**, 372.
Sahade, J.: 1980, *Space Sci. Rev.* **26**, 349.
Sahade, J.: 1986, in *New Insights in Astrophysics*, ESA SP-263, ed(s)., E.J. Rolfe, , 267.
Sahade, J.: 1987, *Comments in Astrophys.* **12**, 11.
Sahade, J.: 1988, in *Atmospheric Diagnostics of Stellar Evolution: Chemical Peculiarity, Mass Loss and Explosion*, IAU Coll. 108, ed(s)., K. Nomoto, 199.
Sahade, J. and Wood, F.B.: 1978, *Interacting Binary Stars*, Pergamon, Oxford.
Schuerman, D.W.: 1972, *Astrophys. Space Sci.* **19**, 351.
Slee, O.B., Nelson, G.J., Stewart, R.T., Wright, A.E., Innis, J.L., Ryan, S.G., and Vaughan, A.E.: 1987, *Monthly Notices Roy. Astron. Soc.* **229**, 659.
Stewart, R.T., Slee, O.B., White, G.L., et al.: 1989, *Astrophys. J.* **342**, 463.
Stothers, R. and Chin, C.-W.: 1985, *Astrophys. J.* **292**, 222.
Strassmeier, K.G., Fekel, F.C., Bopp, B.W., et al.: 1990, *Astrophys. J. Suppl.* **72**, 191.
Struve, O.: 1941, *Astrophys. J.* **93**, 104.
Tomkin, J. and Lambert, D.L.: 1989, *Monthly Notices Roy. Astron. Soc.* **241**, 777.
Van Hamme, W. and Wilson, R.E.: 1990, *Astron. J.* **100**, 1981.
Vanbeveren, D.: 1977, *Astron. Astrophys.* **54**, 877.
Vanbeveren, D.: 1982, *Astron. Astrophys.* **105**, 260.
Vanbeveren, D. and De Loore, C.: 1980, in *Close Binary Stars: Observations and Interpretations*, ed(s)., M.J. Plavec, D.M. Popper, and R.K. Ulrich, D. Reidel Publ. Co., 115.
Vreux, J.M.: 1985, *Astron. Astrophys.* **143**, 209.
Walborn, N.R.: 1976, *Astrophys. J.* **205**, 419.
White, N.E. and Marshall, F.E.: 1983, *Astrophys. J. Letters* **268**, L117.
White, N.E., Culhane, J.L., Parmar, A.N., et al.: 1986, *Astrophys. J.* **301**, 262.
Wilson, R.E.: 1989, *Space Sci. Rev.* **50**, 191.
Wilson, R.E. and Rafert, J.B.: 1981, *Astrophys. Space Sci.* **76**, 23.
Zorec, J. and Niemela, V.: 1980a, *Comptes Rendus Acad. Sci. Paris, Ser. B.* **290**, 67.
Zorec, J. and Niemela, V.: 1980b, *Comptes Rendus Acad. Sci. Paris, Ser.B.* **290**, 95.

INDEX

A 0620-00 277
A stars 357
accreting white dwarfs 248
accretion 249, 329, 331
   column 306
   disk 147, 149, 151, 199, 226, 227, 276, 277, 278, 284, 287, 326, 329, 331, 332
   disk coronae 290
   disks 147, 199, 276, 278, 287
   disks in Algol systems 9
   disks in cataclysmic variables 9
   energy 205
   induced collapse (AIC) 259, 279, 281, 282
   rate 206
active phase 204
AE Aqr 175
AE Aquarii 9
AG Peg 200, 326
air shower 298
AL Scl 318
Algol 78
   binaries 80
   binaries, Algol 80
   binaries, low activity 44
   emission in spectrum of 8
   optical studies 41
   orbital motion of 4
   system 52, 145
   system, active 41
   system, dynamic 40
   system, low activity 41
   system, orbital periods 41
   system, U Cep 79
   systems 52, 145
   systems, instability of secondary components 5, 8
   systems, secondary components of 3
   systems, theory of 8
   -type 39
   -type binaries 52
   variability of 4, 12
Algols 326
   accretion disks 78
   chemical abundance anomalies 47
   gas stream 78
   infrared studies 44
   radio studies 45
   ultraviolet studies 46
   X-ray studies 45

Am stars 318, 319
AM Her 147, 151, 227, 330, 331
   stars 160
AM Herculis 170
   or polar systems 169
amateur observers, contributions of 7
ambipolar diffusion 368
AML (angular momentum loss) model 138
amorphous carbon 222
λ And 323
angular momentum 151, 152, 378
angular momentum loss (AML) 51
AO Cas 22, 324
Ap stars 318
Aql X-1 278
AR Aur 318
AR Psc 321
astrometry of Algol with VLBI 84
atmospheric eclipse 305, 328
ε Aur 326, 327
AW UMa 321
AX Mon 325
ζ Aur 326, 327
ζ Aurigae 305
   binaries 306
ζ Aurigae-type binaries 97

Balmer discontinuity 219
Balmer lines 219
Balmer series 149
Be star 275
BE UMa 329
beat model 290
BG CMi 330
binary 10, 279, 281
   and millisecond pulsars 273
   evolution 313
   protostar 366
   pulsars 266, 272, 291
   stars 360, 379, 399
binary systems, AR Lac 54, 77
   AR Mon 77
   BY Dra 51
   CaII H+K emission 77
   EI Eri 57
   ER Vul 78
   FF Aqr 54
   HD 82558 58
   II Peg 58, 73
   RS CVn 51, 57, 77

RT Lac 77
RZ Cnc 77
SS Cam 77
SZ Psc 77
UX Ari 58, 73, 77
V471 Tau 54
V711 Tau 57, 58, 73
V833 Tau 58
W UMa 51
Z Her 77
black body 216, 222
    expansion parallax 223
    spectra 167
black hole 271, 272, 276, 287, 291, 332, 333
BM Cas 326
BM Ori 313, 314, 382
bolometric correction 219
44 BooA 320
44 BooB 320
bound-free emission 220
boundary layer 306
Bp star 317
bremsstrahlung 220
bursts 278

C+O white dwarfs 259
Čerenkov light 292
Ca II H+K emission 54
capture binaries 282
carbon 209
    burning 251
    deflagration model 251
    grains 223
    oxygen (CO) 211
cataclysmic variable 226, 234, 237, 278, 280, 287
    binaries (CVs) 52
cataclysmic variables as binaries 9
Cen X-3 292, 331
Cen X-4 278
Chandrasekhar 279
    Effect 316, 318, 326, 328
    Limit 235, 271
    mass 259
    mass limit 237
chemical composition 394
Cherenkov 298
chromosphere 199, 307, 327
chromospheric 53
    emission 51
    gas 306
    UV lines, VW Cep 93
chromospherically active binaries 51
chromospherically-active stars, HD 129333 73

CI Cyg 326
Cir X-1 277
circular Stokes parameter 312
circularization 25
circumbinary envelope 18, 21
circumbinary material 225
circumstellar disks 377
circumstellar envelope 18, 19
CK Vul 226
classical CO nova 212
close binary systems 146
clump collision model 377
6-cm radio luminosity 73
CNO cycle 229
CNO reactions 228
CO 216
CO classical nova 211
CO nova 213, 216, 243, 245
collapse 259
color indices 195
column 329
common envelope 251, 275, 279
contact pairs 333
convection 63
convective synchronization 26
cool giants 325
core collapse 275, 277
core-mass luminosity relationship 220
coronal line emission 224
coronal temperatures, Capella 70
coronal temperatures, σ^2 CrB 70
coronal X-ray 51
    activity 86
    emission 68
Crab 289
    pulsar 293
ε CrA 321
β CrB 319
critical Roche lobe 40
cross-correlation measurements 10
crucial azimuthal 367
CTB 109 293
CVs 147, 160
cyclotron 330
    resonances 285
    transitions 295
32 Cyg 326
Cyg X-1 276, 288, 332
Cyg X-2 290, 331
Cyg X-3 278, 283, 289, 291, 292, 298, 331

D-type stars 193, 199, 206
D-types 202
β^+-decay 228, 229
deep convective zones 86

deflagration wave 251
degenerate 228
detectors, non-photographic 10
detonation wave 251
DF Tau 358
differential rotation 53, 61
diffuse enhanced 221
disk 317, 328, 330
    coronae 285
    -envelope 314
    formation 144
    instability model (DIM) 160
α disks 288
distortion 10
    tidal 3
    tidal in U Cep 5
DM sample 359
dMe stars 75
Doppler-effect 149
Doppler imaging 57
Doppler tomography 151
DQ Her 175, 227
dust 192, 328, 330
dwarf nova outbursts 154, 160
dwarf novae (DN) 160, 278

eccentric disk 378
eccentric mass transfer 31
eclipses, CI Cyg 201
eclipsing binaries 148
    Algol 109
    AR Lac 109
    FF Aqr 97, 101, 109
    FF Aqr (=BD-03°5357) 101
    V471 Tau 97, 103, 109
Eddington limit 224, 225
Eddington luminosity 209, 216, 220, 232, 234, 235, 242, 246, 272, 285, 286
EF Eri 151, 330
EK Cep 359, 382
Ekman pumping 28
electric vector 321
electron densities 222
electron scattering 328
    opacity 220
elemental abundances 211
energy balance 152
energy transport 152
enhanced chromospheric 86
envelope 318, 320, 321, 324, 326, 327, 328, 329, 330, 331, 333
ER Vul 321
ζ Eri 319
τ^5 Eri 318
EUV 230

evolutionary stage 313
evolutionary states 311, 328
evolved systems 384
ex-novae as interacting binaries 9
excess TiO absorption 58
EXOSAT 209, 224, 225, 237, 240
extensive air showers 292
EZ CMa 328, 329

F stars 357
Fahlman-Gregory object 293
fast ONeMg 216
fastness parameter 174
FD models 367
finite difference (FD) 364
fission, binary origin by 8
flare activity 51
flares 53
flash 251
flow 329
free-free 218
frictional stress 152

galactic X-ray sources as interacting binaries 9
gamma ray burster 283, 290, 294
gamma-rays 256
gas flow 320
gas streaming 18, 324
gas streams photometric effects of 11
gaseous stream 18
Geminga 289
GG Carinae 399
GK Cep 320
GK Per 147, 175
globular clusters 280
GP Vel 332
GQ Mus 1983 225
GQ Mus 224, 237
grain 209, 210, 222, 223, 325, 330
grain formation 223
gravitational radiation 283
Griffin systems 357
GX 1+4 285
GX 2000+25 277
GX 2030+375 290

Hα profile 221
HD 162679 318
HD 16589 325
HD 44179 314
HD 6619 318
HeI lines 149
helium stars 262
helium stars in binaries 247

Her X-1 276, 277, 279, 292, 331
Hercules X-1 301
hibernation 226
ζ Hor 325
hydrodynamic instabilities 152
hydrogen shell burning 249
hydrostatic equilibrium 152
HZ Her 277

I Aurigae stars as interacting binaries 9
II Peg 321, 323
inclination 332, 333
infrared emission 218, 222, 223
infrared excess 68, 222
infrared spectra 224
infrared studies 223
intensity ratios 197
intermediate polars 172
International Ultraviolet Explorer (IUE) 62
interstellar polarization 313, 314, 318
interstellar vector 313
intersystem gas flows 199
inverse Compton scattering 285, 295
ionization front 221
IP Peg 151
IR 311
iron 285
iron-peak elements 251
ISM (carbon, silicates, SiC, and PAH) 210
ISM 222
IUE 209, 211, 216, 219, 220, 222, 225, 237, 239, 242, 244, 245
IUE observations, V471 Tauri 100
IUE satellite, 90
IUE spectra, VW Cep 90, 93

Jacobi ellipsoids 364
Jeans equilibrium 366
jets 278, 291, 326, 332

K band luminosities 360
K line 285
Kelvin-Helmholz instability 36
Keplerian period 325
Keplerian value 151
kick velocity 267
kinematic viscosity 288
Kurucz spectra 167

Landau levels 295
Large Magellanic Cloud 245, 294
last stable orbit 287
late detonation 254
late detonation models 252
δ Lib 327

light curve 220, 256
light-curves 111, 112, 113, 114, 117, 119, 121, 125, 126
  methods of analysis 10
  synthesis of 10
linear polarization 68
linear Stokes parameter 312
LMC 1988 #1 212, 213, 214, 220, 221
LMC 1990 #1 215, 216, 220
LMC 1990 #2 216, 220, 237, 238, 241
LMC 213, 216, 220, 221, 241, 242, 245, 246, 295
LMC X-1 277, 290, 295
LMC X-3 277, 295
LMC X-4 276, 285
LMX X-1 290
LMXB 147
LMXRB 273, 283, 286, 288
LMXRBs 273, 286, 288
long period systems 388
low-mass X-ray binaries 164, 277, 287, 289
Lowell Observatory 221, 230
luminosity 230
Lyman continuum 216
β Lyr 47, 327
β Lyrae 399
β Lyrae, accretion disk in 7
β Lyrae, application of Kuiper's ideas to 8
β Lyrae, problem of 5, 8
β Lyrae, streams in 7

M15 280
M31 295
M33 295
Maclaurin spheroids 363
magnesium white dwarf 209
magnetic activity 341
  indicators 51
magnetic braking 8, 51
magnetic catalysic variables 185
magnetic CVs 184
magnetic dynamo 51
magnetic environment 319, 323, 331
magnetic field 51, 275, 278, 281, 324, 327, 332, 389
magnetic obliquity 331
magnetosphere 287, 290
magnetospheric models 342
main ionization mechanism 197
main sequence binaries 316
main sequence stars 356
marginally stable orbit 290
mass ejected 210
mass flow 333
mass flux 324

mass loss 324
   mechanisms 23
   rates 22
mass transfer 144, 226
   burst model (MTBM) 161, 165
   model 29
mass-transfer, theoretical computations 8
mass-transferring
   binaries 329
   stream 329
massive X-ray binaries 274
massive XRBs 273, 284
Maxwell-Boltzmann distribution 361
millisecond pulsars 272, 279, 281
Mira shell 326
muons 293
θ Mus 329
MXB 1730-335 286

N49 294
nebular radiation flux 198
1913+16 282
neon 209
neutron star 153, 271, 275, 293, 329, 331, 332
   masses 283
NGC 6624 280
non-electron scatterers 329
non-radial pulsation 328
non-stationary disks 154
non-synchronous polars 184
non-thermal sources of energy 21
nonradial 324
nonsynchronous binaries, mass flow 30
nova 149, 251, 329, 330, 331
nova mechanism 286
Nova DQ Herculis 9
Nova LMC 1990 #2 242
Nova QV Vul 234
Nova RS Oph 1985 239
Nova V693 CrA 237
Nova V827 Her 1987 224
novae 329, 331
novae ejecta 223
NQ Vul 1976 223
nuclear burning, carbon 252
nuclear burning, hydrogen shell 252
nuclear burning, oxygen 252
nuclear burning, silicon 252
nuclear statistical equilibrium (NSE) 263
nucleosynthesis 254, 263, 329

O/B star 147
O+Ne+Mg white dwarfs 259
obscuration phase 305
observation from space 10

old nova, GK Per, 96
old novae 226
ONeMg nova 213, 216, 220, 237, 243, 244, 245
ONeMg nova V693 CrA 1981 213
66 Oph 318
optical depth 333
optical flash phenomenon 295
optical scattering depth 318
orbit 149
orbit-orbit 377
orbital inclination 313, 328, 329, 331, 332
orbital phase 148
δ Ori 324
θ^2 Ori 317
σ Ori 317
σ OriE 317
Orion features 222
Orion spectrum 221
Orion Trapezium 313
OS And 231
OS And 1986 215, 221, 230
overflow 327
oversynchronous stars 29
oxygen 209
   neon, magnesium novae 211
   white dwarf 209
OY car 148, 149, 161

P-Cygni 217, 219
   profile 213, 216
PAH 222
peculiar systems 392
β Per 326, 327
β Persei 399
φ Per 317
o Per 324
period changes 8, 135, 136
period variations 386
PeV gamma rays 292
phase-locked 319, 320, 321, 324, 328, 331, 332
phase locking 317
ζ Phe 318
photometer, visual polarizing 6
photometer, wedge 6
photometric mass-ratios 11
photometric radius 149
photometry 58
   photoelectric 6, 7
   photographic 6
photomultiplier 7
photospheric spots 51
PIC methods 159
planetary nebulae 222, 223, 272

Poisson equation 367
polar jets 326
polarimeters 311
polarization spectra 327
polarization spectrum 325, 326, 328, 329, 330, 332
polars 330
pre-main sequence stars 313, 358
precession 279
presupernova 264
primary eclipse 305
Principal Spectrum 219
process 283
prograde rotation 149
propellor mechanism 276
proton-proton chain 228
protostellar disk 362
protostellar objects 358
pseudo-photosphere 219
PSR 1534+12 267
PSR 1913+16 268
PSR 1957+20 291
pulsar 272, 282, 289
pulsations 324
PW Vul 237
PW Vul 1984 #1 224

Q Cyg 226
QU Vul 1984 #2 216, 218, 222, 224
QU Vul 216, 237
quasars 222
quasi-periodic oscillations (QPOs) 289
quasiequilibrium disk 362
QV Vul 1987
QV Vul 1987 210, 211, 221, 222, 231

R Aqr 326
R Ara 24, 399
radial velocity changes 114, 124
radial velocity curve 201
radiative 197
radii 146
radio 311, 331
radio emission 51, 53, 336
   cataclysmic variables 348
   cataclysmic variables, AM Her binaries 349
   cataclysmic variables, DQ Her Binaries 348
   classical Algols 346
   contact binaries 345
   early-type stars 341
   emission line Algols 347
   emission mechanism 336
   flare (dMe) stars 338
   magnetic Ap/Bp stars 352
   massive stars 351
   multiwavelength observations 338
   pre-main sequence stars 351
   role of binarity 340
   RS CVn binaries 346
   single stars 341, 351, 352
   stellar correlation 337
   symbiotic stars 350
   symbiotics 340
   X-ray binaries 349
radio stars 83
radio studies 218
   of algols 82
radioactive decays 257
radioactive decays of ^{56}Ni, ^{56}Co
rapid burster 290
rapid rotation 86
Rayleigh scattering 332
Rayleigh-Taylor instabilities 262, 290
rectification 11
   (of light-curves) 10
recurrent 211
   nova 209, 238, 245
recycled 282
   pulsars 284
red giant 224, 314, 325
   radiation flux 198
reflection 3, 10
reflection-effect 324
resonant driving 28
RN 213, 216, 220, 226, 234, 237
RN LMC 1990 #2 224
Roche equipotentials 31
Roche geometry 144
Roche lobe 31, 144, 151, 225, 226, 234, 235, 275, 278, 287, 291, 326, 332
   filling of 3
Roche overflow 327
Roche radii 362
Roche radius 145, 163
Rossby number 63
rotation 63, 387
   rates 11
RR Cen 321
RR Pic 211, 329
RS Canum Venaticorum 9
RS Canum Venaticorum systems, recognition of 9
RS CVn 314, 321, 323
   Binaries 54
   stars 52
   system, AR Lac 70
   system, λ And 61
   system, V711 Tau 93

systems, enhanced activity 76
-type 321
RS Oph 1985 240, 241
RS Oph 224, 225, 237, 238, 239, 240, 241, 329, 330
RW Per 42
RW Tauri, emission in spectrum of 5
RW Tri 149
RY Tau 314
RZ Sct 22

S Ant 320
S-stars 193
S-type stars 198, 206
satellite, Einstein 86
satellite, EXOSAT 86
satellite, IUE 86
satellite, ROSAT 86
saturation 70
scatterer 313, 325, 327, 329, 332
scattering 323, 324, 327, 331, 332, 333
    disks 327
    envelope 324
    optical depth 327
Sco X-1 278, 279, 290, 331
sense of orbital revolution 313
SETI 4
shell 316, 318, 324
shock cone 306
shock front 326
shock wave 251
SiC 222
silicates 223
Sirius B 4
slow novae 227
slow ONeMg nova 216
SMC 295
smoothed particle hydrodynamics (SPH) 364
SN 1987A 211
SN II 211
solar-like stars 51
    T Tauri stars 51
solar-type magnetic cycles 58
solar-type stars 51
    chromospherically active 62
solar-type stars, sun 51
spectral syntheses 211
spectroscopy 10
    developments in 7
    limitations of early 6
    of binary systems 5
SPH models 368, 374
spherical symmetry 320
spin-orbit 377
spotted star 323, 325, 333

SS 433 276, 290, 331
stars, distorted, gravity brightening 113, 114, 117, 122
stars, distorted, limb darkening 114, 117, 118
stars, distorted, Lucy law 116, 122
stars, distorted, Von Zeipel law 113, 116, 122
stars, early-type 120, 121, 130, 132, 140
stars, late-type 117, 119, 121, 125, 130, 131, 132
stars, stellar activity 112, 117, 121ff, 138
stars, stellar angular momentum 111, 121, 132, 137, 138, 141, 142
stars, stellar evolution 111, 112, 115, 119, 120, 121, 123, 130,
stars, stellar rotation 111, 114, 116, 117, 130, 131, 132, 136
stars, stellar variability 111, 124, 125, 126, 129, 130, 131, 135,
starspot 52, 53, 57, 86
    model 61
stationary disks 151, 153
stellar activity 51
stellar atmospheres 211
stellar-type flux distributions 166
stellar wind 316, 324, 327, 328
stream 330, 332
    and disk 35
    recombination 332
Strömgren 66
SU UMa stars (SU) 160
sunspots 52, 60
super-Eddington 245
    luminosity 229, 241
supernova 271, 272
    explosions 283
    remnants 293
supernovae, Type Ia 247, 248
supernovae, Type Ib 247
supernovae, Type Ic 247
supernovae, Type II-L 247
supernovae, Type II-P 247
supernovae, Type IIb 247
SX Cas 42
symbiotic binaries 52
symbiotic nova, RR Tel 97
symbiotic stars 153, 164, 189, 251
symbiotic stars, AG Dra 201
symbiotic stars, AG Peg 201, 206
symbiotic stars, AX Per 97, 201
symbiotic stars, CI Cyg 97, 206
symbiotic stars, cool component 189
symbiotic stars, hot component 193
symbiotic stars, photometric IR observations 190
symbiotic stars, T CrB 97

symbiotic stars, temperature 194
symbiotic stars, Z And 97, 206
synchronization 25
SZ Cam 317

ζ Tau 324
T CrB 237, 238, 241, 329
T Pyx 237
T Tauri stars 358
techniques 298
Tev 292
The Red Rectangle 314
thermonuclear 226
   burning 228
   explosions 286
   flashes 200
   runaway 228
   runaway (TNR) 210, 228
Thomson scattering 314, 332
tidal capture 280
time-resolution, observation with high radio observations of TNR 229
TNR 230, 232, 233, 234, 239
toroid 328
TR line fluxes, II Peg 93
transient 288
transition phase 223
transition region 86
transition region (TR) line emission 53
TRO (thermal relaxation oscillation) model 137
47 Tuc 280
turbulence 288
two-dimensional adiabatic flow 152
two-dimensional hydrodynamic simulations 145
TX UMa 47

4U 0115+63 285
U Cep 327
U Cephei 43, 47, 399
   effects of gas-streams in 11
   light-curve of 8
   tidal distortion in 5
U Gem 149, 151
U Gem 151
   stars (UG) 160
U Her 324
U Oph 317
U Sco 1979 241
U Sco 216, 237, 238, 239, 241
U Sge 22, 44
unevolved systems 383
unstable nuclei 228
β^+-unstable nuclei 228, 229, 233

uvby photometry 66
UZ For 181

V1010 Oph stars 107
V1143 Cyg 318
V1343 Aql 332
V1357 Cyg 332
V1370 Aql 1982 222
V1500 Cyg 218, 331
V1500 Cyg 1975 211
V1500 Cyg 218, 220, 227, 331
V1500 Cygni 224
V1668 Cyg 330
V1819 Cyg 1986 224
V1819 Cyg 224
V2214 Oph 1988 224
V356 Sgr 46
V3890 Sgr 238, 239
V394 CrA 238, 241
V453 Scorpii 399
V471 Tau 329, 330
V693 CrA 216
V711 Tau 321
V745 Sco 238, 239
V827 Her 224
V841 Oph 226
V842 Cen 1986 223
V884 Sco 332
Vela 289
Vela X-1 292, 302, 331
velocity 268
very long baseline interferometer (VLBI) 83
viscosity 287, 288
visible-band 311
visual binary 314
Vogt-Russell theorem 309
vortex-flux conservation 34
VV Cep 326, 327
VW Hydri 156
VY Scl stars 160
VZ Scl 149

W-R 103 328
W Serpentis 11, 399
W-type syndrome 112, 116, 117, 119, 120, 121, 126
W UMa system, 44i Boo 93
W UMa system, ER Vul 107
W UMa system, FK Comae 107
W UMa system, VW Cep 87
W UMa-type, common envelopes 113
W UMa-type, frequency of occurrence 111, 124, 130
W UMa-type, period-colour relation 118ff, 129, 132, 134, 138

W UMa-type, period distribution 111, 118, 119, 125, 126, 130, 131
W Ursae Majoris, light curves of 11
W Ursae Majoris systems, mass-ratios of 11
W Ursae Majoris systems, possible origin by fusion 8
W Ursae Majoris systems, theory of 8
W50 294
white dwarf 147, 194, 220, 222, 224, 225, 226, 227, 228, 232, 233, 234, 235, 238, 247, 271, 282, 326, 329, 330, 331
white light faculae 59
wind 22, 317, 318, 323, 324, 325, 326, 332
   collisions 22
   driven accretion 36
   flow 329
winds 318, 323, 324
Wolf-Rayet 328, 329
   stars 247
   stars as binaries 9
WY Sge 226

X-ray 52, 53, 226, 240, 244, 329, 331
   binaries 147, 272
   bursters 284, 285
   bursts 280
   coronal emission 80
   eclipses 277
   emission 178, 230
   flares 81
   flaring 274
   flux 224
   observations 81, 224
   pulsation 275
   transients 274
X-rays 224, 225, 256
XY Leo 321
XZ Cep 317

Y Cyg 317
YY Gem 318

Z Cam stars (ZC) 160
Z Cha 149, 161
Zanstra method 196
Zeeman-Doppler imaging method 77
Zeeman measures 323
Zeemann broadening technique 77

MIX
Papier aus verantwortungsvollen Quellen
Paper from responsible sources
FSC® C105338

If you have any concerns about our products,
you can contact us on
ProductSafety@springernature.com

In case Publisher is established outside the EU,
the EU authorized representative is:
**Springer Nature Customer Service Center GmbH
Europaplatz 3, 69115 Heidelberg, Germany**

Printed by Libri Plureos GmbH
in Hamburg, Germany